Reproductive Biology
and Phylogeny of
Gymnophiona (Caecilians)

Reproductive Biology and Phylogeny Series
Series Editor: Barrie G. M. Jamieson

Published:

Vol. 1 : Reproductive Biology and Phylogeny of Urodela
(Volume Editor: David M. Sever)
Vol. 2 : Reproductive Biology and Phylogeny of Anura
(Volume Editor: Barrie G. M. Jamieson)
Vol. 3 : Reproductive Biology and Phylogeny of Chondrichthyes
(Volume Editor: William C. Hamlett)
Vol. 4 : Reproductive Biology and Phylogeny of Annelida
(Volume Editors: G. Rouse and F. Pleijel)
Vol. 5 : Reproductive Biology and Phylogeny of Gymnophiona
(Caecilians)
(Volume Editor: Jean-Marie Exbrayat)

In press/under preparation:

Vol. 6 : Reproductive Biology and Phylogeny of Birds
(A and B) (Volume Editor: Barrie G. M. Jamieson)
Vol. 7 : Reproductive Biology and Phylogeny of Cetacea
(Volume Editor: D. Miller)

Reproductive Biology and Phylogeny of Gymnophiona (Caecilians)

Volume edited by
JEAN-MARIE EXBRAYAT
Professor
Laboratoire de Biologie générale, Université Catholique de Lyon
and
Laboratoire de Reproduction et Développement des Vertébrés
Ecole Pratique des Hautes Etudes
Lyon, France

Volume 5 of Series:
Reproductive Biology and Phylogeny

Series edited by
BARRIE G.M. JAMIESON
School of Integrative Biology
University of Queensland
St. Lucia, Queensland
Australia

THE UNIVERSITY
OF QUEENSLAND
AUSTRALIA

Science Publishers
Enfield (NH)   Jersey   Plymouth

CIP data will be provided on request.

SCIENCE PUBLISHERS
An Imprint of Edenbridge Ltd., British Isles.
Post Office Box 699
Enfield, New Hampshire 03748
United States of America

Website: *http://www.scipub.net*

sales@scipub.net (marketing department)
editor@scipub.net (editorial department)
info@scipub.net (for all other enquiries)

ISBN (Set)   1-57808-271-4
               978-1-57808-271-1
ISBN (Vol. 5) 1-57808-312-5
               978-1-57808-312-1

© 2006, Copyright reserved

All rights reserved. No part of this publication may be reproduced, stored in a retrieval system, or transmitted in any form or by any means, electronic, mechanical, photocopying or otherwise, without the prior permission.

This book is sold subject to the condition that it shall not by way of trade or otherwise be lent, re-sold, hired out, or otherwise circulated without the publisher's prior consent in any form of binding or cover other than that in which it is published and without a similar condition including this condition being imposed on the subsequent purchaser.

Published by Science Publishers, Enfield, NH, USA
An Imprint of Edenbridge Ltd.
Printed in India

Preface to the Series

This series was founded by the present series editor, Barrie Jamieson, in consultation with Science Publishers, in 2001. The series bears the title 'Reproductive Biology and Phylogeny' and this title is followed in each volume with the name of the taxonomic group which is the subject of the volume. Each publication has one or more invited volume editors and a large number of authors of international repute. The level of the taxonomic group which is the subject of each volume varies according, largely, to the amount of information available on the group, the advice of proposed volume editors, and the interest expressed by the zoological community in the proposed work. The order of publication reflects these concerns, and the availability of authors for the various chapters, and it is not proposed to proceed serially through the animal kingdom in a presumed phylogenetic sequence. Nevertheless, a second aspect of the series is coverage of the phylogeny and classification of the group, as a necessary framework for an understanding of reproductive biology. Evidence for relationships from molecular studies is an important aspect of the chapter on phylogeny and classification. Other chapters may or may not have phylogenetic themes, according to the interests of the authors.

It is not claimed that a single volume can, in fact, cover the entire gamut of reproductive topics for a given group but it is believed that the series gives an unsurpassed coverage of reproduction and provides a general text rather than being a mere collection of research papers on the subject. Coverage in different volumes will vary in terms of topics, though it is clear from the first volumes that the standard of the contributions by the authors will be uniformly high. The stress will vary from group to group; for instance, modes of external fertilization or vocalization, important in one group, might be inapplicable in another.

The first four volumes on Urodela, edited by David Sever, on Anura, edited by myself, Chondrichthyes, edited by William Hamlett, and Annelida, edited by Greg Rouse and Fredrik Pleijel, reflected the above criteria and the interests of certain research teams. This, the fifth volume, has resulted from our good fortune in the acceptance by Professor Jean-Marie Exbrayat of an invitation to edit a volume on Gymnophiona, the third and last volume on the Amphibia. Jean-Marie Exbrayat is an outstanding authority, renowned for the breadth and depth of his studies on caecilians. In contributions to this volume he has demonstrated, again, the highest standards of scholarship and research. His enthusiasm, despite his heavy duties as Dean of his faculty, for the study of these all too neglected animals is apparent in the eight chapters, three with Jeanne Estabel, Elisabeth Anjubault and Souad Hraoui-Bloquet, to which he has contributed. In our choice of topics for the four other chapters we have been able to draw on a most distinguished group of authors: Marvalee Wake, whose works are landmarks in caecilian studies, Mark Wilkinson and Ronald Nussbaum, who have made major contributions to the study of the taxonomy and plylogeny of caecilians, and a group of Indian researchers, Mohammad A. Akbarsha, George M. Jancy, Mathew Smita and Oommen V. Oommen, who are doing cutting-edge work and represent a renaissance in gymnophionan studies for which India was traditionally renowned. Finally, I have joined David Scheltinga, whose work was completed in my laboratory, in a chapter on spermatozoa in which his excellence as an ultrastructural worker and illustrator is apparent.

Other volumes in preparation are on Birds (B.G.M. Jamieson) and Cetacea (D. Miller). While volume editing is by invitation, biologists who consider that a given taxonomic group should be included in the series and may wish to undertake the task of editing a volume should not hesitate to make their views known to the series editor.

My thanks are due to the School of Integrative Biology, University of Queensland, for facilites, and especially to the Executive Dean, Professor Mick McManus, for his continuing encouragement.

I am grateful to the publishers for their friendly support and high standards in producing this series. Sincere thanks must be given to the volume editors and the authors, who have freely contributed their chapters, in very full schedules. The editors and publishers are gratified that the enthusiasm and expertise of these contributors has been reflected by the reception of the series by our readers.

THE UNIVERSITY
OF QUEENSLAND
AUSTRALIA

9 January 2006

Barrie G. M. Jamieson
School of Integrative Biology
University of Queensland
Brisbane

Preface to this Volume

Even today, several animal groups are little known – or even unknown – but need to be deeply studied, especially as they may disappear. Among these little known animals are the gymophionan amphibians, also called caecilians or Apoda. The Gymnophiona is the third amphibian order, Anura and Urodela being the two others. The Gymnophiona contained 170 species according to Taylor (1968), but in their most recent works, Mark Wilkinson and Ronald Nussbaum list 154 species belonging to 34 genera and 6 families (see chapter **2** of this volume).

For many years, studies on the Gymnophiona were disparate and still only a few species have been deeply studied. The result is a fragmented knowledge of these animals. Fortunately, in recent years, some new works have been published on their systematics, using both the classical methods as well as immunology and molecular biology. New data have also been obtained on the biology, life history, reproductive biology, endocrinology and embryonic development of several species. These fascinating aspects of the history of gymnophionan studies are ably reviewed in the first chapter of this volume by Marvalee Wake.

Examining the publications on Caecilians, it appears evident that these publications are very different one from another. Some of works and especially the studies of the 19th century were very exhaustive, concerning a single species deeply studied (i.e. *Ichthyophis glutinosus*, Sarasin and Sarasin 1887-1890), more recently, other works have been devoted to the comparative study of one or several organs in a large number of species (i.e. the papers of M.H. Wake on urogenital organs). In contrast, other publications are short, concerning a special point of anatomy or biology of a single species. Sometimes only a single rare individual has been the subject of a study. Some species, such as *Ichthyophis glutinosus*, *Hypogeophis rostratus*, *Dermophis mexicanus*, *Gymnopis multiplicata* and *Typhlonectes compressicauda* have been deeply studied by several authors. In contrast, other species, such as *Microcaecilia unicolor* or *Grandisonia diminutiva* have been the subject of only a few works.

The topics covered in the various works are also disparate. Phylogenetic systematics has been well studied, with several propositions for classification. Some organs and functions, such as reproductive tracts and breeding, have been the subject of several papers, concerning one or several species with comparative aspects. In contrast, other organs and functions, such as circulatory or respiratory system, have not been exhaustively studied. Some aspects of developmental biology were well studied in 19th century works and, over the last 15 years, these aspects have been the subject of renewed interest. This large disparity in studies is certainly linked to the fact that Gymnophiona are burrowing animals difficult to catch and living in tropical forests so that access of them is often difficult and expensive. These animals are also difficult to breed in aquaria or terraria because the natural conditions to which they are adapted are poorly known, and reproduction is rarely obtained.

This book devoted to reproductive biology and phylogeny of Gymnophiona begins with a chapter on the history of research on this group (M. H. Wake). In the second chapter concerning the phylogeny of Gymnophiona with systematics, M. Wilkinson and R. Nussbaum give the most recent classification of these species with a useful diagnosis. Considering that caecilians belong to a very little known order, even to scientists, chapter 3 contains general observations on anatomy, physiology, with special references to male and female urogenital systems. Because of inevitable delays from the beginning of the preparation to editing of chapters, several new observations and results were published after the general chapters had been written. Therefore chapter 4, devoted to the Mullerian gland in *Uraeotyphlus narayani* or chapter 6 on spermatogenesis, give the most recent results and augment chapter 3 in which more general and classical data about male genital system are given.

In chapter 5, the endocrinology of reproduction in both males and females is covered. Chapter 7 describes the structure of spermatozoa in several species and draws some phylogenetic conclusions. Chapter 8 is devoted to oogenesis and folliculogenesis. Chapter 9 concerns the embryonic development of genital glands, a neglected aspect of caecilian anatomy. Chapter 10 is about parity; chapter 11 investigates the peculiar case of viviparity in *Typhlonectes compressicauda*, a unique feature (to present knowledge) in Gymnophiona. In Chapter 11 concerning fertilization and embryonic development data include normal tables for *Ichthyophis glutinosus, I. kohtaoensis*, and *Typhlonectes compressicauda* with some observations on other species.

I thank Barrie Jamieson who invited me to become volume editor of this work in the series "Reproductive Biology and Phylogeny" of which he is the series editor and I am grateful to him for his advice and practical help throughout all stages of preparation of this volume. I thank also the authors who accepted to contribute to the volume. I thank particularly Marvalee H. Wake whose work constituted one of the main bases for my own studies, and whom I have met several times at congresses. I thank Mark Wilkinson whom I have known for several years and who has spent a time in my own laboratory, Ronald Nussbaum whose works are at the summit of systematics in Gymnophiona. I also thank our Indian colleagues who have made fascinating recent major contributions to the knowledge of these animals. I wish also to thank my colleagues and fellows who contributed to the knowledge of these animals. I cannot forget Jean Lescure (CNRS and Muséum National d'Histoire Naturelle, Paris) with whom I visited French Guyana and collected animals. Lastly, I wish also to particularly thank Michel Delsol, Honorary Professor at the Catholic

University of Lyon and E.P.H.E., who welcomed me into his laboratory, more than 25 years ago, and who initiated my work on Gymnophiona; I dedicate this book to him in recognition of his support and our faithful friendship throughout these long years of common work.

9 January 2006   **Jean-Marie Exbrayat**
                                                                                         Ecully, France

Contents

Preface to the Series – Barrie G. M. Jamieson   v
Preface to this Volume – Jean-Marie Exbrayat   vii

1. A Brief History of Research on Gymnophionan Reproductive Biology and Development   1
   Marvalee H. Wake

2. Caecilian Phylogeny and Classification   39
   Mark Wilkinson and *Ronald A. Nussbaum*

3. Anatomy with Particular Reference to the Reproductive System   79
   Jean-Marie Exbrayat and *Jeanne Estabel*

4. Caecilian Male Mullerian Gland, with Special Reference to *Uraeotyphlus narayani*   157
   Mohammad A. Akbarsha, George M. Jancy, Mathew Smita and *Oommen V. Oommen*

5. Endocrinology of Reproduction in Gymnophiona   183
   Jean-Marie Exbrayat

6. Caecilian Spermatogenesis   231
   Mathew Smita, George M. Jancy, Mohammad A. Akbarsha, Oommen V. Oommen and *Jean-Marie Exbrayat*

7. Ultrastructure and Phylogeny of Caecilian Spermatozoa   247
   David M. Scheltinga and *Barrie G. M. Jamieson*

8. Oogenesis and Folliculogenesis   275
   Jean-Marie Exbrayat

9. Development of Gonads   291
   Elisabeth Anjubault and *Jean-Marie Exbrayat*

10. Modes of Parity and Oviposition   303
    Jean-Marie Exbrayat

11. Viviparity in *Typhlonectes compressicauda*   325
    Jean-Marie Exbrayat and *Souad Hraoui-Bloquet*

12. Fertilization and Embryonic Development 359
    Jean-Marie Exbrayat

Index 387

CHAPTER 1

A Brief History of Research on Gymnophionan Reproductive Biology and Development

Marvalee H. Wake

1.1 INTRODUCTION

Observations on the gymnophione amphibians have been published for more than 250 years. However, knowledge of the diversity of their biology, and even their existence, has been exceptionally limited, even to most biologists. Most of the research published to date deals with fewer than 5% of the approximately 170 species currently described. This paucity of information has begun to change recently, as awareness of the ecological significance of fossorial animals and of the interesting aspects of the biology of the members of the Order Gymnophiona has increased. Currently, as populations and species of caecilians are becoming threatened by changing land use and by disease, there is significant interest among ecologists, systematists, morphologists, physiologists, behaviorists, developmental biologists, paleontologists, and evolutionists in understanding the biology of the caecilian amphibians. This examination of the history of research on the reproductive biology and development of caecilians will of necessity be brief. I will summarize the work on caecilians presented during several 'periods' or 'epochs' of research on the several subareas of caecilian reproduction and development. I append a large Literature Cited of all references cited in this text. I include only references that consider caecilian reproductive biology and development, *sensu lato*, and systematics as it relates to using or understanding that biology. It is not meant to be exhaustive, but to indicate the range of times, styles, and content of publications. In addition, the reader is referred to other bibliographic lists; these several (in English, French, and German), together with those for the chapters in this book, should present an up-to-date expression of the state of knowledge about caecilian reproductive biology, as well as the history of its development.

Department of Integrative Biology and Museum of Vertebrate Zoology, University of California,
  Berkeley CA 94720-3140 USA

This consideration of research on caecilian reproductive biology (including morphology, development, ecology and population biology, and behavior—to the degree that they are known) and development commences with a summary of the earliest research, arbitrarily designated as that published before 1850, and then progresses through the period 1850-1900, a 'golden age' for classification and morphology, including development, as well as exploration of the tropical fauna; 1900-1970, characterized by field work, many species descriptions, and extensive work on morphology; 1970-2000, a period of expansion of interest and research foci; and 2001 on—the future... for the major areas of research on caecilian reproductive biology.

Research on caecilians before 1850 dealt largely with consideration of the affinities of the rare and poorly understood caecilians with other amphibians and with reptiles, as well as a few general descriptions of species and aspects of their morphology, a consequence of collecting in the tropics by adventurous naturalists. Taylor (1968) and Lescure (1986b) briefly reviewed the early history of the systematics of caecilians (see their literature lists for references). Many taxonomies of the period included caecilians with snakes and amphisbaenians. For example, Taylor (1968) noted that Seba (1735) used the name Caecilia to refer to both caecilians and amphisbaenians); most workers seemed to believe that the affinities of caecilians were with reptiles, not amphibians, and that they were 'naked reptiles' (Taylor 1968). Linnaeus (1758) included two species of *Caecilia* (*tentaculata*, which he had described in 1749, and *'glutinosa'*, which he described in 1754) in the 10th Edition of Systema Naturae, so the names gained 'legal' taxonomic status, and were repeatedin the 12th edition (Linnaeus 1766). Daudin (1802-1803) included Caecilia as the 25th genus of reptiles. Oppel (1810) included caecilians among the Batrachia, and recognized a Family Apoda for them, but included the Batrachia among the reptiles. De Blainville (1816) elevated caecilians to the rank of Order, calling them the "Pseudophydiens" of the "Amphibiens," which he considered a separate Class from the Reptiles. However, classifications of caecilians until approximately 1840 usually included caecilians among the reptiles, despite the fact that the identification of caecilians as amphibians was made clear with Müller's (1831, 1831-1832, 1835) and Hogg's (1841) reports of gills in 'young' caecilians (an early contribution of development to systematics). For example, Mayer (1835) directly defied Müller, claiming that caecilians were snakes, based on comparative morphological information. But in 1839, Duméril presented work at the Academy of Sciences in Paris that showed that caecilians must be included among the amphibians, and Duméril and Bibron (1834-1841) explained that caecilians were amphibians based on several aspects of their anatomy. They used Müller's (1831) suggestion that the ordinal name be Gymnophides, and the name is retained today as modified to Gymnophiona (though some dispute the attribution to Müller). As late as the 1880's some researchers still included caecilians with snakes, but debated the issue, though most workers by then accepted that caecilians are amphibians. A number of scientific naturalists in the 18th and early 19th centuries, and even as early as the 17th century (e.g., Marcgrav de Liebstadt 1648), were actively collecting amphibians and reptiles in many tropical countries, and many of their short accounts included comments on the animals in their habitats, as well as descriptions of new species, including several caecilians (e.g., Ihering 1911a, b). The practice of including useful information about morphology in treatises by field biologists primarily interested in species identification, biogeography, ecology, etc., continued (e.g., Parker 1936, 1941) through much of the first half of the 20th century, and beyond (e.g., Taylor 1960, 1965, 1968, 1969a, b).

The Sarasins (1887-1890) included a "Verzeichniss der Originalliteratur ueber die Caeciliiden" in their magnum opus about *Ichthyophis glutinosus*. It is a most useful reference, with literature beginning with Seba (1735) and extending through "31. Januar 1890." Each reference is annotated in terms of content regarding caecilians, often including page numbers. That literature summary is without doubt an unparalleled exposition of work on caecilians from 1735 though early 1890. It includes all of the short descriptive natural history and taxonomic reports, as well as those on morphology and development.

Taylor's (1968) extensive summary of the taxonomy of caecilians, with a great amount of new information about morphology, habits, etc., is a landmark in caecilian biology. In many ways, it stimulated the research that ensued from many workers through the latter decades of the 20th century. It gave morphologists, developmentalists, and ecologists, as well as systematists, a basis for understanding and sorting caecilian taxa in order to do more reliable comparative work. Recent work that makes use of morphological, biochemical, and molecular data is advancing our understanding of the relationships among caecilians and among amphibians (e.g., Feller and Hedges 1998; Gower *et al.* 2002; Hass *et al.* 1993; Hay *et al.* 1995; Hedges and Maxson 1993; Hedges *et al.* 1993; Milner 1988; Nussbaum and Wilkinson 1989; Trueb and Cloutier 1991a, b; Wilkinson *et al.* 2002b; Wake *et al.* 2004). Taylor (1968) also had general comments about the life histories, reproductive biology, and development of caecilians, as well as their general morphology, ecology and biogeography in his Introduction to the compendium. He included a lengthy bibliography as well.

But, necessary as correct identification of species and understanding of their phylogenetic relationships is to study of their reproductive biology, the focus of this review is on the reproductive biology and development of caecilians. Two recent books provide very useful summaries about the biology of caecilians, one in German (Himstedt, 1996) and one in French (Exbrayat 2000). Himstedt provides extensive new information about the biology *of Ichthyophis kohtaoensis*, including aspects of its reproductive biology and development, and places that information in the context of a general review of caecilians. Exbrayat's treatise similarly emphasizes the species that he has studied extensively, *Typhlonectes compressicauda*, and it too is placed in the context of the general biology of caecilians. Because of

4 Reproductive Biology and Phylogeny of Gymnophiona

Fig. 1.1 A. André-Marie-Constant Duméril (1774-1847; Museum National d'Histoire Naturelle, Paris) included comments on the morphology of caecilians in his treatises on amphibian and reptilian taxonomy and relationships. (Please note: Professor Kraig K. Adler supplied all portraits in these figures from his collection, save those sent by three subjects [W. Himstedt, J.-M. Exbrayat, and M. Wilkinson]). **B**. Henri-Marie Ducrotay de Blainville (1777-1850; Museum National d'Histoire Naturelle, Paris) a collaborator of Georges Cuvier's, was among the earliest workers to recognize that caecilians are amphibians, not reptiles, through his work on anatomy and paleontology. **C**. Martin Heinrich Rathke (1793-1860; Gdansk

Fig. 1.1 contd

his research focus, Exbrayat's book includes a great deal of information about the reproductive biology, development, and neuroendocrinology of the species. It is a most useful summary, and includes an extensive list of references to which I refer the reader.

Research on caecilian reproductive biology and development during the period before 1850 was limited, to say the least. At the same time, it is interesting to see what components were considered. Mentioned above are Müller's (1831, 1831-1832, 1835) and Hogg's (1841) comments on the presence of gills in caecilians (Müller's specimen being a southeast Asian *Ichthyophis* in the Leiden Museum). Several of the 'taxonomic' treatments of amphibians that included caecilians made use of information about the anatomy of various species (e.g., Daudin 1802-1803; de Blainville 1816, 1839; Duméril 1839; Duméril and Bibron 1834-1841).

This review will examine the history of research on caecilian reproductive biology for the following topics: general morphology of the female reproductive system; the structure and function of the ovary (including ovum development and structure) and the oviducts; general morphology of the male reproductive system; the structure and function of the testes, the ducts, and the Mullerian gland; reproductive modes in caecilians, including behavior, ecology, and population biology; developmental biology of caecilians, and close with a general conclusion about the prospects for new and exciting research contributions, followed by the extensive but not exhaustive list of references.

1.2 DESCRIPTIONS OF THE REPRODUCTIVE SYSTEM: FEMALE

1.2.1 Introduction

Reproductive systems, both male and female, have received considerable attention, given that caecilians are our focal taxon. Attention has mainly

---

Fig. 1.1 contd

and Kaliningrad) discussed the morphology of caecilian organs, including reproductive, of *Siphonops annulatus* as part of his comparative research. **D.** Franz von Leydig (1852-1921; Jena and Bonn) included caecilians in his extensive comparative anatomical studies. **E.** Johann Wilhelm Spengel (1852-1921; Würzburg and Jena) provided useful description of the urogenital systems of several species of caecilians, especially *Ichthyophis glutinosus*, and was the first to describe spermatogenesis in caecilians, observing it in three species. **F.** Robert Ernst Eduard Wiedersheim (1848-1923; Freiburg-in-Baden and Jena) made the first extensive description of the comparative anatomy of caecilians in his landmark *Die Anatomie der Gymnophionen* (1879); he included the urogenital system and information about development. **G.** Paul Benedikt Sarasin (1856-1929; Basel), with Fritz Sarasin, did seminal work on the morphology, development, and ecology of *Ichthyophis glutinosus* of Sri Lanka. Amateur naturalists, their landmark four-part treatise published in 1887-90, and several other papers, provide significant contributions to our current understanding of caecilian biology. **H.** Fritz (Carl Friedrich) Sarasin (1859-1942; Basel) shared the research contribution on *I. glutinosus* with his cousin Paul, and conducted several biological expeditions to Southeast Asia. **I.** George Albert Boulenger (1858-1937; The Natural History Museum (London) contributed extensively to our understanding of the systematics and biology of many species of caecilians.

Fig 1.2 A. Emmett Reid Dunn (1894-1956; Haverford College) did extensive field work in the New World tropics and dealt with the systematics, including descriptions of morphological and life history features, of caecilians. **B.** Edward Harrison Taylor (1889-1978; University of Kansas) did field work throughout the world's tropics and conducted extensive research on the systematics and morphology of caecilians. His Caecilians of the World A Taxonomic Review (1968) generated new attention to all aspects of the biology of caecilians. **C.** Ronald Archie Nussbaum (1942— ; University of Michigan) does extensive field research on caecilians (and other amphibians and reptiles), and is contributing significantly to our understanding of their systematics, morphology, and biology. **D.** Werner Himstedt (1940— ; Technischen Hochschule Darmstadt) provides significant investigations on the neurophysiology and behavior of caecilians and the ecology and biology of *Ichthyophis kohtaoensis* of Thailand. **E.** Mark Wilkinson (1963—; The Natural History Museum, London) is making major contributions to the ecology of caecilians and to understanding their morphology and their phylogenetic relationships. **F.** Jean-Marie Exbrayat (1952—; Catholic University of Lyon) is doing extensive research on all aspects of the reproductive biology of the viviparous *Typhlonectes compressicauda* of Guyana.

focused on 1) general descriptions of the systems in individual species, and, for females, 2) attention to the comparative morphology of ovaries, ducts, and cloacae, usually with reference to reproductive mode (oviparity or viviparity), with research on males focusing on testis, and to a lesser degree, intromittent organ morphology. Useful descriptive work was published by the outstanding German morphologists of the second half of the 19th Century—

Meckel's (1833) and Mayer's (1835) comparative anatomy treatises included caecilians. They and others presented the descriptions of the internal morphology of caecilians, occasionally including the gonads and ducts (e.g., Rathke 1852; Leydig 1853, 1868). Müller's (1835) paper on the gills of *Coecilia hypocyanea* (now *Ichthyophis hypocyaneus*) included considerable information about general morphology and development, and Bischoff (1838) also commented on the anatomy of that species. Fischer (1843) commented on the caecilian neural system and on aspects of *Siphonops annulatus* morphology. Stannius (1846) included information on *Siphonops annulatus* in his landmark comparative anatomy textbook. Leydig (1853) considered the anatomy of caecilians in his comparative work, and examined caecilians closely (1868), presenting a wealth of information on *Siphonops annulatus* in a short paper (11 pages). Peter (1895a) presented brief comments on the anatomy, including the reproductive organs, of *Scolecomorphus kirkii*, and on other east African caecilians (1908). Peters (1879) discussed the morphology and taxonomic positions of *Rhinatrema* and *Gymnopis*, in addition to his work on development (see below). Rathke (1852), as noted above presented extensive information about the morphology, including the urogenital systems, of *Coecilia annulata* (now *Siphonops annulatus*). Semon (1892) and Spengel (1876) examined in detail the urogenital systems of amphibians, the former concentrating on *Ichthyophis glutinosus*, the latter being more broadly comparative in treating the caecilians he examined. Two brilliant treatises produced during the last part of the 19th century deserve special attention: Wiedersheim's (1879) superb work on the anatomy of gymnophiones, which examined in detail the gross morphology of several species, with some information on histology, including that of urogenital systems, and the tome of the Sarasins (1887-1890), which explored all aspects of the biology of *Ichthyophis glutinosus* in detail. Both (and a number of the other papers mentioned) must be consulted by any serious researcher investigating caecilian biology today.

There was a brief hiatus in publication about caecilians at the turn of the century, but Harry Marcus in Munich and his students quickly filled that void. They contributed a great deal of information about the gross morphology and histology of development and adult structure of *Hypogeophis rostratus*, with some comparison to other species, including *Hypogeophis* (now *Grandisonia*) *alternans*, from 1910 to 1939. Marcus's student Tonutti (1931) contributed extensive and detailed information about all aspects of the reproductive system of *Hypogeophis rostratus*, with some emphasis on the male system, and Marcus (1939) added to that work. Chatterjee (1936) described the viscera of *Uraeotyphlus menoni*, including the urogenital system; Garg and Prasad (1962) considered the female urogenital system of *Uraeotyphlus oxyurus*. Oyama (1952) presented a brief description of the histology of the visceral organs, including ovaries and oviducts, of *Hypogeophis rostratus*. Jared *et al.* (1999) summarized information about the physiology and morphology of caecilians, with particular reference to

Siphonops annulatus, including considerable information about male and female reproductive systems.

1.2.2 Ovaries, Oviducts, and Fat Bodies

Wake (1968) reviewed the literature on the gonads and fat bodies to date, and Exbrayat (2000) that period and the following years. In fact, the extensive work of Exbrayat and his students and colleagues poses an interesting complement and contrast to Wake's approach, the former focusing in depth on a single species, *Typhlonectes compressicauda*, and the latter taking a more comparative and evolutionary approach. Wake (1968) presented a comparison with new information for 39 species on the morphology of the ovaries and ova, and the fat bodies, and their seasonal cycles, based on dissection and histology. Wake (1970) considered urogenital duct morphology for those species, and presented information for several species, with gross and microanatomical data and a consideration of seasonal and evolutionary patterns, with emphasis on the acquisition of viviparity. Wake (1993) and Wake and Dickie (1998) presented extensive gross and histological data and analysis of oviduct structure and function in caecilians, particularly with reference to the evolution of live-bearing modes of reproduction and maternal nutrition of the developing young before birth. Similarly, Exbrayat (1983, 1986) and Exbrayat and Collenot (1983) examined the ovarian cycle, and Exbrayat studied the fat bodies (1988b) and the ostium of the oviduct (1990) in *T. compressicauda*. He followed that with examinations of the cloaca and reproductive apparatus in a sampling of species (1991, 1992a, 1996), and of the endocrine organs associated with reproduction in *T. compressicauda* (1992b). Hraoui-Bloquet *et al.* (1994) examined the ultrastructural changes in the oviductal mucosa during the gestation period in *T. compressicauda*. Exbrayat (1989a) briefly compared the morphology of three species of caecilians in three different families, and concluded that *Microcaecilia unicolor* may be viviparous, based on morphological features of the ovaries and oviducts, and oocyte diameters. Other workers contributed useful information as well. Berois and De Sá (1988) described the ovaries and fat bodies of *Chthonerpeton indistinctum*, and Masood Parveez and Nadkarni (1993a, b) presented examinations of the ovary and the ovarian cycle in the oviparous *Ichthyophis beddomei*.

1.3 DESCRIPTIONS OF THE REPRODUCTIVE SYSTEM: MALE

The male reproductive system has received substantial but rather specific attention over the years. The gross anatomy of the system was included in the work of several of the 19th century authors mentioned with regard to description of the female system; I will not repeat these, but they should also be consulted for information about male urogenital systems. Spengel's (1876) comparative examination of urogenital systems is particularly important, and Semon's (1892) work on comparative development, especially of

Ichthyophis glutinosus, provides useful background. As noted above, Tonutti (1931) contributed extensive and detailed information about all aspects of the reproductive system of *Hypogeophis rostratus*, with some emphasis on the male system, and his professor, Marcus (1939) added to that work. Chatterjee (1936) described the viscera of *Uraeotyphlus menoni*, including the urogenital system. Wake (1968) examined the comparative morphology of the testis in 33 species representing all six families of caecilians, largely via dissection and with many histological preparations, and commented on the patterns of testis structure and its apparent evolution in the clade, duct morphology, the Mullerian glands, the apparent species-specific morphology of the cloaca and the intromittent organ, and aspects of seasonal variation in some species.

1.3.1 Testes and Spermatogenesis

Much of the research on the male system has focused on the testes, and the comparative biology of spermatogenesis. Seshachar's body of work (e.g., 1936, 1937a, b, 1939a, b, 1940, 1942a, b, c, d, 1943a, b, 1945, 1948) has contributed a great deal to our understanding of the comparative biology of spermatogenesis and spermatelosis, and other aspects of testis morphology, as he examined patterns in members of at least six genera in three families. De Sá and Berois (1986) examined testis morphology and spermatogenesis in *Chthonerpeton indistinctum*; Exbrayat (1986) and Exbrayat and Sentis (1982) considered testis morphology and spermatogenic cycles in *Typhlonectes compressicauda*, and van der Horst *et al.* (1991) examined the ultrastructure of *T. natans* sperm. Scheltinga *et al.* (2003) and Scheltinga and Jamieson (Chapter 7 of this volume) have considerably added to our knowledge of gymnophionan spermatozoal ultrastructure and its phylogenetic significance. Exbrayat and Dansard (1992, 1993) examined the ultrastructure of Sertoli cells and the histology of the testis in *T. compressicauda* (Exbrayat and Dansard 1994). Anjubault and Exbrayat (1998) discussed testis structure and seasonal variation in that species, and presented useful new information on the activity of Leydig-like cells and production of testosterone. Wake (1994a) examined the comparative morphology of sperm of 29 species in 22 genera representing all the six families of caecilians at the gross (light microscope) level and she described the spermatogenic cycle of *Dermophis mexicanus*, with its nearly 11 month periodicity (Wake 1995). Most recently, several workers have begun to examine the comparative ultrastructure and light microscopy of spermatogenesis and of sperm (pers. comm.).

1.3.2 The Intromittent Organ

The intromittent organ of caecilians is a uniquely derived structure among vertebrates. All caecilians, so far as has been determined, have internal fertilization, and it is effected by the male everting the rear part of his cloaca, and inserting it into the vent of the female. Fitzinger (1834) first reported a *Caecilia* with 'a penis protruding from the anus'. Duvernoy (1849) described the intromittent organ of *Siphonops annulatus*. Several nineteenth century workers presented observations on the general cloacal morphology of a diversity of species (e.g., Rathke 1852; Spengel 1876; Wiedersheim 1879; Semon 1892). Tonutti extensively investigated caecilian cloacal morphology and evolution (1931, 1932, 1933). He described interspecific differences in morphology, and speculated about the mechanism of eversion of the intromittent organ (phallodeum). He proposed a scenario in which the intromittent organs of all of the amniotes are derived directly from the caecilian phallodeum! Taylor (1965, 1968) suggested that cloacal morphology might be a taxonomic character, but he did not so use it. Various authors (e.g., Noble 1931; Taylor 1968) have illustrated the extruded intromittent organ. More recently, Wake (1972) described the morphology and histology of male and female cloacas, including extruded phallodea, and speculated about the functional morphology of the connective tissue ridges, blind sacs, etc. of the male cloaca and their 'fit' with the structure of the female cloaca. Both males and females appear to have species-specific morphologies. She also discussed possible mechanisms for eversion of the phallodeum, based on the positions and sizes of vascular sinuses, the cloacal morphology, and the potential association of body wall contraction with filling of the sinuses as a mechanism. The cloaca of *Idiocranium* was described later (Wake, 1986b). Wake (1998) also examined histologically the peculiar cartilaginous spicules of the phallodeum of *Scolecomorphus* previously noted by Noble (1931), Taylor (1968), Wake (1972), and Nussbaum (1985), and found them to be composed of chondroid cartilage with a thin mineralized cap on each spicule, the spicules emerging from an extensive cartilaginous base plate. She continued her conjecture about eversion mechanisms, especially for the taxa with an extensive structural plate in the cloaca (functional tests have not yet been performed). Exbrayat (1991, 1992a, 1996) has closely examined the cloacas of *Typhlonectes compressicauda* and other species, including the extruded phallodeum. Most recently, Gower *et al.* (2002) examined the morphology of the cloacae of several species, including many of the taxa that Wake (1972) had examined. With their much greater sample sizes for several taxa (Wake was restricted usually to dissections of single representatives, that work being done in the 'Dark Ages' when series were few, and carefully guarded [wisely] by their museum curators), they were able to comment on the nature of variation of elements of cloacal morphology. They concluded, as had authors before them, that species-specific cloacal morphology, including its individual variation, should be of use in the delineation of additional characters that will be of utility for systematic analysis.

Virtually nothing is known of mate attraction and identification, or courtship, in caecilians, save for the reports of Barrio (1969) on copulation in *Chthonerpeton*, and those of Heinroth (1915), Murphy *et al.* (1977), Sprackland (1982), Exbrayat and Laurent (1983, 1986), Billo *et al.* (1985), Lyman-Henley (1993), and O'Reilly and Ritter (1995) on 'courtship', copulation, and birth in *Typhlonectes*. In these aquatic forms, pairs were observed with the intromittent organ of the male inserted in the female's vent.

1.3.3 The Mullerian Gland

Finally, a nearly unique component of male reproductive morphology in caecilians is that the terminal several millimeters of the embryonic Mullerian duct do not degenerate, as occurs during development in the males of most vertebrate taxa, but they develop into structures lined with large tubular glands that empty into the cloaca. A notable aspect of the adult male reproductive system, consequently, is the presence of "Mullerian glands"—dilated, secretory structures composed of large, tubular glands that are active during spermatogenesis and regressed when the testis is regressed. The structures were described by Tonutti (1931), and extensively investigated by Wake (1981), who reported on their correlation of secretory activity with the period in which the male has mature sperm in his testes and ducts in *Dermophis mexicanus*. Exbrayat (1985) found similar morphology and correlated activity in *T. compressicauda*, an aquatic caecilian whose copulation has been observed (see below). Wake (1981) also characterized the components of the secretion, and found them to be similar to those in the ejaculate of some birds and mammals, and commented on the fact that the mammalian prostate has a central component derived from the Mullerian duct, and she conjectured that the caecilian Mullerian glands might be analogues to or precursors of the prostate gland, enabling nutrition of sperm and a vehicle for their transport during terrestrial copulation and sperm transport. New work is in progress by other authors on the Mullerian gland (see Chapter 4 of this volume).

1.4 ENDOCRINOLOGY AND IMMUNOLOGY OF CAECILIAN REPRODUCTION

The endocrinology and immunology of caecilian reproductive modes have received little attention to date. Both Wake (1968, 1977a, b, 1993) and Exbrayat (Exbrayat and Collenot 1983; Exbrayat and Delsol 1988) presented morphological evidence that corpora lutea persist during the gestation periods of the species they studied, and conjectured that progesterone therefore is involved in maintenance of the pregnancies. Pillay (1957) described the hypthalamo-hypophyseal neurosecretory system of *Gegeneophis carnosus*. Hilscher *et al.* (1994) found that neurohypophysial peptides control oviduct contraction in *Typhlonectes*, and Ebersole *et al.* (1998) provided preliminary information on gonadal steroid production in *T. natans*. Hilscher-Conklin *et al.* (1998) identified and localized neurohypophysial peptides in the brain of *T. natans*. Studies of hypophyseal cellular details indicate that seasonal variation is correlated with reproductive state (Zuber-Vogeli and Doerr-Schott 1981; Doerr-Schott and Zuber-Vogeli, 1984, 1986; Exbrayat, 1989c, 1992a, b; Exbrayat and Morel, 1990-1991. Exbrayat and Morel (1995, 2003) and Exbrayat *et al.* (1997) assessed prolactin receptors in the reproductive organs of certain caecilians. Exbrayat *et al.* (1995) presented preliminary observations on the immunology of the maternal-fetal relationship in *Typhlonectes compressicauda*, finding that spleen cells of a pregnant female were cytotoxic to its larvae, but that maternal serum (centrifuged whole blood with heparin) prevented the toxicity. This information is tantalizing, and I presume that Exbrayat is following up this important area of research.

1.5 LIFE HISTORY INFORMATION—REPRODUCTIVE MODES

Life history 'strategies' among caecilians are very poorly known. Reproductive mode is known for fewer than half of the described species; modes include egg-layers with free-living larvae, direct developers, and live-bearing species. All of the latter, so far as is currently known, provide maternal nutrition to the fetuses after yolk resorption and hatching in the oviducts (summarized in Wake 1977a, b, 1982, 1989a, 1993; Wake and Dickie 1998; Chapter 11 of this volume). Information that is accruing that parental care is frequent in caecilians, whether egg layers or live-bearers. There are several reports that female caecilians that lay eggs guard the clutch, whether their species is oviparous with a free-living larval stage after hatching, or is a direct developer, such as the oviparous-with-larvae *Ichthyophis glutinosus* (Sarasin and Sarasin 1887-1890) and (apparently) *Grandisonia alternans* (Nussbaum 1984), and the direct-developing *Idiocranium russelli* (Sanderson 1937), *Hypogeophis rostratus* (Nussbaum 1984), and *Boulengerula taitana* (Nussbaum and Hinkel 1994). Some species apparently provide maternal nutrition via viviparity with the birth of altricial young that eat maternal skin secretions, such as *Geotrypetes seraphinii* (Reiss and O'Reilly 1999). Following hatching, larvae wriggle from their maternal burrows into nearby streams (Sarasin and Sarasin 1887-1890). Nussbaum (1984) reported that he had collected many clutches of the Seychellean *H. rostratus* and *G. alternans*, and the female always was coiled around her eggs.

Breckenridge (Breckenridge and de Silva 1973; Breckenridge and Jayasinghe 1979; Breckenridge *et al.* 1987) examined clutch size, egg morphology, and egg case structure in *Ichthyophis glutinosus* (as did Sarasin and Sarasin 1887-1890) and Seshachar *et al.* (1982) and Balakrishna *et al.* (1983) in *I. malabarensis*. Seshachar's (1942a) description of the embryos of *Gegeneophis carnosus* illustrates the membrane and case that holds the clutch together, as does Brauer's (1897, 1899) of *Hypogeophis rostratus*. Such data are needed for nearly all species, as Wake (1977a, b, 1992a) noted. Breckenridge and his colleagues also added to information on embryonic and larval development in their discussions. Salthe (1963) alludes to caecilians in his comparison of the egg capsules of frogs and salamanders, and pleads for more information on caecilians.

Peters (1874a) first noted that caecilians could be live-bearers, having found developing young in the oviducts of a *Typhlonectes compressicauda* (then *Caecilia*). Over time, several workers documented the presence of oviductal embryos and fetuses in several species of caecilians. Many discussions of the evolution of viviparity in caecilians (and other amphibians) are present in the literature, including those of Delsol *et al.* (1981, 1986), Duellman and Trueb (1986), Exbrayat and Delsol (1985, 1988), Exbrayat (1989a,b 1992b), Exbrayat *et al.* (1981, 1982, 1983, 1995), Lescure (1986a), Parker (1956), Salthe and Mecham (1974), Wake (1977a, b, 1982, 1989a, 1992a, 1993, 2004), Wake and Dickie (1998). Laurin and Reisz (1997), Laurin and Girondot (1999), and Laurin *et al.* (2000) are engaged in a debate with Wilkinson and Nussbaum (1998) and Wilkinson *et al.* (2002b) about the ancestral amniote reproductive mode, especially embryo retention, and the evolution of amniotes, as well as viviparity. The degree of embryo retention in caecilians is a crucial point in the discussion. It is clear that the most parsimonious (and logical) assessment is that basal caecilians, despite internal fertilization, lack extended embryo retention, as well as being oviparous.

The nature of maternal nutrition in viviparous caecilians has been considered by many authors and for many years. Parker (1956) and Parker and Dunn (1964) noted that viviparous species have a species-specific fetal dentition, but they thought that it was some sort of retention of an ancestral (fish) condition, rather than a functional innovation. Wake (Wake 1968, 1976, 1977a, b, 1980a, 1989a, 1993; Wake and Dickie, 1998) has discussed amphibian and particularly caecilian viviparity extensively, drawing attention to the relatively abundant number of viviparous species of caecilians, in contrast to the paucity of frogs and salamanders that are live-bearers. She has also noted that, to date, it appears that all live-bearing species provide maternal nutrition to the developing young after the yolk is resorbed via a highly secretory oviductal epithelium (matrotrophy), and the young are born fully metamorphosed, usually after a lengthy (7-11 months) gestation period (maternal nutrition and birth after metamorphosis being the two characters Wake requires for the definition of euviviparity in amphibians). This is in contrast to the situation in several frog and salamander species that do not provide maternal nutrition, the yolk being the full extent of maternal provisioning (lecithotrophy) and the young being born at diverse stages of development (lecithotrophy and birth and variable stages of development defining ovoviviparity in amphibians, according to Wake). Wake's data suggest that the developing young explore the oviducts, using their fetal dentitions both to aid in ingestion of the oviductal epithelial secretion and to stimulate mechanically the epithelium, thus increasing cell turnover and secretory activity. Welsch *et al.* (1977) showed that the composition of the oviductal secretion changes during the gestation period in *Chthonerpeton indistinctum*, early in the period having free amino acids and small carbohydrates, and late in gestation being lipid-rich. Wake (1993) and Wake and Dickie (1998; unpublished data) have extensive information that documents the same phenomenon in other viviparous species.

Exbrayat and his colleagues have suggested that *Typhlonectes* not only practices oral ingestion of the maternally secreted nutrient material, but that later in development, the highly expanded, balloon-like gills are used as pseudoplacental surfaces that facilitate uptake of nutrient material

(Delsol et al. 1981, 1986; Exbrayat and Delsol 1985, 1988; Exbrayat et al. 1981, 1983; Exbrayat and Hraoui-Bloquet 1991, 1992; Hraoui-Bloquet and Exbrayat 1994). Toews and MacIntyre (1977) described the blood respiratory properties of fetuses and adults of *Typhlonectes*, one of the exceedingly rare studies of the physiology of an embryo, larva, or fetus of a caecilian.

1.6 LIFE HISTORY INFORMATION—POPULATION STRUCTURE AND REPRODUCTION

Little is known of population structure in caecilians. Accounts range from the anecdotal (e.g., Sanderson 1937, for the direct-developing *Idiocranium*) to the more explicit and quantitative (e.g., Moodie (1978) and Exbrayat (2000) for *T. compressicauda*; Wake (1980b) for *Dermophis mexicanus*). Wake (1977a) presented information on age at first reproduction, juvenile mortality, etc., for *Geotrypetes seraphinii*. The report by the Sarasins (1887-1890) of their field observations in Sri Lanka presents considerable information on the life history of *Ichthyophis glutinosus*, and Himstedt (1996) gives useful data for *I. kohtaoensis*, based on his extensive collecting in Thailand. Bhatta (1999) contributed information on the biology of the Indian species *I. beddomei* and *I. malabarensis*. Taylor (1960, 1968) included limited life history information for several species, including whether they had free-living larvae, were direct-developers, or were live-bearers. Gans (1961) reported that *Siphonops paulensis* is oviparous. Wilkinson (1992) found that *Uraeotyphlus oxyurus* has free-living larvae. Nussbaum and Hinkel (1994) indicated that *Boulengerula*, at least *taitanus*, may be a direct-developer, because Nussbaum found females guarding clutches and observed newly hatched young in the Taita hills. Nussbaum (1984) also summarized information about the life histories of Seychellean caecilians.

Moodie (1978) and Exbrayat (see references below, but especially 2000 for a summary) commented on aspects of population biology in *Typhlonectes compressicauda*. Moodie (1978) found that the diet is largely aquatic invertebrates, and also small fish. The animals live in burrows, and the burrows are often shared. The animals emerge at dusk to feed. Moodie found that birth is synchronous among females, at the end of the dry season when water levels in the Amazon Basin are at their minimum. Clutch size ranged from 1 to 14, with a mode of 10. Clutch size was correlated with maternal size, but not with fetal size, and fetal size was not correlated with clutch size. Skin gland secretions were highly toxic to a local predatory fish. Exbrayat (1983, 1984a, b, 1985, 1986, 1988a, b, 1989c, 1990, 1996, 2000) and Exbrayat and Delsol (1985, 1988), Exbrayat and Flatin (1985) and Exbrayat et al. (1981, 1982, 1995) have studied extensively the reproductive cycle of Guyanean *T. compressicauda*. They found biennial female reproduction, and report similarly on clutch size, etc. The gestation period in the species is 7-9 months, and they present a great deal of information on embryonic and fetal development (see below). Wake (1980b) discussed the structure of a single population of *Dermophis mexicanus*, including data on age at first reproduction (2-3 years), biennial viviparity, gestation period (11 months), clutch size (not correlated with maternal size), high juvenile mortality, embryonic and adult growth, longevity (approximately 14 years), etc. Such data are not yet available in the literature for most other species, though they are slowly accruing. New research is expected to be published by several authors (Wake, pers. comm.).

Recently, field work has produced more explict analyses of the population biology of caecilians. Wake's (1980b) work was based on large samples from a single site collected during different times of the year over a several-year period. Himstedt (1996) added measurably to information about *Ichthyophis* (a hundred-fifty year hiatus from the pioneering work of the Sarasins), and his work on *I. kohtaoensis* allows comparison with the Sarasins' information on *I. glutinosus*. Wilkinson and colleagues now have a magnificent field program in India and other parts of southeast Asia, and are gathering extensive ecological and population data, which has significant bearing on understanding of aspects of reproductive biology. Such work on many taxa is desparately needed, as land use changes and disease are effecting the decline of many caecilian species throughout the world (Wake, unpubl.).

1.7 DEVELOPMENTAL BIOLOGY OF CAECILIANS

Research on caecilian reproduction, especially its morphological basis, has often been part of studies of the development of fertilized ova, embryos, fetuses, and larvae. Because these topics are so often intertwined in the literature, it is relevant to a review of the 'history of research on caecilian reproduction' to introduce the reader to that literature. Of necessity, this summary will be brief and topical, but I hope it will provide a useful source of references, and perhaps stimulate new work on caecilian development and reproduction.

The condition of embryos and larvae of caecilians is reported in some of the earliest descriptions of caecilians. Peters' (1874a, b, c, 1875) brief reports of larvae of *Caecilia* and embryos of *Typhlonectes* are noteworthy. The landmark work of Sarasin and Sarasin (1887-1890, 1887, 1889, 1892) presented a vast amount of information on the developmental biology of *Ichthyophis glutinosus*, with beautiful illustrations of the development of eye, ear, tentacle, skin, scales, and glands, etc. August Brauer (1897, 1899) carefully and thoroughly described the early development of the direct-developing Seychellian *Hypogeophis rostratus*, presenting the equivalent of a normal table of its development (unfortunately, by current standards, devoid of a method of staging), and also presented information on *Grandisonia* (then *Hypogeophis*) *alternans*. Several descriptions of single or a few stages of embryos and larvae are found in the literature, e.g., Balakrishna *et al.* (1983) of the eggs and embryo of *Ichthyophis malabarensis*, Bhatta and Exbrayat (1998) and Exbrayat *et al.* (1998) of *Ichthyophis beddomei*, Breckenridge and

Jayasinghe (1979) and Breckenridge *et al.* (1987) of *Ichthyophis glutinosus*, Exbrayat and colleagues (summarized in Exbrayat 2000) for *Typhlonectes compressicauda*, Gans (1961) and Goeldi (1899) for *Siphonops annulatus*, Largen *et al.* (1972) for *Sylvacaecilia* (then *Geotrypetes*) *grandisonae*, Seshachar (1942a) for *Gegeneophis carnosus*, Taylor (1960, 1968) for *Ichthyophis* and other taxa, and Wake (1967, 1969, 1977a) for various species.

Research on development continues in a mode that is both 'traditional' and modern, as Sammouri *et al.* (1990) presented a normal table of development of *T. compressicauda*, and Dünker *et al.* (2000) developed one for *Ichthyophis kohtaoensis*, and reviewed much of the pertinent literature to date on caecilian embryology. Work on caecilian development becomes ever more comparative, e.g., Barteczko and Jacob's (2002) comparison of the rostral notochord of *I. kohtaoensis* with that of 'higher vertebrates', finding that structure in both is quite similar, and Wrobel and Süss' (2000) discussion of the significance of nephrostomial tubules as observed in *I. kohtaoensis* for the origin of the vertebrate gonad. Wake and Wake's (2000) examination of vertebrogenesis, with emphasis on sclerotome development and resegmentation and a comparison with paleontological data and with current molecular approaches to analysis of skeletogenesis is expected to set the stage for more work on the process of skeletogenesis, in addition to the pattern, in caecilians. More and more researchers are finding caecilians useful organisms for examination of major questions in biology, and the problem orientation is proving productive.

Much of the work to date, though, has dealt with the development of particular elements, most notably skeletal structures. In addition, there is a 'history' of more than 100 years of research on the development of visceral organs, the integument, etc. It is summarized below, to provide ready reference for the reader. Members of the Stellenbosch school did useful work in the 1930's and 1940's on comparative development in caecilians, with an eye to resolve some issues of vertebrate development, evolution, and homology, based on only a few embryos, larvae, and adults of several species (e.g., deVilliers 1936, 1938; de Jager 1939a, b, c, 1947). Later, Stellenbosch workers examined the development of the nasal region (Badenhorst 1978), and compared skull development based usually on a larva or a fetus and an adult (e.g., Els 1963; Visser 1963). Harry Marcus and a number of his students, working in Munich from 1908 to the late 1930's, dissected the development of *Hypogeophis rostratus*, organ by organ, structure by structure, often from developmental series (see below). Ramaswami (1943, 1948) described the head and chondrocranium of *Gegeneophis*. More recently, Wake and Hanken (1982) described the ossification sequence of the skull of the viviparous *Dermophis mexicanus* based on an extensive ontogenetic series, and Wake *et al.* (1985) described the chondrocranium of *Typhlonectes compressicauda*. The latter studies dealt with series, some with few, some with several, stages of development represented.

Research on caecilian development has concentrated on osteology (skull development, including teeth and the hyobranchial apparatus, and vertebrogenesis), and is summarized by Wake (2003). Studies of aspects of skull development include those by de Jager (1939a, b, c, 1947); de Villiers (1936, 1938); Eifertinger (1933) (lower jaw); Els (1963); Gaupp (1906); Gehwolf (1923); Goodrich (1930); Marcus (1908a, b, 1909, 1910, 1922, 1933 [lower jaw]; Meckel (1833); Peter (1898); Peters (1880); Ramaswami (1941a, b, 1942, 1943,1948); Reiss (1996); Sarasin and Sarasin (1887, 1887-1890, 1892); Stadtmüller (1936) (skull and hyobranchial apparatus); Veit (1965); Visser (1963); Wake (1989b) (hyobranchial apparatus), (2003) (all osteology); Wake et al. (1985); Wake and Hanken (1982). Studies of tooth development include those by Clemen and Opolka (1990); Greven and Clemen (1980); Hraoui-Bloquet and Exbrayat (1996); Lawson (1965); Parker (1956); Parker and Dunn (1964); Wake (1976, 1978, 1980a). Studies of vertebrogenesis, including the notochord, are those of: Barteczko and Jacob (2002); Lawson (1966); Marcus (1934a, 1937); Marcus and Blume (1926); Marcus et al. (1933, 1935); Mookerjee (1942); Peter (1894, 1895a, b); Ramaswami (1958); Wake (1970); Wake (1987, 2003); Wake and Wake, (1986, 2000); Welsch and Storch (1971); Wiedersheim (1879); and Williams (1959).

Examination of the development of the caecilian brain and the special sensory organs has a century's history, though few species have been considered. Aspects of brain development have been examined by Estabel and Exbrayat (1998); Estabel et al. (1998); Exbrayat (1989c); Krabbe (1962); Kuhlenbeck (1922); Laubmann (1926, 1927); Olivecrona (1964); Schmidt and Wake (1997, 1998); and Senn and Reber-Leutennger (1986); the pineal was described by Leclercq et al. (1995). The development of the eye has been studied by Arnold (1935); Engelhardt (1924); Himstedt (1995); and Wake (1985); that of the ear by Sarasin and Sarasin (1889) and Fritzsch and Wake (1990); the lateral line organs by Hinsburg (1901); Coggi (1905); Hetherington and Wake (1979); Fritzsch et al. (1985); and Fritzsch and Wake (1986); and the taste buds by Wake and Schwenk (1986); the olfactory and vomeronasal systems by Badenhorst (1978) and Schmidt and Wake (1990). The development and evolution of the tentacle, that chemosensory organ unique to caecilians and composed of structures derived from the visual and olfactory systems, has been studied by Leydig (1868); Wiedersheim (1879, 1880); Sarasin and Sarasin (1889); Engelhardt (1924); Marcus (1930); Badenhorst (1978); Billo (1986); Billo and Wake (1987); and Wake (1992b). The ontogeny of the skin and its glands and scales has been investigated by several workers (e.g., Marcus 1934b; Gabe 1971; Welsch and Storch 1973; Zylberberg et al. 1980; Fox 1983, 1986, 1987; Wake and Nygren 1987; Zylberberg and Wake 1990; and Paillot et al. 1998). The structure and development of the gills has been studied by Fitzinger (1833); Ramaswami (1954); Wake (1967, 1969, 1977b); Welsch (1981); Exbrayat and Delsol (1985, 1988); Sammouri et al. (1990); Hraoui-Bloquet and Exbrayat (1994); Bhatta and Exbrayat (1998); and Dünker et al. (2000); (it should be noted that terrestrial species have triramous gills, whether in larvae or in the oviducts; the typhlonectids have modified the basic triramous structure into large paired sac-like gills). Among the visceral organs the development of the kidney in caecilians has received relatively extensive attention (Brauer 1900, 1902; Field 1891; Fox 1963; Kozlik 1940; Semon 1892; Spengel 1876; Wake 1968). Welsch *et al.* (1974) examined thyroid development and Welsch (1982) that of the thymus; lung development was studied by Marcus (1923, 1927) and by Welsch (1981). The development of the heart was examined by Schilling (1935) and by Marcus (1935b), the intestine by Marcus (1932) and Hraoui-Bloquet and Exbrayat (1992), and the tongue by Marcus (1930), Tiepel (1932), and Hraoui-Bloquet and Exbrayat (1997a,b).

Because few species have been examined, rarely comparatively and never, to date, using modern genetic and molecular methods, the developmental biology of caecilians is an open an exciting arena for new discovery about pattern and process of development and evolution.

1.8 METAMORPHOSIS

Amphibian metamorphosis has been the subject of many papers and several books. However, only recently have authors considered caecilians in their treatments, focusing on the better-known frogs and salamanders. Fritzsch (1990) is one exception to this; he compared the metamorphosis of components of the sensory and nervous systems, and found that salamanders and caecilians share similarities that differ profoundly from the neural metamorphosis of frogs. Wake (1994b, 2004) also compared members of the three amphibian orders to evaluate the patterns and processes of metamorphosis, and found the events of caecilian metamorphosis much more protracted, especially in viviparous taxa, than in other amphibians. Reiss (2002) did a phylogenetic study of the events of amphibian metamorphosis in a presentation that is sure to stimulate new research. However, some aspects of caecilian metamorphosis have been studied specifically in members of the order. Breckenridge and Jayasinghe (1979) and Breckenridge *et al.* (1987) commented that metamorphosis *in Ichthyophis glutinosus* appeared to be 'protracted' (though they did not compare and analyse specific features). Wake and Hanken (1982) discussed aspects of osteological and soft tissue (gills, skin, etc.) metamorphosis in *Dermophis mexicanus*, and Reiss (1996) evaluated metamorphic events of the palate in *Epicrionops*. Fox (1987) commented on metamorphosis in his comparison of larval and adult skin in *Ichthyophis*. Several authors (mentioned above) who have investigated lateral line organs have mentioned their loss at metamorphosis (and the possible retention of ampullary organs in adult typhlonectids). Exbrayat and Hraoui-Bloquet (1994, 1995) evaluated metamorphosis in four species of caecilians for which they had information, and concluded that heterochronous patterns were apparent, and that there may be a tendency to a reduction of metamorphic time in amphibians (this is apparently based on their assumption that caecilians are characterized by an ancestral mode, and frogs are derived—a within-clade analysis and then comparison might refute this). Significantly, there has been no experimental work to date on caecilian metamorphosis, so mechanisms that are involved in differences in patterns of metamorphosis are unknown. There clearly is a great deal of interesting work available to experimental developmentalists, endocrinologists, and physiologists!

1.9 CONCLUSION

Research on the reproductive biology of caecilians has taken place for more than 150 years. It has contributed to a larger understanding of the relationships of caecilians to other amphibians and to other vertebrates, to the provision of considerable data about several aspects of caecilian reproductive biology (but only for a handful of species in any relative depth), and it remains largely morphological. Further, much of the morphology is done at the whole organism (dissection) and light microscope (tissue/organ) levels, with limited ultrastructural and molecular analysis. Some has occurred, as indicated in this review, but the need for 'fresh' material has been a frustrating limiting factor for studies that have begged for it [e.g., Wake, unpubl.]). Fortunately, material for a few species is becoming available as more researchers are able to do field work. Further, attention to physiological and endocrinological aspects of reproduction in caecilians, to the ecological and population-biology aspects of their reproductive biology, to the behavioral components of courtship and reproduction, (now especially in terrestrial taxa), and to studies of mechanisms in reproductive biology at many levels (e.g., ranging from fertilization mechanisms to courtship to development to evolution) is beginning to engage a new generation of researchers. Needless to say, I trust, I applaud the burgeoning interest in the biology of caecilians finally being exhibited by a number of researchers throughout the world. Understanding of the phylogenetic relationships is under study in several laboratories, using both molecular and morphological techniques, more extensive ecological research is being done (and published), and some physiological and developmental research is under way. Caecilians are excellent subjects for studies of biology, not simply because they exist and are little-known, but because their features of limb loss and body elongation, modifications of sensory and reproductive modes, etc., and their adaptive radiation make them interesting species for analysis of pattern and process of evolution. I hope that we are not engaging in examination of the biology and evolution of caecilians too late—it will be a tragedy to simply document their decline, as we try to understand their essence.

1.10 ACKNOWLEDGEMENTS

I thank Jean-Marie Exbrayat for the invitation to contribute to this volume, and for many fruitful discussions about the reproductive biology and development of caecilians. Barrie Jamieson is making a major contribution to biology through his interest in the reproductive biology, especially sperm structure, in all animals, and through his editing of these volumes on animal reproductive biology. I especially thank many colleagues and students at Berkeley and throughout the world for discussions of caecilian biology; I hope that this has stimulated further interest in research on those organisms. I warmly acknowledge Kraig Adler's generosity in supplying portraits of many scientists who have worked on caecilians, and three colleagues for sending their images. I much appreciate the space and facilities, and the community of scholars, provided by the Radcliffe Institute for Advanced Studies, where writing of this review was initiated while on sabbatical leave from Berkeley. I am grateful to the National Science Foundation for supporting much of my research on caecilians, and the problems and questions that their biology poses for evolutionary morphologists, currently under award #IBN 02 2012

1.11 LITERATURE CITED

Arnold, W. 1935. Das Auge von *Hypogeophis*. Beitrag zur Kenntnis der Gymnophionen XXVII. Morphologisches Jahrbuch 76: 589-625.

Anjubault, E. and Exbrayat, J.-M. 1998. Yearly cycle of Leydig-like cells in testes of *Typhlonectes compressicaudus* (Amphibia, Gymnophiona). Pp. 53-58. In C. Miaud and R. Guyetant (eds), *Current Studies in Herpetology*, Proceedings of the 9th General Meeting of the Societas Europaea Herpetologica, Le Bourget du Lac, France.

Badenhorst, A. 1978. The development and the phylogeny of the organ of Jacobson and the tentacular apparatus of *Ichthyophis glutinosus* (Linn.). Annals of the University of Stellenbosch, Series A 2, 1: 1-26.

Balakrishna, T. A., Gundappa, K. R., and Shakuntala K. A. 1983. Observations on the eggs and embryo of *Ichthyophis malabarensis* (Taylor) (Apoda: Amphibia). Current Science, Bangalore 52: 990-991.

Barrio, A. 1969. Observaciones sobre *Chthonerpeton indistinctum* (Gymnophiona, Caeciliidae) y su reproduccion. Physis 28: 499-503.

Barteczko, K. and Jacob, M. 2002. The morphology of the rostral notochord in embryos of *Ichthyophis kohtaoensis* (Amphibia, Gymnophiona) is comparable to that of higher vertebrates. Anatomy and Embryology 205: 99-112.

Berois, N. and de Sá, R. 1988. Histology of the ovaries and fat bodies of *Chthonerpeton indistinctum*. Journal of Herpetology 22: 146-151.

Bhatta, G. 1999. Some aspects of general activity, foraging and breeding in *Ichthyophis beddomei* (Peters) and *Ichthyophis malabarensis* (Taylor) (Apoda: Ichthyophiidae) in captivity. Zoos' Print Journal 14: 23-36.

Bhatta, G. K. and Exbrayat, J.-M. 1998. Premières observations sur le developpement embryonnaire d'*Ichthyophis beddomei*, Amphibien Gymnophione ovipare. Bulletin de la Société Zoologique de France 124: 117-118.

Billo, R. 1986. Tentacle apparatus of caecilians. Mémoires de la Société Zoologique de France 43: 71-75.

Billo, R., and Wake, M. H. 1987. Tentacle development in *Dermophis mexicanus* (Amphibia; Gymnophiona) with an hypothesis of tentacle origin. Journal of Morphology 192: 101-111.

Billo, R. R., Straub, J. O. and Senn, D. G. 1985. Vivipare Apoda (Amphibia: Gymnophiona), *Typhlonectes compressicaudus* (Duméril et Bibron, 1841): Kopulation, Tragzeit, und Geburt. Amphibia-Reptilia 6: 1-9.

Bischoff, T. 1838. Anatomisch-physiologische Bermerkungen. Archiv für Anatomische und Physiologie und wissenschaft Medicin 838: 353.

Brauer, A. 1897. Beitrage zur Kenntnis der Entwichlungsgeschichte und der Anatomie der Gymnophionen. Zoologische Jahrbucher. Abteilung für Anatomie und Ontogenie der Thiere 10: 389-472.
Brauer, A. 1899. Beiträge zur Kenntniss der Entwicklung und Anatomie der Gymnophionen. Zoologischer Jahrbucher. Abteilung für Anatomie und Ontogenie der Thiere 12: 477-508.
Brauer, A. 1900. Zur Kenntnis der Entwicklung der Excretionsorgane der Gymnophionen. Zoologischer Anzeiger 23: 353-358.
Brauer, A. 1902. Beitrag zur Kenntnis der Entwicklung und der Anatomie der Gymnophionen. III. Entwicklung der Excretionsorgane. Zoologischer Jahrbucher. Abteilung für Anatomie und Ontogenie der Thiere 3: 1-176.
Breckenridge, W. R. and de Silva, G. I. S. 1973. The egg case of *Ichthyophis glutinosus* (Amphibia, Gymnophiona). Ceylon Association for the Advancement of Science 29: 107.
Breckenridge, W. R. and Jayasinghe, S. 1979. Observations on the eggs and larvae of *Ichthyophis glutinosus*. Ceylon Journal of Science (Biological Sciences) 13: 87-202.
Breckenridge, W. R., Nathanael, S. and Pereira, L. 1987. Some aspects of the biology and development of *Ichthyophis glutinosus* (Amphibia: Gymnophiona). Journal of Zoology, London 211: 437-499.
Clemen, G. and Opolka, A. 1990. Dental laminae and teeth of embryonic *Ichthyophis glutinosus* (L.) (Amphibia: Gymnophiona). Anatomischer Anzeiger 170: 111-117.
Chatterjee, B. K. 1936. The anatomy of *Uraeotyphlus menoni* Annandale. Part I. The digestive, circulatory, respiratory, and urino-genital systems. Anatomischer Anzeiger 81: 393-414.
Coggi, A. 1905. Le ampolle di Lorenzini nei Gimnofioni. Monitore Zoologica Italiano 16: 49-56.
Daudin, F. M. 1802-1803. Histoire naturelle, générale et particulière des reptiles; ouvrage faisant suite à l'histoire naturelle, générale et particulière composée par Leclerc de Buffon, et rédigée par S. C. Sonnino, Paris 7: 411-429.
de Blainville, H. M. 1816. Prodrome d'une nouvelle distribution systématique du règne animal. Bulletin des Sciences, Société Philomatique de Paris 1816: 111.
de Blainville, H. M. 1939. Notice historique sur la place assignée aux Cécilies dans la série Zoologique. Comptes Rendus des Séances de l'Académie des Sciences de Paris 1839: 1-15, and Annales des Sciences Naturelles, Serie 2, 12 Zoologie: 360-367.
de Jager, E. 1939a. Contributions to the cranial anatomy of the Gymnophiona. Further points regarding the cranial anatomy of the genus *Dermophis*. Anatomischer Anzeiger 88: 193-222.
de Jager, E. 1939b. The cranial anatomy of *Coecilia ochrocephala* Cope (further contributions to the cranial morphology of the Gymnophiona). Anatomischer Anzeiger 88: 433-469.
de Jager, E. 1939c. The gymnophione quadrate and its processes, with special reference to the processus ascendens in a juvenile *Ichthyophis glutinosus*. Anatomischer Anzeiger 88: 223-232.
de Jager, E. 1947. Some points in the development of the stapes of *Ichthyophis glutinosus*. Anatomischer Anzeiger 96: 203-210.
Delsol, M., Exbrayat, J.-M., Flatin, J. and Gueydan-Baconnier, M. 1986. Nutrition embryonnaire chez *Typhlonectes compressicaudus* (Duméril et Bibron, 1841), amphibien apode vivipare. Mémoires de la Société Zoologique de France 43: 39-54.

Delsol, M., Flatin, J., Exbrayat, J.-M. and Bons, M. 1981. Développement de *Typhlonectes compressicaudus*, Amphibien Apode vivipare. Hypothéses sur sa nutrition embryonnaire et larvaire par un ectotrophoblaste. Comptes Rendus des Séances de l'Académie des Sciences de Paris 293: 281-285.

de Sá, R. and Berois, N. 1986. Spermatogenesis and histology of the testes of the caecilian *Chthonerpeton indistinctum*. Journal of Herpetology 20: 510-514.

de Villiers, C. G. S. 1936. Some aspects of the amphibian suspensorium, with special reference to the paraquadrate and quadratomaxillary. Anatomischer Anzeiger 81: 225-247.

de Villiers, C. G. S. 1938. A comparison of some cranial features of the East African Gymnophiones *Boulengerula boulengeri*, Tornier and *Scolecomorphus uluguruensis* Boulenger. Anatomischer Anzeiger 86: 1-26.

Doerr-Schott, J. and Zuber-Vogeli, M. 1984. Immunohistochemical study of the adenohypophysis of *Typhlonectes compressicaudus* (Amphibia, Gymnophiona). Cell and Tissue Research 235: 211-214.

Doerr-Schott, J. and Zuber-Vogeli, M. 1986. Cytologie et immunocytologie de l'hypophyse de *Typhlonectes compressicaudus*. Mémoires de la Société Zoologique de France 43: 77-79.

Duellman, W. E. and Trueb, L. 1986. *Biology of Amphibians*. John Hopkins University Press, Baltimore and London.

Duméril, A. M. C. 1839. Mémoire sur la classification des Ophiosomes ou Ceciloides, familie de Reptiles qui participent des Ophidiens et des Batraciens, relativement a la forme et à l'organisation. Comptes Rendus des Séances de l'Académie des Sciences de Paris 9: 581-587.

Duméril, A. M. C. and Bibron, G. 1834-1841. Erpétologie generale ou histoire naturelle complète des Reptiles. Librairie encyclopédique de Roret, Paris.

Dünker, N., Wake, M. H. and Olson, W. M. 2000. Embryonic and larval development in the caecilian *Ichthyophis kohtaoensis* (Amphibia, Gymnophiona): a staging table. Journal of Morphology 243: 3-34.

Duvernoy, G. L. 1849. Cours d'histoire naturelle des corps organisès professé au college de France. Revue et Magazine de Zoologie 1849: 179-189.

Ebersole, T. J., Goetz, F. W. and Boyd, S. K. 1998. Gonadal steroid production in the viviparous caecilian amphibian *Typhlonectes natans*. American Zoologist 38: 21A.

Eifertinger, L. 1933. Die Entwicklung des knöchernen Unterkiefers von *Hypogeophis*. (Beitrag zur Kenntnis der Gymnophionen. XX.) Zeitschrift für Anatomie und Entwicklungsgeschichte 101: 534-552.

Els, A. J. 1963. Contributions to the cranial morphology of *Schistometopum thomensis* (Bocage). Annals of the University of Stellenbosch Series A 38: 39-64.

Engelhardt, F. 1924. Tentakelapparat und Auge von *Ichthyophis*. Jena Zeitschrift für Naturwissenschaft 60: 241-305.

Estabel, J. and Exbrayat, J.-M. 1998. Brain development of *Typhlonectes compressicaudus*. Journal of Herpetology 32: 1-10.

Estabel, J., Bhatta, G. K. and Exbrayat, J.-M. 1998. Comparison of brain development in two Caecilians *Typhlonectes compressicaudus* and *Ichthyophis beddomei*. Pp. 105-111. In C. Miaud and R. Guyetant (eds), *Current Studies in Herpetology*, Proceedings of the 9th General Meeting of the Societas Europaea Herpetologica, Le Bourget du Lac, France.

Exbrayat, J.-M. 1983. Premiéres observations sur le cycle annuel de l'ovaire de *Typhlonectes compressicaudus* (Duméril et Bribron, 1841), Amphibien Apode. Comptes Rendus des Séances de l'Académie des Sciences de Paris 296: 493-498.

Exbrayat, J.-M. 1984a. Cycle sexuel et reproduction chez un Amphiien Apode: *Typhlonectes compressicaudus* (Duméril et Bibron, 1841). Bulletin de la Société Herpétologique de France 32: 31-35.

Exbrayat, J.-M. 1984b. Quelques observations sur l'evolution des voies genitales femelles de *Typhlonectes compressicaudus* (Duméril et Bibron, 1841), Amphibien Aponde vivipare, au cours du cycle de reproduction. Comptes Rendus des Séances de l'Académie des Sciences de Paris 298: 13-18.

Exbrayat, J.-M. 1985. Cycle des canaux de Müller chez le mâle adulte de *Typhlonectes compressicaudus* (Duméril et Bibron, 1841), Amphibien Apode. Comptes Rendus des Séances de l'Académie des Sciences de Paris 301: 507-512.

Exbrayat, J.-M. 1986. Le testicule de *Typhlonectes compressicaudus*: Structure, ultrastructure, croissance et cycle de reproduction. Mémoires de la Société Zoologique de France 43: 121-132.

Exbrayat, J.-M. 1988a. Croissance et cycle des voies genitales femelles de *Typhlonectes compressicaudus* (Duméril et Bibron, 1841), amphibien apode vivipare. Amphibia-Reptilia 9: 117-134.

Exbrayat, J.-M. 1988b. Variations pondérales des organes de reserve (corps adipeux et foie) chez *Typhlonectes compressicaudus*, Amphibien Apode vivipare au cours des alternances saisonnières et des cycles de reproduction. Annales des Sciences naturelles, Zoologie, Paris 13ème série 9: 45-53.

Exbrayat, J.-M. 1989a. Quelques observations sur les appareils génitaux de trois gymnophiones; hypothèses sur le mode de reproduction de *Microcaecilia unicolor* (Amphibia, Gymnophiona). Bulletin de la Société Herpétologique France 52: 34-44.

Exbrayat, J.-M. 1989b. Quelques aspects de l'évolution de la viviparité chez les Vertébrés. Pp. 143-169 In J. Bons and M. Delsol (eds), *Evolution Biologique. Quelques donnés actuelles*. Boubée, Paris et Lyon.

Exbrayat, J.-M. 1989c. The cytological modifications of the distal lobe of the hypophysis in *Typhlonectes compressicaudus* (Duméril and Bibron, 1841), Amphibia, Gymnophiona, during the cycles of seasonal activity. I. In adult males. Biological Structures and Morphogenesis 2: 117-123.

Exbrayat, J.-M. 1990. Quelques observations sur l'ostium et la paroi coelomique chez deux Amphibiens Gymnophiones femelles au cours de différentes périodes du cycle de reproduction. Bulletin de la Société Zoologique de France 115: 199-200.

Exbrayat, J.-M. 1991. Anatomie du cloaque chez quelques Gymnophiones. Bulletin de la Société Herpétologique de France 58: 31-43.

Exbrayat, J.-M. 1992a. Appareils génitaux et reproduction chez les Amphibiens Gymnophiones. Bulletin de la Société Zoologique de France 117: 291-296.

Exbrayat, J.-M. 1992b. Reproduction et organes endocrines chez les femelles d'un Amphibien Gymnophione vivipare, *Typhlonectes compressicaudus*. Bulletin de la Société Herpétologique de France 64: 37-50.

Exbrayat, J.-M. 1996. Croissance et cycle du cloaque chez *Typhlonectes compressicaudus* (Duméril et Bibron, 1841), Amphibien Gymnophione. Bulletin de la Société Zoologique de France 121: 93-98.

Exbrayat, J.-M. 2000. *Les Gymnophiones Ces curieux Amphibiens*. Société Nouvelle des Editions Boubee, Paris. Pp. 443.

Exbrayat, J.-M. and Collenot, G. 1983. Quelques aspects de l'evolution de l'ovaire de *Typhlonectes compressicaudus* (Duméril et Bibron, 1841), Batracien Apode vivipare. Etude quantitative et histochimique des corps jaunes. Reproduction, Nutrition, Développement 23: 889-898.

Exbrayat, J.-M. and Dansard, C. 1992. Ultrastructure des cellules de Sertoli chez *Typhlonectes compressicaudus*, Amphibien Gymnophione. Bulletin de la Société Zoologique de France 117: 166-167.

Exbrayat, J.-M. and Dansard, C. 1993. An ultrastructural study of the evolution of Sertoli cells in a Gymnophionan Amphibia. Biology of the Cell 79: 90.

Exbrayat, J.-M. and Dansard, C. 1994. Apports de techniques complémentaires à la connaissance de l'histologie du testicule d'un Amphibien Gymnophione. Revue Française d'Histotechnologie 7: 19-26.

Exbrayat, J.-M. and Delsol, M. 1985. Reproduction and growth of *Typhlonectes compressicaudus*, a viviparous Gymnophione. Copeia 1985: 950-955.

Exbrayat, J.-M. and Delsol, M. 1988. Oviparité et développement intra-utérin chez les Gymnophiones. Bulletin de la Société Herpétologique de France 45: 27-36.

Exbrayat, J.-M. and Flatin, J. 1985. Les cycles de reproduction chez les Amphibiens Apondes. Influence des variations saisonnières. Bulletin de la Société Zoologique de France 110: 301-305.

Exbrayat, J.-M. and Hraoui-Bloquet, S. 1991. Morphologie de l'épithélium branchial des embryons de *Typhlonectes compressicaudus* (Amphibien Gymnophione) etudié en microscopie électronique a balayage. Bulletin de la Société Herpétologique de France 57: 45-62.

Exbrayat, J.-M. and Hraoui-Bloquet, S. 1992. Evolution de la surface branchiale des embryons de *Typhlonectes compressicaudus*, Amphibien Gymnophione vivipare, au cours du développement. Bulletin de la Société Zoologique de France 117: 340.

Exbrayat, J.-M. and Hraoui-Bloquet, S. 1994. Un exemple d'hétérochronie: la metamorphose chez les gymnophiones. Bulletin de la Société Zoologique de France 119: 117-126.

Exbrayat, J.-M, and Hraoui-Bloquet, S. 1995. Evolution of reproductive patterns in Gymnophiona Amphibia. Pp. 48-52. In G. A. Llorente, A. Montori, X. Santos and M. A. Carretero (eds), *Scientia Herpetologica*, Proceedings of the 7th General Meeting of the Societas Europaea Herpetologica, Barcelona, Spain.

Exbrayat, J.-M. and Laurent, M.-T. 1983. Quelques observations concernant le maintien en élevage de deux amphibians apodes: *Typhlonectes compressicaudus* et un *Ichthyophis*. Reproduction de *Typhlonectes compressicaudus*. Bulletin de la Société Herpétologique de France 26: 25-26.

Exbrayat, J.-M. and Laurent, M.-T. 1986. Quelques observations sur la reproduction en élevage de *Typhlonectes compressicaudus*, Amphibien Apode vivipare. Possibilité de rythmes endogènes. Bulletin de la Société Herpétologique de France 40: 52-62.

Exbrayat, J.-M. and Morel, G. 1990-1991. The cytological modifications of the distal lobe of the hypophysis in *Typhlonectes compressicaudus* (Duméril et Bibron, 1841), Amphibia Gymnophiona, during the cycles of seasonal activity. II. In adult females. Biological Structures and Morphogenesis 3: 129-138.

Exbrayat, J.-M. and Morel, G. 1995. Prolactin (PRL)-coding mRNA in *Typhlonectes compressicaudus*, a viviparous gymnnophionan amphibian: an in situ hybridization study. Cell and Tissue Research 280: 133-138.

Exbrayat, J.-M. and Morel, G. 2003. Visualization of prolactin-receptors (PRL-R) by in situ hybridization in reproductive organs of *Typhlonectes compressicauda*, a hymnophionan amphibian. Cell and Tissue Research 312: 361-367.

Exbrayat, J.-M. and Sentis, P. 1982. Homogénéite du testicule et cycle annuel chez *Typhlonectes compressicaudus* (Dumeril et Bibron, 1841), amphibien apode vivipare. Comptes Rendus des Séances de l'Académie des Sciences de Paris 294: 757-762.

Exbrayat, J.-M., Delsol, M. and Flatin, J. 1981. Première remarques sur la gestation chez *Typhlonectes compressicaudus* (Duméril et Bibron, 1841), Amphibien Aponde vivipare. Comptes Rendus des Séances de l'Académie des Sciences de Paris 292: 417-420.

Exbrayat, J.-M., Delsol, M. and Flatin, J. 1982. Observations concernant la gestation de *Typhlonectes compressicaudus* (Dumeril et Bibron, 1841), Amphibien Apode vivipare. Bulletin de la Société Zoologique de France 107: 486.

Exbrayat, J.-M., Delsol, M. and Flatin, J. 1995. Amphibiens Gymnophiones, organes génitaux et biologie de la reproduction. Pp. 1287-1298. In P. P. Grassé and M. Delsol (eds), *Traité de Zoologie*, Volume 4, Amphibiens. Masson, Paris.

Exbrayat, J.-M., Delsol, M. and Lescure, J. 1983. La viviparite chez *Typhlonectes compressicaudus*, amphibien apode. Bulletin de la Société Herpétologique de France 26: 23-24.

Exbrayat, J.-M., Ouhtit, A., and Morel G. 1997. Visualization of gene expression of prolactin receptors (PRL-R) by in situ hybridization, in *Typhlonectes compressicaudus*, a gymnophionan amphibian. Life Sciences 61: 1915-1928.

Exbrayat, J.-M., Pujol, P., and Hraoui-Bloquet, S. 1995. First observations on the immunological materno-foetal relationships in *Typhlonectes compressicaudus*, a viviparous gymnophionan Amphibia. Pp. 271-273. In G. A. Llorente, A. Montori, X. Santos and M. A. Carretero (eds), *Scientia Herpetologica*, Proceedings of the 7th General Meeting of the Societas Europaea Herpetologica, Barcelona, Spain.

Exbrayat, J.-M., Bhatta, G. K., Estabel, J. and Paillot, R. 1998. First observations on embryonic development of *Ichthyophis beddomei*, an oviparous Gymnophionan Amphibia. Pp. 113-120. In C. Miaud and R. Guyetant (eds), *Current Studies in Herpetology*, Proceedings of the 9th General Meeting of the Societas Europaea Herpetologica, Le Bourget du Lac, France.

Feller, A. and Hedges, S. B. 1998. Molecular evidence for the early history of living amphibians. Molecular Phylogeny and Evolution 9: 509-516.

Field, H. H. 1891. Development of the pronephros and segmental duct in amphibians. Bulletin of the Museum of Comparative Zoology, Harvard University 21: 201-340.

Fischer, J. G. 1843. Amphibiorum nudorum neurologiae specimen primum. Berolini 1843: 40-75.

Fitzinger, W. 1833. Ueber die Kiemenlöcher der Coecilien. Isis von Oken: 1833: 380.

Fitzinger, W. 1834. Ueber ein Exemplar einer *Caecilia sp.*, mit aus dem After hervorragendem Penis. Isis von Oken 1834: 695.

Fox, H. 1963. The amphibian pronephros. Quarterly Review of Biology 38: 1-25.

Fox, H., 1983. The skin of *Ichthyophis* (Amphibia: Caecilia): An ultrastructural study. Journal of Zoology London 199: 223-248.

Fox, H. 1985. The tentacles of *Ichthyophis* (Amphibia: Caecilia) with special reference to the skin. Journal of Zoology London 205: 223-234.

Fox, H. 1986. Early development of caecilian skin with special reference to the epidermis. Journal of Herpetology 20: 154-167.

Fox, H. 1987. On the fine structure of the skin of larval juvenile and adult *Ichthyophis* (Amphibia: Caecilia). Zoomorphologie 107: 67-76.

Fritzsch, B. 1990. The evolution of metamorphosis in amphibians. Journal of Neurobiology 21: 1011-1021.

Fritzsch, B. and Wake, M. H. 1986. The distribution of ampullary organs in Gymnophiona. Journal of Herpetology 20: 90-93.

Fritzsch, B., and Wake, M. H. 1990. The inner ear of gymnophione amphibians and its nerve supply: A comparative study of regressive events in a complex sensory system (Amphibia, Gymnophiona). Zoomorphology. 108: 201-217.

Fritzsch, B., Wahnschaffe, U., Crapon de Caprona, M. D. and Himstedt, W. 1985. Anatomical evidence for electroreception in larval *Ichthyophis kohtaoensis*. Naturwissenschaften 72: 102-104.

Gabe, M. 1971. Données histologiques sur le tegument d'*Ichthyophis glutinosus* L. (Batracien, Gymnophione). Annales des Sciences Naturelles, Zoologie, Paris 13: 573-608.

Gans, C. 1961. The first record of egg laying in the caecilian *Siphonops paulensis* Boettger. Copeia 1961: 490-491.

Garg, B. L. and Prasad, J. 1962. Observations of the female urogenital organs of limbless amphibians *Uraeotyphlus oxyurus*. Journal of Animal Morphology and Physiology 9: 154-156.

Gaupp, E. 1906. Die Entwicklung des Kopfskelettes. Pp. 570-890. In O. Hertwig (ed.), *Handbuch der Vergleichenden und Experimentellen Entwicklungslehre der Wirbeltiere*, Vol. 3, Pt. 2. Gustav Fischer, Vienna.

Gehwolf, S. 1923. Die Kehlkopf bei *Hypogeophis*. Zeitschrift für Anatomie und Entwicklungsgeschichte 68: 433-454.

Goeldi, E. A. 1899. Uber die Entwicklung von *Siphonops annulatus*. Zoologische Jahrbucher Systematische 12: 170-173.

Goodrich, E. S. 1930. *Studies on the Structure and Development of Vertebrates*. Macmillan and Co., London.

Goncalves, A. A. 1977. Dimorfismo sexual de *Typhlonectes compressicaudus* (Amphibia Apoda). Boletin do Fisiologie Animal, Universidad Sao Paulo. 1: 141-142.

Gower, D. J. and Wilkinson, M. 2002. Phallus morphology in caecilians (Amphibia, Gymnophiona) and its systematic utility. Bulletin of the Natural History Museum London (Zoology) 68: 143-154.

Gower, D. J., Kupfer, A., Oommen, O. V., Himstedt, W., Nussbaum, R. A., Loader, S. P., Presswell, B., Muller, H., Krishna, S. B., Boistel, R. and Wilkinson, M. 2002. A molecular phylogeny of ichthyophiid caecilians (Amphibia: Gymnophiona: Ichthyophiidae): Out of India or out of southeast Asia? Proceedings of the Royal Society B 269: 1563-1569.

Greven, H. and Clemen, G. 1980. Beobachtungen an den Zähnen und Zahnleisten des Munddaches von *Siphonops annulatus* (Mikan) (Amphibia: Gymnophiona). Anatomischer Anzeiger 147: 270-279.

Hass, C. A., Nussbaum, R. A. and Maxson, L. R. 1993. Immunological insights into the evolutionary history of caecilians (Amphibia: Gymnophiona): relationships of the Seychellean caecilians and a preliminary report on family-level relationships. Herpetological Monographs 6: 56-63.

Hay, J. M., Ruvinsky, I., Hedges, S. B. and Maxson, L. R. 1995. Phylogenetic relationships of amphibian families inferred from DNA sequences of mitochondrial 12S and 16S ribosomal RNA genes. Molecular Biology and Evolution 12: 928-937.

Hedges, S. B. and Maxson, L. R. 1993. A molecular perspective on lissamphibian phylogeny. Herpetological Monographs 6: 27-41.

Hedges, S. B., Nussbaum, R. A., and Maxson, L. R. 1993. Caecilian phylogeny and biogeography inferred from mitochondrial DNA sequences of the 12S rRNA and 16S rRNA genes (Amphibia: Gymnophiona). Herpetological Monographs 6: 64-76.

Heinroth, O. 1915. Geburt von *Typhlonectes natans* (Blindwuhle) im Aquarium. Blatter für Aquarien-Terrarienkunde 26: 34-35.

Hetherington, T. E. and Wake, M. H. 1979. The lateral line system in larval *Ichthyophis* (Amphibia: Gymnophiona). Zoomorphologie 93: 209-225.

Hilscher, C. A., Conklin, D. J. and Boyd, S. K. 1994. Neurohypophysial peptides control of oviduct contraction in a caecilian amphibian. American Zoologist 34: 20A.
Hilscher-Concklin, C., Conlon, J. M., and Boyd, S. K. 1998. Identification and localization of neurohypophysial peotides in the brain of a caecilian amphibian, *Typhlonectes natans* (Amphibia: Gymnophiona). Journal of Comparative Neurology 394: 139-151.
Himstedt, W. 1995. Structure and function of the eyes in the caecilian *Ichthyophis kohtaoensis* (Amphibia, Gymnophiona). Zoology 99: 81-94.
Himstedt, W. 1996. *Die Blindwühlen*. Westarp-Wissenschaft Magdeburg. Pp. 160.
Himstedt, W. and Fritzsch, B. 1990. Behavioral evidence for electroreception in larvae of *Ichthyophis kohtaoensis* (Amphibia; Gymnophiona). Herpetological Journal 5: 266-270.
Hinsburg, V. 1901. Die Entwicklung der Nasenhöhle bei Amphibien. Teil III: Gymnophionen. Archiv für Microskopische Anatomie 60: 369-385.
Hogg, J. 1841. On the existence of branchiae in the young caeciliae, and on a modification and extension of the branchial classification of the Amphibia. Annals and Magazine of Natural History 45: 353-363.
Hraoui-Bloquet, S. and Exbrayat, J.-M. 1992. Développement embryonnaire du tube digestif chez *Typhlonectes compressicaudus* (Duméril et Bibron, 1841), Amphibien Gymnophione vivipare. Annales des Sciences Naturelles, Zoologie, Paris 13^{éme} série, 13: 11-23.
Hraoui-Bloquet, S. and Exbrayat, J.-M. 1994. Développement des branchies chez les embryons de *Typhlonectes compressicaudus*, Amphibien Gymnophione vivipare. Annales des Sciences Naturelles, Zoologie, Paris, 13ème série, 15: 33-46.
Hraoui-Bloquet, S. and Exbrayat, J.-M. 1996. Les dents de *Typhlonectes compressicaudus* (Amphibia, Gymnophiona) au cours du développement. Annales des Science Naturelle, Zoologie, Paris, 13ème série 17: 11-23.
Hraoui-Bloquet, S. and Exbrayat, J.-M. 1997a. Développement embryonnaire de la langue de *Typhlonectes compressicaudus*, Amphibien Gymnophione vivipare. Bulletin de la Société Zoologique de France 122: 452.
Hraoui-Bloquet, S. and Exbrayat, J.-M. 1997b. Développement de la langue de *Typhlonectes compressicaudus* (Dumeril et Bibron, 1841), amphibien gymnophione vivipare. Bulletin de la Société Herpétologique de France 82-83: 39-46.
Hraoui-Bloquet, S., Escudie, G. and Exbrayat, J.-M. 1994. Aspects ultrastructuraux de l'évolution de la muqueuse utérine au cours de la gestation chez *Typhlonectes compressicaudus*, Amphibien Gymnophione vivipare. Bulletin de la Société Zoologique de France 119: 237-242.
Ihering, R. 1911a. Cobras e amphibios das Ilhotas de "Aguape". Revista do Museo Paulista 3: 454-461.
Ihering, R. 1911b. Os amphbios do Brasil. Ia. Ordem: Gymnophiona. Revista Museo Paulista 8: 89-91.
Jared, C., Navas, C. A. and Toledo, R. C. 1999. An appreciation of the physiology and morphology of caecilians (Amphibia: Gymnophiona). Comparative Biochemistry and Physiology, Part A, 123: 313-328.
Kozlik, M. 1940. Das Nephron der Gymnophiona. 3 Mitteilung zu: Über den Bau des Nierenkanälchens. Zeitschrift für Anatomie und Entwicklungsgeschichte 110: 767-783.
Krabbe, K. H. 1962. *Studies on the morphogenesis of the brain in some urodeles and gymnophions ("Amphibians") (Morphogenesis of the vertebrate brain IX)*. Ejnar Munksgaard, Copenhagen. Pp. 50 + 100 plates.

Kuhlenbeck, H. 1922. Zur Morphologie des Gymnophionengehirns. Jena Zeitschrift für Naturwissenschaft 58: 453-484.
Largen, M. J., Morris, P. A. and Yalden, D. W. 1972. Observations on the caecilian *Geotrypetes grandisonae* Taylor (Amphibia: Gymnophiona) from Ethiopia. Monitore Zoologico Italiano Supplemento IV 8: 185-205.
Laubmann, W. 1926. Die Entwicklung der Hypophyse bei *Hypogeophis rostratus*. Zeitschrift für Anatomie und Entwicklungsgeschichte 80: 79-103.
Laubmann, W. 1927. Über die Morphogenese vom Gehirn und Geruchsorgan der Gymnophionen. Zeitschrift für Anatomie und Entwicklungsgeschichte 84: 597-637.
Laurin, M. and Girondot, M. 1999. Embryo retention in sarcopterygians, and the origin of the extra-embryonic membranes of the amniotic egg. Annales des Sciences Naturelles, Zoologie, Paris 3: 99-104.
Laurin, M. and Reisz, R. R. 1997. A new perspective on tetrapod phylogeny. Pp. 9-59. In S. Sumida and K. L. Martin (eds), *Amniote Origins: Completing the Transition to Land*. San Diego: Academic Press.
Laurin, M., Reisz, R. R. and Girondot, M. 2000. Caecilian viviparity and amniote origins: a reply to Wilkinson and Nussbaum. Journal of Natural History 34: 311-315.
Lawson, R. 1965. The development and replacement of teeth in *Hypogeophis rostratus* (Amphibia, Apoda). Journal of Zoology 147: 352-362.
Lawson, R. 1966. The development of the centrum of *Hypogeophis rostratus* (Amphibia, Apoda) with special reference to the notochordal (intravertebral) cartilage. Journal of Morphology 118: 137-148.
Leclercq, B., Martin-Bouyer, L. and Exbrayat, J.-M. 1995. Embryonic development of pineal organ in *Typhlonectes compressicaudus* (Dumeril and Bibron, 1841), a viviparous Gymnophionan Amphibia. Pp. 107-111. In G. A. Llorente, A. Montori, X. Santos and M. A. Carretero (eds), *Scientia Herpetologica*, Proceedings of the 7th General Meeting of the Societas Europaea Herpetologica, Barcelona, Spain.
Lescure, J. 1986a. Modes particuliers de reproduction. Pp. 429-456. In P. P. Grasse and M. Delsol (eds), *Traité de Zoologie*, Amphibiens, Tome XIV. Masson, Paris.
Lescure, J. 1986b. Histoire de la classification des Céciles (Amphibia, Gymnophiona). Mémoires de la Société Zoologique de France 43: 11-19.
Leydig, F. 1853. *Anatomisch—Histologische Untersuchungen über Fische und Reptilien*. Georg Riemer, Berlin.
Leydig, F. 1868. Über die Schleichenlurche (Coeciliae). Ein Beitrag zur anatomischen Kenntnis der Amphibien. Zeitschrift für Wissenschaftliche Zoologie 18: 283-286, 291-297.
Linnaeus, C. 1758. Systema naturae per regna tria naturae secundum classes, ordines, genera, species, cum characteribus, differentiis, synonymis, locis. Editio decima, Holmiae.
Linnaeus, C. 1766. Systema naturae per regna tria naturae secundum classes, ordines, genera, species, cum characteribus, differentiis, synonymis, locis. Editio duodecima, Holmiae.
Lyman-Henley, L. P. 1993. Observations on a captive-bore litter of typhlonectid caecilians (Amphibia: Gymnophiona). Herpetological Review 24: 146-147.
Marcus, H. 1908a. Beiträge zur Kenntnis der Gymnophionen. I. Über das Schlundspaltengebiet. Archiv für mikroskopische Anatomie 71: 695-744.
Marcus, H. 1908b. Ueber Mesodermbildung im Gymnophionenkopf. Gesellschaft für morphologische und physiologische Sitzungberichte 24: 79-89.
Marcus, H. 1909. Beitrag zur Kenntnis der Gymnophionen. III. Zur Entwicklungsgeschichte des Kopfes. Morphologisches Jahrbuch 40: 105-183.

Marcus, H. 1910. Beitrag zur Kenntnis der Gymnophionen. IV. Zur Entwicklungsgeschichte des Kopfes. Festschrift für Robert Hertwig, Jena 2: 373-462.

Marcus, H. 1922. Der Kehlkopf bei *Hypogeophis*. Vehrhandlung Anatomisches Gesellschaft Jena 31: 188-202.

Marcus, H. 1923. Beitrag zur Kenntnis der Gymnophionen. VI. Über den Übergang von der Wasser-zur Luftatmung mit besonderer Berücksictugung des Atemmechanismus von *Hypogeophis*. Zeitschrift für Anatomie und Entwicklungsgeschichte 69: 328-343.

Marcus, H. 1927. Lungenstudien. Morphologisches Jahrbuch 58: 100-127.

Marcus, H. 1930. Beitrag zur Kenntnis der Gymnophionen. XIII. Über die Bildung von Geruchsorgan, Tentakel und Choanen bei *Hypogeophis*, nebst Vergleisch mit Dipnoern und Polypterus. Zeitschrift für Anatomie und Entwicklungsgeschichte 91: 657-691.

Marcus, H. 1932. Weitere Versuche und Beobachtungen über die Vorderdarmentwicklung bei den Amphibien. Zoologische Jarhbucher Abteilung Anatomie 55: 581-602.

Marcus, H. 1933. Beitrag zur Kenntnis der Gymnophionen. XX. Zur Entstehung des Unterkiefers von *Hypogeophis*. Anatomischer Anzeiger 77: 178-184.

Marcus, H., 1934a. Über den Einfluss des Kriechens auf Wirbelzahl und Organgestalt bei Apoden. Biologischer Zentralblatt 54: 518-523.

Marcus, H. 1934b. Beitrag zur Kenntnis der Gymnophionen. XXI. Das Integument. Zeitschrift für Anatomie und Entwicklungsgeschichte 103: 189-234.

Marcus, H. 1935a. Zur Entstehung der Stapesplatte bei *Hypogeophis*. Anatomische Anzeiger 80: 81.

Marcus, H. 1935b. Zur Stammengeschichte der Herzens. Morphologisches Jahrbuch 76: 92-103.

Marcus, H. 1937. Beitrag zur Kenntnis der Gymnophionen XXX. Über Myotome Horizontal septum und Rippen bei *Hypogeophis* und Urodelen. Zeitschrift für Anatomie und Entwicklungsgeschichte 107: 531-552.

Marcus, H. 1939. Über Keimbahn, Keimdrüsen, Fettkörper und Urogenitalverbindung bei *Hypogeophis*. Bio-Morphosis 1: 360-384.

Marcus, H. und Blume, W. 1926. Beitrag zur Kenntnis der Gymnophionen VII. Über Wirbel und Rippen bei *Hypogeophis* nebst Bemerkungen über *Torpedo*. Zeitschrift für Anatomie und Entwicklungsgeschichte 80: 1-78.

Marcus, H., Stimmelmayr, E. und Porsch, G. 1935. Beitrag zur Kenntnis der Gymnophionen XXV. Die Ossification des *Hypogeophis* schädels. Morphologisches Jahrbuch 76: 375-420.

Marcus, H. Winsauer, O. und Hueber, A. 1933. Beitrag zur Kenntnis der Gymnophionen XVIII. Der Kinetische Schädel von *Hypogeophis* und die Gehörknöschelchen. Zeitschrift für Anatomie und Entwicklungsgeschichte 100: 149-193.

Marcgrav de Liebstadt, G. 1648. *Historiae rerum naturalium Brasiliae*. Amsterdam folio.

Masood Parveez, U. and Nadkarni, V. B. 1991. Morphological, histological, histochemical, and annual cycle of the oviduct in *Ichthyophis beddomei* (Amphibia: Gymnophiona). Journal of Herpetology 25: 234-237.

Masood Parveez, U. and Nadkarni, V. B. 1993a. The ovarian cycle in an oviparous gymnophione amphibian, *Ichthyophis beddomei* (Peters). Journal of Herpetology 27: 59-63.

Masood Parveez, U. and Nadkarni, V. B. 1993b. Morphological, histological and histochemical studies on the ovary of an oviparous caecilian, *Ichthyophis beddomei* (Peters). Journal of Herpetology 27: 63-69.
Mayer, A. F. J. C. 1835. *Analecten für vergleichende Anatomie*. Bonn.
Meckel, J. F. 1833. *System der vergleichenden Anatomie, sechter Theil*. Halle.
Milner, A. R. 1988. The relationships and origin of living amphibians. Pp. 59-102. In M. J. Benton (ed), *The Phylogeny and Classification of the Tetrapods*. Vol. 1. Amphibians, Reptiles, Birds, Clarendon Press, Oxford.
Moodie, G. E. E. 1978. Observations on the life history of *Typhlonectes compressicaudus* (Duméril and Bibron) in the Amazon basin. Canadian Journal of Zoology 56: 1005-1008.
Mookerjee, H. K. 1942. On the development of the vertebral column in Gymnophiona. Proceedings of the 29th Indian Science Congress 159: 256-258.
Müller, J. 1831. Kiemenlöcher an einer jungen *Coecilia hypocyanea* im Museum der Naturgeschichte zu Leiden beobachtet. Isis von Oken 1831: 709-711.
Müller, J. 1831/32. Beitrage zur Anatomie und Naturgeschichte der Amphibien. Zeitschrift für Physiologie 4: 190-222.
Müller, J. 1835. Über die Kiemenlöcher der jungen *Coecilia hypocyanea*. Archiv für Anatomie, Physiologie und wissenschaft Medicin 1835: 391-397
Murphy, J. D., Quinn, H. and Campbell, J. A. 1977. Observations on the breeding habits of the aquatic caecilian *Typhlonectes compressicaudus*. Copeia 1977: 66-69.
Noble, G. K. 1931. *The Biology of the Amphibia*. McGraw-Hill, New York.
Nussbaum, R. A. 1984. Amphibians of the Seychelles. Pp. 379-415. In D. R. Stoddart (ed), *Biogeography and Ecology of the Seychelles Islands*, Dr. W. Junk Publication, The Hague.
Nussbaum, R. A. 1985. Systematics of the caecilians (Amphibia: Gymnophiona) of the family Scolecomorphidae. Occasional Papers of the Museum of Zoology, University of Michigan 682:1-30.
Nussbaum, R. A. and Hinkel, H. 1994. Revision of East African caecilians of the genera *Afrocaecilia* Taylor and *Boulengerula* Tornier (Amphibia: Gymnophiona: Caeciliaidae). Copeia 1994: 750-760.
Nussbaum, R. A. and Wilkinson, M. 1989. On the classification and phylogeny of caecilians (Amphibia, Gymnophiona), a critical review. Herpetological Monographs 3: 1-42.
Olivecrona, H. 1964. Notes on forebrain morphology in the Gymnophion (*Ichthyophis glutinosus*). Acta Morphologica Neerlando-Scandinavica 6:45-53.
Oppel, M. 1810. Second memoire sur la classification des Reptiles. Annales du Museum National d'Histoire Naturelle 16:394-418.
O'Reilly, J. C. and Ritter, D. A. 1995. Observations on the birth of a caecilian (Amphibia: Gymnophiona). Herpetological Natural History 3: 199-202.
Oyama, J. 1952. A microscopic study of the visceral organs of a Gymnophiona, *Hypogeophis rostratus*. Kumamoto Journal of Science 1B: 117-125.
Paillot, R., Estabel, J. and Exbrayat, J.-M. 1998. Integument development in *Typhlonectes compressicaudus* (Amphibia, Gymnophiona). Pp. 357-362. In C. Miaud and R. Guyetant (eds), *Current Studies in Herpetology*, Proceedings of the 9th General Meeting of the Societas Europaea Herpetologica, Le Bourget du Lac, France.
Parker, H. W. 1936. The amphibians of the Mamfe Division, Cameroons. I. Zoogeography and systematics. Proceedings of the Zoological Society of London. Part I. 1936: 64-163.

Parker, H. W. 1941. The caecilians of the Seychelles. Annals and Magazine of Natural History, Series 11. 7: 1-17.
Parker, H. W. 1956. Viviparous caecilians and amphibian phylogeny. Nature 178: 250-252.
Parker, H. W. and Dunn, E. R. 1964. Dentitional metamorphosis in the Amphibia. Copeia 1964: 75-85.
Peter, K. 1894. Die Wirbelsäule der Gymnophionen. Berichte naturforschung Gesellschaft Freiburg 9: 35-58.
Peter, K. 1895a. Der Anatomie von *Scolecomorphus kirkii*. Berichte naturforschung Gesellschaft Freiburg 9: 183-193.
Peter, K. 1895b. Ueber die Bedeutung des Atlas der Amphibien. Anatomischer Anzeiger 10: 565-574.
Peter, K. 1898. Die Entwicklung und funktionelle Gestaltung des Schädels von *Ichthyophis glutinosus*. Morphologisches Jahrbuch 25: 1-78.
Peter, K. 1908. Zur Anatomie eines ost-afrikansiches apoden nebst Bemerkungen Uber die Einteilung dieser Gruppe. Zoologische Jahrbucher Abteilung für Anatomie 26: 527-536.
Peters, W. 1874a. Observations sur le developpement du *Caecilia compressicauda*. Annales des Science Naturelle Zoologie, Paris, 5ème série 19, art. 13.
Peters, W. 1874b. Derselbe das ferner ueber die Entwicklung der Caecilien und besonders der *Caecilia compressicauda*. Monatsberichte der Akademische Wissenschaft Berlin 1874: 45-49.
Peters, W. 1874c. Über die Entwicklung der Caecilien und besonders der *Caecilia compressicauda*. Monatsberichte der Akademische Wissenschaft Berlin 1874: 45-49.
Peters, W. 1875. Über die Entwicklung der Caecilien. Monatsberichte der Akademie Wissenschaftliche Berlin 1875: 483-486.
Peters, W. 1879. Über die Eintheilung der Caecilien und insbesondere über die Gattungen *Rhinatrema* und *Gymnopis*. Monatsberichte Konigl. Preussische Akademie Berlin 1879: 924-943.
Peters, W. 1880. Über Schädel von zwei Cäcilien, *Hypogeophis rostratus* und *H. seraphini*. Sitzber. Gesellschaft für naturforschung Freiburg 1880: 53-56.
Pillay, K. V. 1957. The hypothalamo-hypophyseal neurosecretary system of *Gegenophis carnosus* Beddome. Zeitschrift für Zellforschung 46: 577-582.
Ramaswami, L. S. 1941a. Some aspects of the cranial morphology of *Uraeotyphlus narayani* Seshachar (Apoda). Records of the Indian Museum 43: 143-207.
Ramaswami, L. S. 1941b. Pterygoquadrate connexions in the embryo of *Ichthyophis glutinosus* (Linné) (Apoda). Nature 148: 470.
Ramaswami, L. S. 1942. The stapedial connexions in *Ichthyophis glutinosus* Linné (Apoda). Current Science Bangalore 11: 106-7.
Ramaswami, L. S. 1943. An account of the head morphology of *Gegenophis carnosus* (Beddome), Apoda. Half-yearly Journal of Mysore University 3: 205-220.
Ramaswami, L. S. 1948. The chondrocranium of *Gegenophis* (Apoda, Amphibia). Proceedings of the Zoological Society 118: 752-760.
Ramaswami, L. S. 1954. The external gills of *Gegenophis* embryos. Anatomischer Anzeiger 101: 120-123.
Ramaswami, L. S. 1958. The development of the apodan vertebral column. Zoologischer Anzeiger 161: 271 280.
Rathke, H. 1852. Bemerkungen über mehrere Körpertheile der *Coecilia annulata*. Müllers Archiv 1852: 334-350.

Reiss, J. O. 1996. Palatal metamorphosis in basal caecilians (Amphibia: Gymnophiona) as evidence for lissamphibian monophyly. Journal of Herpetology 30: 27-39.
Reiss, J. O. 2002. The phylogeny of amphibian metamorphosis. Zoology (Jena) 105: 85-96.
Reiss, J. O. and O'Reilly, J. C. 1999. Skull development in the West African caecilian *Geotrypetes seraphini* (Gymnophiona: Caeciliidae). American Zoologist 39: 82-83A.
Salthe, S. N. 1963. The egg capsules in the Amphibia. Journal of Morphology 113: 161-171.
Salthe, S. N. and Mecham, J. S. 1974. Reproductive and courtship patterns. Pp. 309-521, In B. Lofts (ed), *Physiology of Amphibia*, vol. 2, Academic Press, New York.
Sammouri, R., Renous, S., Exbrayat, J.-M., and Lescure, J. 1990. Développement embryonnaire de *Typhlonectes compressicaudus* (Amphibia, Gymnophiona). Annales des Sciences Naturelle, Zoologie, Paris, 13ème série 11: 135-163.
Sanderson, I. T. 1937. *Animal Treasure*. Viking Press, New York. Pp. 221-224.
Sarasin, P. and Sarasin, F. 1887. Einige Punkte aus der Entwicklungsgeschichte von *Ichthyophis glutinosus*. Zoologischer Anzeiger 19: 194.
Sarasin, P. and F. Sarasin. 1887-1890. *Ergebnisse naturwissenschaftlichen. Forschungen auf Ceylon in den Jahren 1884-1886. Zur Entwicklungsgeschichte und Anatomie der ceylonischen Blindwühle Ichthyophis glutinosus*. C. W. Kriedel's Verlag, Wiesbaden.
Sarasin, P. and Sarasin, F. 1889. Über das Gehörorgan und Tentacle von *Ichthyophis glutinosus*. Sitzungberichte Gesellschaft für Naturwissen Freiborg 1889: 137, 147.
Sarasin, P. and Sarasin, F. 1892. Über das Gehörorgan der Caecilien. Anatomischer Anzeiger 7: 812-815.
Scheltinga, D. M., Wilkinson, M., Jamieson, B. G. M. and Oommen, O. V. 2003. Ultrastructure of the mature spermatozoa of caecilians (Amphibia: Gymnophiona). Journal of Morphology (In press).
Schilling, C. 1935. Das Herz von *Hypogeophis* und seine Entwicklung. Morphologische Jahrbucher 76: 52-91.
Schmidt, A. and Wake, M. H. 1990. Olfactory and vomeronasal systems of caecilians (Amphibia: Gymnophiona). Journal of Morphology 205: 255-268.
Schmidt, A. and Wake, M. H. 1997. Celllular migration and morphological complexity in the caecilian brain. Journal of Morphology 231: 11-27.
Schmidt, A. and Wake, M. H. 1998. Development of the tectum in gymnophiones, with comparison to other amphibians. Journal of Morphology 236: 233-246.
Seba, A. 1735. Locupletissimi rerum naturalium thesauri accurate descriptio. Amstelaedami 2: 26.
Semon, R. 1892. Studien über den Bauplan des Urogenitalsystems der Wirbeltiere. Dargelegt an der Entwicklung dieses Organsystems bei *Ichthyophis glutinosus*. Zeitschrift für Naturwissenschaft 26: 80-203.
Senn, D. G. and Reber-Leutenegger, S. 1986. Notes on the brain of Gymnophiona. Mémoires de la société Zoologique de France 43: 65-66.
Seshachar, B. R. 1936. The spermatogenesis of *Ichthyophis glutinosus* (Linn.) Part I. The spermatogonia and their division. Zeitschrift für Forschungsgemeinschaft 24: 662-706.
Seshachar, B. R. 1937a. The spermatogenesis of *Ichthyophis glutinosus* (Linn.). Part II. The meiotic divisions. Zeitschrift für Forschungsgemeinschaft 27: 133-158.
Seshachar, B. R. 1937b. Germ-cell origin in the adult caecilian *Ichthyophis glutinosus* (Linn.). Zeitschrift für Zellforschung Mikroskopische Anatomie 26: 293-304.
Seshachar, B. R. 1939a. The spermatogenesis of *Uraeotyphlus narayani* Seshachar. La Cellule 48: 63-76.
Seshachar, B. R. 1939b. Testicular ova in *Uraeotyphlus nararyani* Seshachar. Proceedings of the Indian Academy of Science 10B: 213-217.

Seshachar, B. R. 1940. The apodan sperm. Current Science, Bangalore 9: 464-468.
Seshachar, B. R. 1942a. The eggs and embryos of *Gegenophis carnosus*. Bedd. Current Science, Bangalore 1942: 439-441.
Seshachar, B. R. 1942b. Stages in the spermatogenesis of *Siphonops annulatus* Mikan, and *Dermophis gregorii* Blgr. (Amphibian: Apoda). Proceedings of the Indian Academy of Science 15B: 263-277.
Seshachar, B. R. 1942c. Origin of the intralocular oocytes in male Apoda. Proceedings of the Indian Academy of Science 15B: 278-279.
Seshachar, B. R. 1942d. The Sertoli cells in Apoda. Half-yearly Journal of the University of Mysore 3B: 65-71.
Seshachar, B. R. 1943a. The spermatogenesis of *Ichthyophis glutinosus* (Linn.), Part 3, spermatelosis. Proceedings of the National Institute of Science, India 9: 271-286.
Seshachar, B. R. 1943b. The amphibian sperm. Current Science, Bangalore 12: 247-249.
Seshachar, B. R. 1945. Spermatelosis in *Uraeotyphlus narayani* Seshachar and Gegenophis carnosus Beddome (Apoda). Proceedings of the National Institute of Science, India 11: 336-340.
Seshachar, B. R. 1948. The nucleolus of the apodan Sertoli cell. Nature 161: 558-559.
Seshachar, B. R., Balakrishna, T. A. Shakuntala, K. and Gundpapa, K. R. 1982. Some unique features of egg laying and reproduction in *Ichthyophis malabarensis* (Taylor) (Apoda: Amphibia). Current Science, Bangalore 51: 32-34.
Spengel, J. W. 1876. Das Urogenitalsystem der Amphibien. I. Theil. Der anatomische Bau des Urogenitalsystems. Arbeit der zoologische- Zootomie Institut, Würzburg 2: 195-509.
Sprackland, R. G. 1982. *Typhlonectes compressicaudus* (aquatic caecilian) reproduction. Herpetological Review 13: 94.
Stadtmüller, F. 1936. Kranium und Visceralskelett der Stegocephalien und Amphibien. In Bolk *et al.* (eds.), *Handbuch der vergleichende Anatomie der Wirbeltiere*, 4th edition.
Stannius, H. 1846. *Lehrbuch der vergleichenden Anatomie der Wirbelthiere*. Berlin.
Taylor, E. H. 1960. On the caecilian species *Ichthyophis monochrous* and *Ichthyophis glutinosus* with descriptions of related species. University of Kansas Science Bulletin 40: 37-120.
Taylor, E. H. 1965. New Asiatic and African caecilians with a redescription of certain other species. University of Kansas Science Bulletin 46: 253-302.
Tayor, E. H. 1968. *The Caecilians of the World—A Taxonomic Review*. University of Kansas Press, Lawrence, KS. Pp. 848.
Taylor, E. H. 1969a. Skulls of Gymnophiona and their significance in the taxonomy of the group. University of Kansas Science Bulletin 48: 585-687.
Taylor, E. H. 1969b. A new family of African Gymnophiona. University of Kansas Science Bulletin 48: 297-305.
Tiepel, H. 1932. Beitrag zur Kenntnis der Gymnophionen. XV.I. Die Zunge. Zeitschrift für Anatomie und Entwichlungsgeschichte 98: 726-746.
Toews, D. and MacIntyre, D. 1977. Blood respiratory properties of a viviparous amphibian. Nature 266: 464-465.
Tonutti, E. 1931. Beitrag zur Kenntnis der Gymnophionen. XV. Das Genitalsystem. Morphologische Jahrbuch 68: 151-292.
Tonutti, E. 1932. Vergleichende morphologische Studien über Enddaarm und Kopulationsorgane bis weiteren Gymnophionen. Morphologisches Jahrbuch 70: 101-130.
Tonutti, E. 1933. Beitrag zur Kenntnis der Gymnophionen. XIX. Kopulationsorgane bis weiteren Gymnophionenarten. Morphologisches Jahrbuch 72: 155-211.

Trueb, L. and Cloutier, R. 1991a. Toward an understanding of the amphibians: two centuries of systematic history. Pp. 175-193. In H.-P. Schultze and L. Trueb (eds), *Origins of the Higher Groups of Tetrapods Controversy and Consensus*, Comstock, Ithaca.

Trueb, L. and Cloutier, R. 1991b. Phylogenetic investigation of the inter- and intrageneric relationships of the Lissamphibia (Amphibia: Temnospondyli). Pp. 223-313. In H.-P. Schultze and L. Trueb (eds), *Origins of the Higher Groups of Tetrapods Controversy and Consensus*, Comstock, Ithaca.

Van der Horst, G., Visser, J. and van der Merwe, L. 1991. The ultrastructure of the spermatozoan of *Typhlonectes natans* (Gymnophiona: Typhlonectidae). Journal of Herpetology 25: 441-447.

Veit, O. 1965. Beiträge zur Kenntnis des Kopfes der Wirbeltiere IV. Beobachtungen über eine Chorda rostralis bei Hypogeophis. Morphologisches Jahrbuch 107: 11-41.

Visser, M. H. C. 1963. The cranial morphology of *Ichthyophis glutinosus* (Linné) and *Ichthyophis monochrous* (Bleeker). Annals of the University of Stellenbosch, Series A 38: 67-102.

Wake, D. B. 1970. Aspects of vertebral evolution in the modern Amphibia. Forma et Functio 3: 33-60.

Wake, D. B. and Wake, M. H. 1986. On the development of vertebrae in gymnophione amphibians. Mémoirs de la Société Zoologique de France 43: 67-70.

Wake, M. H. 1967. Gill structure in the caecilian genus *Gymnopis*. Bulletin of the Southern California Academy of Sciences 66: 109-116.

Wake, M. H. 1968. Evolutionary morphology of the caecilian urogenital system. Part I. The gonads and fat bodies. Journal of Morphology 126: 291-332.

Wake, M. H. 1969. Gill ontogeny in embryos of *Gymnopis* (Amphibia; Gymnophiona). Copeia 1969: 183-184.

Wake, M. H. 1970. Evolutionary morphology of the caecilian urogenital system. Part II. The kidneys and urogenital ducts. Acta Anatomica 75: 321-358.

Wake, M. H. 1972. Evolutionary morphology of the caecilian urogenital system. IV. The cloaca. Journal of Morphology 136: 353-366.

Wake, M. H. 1976. The development and replacement of teeth in viviparous caecilians. Journal of Morphology 148: 33-64.

Wake, M. H. 1977a. The reproductive biology of caecilians: an evolutionary perspective. Pp. 73-102. In D. H. Taylor and S. I. Guttman (eds), *The Reproductive Biology of Amphibians*, Plenum Publishers, New York.

Wake, M. H. 1977b. Fetal maintenance and its evolutionary significance in the Amphibia: Gymnophiona. Journal of Herpetology 11: 379-386.

Wake, M. H. 1978. Ontogeny of *Typhlonectes obesus*, with emphasis on dentition and feeding. Papéis Avulsos Zoologia 12: 1-13.

Wake, M. H. 1980a. Fetal tooth development and adult replacement in *Dermophis mexicanus* (Amphibia: Gymnophiona): fields versus clones. Journal of Morphology 166: 203-216.

Wake, M. H. 1980b. Reproduction, growth, and population structure of the Central American caecilian *Dermophis mexicanus*. Herpetologica 36: 244-256.

Wake, M. H. 1981. Structure and function of the male Müllerian gland in caecilians (Amphibia: Gymnophiona), with comments on its evolutionary significance. Journal of Herpetology 15: 17-22.

Wake, M. H. 1982. Diversity within a framework of constraints: Reproductive modes in the Amphibia. Pp. 87-106. In D. Mossakowski and G. Roth (eds), *Environmental Adaptation and Evolution. A Theoretical and Empirical Approach*. Gustav Fischer Verlag, Stuttgart.

Wake, M. H. 1985. The comparative morphology and evolution of the eyes of caecilians (Amphibia: Gymnophiona). Zoomorphology 105: 277-295.
Wake, M. H. 1986a. A perspective on the systematics and morphology of the Gymnophiona (Amphibia). Mémoires de la Société Zoologique de France 43: 21-38.
Wake, M. H. 1986b. The morphology of *Idiocranium russelli* (Amphibia: Gymnophiona), with comments on miniaturization through heterochrony. Journal of Morphology 189: 1-16.
Wake, M. H. 1987. Haemal arches in amphibians: a problem in homology and phylogeny. American Zoologist 27: 33A.
Wake, M. H. 1989a. Phylogenesis of direct development and viviparity in vertebrates. Pp. 235-250. In D. B. Wake and G. Roth (eds), *Complex Organismal Functions: Integration and Evolution in Vertebrates*, John Wiley, Chichester.
Wake, M. H. 1989b. Hyobranchial metamorphosis in *Epicrionops* (Amphibia: Gymnophiona: Rhinatrematidae): Replacement of bone by cartilage. Annales des Sciences Naturelles, Zoologie, Paris 10: 171-182.
Wake, M. H. 1992a. Reproduction in caecilians. Pp. 112-120 in W. C. Hamlett (ed), *Reproductive Biology of South American Vertebrates*. Springer Verlag, New York.
Wake, M. H. 1992b. "Regressive" evolution of special sensory organs in caecilians (Amphibia: Gymnophiona): opportunity for morphological innovation. Zoologisches Jahrbuch 122: 325-329.
Wake, M. H. 1993. The evolution of oviductal gestation in amphibians. Journal of Experimental Zoology 266: 394-413.
Wake, M. H. 1994a. Comparative morphology of caecilian sperm (Amphibia: Gymnophiona). Journal of Morphology 221: 261-276.
Wake, M.H. 1994b. The concept of metamorphosis in viviparous vs oviparous amphibians. American Zoologist 34(5): 93A.
Wake, M. H. 1995. The spermatogenic cycle of *Dermophis mexicanus* (Amphibia: Gymnophiona). Journal of Herpetology 29: 119-122.
Wake, M. H. 1998. Cartilage in the cloaca: phallodeal spicules in caecilians (Amphibia: Gymnophiona). Journal of Morphology 237: 177-186.
Wake, M. H. 2003. The osteology of caecilians. Pp. 1811-1878. In H. Heatwole and M. Davies (eds). Amphibian Biology, vol. 5, Osteology. Surrey Beatty and Sons, Chipping Norton, Australia.
Wake, M. H. 2004. Embryonization and the evolution of viviparity. Pp. 151-169. In B. K. Hall, G. Mueller and R. D. Pearson (eds), *The Environment and the Evolution of Development*, MIT Press, Boston, MA.
Wake, M. H. and Dickie, R. 1998. Oviduct structure and function and reproductive modes in amphibians. Journal of Experimental Zoology 282: 477-506.
Wake, M. H. and Hanken, J. 1982. The development of the skull of *Dermophis mexicanus* (Amphibia: Gymnophiona), with comments on skull kinesis and amphibian relationships. Journal of Morphology 173: 203-223.
Wake, M. H. and Nygren, K. M. 1987. Variation in scales in *Dermophis mexicanus* (Amphibia: Gymnophiona: Caeciliidae). Fieldiana, new series 36: 1-8.
Wake, M. H. and Schwenk, K. 1986. A preliminary report on the morphology and distribution of taste buds in gymnophiones, with comparison to other amphibians. Journal of Herpetology 20: 254-256.
Wake, M. H. and Wake, D. B. 2000. Early developmental morphology of vertebrae in caecilians (Amphibia: Gymnophiona): resegmentation and phylogenesis. Zoology Analysis of Complex Systems 103: 68-88.

Wake, M. H., Exbrayat, J.-M. and Delsol, M. 1985. The development of the chondrocranium of *Typhlonectes compressicaudus* (Gymnophiona), with comparison to other species. Journal of Herpetology 19: 68-77.
Wake, M. H., Parra Olea, G. and Sheen, J. P. 2004. Biogeography and molecular phylogeny of certain New World caecilians. In M. A. Donnelly, B. I. Crother, C. Guyer, M. H. Wake and M. White (eds*), Ecology and Evolution in the Tropics: A Herpetological Perspective.* University of Chicago Press, Chicago. (In press).
Welsch, U. 1981. Fine structural and enzyme histochemical observations on the respoiratory epithelium on the caecilian lungs and gills. A contribution to the understanding of the evolution of the vertebrate respiratory epithelium. Archives of Histology Japan 44: 117-133.
Welsch, U. 1982. Morphologische Beobachtungen am thymus larvaler und adultery Gymnophionen. Zoologische Jahrbucher Abteilung für Anatomie 107: 288-305.
Welsch, U. and Storch, V. 1971. Fine structural and enzymehistochemical observations on the notochord of *Ichthyophis glutinosus* and *Ichthyophis kohtaoensis* (Gymnophiona, Amphibia). Zeitschrift für Zellforschung 117: 443-450.
Welsch, U. and Storch, V. 1973. Die Feinstruktur verhornter und nichverhornter ektodermaler Epithelien und der Hautdrusen, embryonaler und adulter Gymnophionen. Zoologische Jahrbuch Anatomie 90: 323-342.
Welsch, U., Müller, W. and Schubert, C. 1977. Electron-microscopical and histochemical observations on the reproductive biology of viviparous caecilians (*Chthonerpeton indistinctum*). Zoologische Jahrbuch Anatomie 97: 532-549.
Welsch, U., Schubert, C. and Storch, V. 1974. Investigations on the thyroid gland of embryonic, larval and adult *Ichthyophis glutinosus* and *Ichthyophis kohtaoensis* (Gymnophiona, Amphibia). Histology, fine structure and studies with radioactive iodide (I 3I). Cell and Tissue Research 155: 245-268.
Wiedersheim, R. 1879. *Die Anatomie der Gymnophionen*. Gustav Fischer Verlag, Jena.
Wiedersheim, R. 1880. Uber den sogenannten Tentakel der Gymnophionen. Zoologischer Anzeiger (Kleinere Mitteilungen) 3: 493-495.
Wilkinson, M. 1989. On the status of *Nectocaecilia fasciata* Taylor, with a discussion of the phylogeny of the Typhlonectidae (Amphibia: Gymnophiona). Herpetologica 45: 23-36.
Wilkinson, M. 1992. On the life history of the caecilian genus *Uraeotyphlus* (Amphibia: Gymnophiona). Herpetological Journal 2: 121-124.
Wilkinson, M. and Nussbaum, R. A. 1998. Caecilian viviparity and anmiote origins. Journal of Natural History 32: 1403-1409.
Wilkinson, M., Richardson, M. K., Gower, D. J. and Oommens, O. V. 2002a. Extended embryo retention, caecilian oviparity and amniote origins. Journal of Natural History 36: 2185-2198.
Wilkinson, M., Sheps, J. A., Oommen, O. V. and Cohen, B. L. 2002b. Phylogenetic relationships of Indian caecilians (Amphibia: Gymnophiona) inferred from mitochondrial rRNA gene sequences. Molecular Phylogenetics and Evolution 23: 401-407.
Williams, E. E. 1959. Gadow's arcualia and the development of tetrapod vertebrae. Quarterly Review of Biology 34: 1-32.
Wrobel, K.-H. and Süss, F. 2000. The significance of rudimentary nephrostomial tubules for the origin of the vertebrate gonad. Anatomy and Embryology 201: 273-290.
Zuber-Vogeli, M. and Doerr-Schott, J. 1981. Description morphologique et cytologique de l'hypophyse de *Typhlonectes compressicaudus* (Duméril et Bibron)

(Amphibien Gymnophione de Guyane francaise). Comptes Rendus des Séances de l'Académie des Sciences de Paris, Series III 292: 503-506.

Zylberberg, L. and Wake, M. H. 1990. Structure of the scales of *Dermophis* and *Microcaecilia* (Amphibia: Gymnophiona), and a comparison to dermal ossifications of other vertebrates. Journal of Morphology 205: 255-268.

Zylberberg, L., Castanet, J. and de Ricqles, A. 1980. Structure of the dermal scales in Gymnophiona (Amphibia). Journal of Morphology 165: 41-54.

CHAPTER 2

Caecilian Phylogeny and Classification

Mark Wilkinson[1*] and Ronald A. Nussbaum[2]

*This work is dedicated to the memories of John Eric Wilkinson and Annie Wilkinson.

2.1 INTRODUCTION

Fifteen years ago, we published a critical review of caecilian phylogeny and classification (Nussbaum and Wilkinson 1989). We hoped to establish some stability in caecilian classification in the face of some highly divergent phylogenetic hypotheses and alternative taxonomic treatments (Wake and Campbell 1983; Duellman and Trueb 1986; Lescure et al. 1986; Laurent 1986). We concluded that caecilian phylogeny was too poorly known to provide the basis for a working phylogenetic classification that recognized only well-founded monophyletic groups. Instead, we provided an interim, conservative classification in which 154 nominate species were partitioned into six families with no sub- or suprafamilial ranks other than 34 genera. Up until 1968, all caecilians were placed into a single family, the Caeciliidae, and we recognized that the subsequent removal of distinctive subsets of species in the establishment of additional families had left the Caeciliidae a most likely paraphyletic assemblage of caecilians that did not fit into one of the better circumscribed families. This is reflected in the relative numbers of taxa: well over half of all recognized caecilian species and genera are caeciliids.

Some important milestones have appeared over the last 15 years in caecilian phylogenetics. Since our previous review, the first phylogenetic study of the interrelationships of caecilians based on DNA sequence data was published. Hedges et al. (1993) analyzed partial 16S and 12S mt rDNA sequences for 13 caecilian species in 9 genera, including members of four of the family-level taxa recognized in our 1989 classification. Recently, the taxonomic coverage for these molecular markers has begun to expand, so that comparative sequence data are now available for 23 species, 16 genera

---
[1]Department of Zoology, The Natural History Museum, London SW7 5BD, United Kingdom
[2]Division of Amphibians and Reptiles, Museum of Zoology, The University of Michigan, Ann Arbor, Michigan, 48109-1079, USA

and for representatives of all six families (Wilkinson *et al.* 2002; 2003b). Gower *et al.* (2002), in a study focusing on ichthyophiid caecilians, demonstrated the potential for sequence data from ribosomal and protein coding (cytochrome B) mt DNA to help resolve low level taxonomic problems when the taxonomic sampling is sufficiently dense, and Gower *et al.* (2005) tentatively identified an undescribed cryptic species of Sri Lankan *Ichthyophis* using molecular data. Recently, San Mauro *et al.* (2004) addressed relationships among single representatives of each of the six families with a combination of complete mitochondrial genomes and RAG-1 nuclear gene sequences.

Morphological data sets have also been expanded in terms of taxa and through the discovery of additional characters, and previously assembled data have been critically reviewed and revised to reduce errors (Naylor and Nussbaum 1980; Nussbaum and Naylor 1982; Scheltinga *et al.* 2003, see also Chapter 7 of this volume; Wilkinson and Nussbaum 1996; Wilkinson 1996a, 1997). There have also been a few phylogenetic studies of monophyletic subgroups at the genus- or species-level using morphological data (Nussbaum and Hinkel 1994; Wilkinson and Nussbaum 1999; Wilkinson *et al.* 2004).

In our 1989 classification, we provided diagnoses of caeciliid genera based on a core set of characters. Although each genus was understood to have a unique combination of characters, uniquely derived characters supporting the monophyly of most caeciliid genera were simply unknown, and knowledge of the diversity within the more speciose nominate genera was limited to one or a few species. Since 1989, a single genus and 16 species (one of which we consider invalid) have been newly described, and 5 species have been removed from synonymy. In the same period, two genera and 5 species have been lost to synonymy (in addition to those we excluded from our treatment and subsequently synonymized). Little else has changed, and the limited low-level taxonomic activity belies the fact that taxonomy at the species- and genus-level remains in need of careful study and stabilisation. The new genus, *Atretochoana* was established to receive a single species of typhlonectid caecilian with a radically divergent morphology discovered in the course of routine taxonomic work (Nussbaum and Wilkinson 1995). *Atretochoana* is the largest lungless tetrapod and the only known lungless caecilian, and it possesses many unique features associated with a novel cranial architecture (Wilkinson and Nussbaum 1997). Its discovery represents a substantial increase in the perceived diversity of caecilians, and of tetrapods (Donoghue and Alverson 2000), and it serves to emphasise the limited knowledge of caecilian biodiversity.

Overall, taxonomic coverage has remained patchy in both morphological and molecular phylogenetic studies. Consequently, even where inferred relationships for the subset of sampled taxa are well-supported, they are not readily translated into a phylogenetic classification of the entire Order. This is exacerbated by the low-level taxonomic uncertainties that are currently a major obstacle to progress in caecilian systematics and for caecilian biology more generally. However, some phylogenetic relationships have been confidently established, some more tentatively so, and, importantly, we have a clearer picture of what remains to be done. In this chapter, we present an overview of current understanding of caecilian phylogeny and an update of our 1989 classification.

2.2   CAECILIAN PHYLOGENY

Caecilians constitute one of three extant orders of the amphibian subclass Lissamphibia, which includes all of the extant Amphibia. The caecilians, frogs and toads, salamanders and newts are generally believed to comprise a monophyletic group based on a variety of presumed shared, derived character states such as smooth (externally scaleless) epidermis and the presence of gonadal fat bodies (e.g., Parsons and Williams 1963). That Lissamphibia is monophyletic with respect to extant taxa is strongly supported by molecular data which also tend to support either a sister group relationship between caecilians and salamanders or between caecilians and salamanders plus frogs (e.g., Feller and Hedges 1998; San Mauro et al. 2004). However, various hypotheses of polyphyletic origins of the three orders from different fossil taxa have been proposed, and monophyly with respect to several extinct groups of Amphibia is far from uniformly accepted (e.g., Milner 1993; Schoch and Milner 2004).

Regardless of origins and relationships, the three extant orders (Gymnophiona = caecilians; Caudata = salamanders; Anura = frogs) are readily distinguished. Many derived characteristics unambiguously identify all species of Gymnophiona (see Diagnosis below) including the presence of a unique, dual, jaw-closing mechanism (Nussbaum 1977, 1983; Fig. 2.1). The presence of paired tentacular sensory organs on the snout on the edges of, or anterior to, the eyes is unique and can be readily determined with a hand lens even for very small specimens, although in some taxa they become apparent only at or close to metamorphosis and they are not present in larvae. Of course, caecilians also have characteristically elongate, snake-like bodies and completely lack limbs or girdles.

In 1989 we considered caecilian phylogeny to be poorly understood but not completely unknown. Nussbaum (1977, 1979, 1985) had provided good morphological support for the monophyly of two families, the Rhinatrematidae and Scolecomorphidae, and for two hypotheses of interfamilial relationships: that the Rhinatrematidae is the sister taxon of all other caecilians, and that the Scolecomorphidae, Typhlonectidae and Caeciliidae comprise a monophyletic group. We also accepted the monophyly of the Typhlonectidae and had a fairly well-supported phylogeny for the four then recognised typhlonectid genera (Wilkinson 1989). We considered the caeciliids of the Seychelles archipelago to be a monophyletic group on the basis of cytogenetic data (Nussbaum and Ducey 1988); and we considered some pairs of caeciliid genera, such as the East

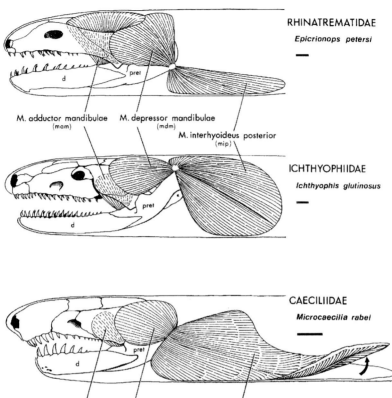

Fig. 2.1. Uniquely derived, dual jaw-closing mechanism, which is diagnostic of caecilians (Nussbaum 1977). The two parts consist of the ancestral jaw-closing mechanism, common to all vertebrates, and a novel component. In the ancestral mechanism, the *m. adductor mandibulae* (*mam*) pulls up on the lower jaw (d, dentary) in front of the articulation of the lower jaw with the skull. In the novel component, the *m. interhyoideus posterior* (*mip*) pulls down on a process of the dentary (*pret*, or *processus retroarticularis*) that projects posteriorly from the jaw articulation, causing the lower jaw to swing up. The *mip*, normally a throat constrictor, takes on a new function of jaw-closing in caecilians. The *m. depressor mandibulae* (*mdm*) serves to open the jaws in all caecilians by pulling up on the *pret*. Rhinatrematids have the presumed ancestral condition in which the ancestral jaw-closing mechanism dominates. Ichthyophiids, caeciliids, and scolecomorphids demonstrate progressively increased dominance of the novel component. The lower jaw becomes progressively shorter and the *pret* progressively longer and more curved dorsally in the same evolutionary sequence. The horizontal bar = 1 mm. Adapted from Nussbaum, R. A. 1983. Journal of Zoology, London 199: 545-554, Fig. 2.

African *Afrocaecilia* and *Boulengerula*, Central American *Dermophis* and *Gymnopis*, and South American *Caecilia* and *Oscaecilia*, to be obviously closely related.

In this section we review current understanding of caecilian phylogeny. Where possible we identify uniquely derived features that unambiguously support hypothesised monophyletic groups of caecilians. Note that inferred relationships of families and of genera that are not monotypic are mostly based on observations of only a subset of the constituent species. Detailed morphological observations and molecular data are lacking for the vast majority of caecilian species. Thus, where we list uniquely derived features supporting the monophyly of a particular group, we do so with the strong caveat that these features may be unknown for some or most of the included species. The relationships we discuss are summarised in Figure 2.2

2.2.1 Rhinatrematidae

Nussbaum (1977) listed six uniquely derived features that support the monophyly of the Rhinatrematidae, a small Neotropical family including the two genera *Rhinatrema* and *Epicrionops* and nine recognised species. Most distinctive of the supporting conditions is the presence of a posterior notch in the squamosal that accommodates a distinct process of the *os basale*. Apart from the lack of a distinct basipterygoid process, Nussbaum's (1977) other derived features of rhinatrematids, the reduction or absence of ceratobranchials 2 and 3; the larynx posterior to the glossal skeleton; and the absence of the *musculus subarcualis rectus* II and III, all relate to the reduction of the posterior hyobranchial apparatus and are unlikely to be completely independent. A *musculus subarcualis rectus* II is now known to be absent in typhlonectids also, although in typhlonectids its absence is associated with the elaboration, rather than reduction, of the buccopharyngeal pump (Wilkinson and Nussbaum 1997), and is surely convergent with the condition in rhinatrematids. Wilkinson (1996a) identified two derived cardiovascular features that also support rhinatrematid monophyly, namely the partial division of the normally undivided sinuatrial aperture and the left pulmonary artery supplying the oesophagus rather than the left lung.

None of these features has been documented in all the currently recognised rhinatrematid species, most of which remain unstudied in any detail, but most have been documented for the monotypic *Rhinatrema* and for one or more species of *Epicrionops*. Based on the general similarity of all known rhinatrematids, we do not expect these morphological features to vary much within the family. Rhinatrematid sampling in Wilkinson's (1997) morphological phylogenetic analysis and in the molecular study of Gower *et al.* (2002) was limited to only two species of *Epicrionops* in the former and one *Epicrionops* and the monotypic *Rhinatrema* in the latter. In both analyses, the results were consistent with rhinatrematid monophyly, and the separation of the rhinatrematids from all other caecilians was well supported. Phylogenetic relationships within the Rhinatrematidae are as yet unstudied. *Epicrionops* differ from *Rhinatrema* primarily in having a longer tail with well-developed haemal arches, a more longitudinal vent, and in retaining a small ceratobranchial 3, all of which are probably ancestral character states. Derived features that support the monophyly of *Epicrionops* are currently unknown.

2.2.2 Non-rhinatrematids — Neocaecilia

The discredited phylogenies of Laurent (1986) and Lescure *et al.* (1986) notwithstanding, since Nussbaum's (1977, 1979) pioneering morphological phylogenetic studies, the basal split in the caecilian tree has been thought to be between the rhinatrematids and all other caecilians. Nussbaum (1977) identified 13 morphological features of rhinatrematids that he considered ancestral and unique within caecilians, with the corresponding derived character states supporting the monophyly of all non-rhinatrematid caecilians. Of these, we currently view five as uniquely derived: sides of parasphenoid converge anteriorly; no contact between quadrate and maxillopalatine; long and recurved retroarticular processes; *musculus interhyoideus posterior* has no insertion on the ceratohyal; and *musculi adductores mandibulae externi* do not meet mid-dorsally. The presence of two rows of trunk neuromasts in larvae is interpreted as an additional unique and ancestral feature of rhinatrematids (Wilkinson 1992a), and Wilkinson (1996a) identified five uniquely derived cardiovascular features that also support the monophyly of the non-rhinatrematids: ventricle relatively narrow and elongate; a bipartite *sinus venosus*; *conus arteriosus* not bent to the left; an elongate *truncus arteriosus*; and common systemicocarotid arteries.

Until recently, molecular data have been mostly consistent with, but had not provided compelling support for, the basal split between rhinatrematids and all other caecilians (Hedges *et al.* 1993). Most recent analyses, both morphological (Wilkinson and Nussbaum 1996; Wilkinson 1997) and molecular (Gower *et al.* 2002; Wilkinson *et al.* 2002, 2003b) data have assumed that the Rhinatrematidae is the sister group of all other caecilians in order to root the caecilian tree and have not provided any further test of this hypothesis. Although taxon sampling is limited, the extensive mitochondrial and nuclear gene data of San Mauro *et al.* (2004) provide very strong support for non-rhinatrematid monophyly, and we consider it to be a very well supported hypothesis on the basis of diverse morphological and molecular data.

Canatella and Hillis (1993) coined the term Stegokrotaphia for the clade including all non-rhinatrematids. However, this name belies a great diversity of zygokrotaphic scolecomorphids, caeciliids and typhlonectids within the group. For this reason we prefer the anatomically neutral Neocaecilia as a rankless epithet for the suprafamilial clade including all caecilians with jaw-closing muscles that do not extend onto the top of the skull from the adductor chamber.

2.2.3 Ichthyophiidae

The Ichthyophiidae comprises some 39 species in the two genera *Caudacaecilia* and *Ichthyophis*. *Caudacaecilia* is restricted to South East Asia whereas *Ichthyophis* also has representatives in South Asia. In Nussbaum's (1979) analyses, *Caudacaecilia* and Ichthyophiidae were closely related, but lacked any uniquely derived features that supported their monophyly. In fact, despite the external similarity of all ichthyophiids, and practical difficulty of distinguishing ichthyophiid species (Nussbaum and Gans 1980), only a single feature that supports ichthyophiid monophyly has been reported previously, the presence of angulate annuli on the anteroventral surface (Nussbaum 1977; Wilkinson and Nussbaum 1996). An additional supporting derived feature, a short parasphenoid that does not extend as far anteriorly as the posterior margin of the choanae, is characteristic of all ichthyophiid skulls and of no non-ichthyophiid skulls that we have examined.

Wilkinson (1997) and Wilkinson et al. (2002) both recovered the very few ichthyophiids included in their morphological and molecular phylogenetic analyses (three and two species of *Ichthyophis* respectively) as a reasonably well-supported monophyletic group. In contrast, the molecular study of Gower et al. (2002) included a broader range of *Ichthyophis* species and yielded optimal trees in which the Ichthyophiidae is paraphyletic with respect to the Uraeotyphlidae. Although this result was not significantly better supported than alternative trees in which the Ichthyophiidae is monophyletic, and thus does not justify any taxonomic changes at this time, it raises the possibility of ichthyophiid paraphyly and indicates the need for further phylogenetic study and additional character data. Phylogenetic relationships among ichthyophiids are poorly understood, but Gower et al. (2002) suggested that the *Ichthyophis* of Sri Lanka, and those of South East Asia comprise distinct monophyletic groups (see also Bossuyt et al. 2004; Gower et al. 2005). All recent work supports the idea that the ichthyophiids were present on the Indian plate prior to its collision with Laurasia, and that South East Asian ichthyophiids result from one or more dispersals out of India (Gower et al. 2002; Wilkinson et al. 2002).

Caudacaecilia and *Ichthyophis* are differentiated by the absence or presence of splenial teeth in adults respectively. Absence of splenial teeth is considered derived within caecilians (Nussbaum 1979) but appears to be quite homoplastic (Nussbaum and Wilkinson 1989). Thus there is only very weak evidence known to support the monophyly of *Caudacaecilia*, no known derived features supporting the monophyly of *Ichthyophis*, and a strong possibility that *Ichthyophis* is paraphyletic with respect to *Caudacaecilia*. No *Caudacaecilia* have been included in any numerical phylogenetic analysis since Nussbaum (1979).

2.2.4 Uraeotyphlidae

The genus *Uraeotyphlus* comprises five nominate species from southern peninsular India. The genus was included in the Caeciliidae by Taylor

(1968, 1969) but transferred to its own subfamily within the Ichthyophiidae by Nussbaum (1979). Duellman and Trueb (1986) elevated Nussbaum's Uraeotyphlinae to family level because their phylogenetic analyses suggested that inclusion of the Uraeotyphlinae within Ichthyophiidae rendered the latter paraphyletic. Uraeotyphlids share a combination of ancestral and derived features, but there are no known uniquely derived features that support the monophyly of the genus and family, a consequence of extensive convergence between uraeotyphlids and various caeciliids. Gower et al. (2002) included three uraeotyphlid species in their molecular phylogenetic analysis and obtained strong support for monophyly of the group as a whole as well as for relationships among the species. The anterior tentacles, dorsal nares, and recessed subterminal mouths, though not unique, are probably derived within the Ichthyophiidae plus Uraeotyphlidae, providing qualified support for the monophyly of *Uraeotyphlus* (Wilkinson and Nussbaum 1996).

2.2.5 Ichthyophiidae + Uraeotyphlidae — Diatriata

Nussbaum (1979), and subsequently both Duellman and Trueb (1986) and Hillis (1991), recovered the Uraeotyphlidae as more closely related to caeciliids, scolecomorphids and typhlonectids than to ichthyophiids. The major change in our understanding of higher caecilian phylogeny has been the adoption of the alternative hypotheses that the Uraeotyphlidae plus Ichthyophiidae are a clade. This hypothesis was well supported by morphological data that incorporated newly discovered cardiovascular characters (Wilkinson and Nussbaum 1996; Wilkinson 1997), and it has subsequently received strong support from analyses of gene sequence data (Wilkinson et al. 2002, 2003b; San Mauro et al. 2004). We currently interpret five uniquely derived features to support the Uraeotyphlidae + Ichthyophiidae clade: circumorbital bone (often termed postfrontal) present, external division of the atrium, an elongate anterior pericardial space, two posterior internal flexures in the *musculus rectus lateralis*, and an internal flexure in the *musculus subvertebralis*. The first of these is somewhat variable in that it may be more or less fused to adjacent elements, and it has been previously considered primitive, but, at least within caecilians, the reverse polarity is more parsimonious. We find the strong and congruent support from morphology and from molecules for this phylogenetic hypothesis to be compelling. We propose Diatriata, as a suitable rankless name for this suprafamilial clade comprising those caecilians with partial external division of their atrium.

2.2.6 The Higher Caecilians — Teresomata

The sister group of the Uraeotyphlidae + Ichthyophiidae, is a clade comprising the Scolecomorphidae, Typhlonectidae and the Caeciliidae, and informally termed the advanced (Nussbaum 1991) or higher (e.g., San Mauro et al. 2004) caecilians. We know of only two putatively unique and derived features of the advanced caecilians, the absence of a true tail, with a true tail being defined compositely as a tapering postcloacal region with internal (vertebrae) and external segmentation (annuli), and the absence of internal flexures in the *musculus rectus lateralis* (Nussbaum and Naylor 1982; Wilkinson 1997). However, this grouping has been recovered in all numerical phylogenetic analyses of morphology or molecules, and with strong support. Compared to its sister clade, the advanced caecilians appear to be more speciose, and the group is much more diverse in morphology, ecology and life history. Its distribution is more cosmopolitan, with representatives in all areas where caecilians are found except South East Asia. The informally named 'advanced caecilians' has stood the test of time, and we suggest Teresomata as a rankless name for this suprafamilial clade, which encompasses the Scolecomorphidae, Typhlonectidae, and paraphyletic Caeciliidae. These caecilians lack true tails and have, for the most part, more rounded (teres) ends to their bodies (soma) than rhinatrematids, ichthyophiids and uraeotyphlids.

2.2.7 Scolecomorphidae

The Scolecomorphidae was established by Taylor (1969) for a few distinctive African caecilians. It currently comprises the West African *Crotaphatrema* and East African *Scolecomorphus*, each with three nominate species. These genera share many distinctive, derived morphological features that provide strong support for scolecomorphid monophyly. These include the absence of stapes and *foramina ovales*, absence of internal processes on lower jaws, a transverse bar extending between the posteromedial edges of the posteriormost ceratobranchial elements of the glossal skeleton (Nussbaum 1977), and a mobile eye attached to the base of the tentacle (Taylor 1968; Nussbaum 1981, 1985; O'Reilly et al. 1996). No *Crotaphatrema* have been included in any numerical phylogenetic analyses, but Wilkinson (1997) included all three *Scolecomorphus* species in his morphological phylogenetic analysis, and two species were included in the Wilkinson et al. (2003b) molecular phylogenetic study. In both cases there was strong support for the monophyly of *Scolecomorphus*, which is supported by the presence of a uniquely large diastema between the vomerine and palatine dental series. Monophyly of *Crotaphatrema* is supported by the particular form of stegokrotaphy in which the upper temporal fossa is obliterated by an outgrowth of the parietal (Nussbaum 1985). Wake (1998) suggested a close relationship between *S. kirkii* and *S. vittatus* on the basis of similar phallus morphology.

2.2.8 Typhlonectidae

The Typhlonectidae was established by Taylor (1968) for a group of Neotropical caecilians that he believed were aquatic and which are now considered to be either aquatic or semi-aquatic (Nussbaum and Wilkinson 1987; 1989). As currently conceived, the family includes five genera, three of which are monotypic. A highly distinctive derived feature that supports typhlonectid monophyly is the fused, sac-like form of the foetal gills (Wilkinson 1989; Wilkinson and Nussbaum 1999). This feature is known in all typhlonectids for which fetuses have been examined but remains unknown in *Potomotyphlus* and *Atretochoana*. Wilkinson and Nussbaum (1999) identified six additional features that appeared to be unique and derived in typhlonectids. These are: small tentacular apertures and (non-protrusible) tentacles, relatively dorsally oriented occipital condyles, a ventral process of the squamosal bracing against the maxillopalatine, M-shaped ceratohyals, a sliding articulation between the third and fourth ceratobranchials, and the *musculus subvertebralis pars ventralis* with a scalloped origin.

Wilkinson's (1997) morphological phylogenetic analysis found strong support for the pairing of the only two typhlonectids, *Chthonerpeton indistinctum* and *Typhlonectes natans*, that it included, and is the only study to have provided a numerical phylogenetic test of typhlonectid monophyly (which has never been seriously questioned). Well-supported relationships within the Typhlonectidae have been inferred on the basis of extensive morphological data (Wilkinson and Nussbaum 1999; see Fig. 2.2). Although monophyly of *Chthonerpeton* has not been established, the generotype, *C. indistinctum*, lies outside a group including all other typhlonectid genera. *Nectocaecilia*, which is believed to be semi-aquatic, is the sister group of a clade of fully aquatic, finned caecilians that comprises *Atretochoana*, *Potomotyphlus*, and *Typhlonectes*. *Potomotyphlus* appears to be the sister genus of the lungless *Atretochoana* on the basis of a number of features associated with a reduction in pulmonary respiration.

2.2.9 Caeciliidae

As currently conceived, the Caeciliidae appears to be a relatively heterogeneous and paraphyletic assemblage comprising all those caecilians that have never been removed to another family. Molecular data strongly support the paraphyly of the Caeciliidae with respect to the Typhlonectidae, recovering *Caecilia* as more closely related to *Typhlonectes* than to a broad range of other caeciliids (Hedges *et al.* 1993; Wilkinson *et al.* 2002, 2003b). Hedges *et al.* (1993) proposed removing caeciliid paraphyly by recognising the typhlonectids at the sub-familial rather than familial level, but in the absence of a better understanding of the relationships among higher caecilians this action would only shift the problem of paraphyly to a different taxonomic level. Paraphyly of the Caeciliidae with respect to the Scolecomorphidae is also suggested by the most recent molecular phylogenetic study (Wilkinson *et al.* 2003b). Phylogenetic analyses based on morphology are less clear cut, with caeciliid monophyly or paraphyly with respect to both the Typhlonectidae and Scolecomorphidae achieved under alternative weighting schemes (Wilkinson 1997) and thus not well-supported. A single derived feature, an elongate *musculus interhyoideus posterior* (Fig. 2.1), supports the monophyly of the Caeciliidae. However, given the strong molecular support for caeciliid paraphyly, the shorter form of the muscle in typhlonectids must be presumed to be due to reversal, presumably associated with their zygokrotaphy.

We have only limited understanding of the relationships among caeciliids. Some relationships are suggested by taxonomic history. *Oscaecilia* was established by Taylor (1968) through the partitioning of *Caecilia* on the basis of a single difference (the eye covered with bone or not) of dubious value. Although both genera are thus of uncertain monophyly, we are content to assume that they are jointly monophyletic based on their overall similarity in external morphology and cranial architecture. Similarly, *Luetkenotyphlus* was established through partitioning of *Siphonops*, and these genera are presumed to be closely related to each other and to *Mimosiphonops* (Wilkinson and Nussbaum 1992), although we know of no uniquely derived characters of this group or of any of its constituent genera. The Central American species of the genera *Dermophis* and *Gymnopis* are presumed to be jointly monophyletic on the basis of their overall similarity in morphology and reproduction. They differ in the presence or absence of splenial teeth and whether the eye is covered with bone, features which appear highly homoplastic in caecilians and which do not convince us of the monophyly of either genus. Of the caeciliid taxa studied thus far, molecular data strongly support a close relationship of *Dermophis* (and by implication *Gymnopis*) with the African genus *Schistometopum*, reflecting a previous taxonomic association.

Among Old World caeciliids, molecular evidence supports the monophyly of the caeciliids of the Seychelles, which Nussbaum and Ducey (1988) had argued previously on the basis of cytological data. All species of the Seychellean caeciliid clade have been included in molecular phylogenetic analyses, an atypical level of taxonomic coverage for caecilians. These data convincingly identify *Praslinia* as the sister taxon to a *Grandisonia* + *Hypogeophis* clade, without resolving the relationships within this latter clade. They suggest that *Grandisonia*, which (like the Seychellean clade as a whole) lacks any known uniquely phenotypic derived traits, is paraphyletic with respect to *Hypogeophis*. Among the sampled taxa, the molecular data also provide strong support for the Indian caeciliid *Gegeneophis* being the sister-group of the Seychelles caeciliids (Wilkinson *et al.*, 2002; 2003b), suggesting the possibility of an Indo-Seychellean caeciliid clade that we might expect to also include the thus far unstudied Indian caeciliid *Indotyphlus*. An Indo-Seychellean connection for caecilians was predicted on the basis of biogeographic and plate tectonics considerations (Nussbaum 1984). Molecular data also support, albeit not strongly, the pairing of the East African *Boulengerula* and West African *Herpele*.

Wilkinson (1997) considered the available morphological data to be insufficient to unravel relationships within the higher caecilians. We interpret the failure of analyses based on morphology to support those relationships that are well supported by molecular data, including the monophyly of the Seychellean caeciliids, the Indo-Seychellean clade, and the *Dermophis-Schistometopum* grouping, to further indicate the current limitations of the morphological data rather than undermining the molecular results. Morphological data have provided very useful phylogenetic characters, but broader and deeper sampling of characters and taxa are needed, particularly within the higher caecilians. Although relationships within the higher caecilians are poorly resolved in our relatively conservative consensus phylogeny (Fig. 2.2), sampling of taxa for both morphological and molecular phylogenetic study is improving.

2.2.10 Prospects

The relatively small number of species of Gymnophiona means that a fairly comprehensive phylogeny for the major lineages of caecilians is a realistic short-term goal that we believe is within reach, and will be reached in the near future. At lower taxonomic levels the systematic foundations are not so good. Until about 1972, caecilian taxonomy was dominated by E. H. Taylor who described many species that have not withstood subsequent scrutiny

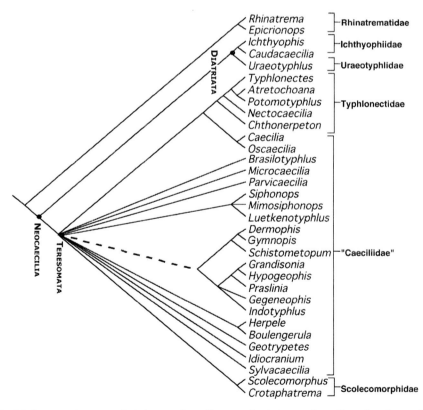

Fig. 2.2. Summary (consensus) phylogeny of caecilians constructed manually and based on the results of previous numerical phylogenetic analyses and inferences from taxonomy as described in the text. The dashed line indicates particularly uncertain monophyly highlighting the potential for the branches from the main polytomy, including many unstudied taxa, to perhaps lie within this group. Note that monophyly of many genera is at best uncertain and that there is a need for much more detailed, low-level, taxonomic work. Original.

and have been lost to synonymy. The reduction in numbers of recognised species has more or less balanced the description of new species since Taylor's time, and the work of checking and testing Taylor's species-level taxonomic work remains largely incomplete. Most caecilian species are poorly known and poorly circumscribed, impinging on all aspects of their biology (e.g., Gower and Wilkinson, 2005), and taxonomic uncertainty and instability can be expected to continue for some time. However, molecular phylogenetic studies at low taxonomic levels are proving useful in helping to delimit and distinguish morphologically similar species in genera such as *Ichthyophis*, where identification to species is notoriously difficult and the existing taxonomy exceedingly problematic. Molecular studies should facilitate taxonomic work while providing the low-level phylogenies needed to study caecilian evolution in detail. Ultimately any understanding of caecilian species, be it derived from molecules or morphology or (preferably) both, will depend on collecting sufficient (i.e., much more) new material across many taxa. Traditionally, caecilians have been poorly sampled in the field, but this is improving. We think it likely that many new caecilian species will be discovered, both in research collections and through additional collecting. It is noteworthy that the majority of new species described in the 15 years since our last review have been described in the last five years, perhaps indicating a rise in interest and the beginnings of a new era of taxonomic discovery (e.g., Gower *et al.*, 2004). As caecilian taxonomy stabilises, and the rate of losses to synonymy decreases, we expect the overall number of recognised caecilian species to increase, and we suspect that the number of currently recognised species is a considerable underestimate of their actual diversity.

2.3 CLASSIFICATION

The classification presented here includes no major changes from our previous summary (Nussbaum and Wilkinson 1989), because we think none is warranted by the current state of knowledge. We draw attention to three important studies of the taxonomy of regional caecilian faunas by Pillai and Ravichandran (1999), Savage and Wake (2001) and Lynch (1999) that provide recent keys. We prefer to accept the paraphyly of the Caeciliidae for the time being, in the belief that the best way of removing it will only become apparent with a more comprehensive understanding of the relationships of caeciliid genera, particularly those that remain unstudied phylogenetically. Formal taxonomic revision aimed at removing the paraphyly now, although well-intended, would be incomplete and unlikely to promote stability in the meanings of names. Thus we use the same six-family system and same format as in 1989, but with updated species lists and synonymies. Where possible, we have indicated derived features for genera that are not monotypic and where this has not been discussed above, but our diagnoses rely upon unique combinations of features, not all of which are themselves unique. Reference to a pseudoectopterygoid (Wilkinson and Nussbaum 1992) is to the separate palatal bone that lies between the pterygoid (process of the quadrate) and the maxillopalatine in some caeciliids. We also prefer to use inner mandibular for those teeth usually referred to as splenials. In both cases our usage is intended to avoid asserting homologies for which there is no good evidence.

ORDER **GYMNOPHIONA** RAFINESQUE **1814**

Diagnosis. Lissamphibia without limbs and girdles; with paired sensory tentacles on the snout; with a dual jaw-closing mechanism consisting of the ancestral component (*musculus adductor mandibulae* pulling up on ramus of lower jaw) and a unique, novel component (*musculus interhyoideus posterior* pulling back and down on retroarticular process of lower jaw); with an eversible phallus in males formed by the posterior part of the cloaca; with an *os basale*; trunk vertebrae with enlarged basipophyseal processes; atlas without a *tuberculum interglenoideum*.

Content. 6 families, 33 genera, 170 species.

Distribution. Pantropical, except for Madagascar and southeast of Wallace's Line. Distribution includes some subtropical areas.

Remarks. Jenkins and Walsh (1993) described a 'caecilian' fossil, *Eocaecilia*, from the Jurassic of North America and more recently Evans and Sigogneau-Russell (2001) described a 'stem-group caecilian' *Rubricacaecilia* from the Lower Cretaceous of North Africa. This former taxon, based on a brief description that has yet to be significantly expanded, has seemingly been universally accepted as a caecilian with little, if any, critical discussion. Unlike living caecilians, *Eocaecilia* has small limbs, and thus, by the diagnosis given above, it is not a gymnophionan (= caecilian). The presence or absence of limbs in the fragmentary *Rubricacaecilia* is unclear, but it has other features, notably the atlas has a *tuberculum interglenoideum*, which places it outside our Gymnophiona. Evans and Sigogneau-Russell (2001) supported Trueb and Cloutier's (1991) use of "Apoda Oppel 1811 for the crown group alone and Gymnophiona Rafinesque for the clade comprising stem-group taxa + Apoda", a use also adopted by Schoch and Milner (2004). Our previous and present diagnoses of Gymnophiona are based on the living species, and they exclude what Evans and Sigogneau-Russell view as stem-group caecilians. Neontologists use the term Gymnophiona to convey generalities about the living caecilians that may not hold or may not be known in the stem group. Changing the meaning of Gymnophiona means that general statements in the literature, such as 'Gymnophiona are legless', become technically incorrect, and neontologists, who comprise the bulk of caecilian researchers would have to learn to adopt Apoda in order to make generalisations about the living caecilians. The latter name is also problematic because zoological classification is festooned with other uses of it, most importantly the homonymy with *Apoda* Haworth 1809, a genus of moth. We believe clarity would be best served if the use of Apoda in caecilian classification were completely abandoned. We see no good reason to rediagnose Gymnophiona in order to accommodate relatively poorly known fossil taxa that, while possibly closely related to caecilians, are not caecilians in the sense in which the term is generally used.

A. Family Rhinatrematidae Nussbaum 1977 (Fig. 2.3 A, B)

Diagnosis. Gymnophiona with true tails consisting of a postcloacal segment with vertebrae, myomeres, and complete skin annuli; primary annuli divided by secondary and tertiary grooves; all annular grooves orthoplicate; numerous scales in all annular grooves and in some of the dorsal grooves of the collars; strongly zygokrotaphic skulls with the *musculi adductores mandibulae externi* passing through the temporal fossae to meet at the midline of the skull along the interparietal suture; maxillopalatine in contact with the quadrate; squamosal widely separated from the frontal, notched posteriorly, the notch opposing a dorsolateral process of the *os basale*; premaxillae and nasals present as separate bones; mouth terminal; retroarticular process of lower jaw short and not curved dorsally; *musculus interhyoideus posterior* short; stapes pierced by stapedial artery; tentacle immediately anterior to or on the anterior edge of eye; eyes visible externally, in a socket in the maxillopalatine; hyobranchium of adults with only three ceratobranchial elements decreasing in size posteriorly, with the larynx situated posterior to the hyobranchium (not enclosed between the two arms of the posteriormost ceratobranchials); hyobranchial elements of larvae mineralized, hyobranchium of metamorphosed individuals cartilaginous; *truncus arteriosus* short; atrium undivided externally.

Content. 2 genera, 9 species.

Distribution. Northern South America.

Remarks. There has been no change in the taxonomy of the Rhinatrematidae since our review of 1989, and no species description or other taxonomic actions since Taylor (1968).

1. *Epicrionops* Boulenger 1883

Type species: *Epicrionops bicolor* by original monotypy.

Diagnosis. Rhinatrematids with three ceratobranchial arches in adults; a longitudinal cloacal opening; relatively long tail consisting of more than 11 postcloacal annuli; more than one row of scales per annular groove.

Content. 8 species: *bicolor, columbianus, lativittatus, marmoratus, niger, parkeri, peruvianus, petersi.*

Distribution. Colombia, Ecuador, Peru, and Venezuela.

2. *Rhinatrema* Duméril & Bibron, 1841

Type species: *Caecilia bivittata* Guérin-Méneville 1829, by monotypy.

Diagnosis. Rhinatrematids with two ceratobranchial arches in adults; more transverse or subcircular cloacal opening; relatively short tail consisting of 11 or fewer postcloacal annuli; a single row of scales per annular groove.

Content. 1 species: *bivittatum.*

54 Reproductive Biology and Phylogeny of Gymnophiona

Fig. 2.3 contd

Distribution. Brazil, French Guiana, Guyana, and Surinam.

Remarks. Nussbaum and Hoogmoed (1979) noted the rarity of *Rhinatrema bivittatum* in scientific collections. Recent fieldwork in French Guiana yielded considerable additional material suggesting the species is not particularly rare in the wild (Wilkinson, Gower and Kupfer, unpublished).

B. Family Ichthyophiidae Taylor 1968 (Figs. 2.3C, D; 2.6A, B)

Diagnosis. Gymnophiona with true tails; skull stegokrotaphic; *musculi adductores mandibulae externi* confined beneath the skull roof, not meeting middorsally; distinct septomaxillae, premaxillae, nasals, and prefrontals; circumorbitals (postfrontals) distinct or partially or entirely fused to maxillopalatine or squamosal; frontal and squamosal in contact; no dorsolateral process on *os basale*; no posterior notch in squamosal; quadrate and maxillopalatine broadly separated; stapes pierced by stapedial artery; mouth nearly terminal; retroarticular process of lower jaw curved dorsally; *musculus interhyoideus posterior* short; tentacular opening between the eye and nostril, usually closer to the eye and below the eye-nostril line; ceratohyal arch U-shaped; four ceratobranchial arches in larvae, arches 3 and 4 fused in adults; larynx positioned between the distal ends of fused arches 3 and 4; all primary annuli subdivided by secondary and tertiary grooves in metamorphosed individuals; annular grooves angulate ventrally over most of the body, orthoplicate posteriorly only; numerous scales present in all but perhaps a few anterior annular grooves of adults; aortic arches proximal to the heart fused into an elongate *truncus arteriosus*; atrium partially divided externally.

Content. 2 genera, 37 species.

Distribution. India, Sri Lanka, and southeast Asia including southern Philippines and Indo-Malaysian Archipelago northwest of Wallace's Line.

1. Caudacaecilia Taylor 1968

Type species. *Ichthyophis nigroflavus* Taylor 1960, by original designation.

Diagnosis. Ichthyophiids without inner mandibular teeth.

Content. 5 species: *asplenia, larutensis, nigroflava, paucidentula, weberi*.

Distribution. Borneo, Malay Peninsular, Philippines, Sumatra.

Remarks. Taylor (1968) and Nussbaum and Gans (1980) examined specimens of *Caudacaecilia* (cf. *asplenia*) from Sri Lanka in museums, but concluded that there was some doubt about the collection data. Recent caecilian surveys in Sri Lanka have not revealed additional specimens attributable to this genus.

2. Ichthyophis Fitzinger 1826

Fig. 2.3 contd

Fig. 2.3 A-B. Adult *Rhinatrema bivittatum* (Rhinatrematidae) from French Guiana; the tentacle (small white spot) is on the anterior edge of the eye; photo by Peter Stafford. **C-D.** Adult *Ichthyophis bannanicus* (Ichthyophiidae) from Mengla, Yunnan, China; the tentacle is just above the mouth and closer to the eye than to the naris; photo by Edmund D. Brodie, Jr.

Type species. *Caecilia glutinosa* Linnaeus 1758, by original monotypy.

Diagnosis. Ichthyophiids with splenial teeth.

Content. 34 species: *acuminatus, atricollaris, bannanicus, beddomei, bernisi, biangularis, billitonensis, bombayensis, dulitensis, elongatus, garoensis, glandulosus, glutinosus, humphreyi, husaini, hypocyaneus, javanicus, kohtaoensis, laosensis, longicephalus, malabaricus, mindanaoensis, monochrous, orthoplicatus, paucisulcus, peninsularis, pseudangularis, sikkimensis, singaporensis, subterrestris, sumatranus, supachaii, tricolor, youngorum.*

Distribution. South East Asia, India, Sri Lanka, southern Philippines, western Indo-Australian Archipelago.

Remarks. Two new species of *Ichthyophis* (*husaini* and *garoensis*) were described by Pillai and Ravichandran (1999), and Gower et al. (2005) suggested the presence of an undescribed cryptic species in Sri Lanka. Kupfer and Müller (2004) provided a rediagnosis of *I. supachii*. *I. longicephalus* (Pillai 1986) was overlooked and not included in our previous treatment (Nussbaum and Wilkinson 1989).

C. Family Uraeotyphlidae Nussbaum 1979 (Fig. 2.4 A, B)

Diagnosis. Gymnophiona with true tails; weakly stegokrotaphic skulls; *m. adductor mandibulae externi* confined beneath the skull roof but may be visible through a small opening between the squamosal and parietal; number and arrangement of skull and lower jaw bones and configuration of the hyobranchium as in the Ichthyophiidae; Stapes imperforate; *m. interhyoideus posterior* short; mouth recessed or subterminal; tentacular opening far forward, below nostril; external nares relatively dorsal, most primary annuli divided by secondary grooves, a few anterior primary annuli may not be subdivided, or primary and higher-order annuli indistinguishable externally; annular grooves do not completely encircle the body; scales present; aortic arches proximal to the heart fused into an elongate *truncus arteriosus*; atrium partially divided externally.

Content. 1 genus, 5 species.

Distribution. Southern peninsular India.

Remarks. The diagnosis (of Nussbaum and Wilkinson 1989) has been modified to account for new information on the diversity of uraeotyphlid annulation patterns (Gower and Wilkinson in prep.; Nussbaum, pers. obs.).

1. *Uraeotyphlus* Peters 1879

Type species. *Coecilia oxyura* Duméril and Bibron 1841, by subsequent designation of Noble (1924).

Diagnosis. As for the family.

Content. 5 species: *interruptus, malabaricus, menoni, narayani, oxyurus.*

Distribution. Southern peninsular India, Kerala, Karnataka and Tamil Nadu.

Remarks. Pillai and Ravichandran (1999) described *U. interruptus* from Kerala since our (Nussbaum and Wilkinson 1989) review.

D. Family Scolecomorphidae Taylor 1969 (Fig. 2.5 C, D, E)

Diagnosis. Gymnophiona that lack stapes and *foramina ovales*; septomaxillae and prefrontals present; no internal process on the pseudoangular bone; no *m. levator quadrati*; *m. interhyoideus posterior* short; a distinctive hyobranchium in which the flattened distal ends of the fourth branchial arch are connected by a transverse bar above the larynx; all primary annuli undivided; aortic arches proximal to the heart fused into an elongate *truncus arteriosus*; atrium undivided externally.

Content. 2 genera, 6 species.

Distribution. East and equatorial West Africa.

1. *Crotaphatrema* Nussbaum 1985

Type species. *Herpele bornmuelleri* Werner, 1899, by original designation.

Diagnosis. Scolecomorphids without temporal fossae; without diastemata between the vomerine and palatine series of teeth; and with the maxillary series extending further posteriorly than the palatine series.

Content. 3 species: *bornmuelleri, lamottei, tchabalmbaboensis*.

Distribution. Cameroon.

Remarks. Lawson (2000) added *C. tchabalmbaboensis* to the known species of *Crotaphatrema*.

2. *Scolecomorphus* Boulenger 1883

Type species. *Scolecomorphus kirkii* Boulenger 1883, by original monotypy.

Diagnosis. Scolecomorphids with temporal fossae; with diastemata between the vomerine and palatine series of teeth; and with all or most of the palatine teeth posterior to the maxillary teeth.

Content. 3 species: *kirkii, uluguruensis, vittatus*.

Distribution. East Africa: Malawi, and Tanzania.

Remarks. The phallus of all *Scolecomorphus* species is equipped with cartilaginous spines or spicules (Taylor 1968; Wake 1998). This may be a uniquely derived character of the genus but the condition in *Crotaphatrema* is undocumented.

E. Family Caeciliidae Rafinesque 1814 (Figs. 2.4 C, D; 2.5 A; 2.6 C-F)

Diagnosis. Gymnophiona with nasal and premaxilla fused; septomaxilla, prefrontal, postfrontal and pterygoid lost or fused to adjacent bones; pseudoectopterygoid present or not; maxillopalatine widely separated from quadrate; temporal fossae usually absent, if present, *m. adductor mandibulae externi* do not pass dorsally through the fossae; *M. interhyoideus posterior* long, with posterior portion extending as far as the sixth trunk myomere; M-shaped ceratohyal arch; larynx between distal ends of fused third and fourth ceratobranchials; no tail; some, none, or all primary annuli subdivided by secondary grooves; no tertiary grooves; scales present or absent; external gills of embryos in three rami (one ramus may be reduced or vestigial), not fused and sac-like; aortic arches proximal to the heart fused into an elongate *truncus arteriosus*; atrium undivided externally.

Content. 21 genera, 98 species.

58 Reproductive Biology and Phylogeny of Gymnophiona

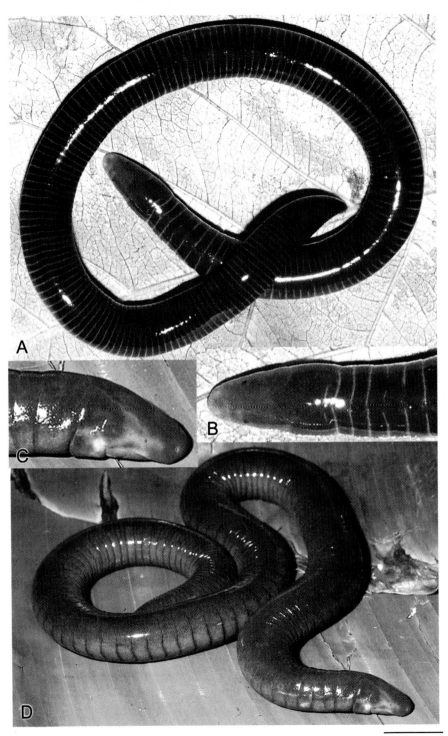

Fig. 2.4 contd

Distribution. Tropical Central and South America, equatorial East and West Africa, islands of the Gulf of Guinea, Seychelles Archipelago, and India.

Remarks. During the late 1980's and early 1990's, the name "Caeciliaidae" was used for this family in an attempt to remove the homonymy of Caeciliidae Rafinesque, 1814 (Amphibia) with Caeciliidae Kolbe, 1880 (Insecta). The problem is outlined in Moore *et al.* (1984). In 1996, the Commission on Zoological Nomenclature, under its plenary powers, ruled (Opinion 1830, BZN 53(1):68-69) that Caeciliidae Rafinesque, 1814 is the valid amphibian name, and the insect name was changed to Caeciliusidae Kolbe, 1880.

1. *Boulengerula* Tornier 1897

Type species. *Boulengerula boulengeri* Tornier 1897, by monotypy.

Diagnosis. Caeciliids with eye (if present) under bone; no temporal fossae; mesethmoid exposed between frontals or not; inner mandibular teeth present or not; no secondary grooves; no scales; tentacular opening nearer to eye than to external naris; an unsegmented terminal shield; no narial plugs; a strong diastema between the vomerine and palatine teeth present or not; a vertical keel on the end of the terminal shield.

Content. 6 species: *boulengeri, changamwensis, denhardti, fischeri, taitanus, uluguruensis.*

Distribution. Kenya, Malawi, Tanzania and Rwanda.

Remarks. Nussbaum and Hinkel (1994) placed *Afrocaecilia* in the synonymy of *Boulengerula* and described *B. fischeri*. Wilkinson *et al.* (2004) removed *B. denhardti* from the synonymy of *Schistometopum gregorii*. Their phylogenetic analyses using Nussbaum and Hinkel's morphological data were unable to resolve well-supported relationships within the genus. Monophyly of the genus is not seriously in question. Many of the diagnostic features are derived within the Neocaecilia although none of them uniquely so, and in molecular analyses, grouping of the two species of *Boulengerula* included thus far is well supported (Wilkinson *et al.* 2003b).

2. *Brasilotyphlus* Taylor 1968

Type species. *Gymnopis braziliensis* Dunn 1945, by original designation and monotypy.

Diagnosis. Caeciliids with eye under bone; no temporal fossae; mesethmoid covered by frontals; no splenial teeth; secondary grooves present; scales present; tentacular opening closer to eye than to external naris; no terminal

---

Fig. 2.4 contd

Fig. 2.4 A-B. Adult *Uraeotyphlus* sp. (Uraeotyphlidae) from the Western Ghats, peninsular India; note the dorsal orientation of the eyes and external nares, and the undivided primary annuli on the anterior portion of the body; photo by John Measey. **C-D**. Adult *Caecilia* cf. *tentaculata* (Caeciliidae) from South America; note the lack of undivided primary annuli over most of the body, but with some subdivided primaries posteriorly; the tentacle cannot be seen, because it is directed ventrally from the "shelf" below the external naris; photo by Peter Stafford.

60 Reproductive Biology and Phylogeny of Gymnophiona

Fig. 2.5 contd

shield; a very short series of premaxillary-maxillary teeth, not extending posterior of the choanae; a strong diastema between the vomerine and palatine teeth; a vertical keel on the body terminus.

Content. 1 species: *braziliensis*.

Distribution. Brazil.

Remarks. The affinities of this genus, which has similarities to *Boulengerula* in the Old-World and *Microcaecilia* in the New, are quite unclear at present (Nussbaum and Hinkel 1994).

3. *Caecilia* Linnaeus 1758

Type species. *Caecilia tentaculata* Linnaeus 1758, by subsequent designation of Dunn (1942).

Diagnosis. Caeciliids with eye not covered with bone; no temporal fossae; mesethmoid exposed between frontals; inner mandibular teeth present; secondary grooves present or absent; scales present or absent; subdermal scales present or absent; tentacular opening directly below external naris, closer to naris than to eye; unsegmented terminal shield present or not; narial plugs present; no diastema between vomerine and palatine teeth; no terminal keel; teeth relatively few and large, usually replaced alternately in groups; vomeropalatine tooth row displaced posteriorly, not parallel to premaxillary-maxillary tooth row, diverging from the latter anteriorly forming an angle where the two rows meet rather than a semicircle.

Content. 33 species: *abitaguae, albiventris, antioquiaensis, armata, attenuata, bokermanni, caribea, corpulenta, crassisquama, degenerata, disossea, dunni, flavopunctata, gracilis, guntheri, inca, isthmica, leucocephala, marcusi, mertensi, nigricans, occidentalis, orientalis, pachynema, perdita, pressula, subdermalis, subnigricans, subterminalis, tentaculata, tenuissima, thompsoni, volcani*.

Distribution. Eastern Panama and northern and central South America.

Remarks. *Caecilia* is the largest genus of caecilians in the New World and has a broad distribution. Over half the species were described by E. H. Taylor, and most are poorly characterised and delimited. Surprisingly, there have been no new species described and little taxonomic work on the group since 1989 despite the clear need for the latter. A helpful treatment of the *Caecilia* of Colombia is given by Lynch (1999). Summers and Wake (2001) redescribed

---

Fig. 2.5 contd

Fig. 2.5 A. Adult female *Schistometopum thomense* (Caeciliidae) and new-born from Ihla São Tomé, Gulf of Guinea; the species is viviparous; photo by Ronald A. Nussbaum. **B**. Adult *Chthonerpeton indistinctum* (Typhlonectidae) from southern South America; this species is terrestrial/semi-aquatic, whereas other genera of the family are more fully aquatic; photo by John Measey. **C-E**. Adult *Scolecomorphus kirkii* (Scolecomorphidae) from Tanzania; the eye, which rides on the base of the tentacle, is shown under the skin and skull bones in the nearly resting position (C), in the nearly maximally protruded position (D), and in the fully protruded position (E) in which the eye is carried completely outside of the skull on the base of the tentacle; there is a pigmentless area of the skin over the track of the eye as it moves back and forth with the tentacle, which presumably allows light to pass through the skin and stimulate the retina; see O'Reilly *et al.* (1996) for details; photos by Daniel Boone.

Fig. 2.6 contd

the holotype of *C. volcani*. *C. isthmica* was accidentally omitted from our previous treatment (although counted in the total number of species).

4. *Dermophis* Peters 1879

Type species. *Siphonops mexicanus* Duméril & Bibron 1841, by subsequent designation of Noble (1924).

Diagnosis. Caeciliids with eye not covered with bone; no temporal fossae; mesethmoid covered or exposed; no inner mandibular teeth; secondary grooves present; scales present; tentacular opening closer to eye than to external naris; no unsegmented terminal shield; no narial plugs; no diastema between vomerine and palatine teeth; no terminal keel.

Content. 7 species: *costaricensis, glandulosus, gracilior, mexicanus, oaxacae, occidentalis, parviceps*.

Distribution. Southern Mexico south to northwestern Colombia.

Remarks. Following our 1989 comment that some of the species of *Dermophis* considered invalid by Savage and Wake (1972) were valid, Savage and Wake (2001) resurrected four species from the synonymies they had previously proposed.

5. *Gegeneophis* Peters 1879

Type species: *Epicrium carnosum* Beddome 1870, by original monotypy.

Diagnosis. Caeciliids with eye under bone; no temporal fossae; mesethmoid not exposed dorsally; inner mandibular teeth present; secondary grooves present; scales present; tentacular opening midway between eye and external naris; no unsegmented terminal shield; narial plugs on tongue; no diastema between vomerine and palatine teeth; terminal keel present or absent.

Content. 8 species: *carnosus, danieli, fulleri, krishni, madhavai, nadkarnii, ramaswamii, seshachari*.

Distribution. India.

Remarks. Five new species of *Gegeneophis* have recently been described from Maharashtra (Ravichandran *et al.* 2003; Giri *et al.* 2003) Karnataka (Pillai and Ravichandran 1999; Bhatta and Srinivasa 2004) and Goa (Bhatta and Prasanth 2004). Giri *et al.* (2003) also revised the generic diagnosis. No uniquely derived traits are known for this genus.

6. *Geotrypetes* Peters 1880

Fig. 2.6 contd

Fig. 2.6 A-B. Adult female *Ichthyophis kohtaoensis* (Ichthyophiidae) from Thailand guarding her clutch of early-stage embryos; the species is oviparous with indirect development; the nest is terrestrial; the hatchling larvae make their way to nearby streams where they grow and eventually metamorphose into terrestrial subadults; photo by Alexander Kupfer. **C-D.** Adult female *Boulengerula boulengeri* (Caeciliidae) from Tanzania guarding her early-stage embryos in a terrestrial nest; the species is oviparous with direct development (no larval stage); photo by Alexander Kupfer. **E-F.** Adult *Boulengerula taitanus* (Caeciliidae) from the Taita Hills, Kenya, guarding her early-stage embryos in a terrestrial nest; the species is oviparous with direct development; photo by Alexander Kupfer.

Type species. *Caecilia seraphini* Duméril 1859, by original monotypy.

Diagnosis. Caeciliids with eye not covered with bone; temporal fossae present; mesethmoid exposed dorsally; inner mandibular teeth present; secondary grooves present; scales present; tentacular opening closer to external naris than to eye; no unsegmented terminal shield; narial plugs present on tongue; no diastema between vomerine and palatine teeth; no terminal keel.

Content. 3 species: *angeli, pseudoangeli, seraphini*.

Distribution. Equatorial West Africa and Bioko Island.

Remarks. Nussbaum and Pfrender (1998) noted that *Schistometopum garzonheydti* from Bioko is a junior synonym of *Geotrypetes seraphini*. *Geotrypetes seraphini* have a distinctively shaped *os basale* (pers. obs.) and unique arrangements of cranial muscles and anterior trunk muscles (Sheps et al. 1997; Wilkinson, unpublished) which, if present in the other species, would provide strong support for the monophyly of the genus.

7. *Grandisonia* Taylor 1968

Type species: *Hypogeophis alternans* Stejneger 1893, by original designation.

Diagnosis. Caeciliids with eye not covered with bone; no temporal fossae; mesethmoid not exposed dorsally; inner mandibular teeth present; secondary grooves present on more than half of the primary annuli, may be missing on some anterior primaries; scales present; tentacular opening variable in position, may be closer to external naris than to eye, nearly midway between eye and external naris, or slightly closer to eye; no unsegmented terminal shield; narial plugs present on tongue; no diastema between vomerine and palatine teeth; no terminal keel.

Content. 4 species: *alternans, brevis, larvata, sechellensis*.

Distribution. Seychelles Archipelago.

Remarks. *Grandisonia diminutiva* is based on juvenile specimens of *G. sechellensis* (Nussbaum, unpublished). The small "tail" that Taylor (1968) considered to be diagnostic of *G. diminutiva* does not exist, and the rest of the characteristics of the type series fall well within the range of *G. sechellensis*.

8. *Gymnopis* Peters 1874

Type species: *Gymnopis multiplicata* Peters 1874, by original monotypy.

Diagnosis. Caeciliids with eye covered by bone; no temporal fossae; mesethmoid not exposed dorsally; inner mandibular teeth present; secondary grooves present; scales present; tentacular opening closer to eye than to external naris; no unsegmented terminal shield; no narial plugs; no diastema between vomerine and palatine teeth; no terminal keel.

Content. 2 species: *multiplicata, syntrema*.

Distribution. Guatemala south to Panama.

Remarks. The complicated history of the taxonomy of *Gymnopis syntrema* was reviewed by Nussbaum (1988). Wake (in Savage and Wake 2001: 52) indicated that she 'disagrees with Nussbaum's concept of *Gymnopis* as it relates to *G. syntremus* [sic]' and would treat the issue elsewhere.

9. *Herpele* Peters 1879

Type species: *Caecilia squalostoma* Stutchbury 1834, by original monotypy.

Diagnosis. Caeciliids with eye under bone; no temporal fossae; mesethmoid slightly visible or not dorsally; inner mandibular teeth present; secondary grooves present; scales present; tentacular opening closer to external naris than to eye; no unsegmented terminal shield; narial plugs present; no diastema between vomerine and palatine teeth; no terminal keel.

Content. 2 species: *multiplicata, squalostoma.*

Distribution. Equatorial West Africa, including the Gulf of Guinea island, Bioko (Fernando Po).

Remarks. The status of *Herpele multiplicata*, a species known only from a holotype specimen that is now lost, was recently reviewed by Wilkinson *et al.* (2003a) who wrongly gave the date of description of *H. squalostoma* as 1859 instead of 1834. *H. squalostoma* has an unusual arrangement of its systemic arches that is unique among vertebrates (Wilkinson 1992b), which, if present also in *H. multiplicata*, would provide good evidence of monophyly of the genus.

10. *Hypogeophis* Peters 1879

Type species: *Coecilia rostrata* Cuvier 1829, by subsequent designation of Parker (1958).

Diagnosis. Caeciliids with eye in socket, not under bone; no temporal fossae; mesethmoid not exposed dorsally; secondary grooves present, confined to posterior third or less of body; scales present; tentacular opening far forward, closer to external naris than to eye; no unsegmented terminal shield; narial plugs present; no diastema between vomerine and palatine teeth; no terminal keel.

Content. 1 species: *rostratus.*

Distribution. Seychelles Archipelago.

11. *Idiocranium* Parker 1936

Type species: *Idiocranium russelli* Parker 1936, by original designation and monotypy.

Diagnosis. Caeciliids with eye not under bone; no temporal fossae; mesethmoid widely exposed dorsally; frontals reduced, not in contact with maxillaries; nasal in contact with squamosal; inner mandibular teeth present; secondary grooves present; scales present; tentacular opening closer to external naris than to eye; no unsegmented terminal shield; narial plugs present; no diastema between vomerine and palatine teeth; no terminal keel.

Content. 1 species: *russelli.*

Distribution. Cameroon.

12. *Indotyphlus* Taylor 1960

Type species: *Indotyphlus battersbyi* Taylor 1960, by original designation and monotypy.

Diagnosis. Caeciliids with eye not under bone; no temporal fossae; mesethmoid not exposed dorsally; inner mandibular teeth present; secondary grooves present; scales present; tentacular opening closer to eye than to external naris; no unsegmented terminal shield; narial plugs on tongue; no diastema between vomerine and palatine teeth; no terminal keel.

Content. 2 species: *battersbyi, maharashtraensis*.

Remarks. Giri *et al.* (2004) recently described the second known species of *Indotyphlus*, noting that the narial plugs are not particularly small in either species, a feature we previously included in the generic diagnosis. We know of no uniquely derived traits.

Distribution. India, northern Western Ghats.

13. *Luetkenotyphlus* Taylor 1968.

Type species: *Siphonops brasiliensis* Lütken, 1852, by original designation and monotypy.

Diagnosis. Caeciliids with eye not under bone; no temporal fossae; dorsal exposure of mesethmoid unknown; no inner mandibular teeth; no secondary grooves; no scales; tentacular opening closer to eye than to external naris; an unsegmented terminal shield; no narial plugs; premaxillary-maxillary series of teeth short, not extending posterior to the choanae; no diastema between vomerine and palatine teeth; a diastema between anterior ends of the two series of vomerine teeth in adults; no terminal keel.

Content. 1 species: *brasiliensis*.

Distribution. Argentina, Brazil, Paraguay.

Remarks. Nussbaum (1986) and Nussbaum and Wilkinson (1989) used the spelling "Lutkenotyphlus" for this genus, because we assumed that Lütken is not a German word. This assumption was based on the fact that Christian Frederik Lütken, for whom the genus is named, is Danish, and his ancestry is also Danish. However, under a strict interpretation of the rules of zoological nomenclature—when it doubt consider a word with an umlaut to be a German word—our earlier spelling appears to have been an unjustified emendation.

14. *Microcaecilia* Taylor 1968

Type species: *Dermophis albiceps* Boulenger 1882, by original designation.

Diagnosis. Caeciliids with eye under bone; no temporal fossae; mesethmoid not exposed dorsally; no inner mandibular teeth; secondary grooves usually present, absent in one species; scales present; tentacular opening closer to eye than to external naris; no unsegmented terminal shield; no narial plugs; no diastema between vomerine and palatine teeth; terminal keel present or absent.

Content. 5 species: *albiceps, rabei, supernumeraria, taylori, unicolor*.

Distribution. Ecuador, French Guiana, Guyana, Surinam, Venezuela.

Remarks. *Microcaecilia* lacks any known uniquely derived features supporting its monophyly. Our previous diagnosis included the absence of a terminal keel, but this is present in at least *M. unicolor* (Wilkinson, unpublished). Nussbaum (unpublished) has examined specimens from northern Brazil that appear to be assignable to *Microcaecilia*.

15. *Mimosiphonops* Taylor 1968

Type species: *Mimosiphonops vermiculatus* Taylor 1968, by original designation and monotypy.

Diagnosis. Caeciliids with eye in socket, not under bone; presence or absence of temporal fossae and dorsal exposure of mesethmoid unknown, probably as in *Siphonops*; inner mandibular teeth present; no secondary grooves; no scales; tentacular opening nearly equidistant between eye and external naris; an unsegmented terminal shield; no narial plugs; a diastema between vomerine and palatine teeth; no terminal keel.

Content. 2 species: *reinhardti, vermiculatus*.

Distribution. southern Brazil.

Remarks. Wilkinson and Nussbaum (1992) placed *Pseudosiphonops* in the synonymy of *Mimosiphonops* and *P. ptychodermis* in the synonymy of *M. vermiculatus*. Wake (2003) gave separate accounts for *Pseudosiphonops* and *Mimosiphonops* in her review of caecilian osteology but did not comment on the earlier proposed synonymy and presented no evidence that would count against it. *Mimosiphonops* lacks known uniquely derived traits. Within the siphonoforms (*Mimosiphonops, Luetkenotyphlus, Siphonops*) the relatively anterior tentacle position, more strongly recessed mouths, and strong diastema between the vomerine and palatine teeth are probably derived.

16. *Oscaecilia* Taylor 1968

Type species: *Caecilia ochrocephala* Cope 1866, by original designation.

Diagnosis. Caeciliids with eye under bone; no temporal fossae; mesethmoid exposed dorsally; inner mandibular teeth present; secondary grooves present; scales present; subdermal scales present or absent; tentacular opening directly below external naris, closer to naris than to eye; no unsegmented terminal shield; narial plugs present; no diastema between vomerine and palatine teeth; no terminal keel; teeth relatively few and large, replaced alternately in groups; vomeropalatine tooth row displaced posteriorly, not parallel to premaxillary-maxillary tooth row, diverging from the latter anteriorly forming an angle where the two rows meet rather than a semicircle.

Content. 9 species: *bassleri, elongata, equatorialis, hypereumeces, koepckeorum, ochrocephala, osae, polyzona, zweifeli*.

Distribution. Southern central and northern South America, Central America (Costa Rica).

Remarks. Lahanas and Savage (1992) described *O. osae* from Costa Rica since our last summary (Nussbaum and Wilkinson 1989).

17. *Parvicaecilia* Taylor 1968

Type species: *Gymnopis nicefori* Barbour 1924, by original designation.

Diagnosis. Caeciliids with eye not under bone; presence or absence of a temporal fossae and dorsal exposure of mesethmoid unknown; no inner mandibular teeth; secondary grooves present; scales present; tentacular opening closer to eye than to external naris; no unsegmented terminal shield; no narial plugs; premaxillary-maxillary series of teeth short, not extending posterior to the choanae; no diastema between the vomerine and palatine teeth; no terminal keel.

Content. 2 species: *nicefori, pricei.*

Distribution. Colombia.

Remarks. We know of no uniquely derived traits of this poorly known genus.

18. *Praslinia* Boulenger 1909

Type species: *Praslinia cooperi* Boulenger 1909, by monotypy.

Diagnosis. Caeciliids with eye not under bone; no temporal fossae; mesethmoid not exposed dorsally; inner mandibular teeth present; teeth small, uniform in size, more than 50 per row, except for inner mandibulars; mouth terminal; secondary grooves present; scales present; tentacular opening adjacent to anterior edge of eye; no unsegmented terminal shield; no narial plugs; no diastema between vomerine and palatine teeth; no terminal keel.

Content. 1 species: *cooperi.*

Distribution. Seychelles Archipelago.

19. *Schistometopum* Parker 1941

Type species: *Dermophis gregorii* Boulenger 1894, by original designation.

Diagnosis. Caeciliids with eye in socket, not under bone; no temporal fossae; mesethmoid exposed dorsally; inner mandibular teeth present; secondary grooves present; scales present; tentacular opening closer to eye than to external naris; no unsegmented terminal shield; no narial plugs; no diastema between vomerine and palatine teeth; no terminal keel.

Content. 2 species: *gregorii, thomense.*

Distribution. Kenya, Tanzania, Gulf of Guinea islands.

Remarks. Nussbaum and Pfrender (1998) noted that *S. ephele* is a geographic variant of *S. thomense*; *S. brevirostre* is a junior synonym of *S. thomense*; and *S. garzonheydti* is a junior synonym of *Geotrypetes seraphini*. Gower and Wilkinson (2002) suggested that the species of *Schistometopum* share a uniquely derived phallus ornamentation, and there is strong support for monophyly from molecular data (Wilkinson *et al.*, 2003b).

20. *Siphonops* Wagler 1828

Type species: *Caecilia annulata* Mikan 1820, by original monotypy.

Diagnosis. Caeciliids with eye in socket, not under bone; no temporal fossae; mesethmoid exposed dorsally; no inner mandibular teeth; no secondary grooves; no scales; tentacular opening closer to eye than to external naris; an unsegmented terminal shield; no narial plugs; no diastema between vomerine and palatine teeth; no terminal keel.

Content. 5 species: *annulatus, hardyi, insulanus, leucoderus, paulensis.*

Distribution. Argentina, Bolivia, Brazil, Colombia, Ecuador, Guyana, Paraguay, Peru, Venezuela, and probably Uruguay, Surinam, and French Guiana.

Remarks. We know of no uniquely derived traits. The three species *annulatus, leucoderus, paulensis* have a distinctive and presumably derived colour pattern with a blue background and whitish annular ring, but this is shared with the species of *Mimosiphonops* suggesting that *Siphonops* may be paraphyletic.

21. Sylvacaecilia Wake 1987

Type species: *Geotrypetes grandisonae* Taylor 1970, by original designation and monotypy.

Diagnosis. Caeciliids with eye not under bone; no temporal fossae; mesethmoid not exposed dorsally; inner mandibular teeth present; secondary grooves present; scales present; tentacular opening closer to eye than to external naris; no unsegmented terminal shield; narial plugs present; no diastema between vomerine and palatine teeth; no terminal keel.

Content. 1 species: *grandisonae.*

Distribution. Ethiopia.

F. Family Typhlonectidae Taylor 1968 (Fig. 2.5 B)

Diagnosis. Gymnophiona with the same number and arrangement of skull bones as caeciliids, but pseudoectopterygoid never present and mesethmoid always covered by frontals; temporal fossae present; eye never under bone; inner mandibular teeth present; teeth monocuspid; *m. interhyoideus posterior* short; tentacle small; choanae large with well developed valves; narial plugs present; relatively large cloacal disk; viviparous; embryonic gills fused into a large, sac-like structure on each side; undivided primary annuli only, or some primary annuli with pseudosecondary grooves; aortic arches proximal to the heart fused into an elongate *truncus arteriosus*; atrium partially divided externally.

Content. 5 genera and 12 species.

Distribution. South America.

1. *Atretochoana* Nussbaum and Wilkinson 1995.

Type species: *Typhlonectes eiselti* Taylor 1968, by original designation and monotypy.

Diagnosis. Typhlonectids with sealed choanae; no lungs; no pulmonary blood vessels; postcranial jaw articulation; posteriorly directed and elongate stapes; novel stapedial muscle; the tentacular aperture intermediate in position between eye and external naris; tentacular groove not covered with bone; body laterally compressed with a middorsal fin; unknown habit, but suspected to be lotic-torrential.

Content. 1 species: *eiselti*.

Distribution. 'South America'.

Remarks. The holotype of *A. eiselti* lacks detailed locality data, but is labelled "South America". Wilkinson *et al.* (1998) reported the second known specimen of *A. eiselti*, which is also without data but suspected to be from Brazil.

2. *Chthonerpeton* Peters 1879

Type species: *Siphonops indistinctus* Reinhardt and Lütken 1861, by monotypy.

Diagnosis. Typhlonectids with the tentacular aperture intermediate in position between eye and external naris; tentacular groove not covered with bone; foetal gills attaching laterally to the nuchal region, the two gill bases well separated dorsally; no lateral compression of the body; no middorsal ridge or free fold (fin); left lung rudimentary; external naris ovate; choanal valve aperture along entire length of valve; cloacal disk subcircular; semiaquatic habit.

Content. 8 species: *arii, braestrupi, exile, indistinctum, noctinectes, onorei, perissodus, viviparum*.

Distribution. Northern Argentina, Brazil, Ecuador, Uruguay.

Remarks. The two species *C. arii* (Cascon and Lima-Verde 1994) and *C. noctinectes* (Da Silva *et al.* 2003) were described since 1989. Based on specimens in Museum collections, there appears to be a number of undescribed *Chthonerpeton* (Wilkinson, unpublished). Uniquely derived traits of *Chthonerpeton* are unknown, but the type species, *C. indistinctum* lacks a distinct *m. rectus lateralis*, a highly unusual condition unknown in other caecilians. If true of the other species this would provide strong support for the monophyly of the genus.

3. *Nectocaecilia* Taylor 1968

Type species: *Chthonerpeton petersii* Boulenger 1882, by original designation.

Diagnosis. Typhlonectids with the tentacular aperture close behind external naris; tentacular groove partially roofed by bone in adults; fetal gills attaching dorsolaterally, the two gill bases slightly separated middorsally; no lateral compression of the body; no dorsal free fold or ridge; left lung well developed; subcircular cloacal disk; external naris subtriangular; choanal valve aperture along entire length of the valve; semiaquatic habit.

Content. 1 species: *petersii*.

Distribution. Venezuela.

Remarks. Wilkinson (1996b) placed *N. haydee* (Roze 1963) in the synonymy of *Typhlonectes natans*.

4. *Potomotyphlus* Taylor 1968

Type species: *Caecilia kaupii* Berthold 1859, by original designation.

Diagnosis. Typhlonectids with the tentacular aperture close behind external naris; tentacular groove partially roofed with bone in adults; foetal gills unknown; body laterally compressed; middorsal free fold or ridge present; left lung well developed, dilated, much wider than right lung; head small relative to body; cloacal disk subcircular posteriorly with a narrower anterior portion; external naris subtriangular; choana extremely large; choanal valve with aperture restricted to a small funnel-like flap; aquatic habit.

Content. 1 species: *kaupii*.

Distribution. Brazil, Ecuador, French Guiana, Peru, Venezuela.

5. *Typhlonectes* Peters 1879

Type species: *Caecilia compressicauda* Duméril and Bibron 1841, by subsequent designation of Dunn (1942).

Diagnosis. Typhlonectids with tentacular aperture close behind external naris; tentacular groove partially roofed by bone in adults; foetal gills attaching dorsally, the two gill bases fused with no separation; body laterally compressed, at least posteriorly, in adults; a middorsal ridge or free fold present; left lung well developed; cloacal disk, subcircular; external naris subtriangular; choanal valve aperture along full length of valve; habit aquatic.

Content. 2 species: *compressicauda, natans*.

Distribution. Colombia, Peru, Venezuela, French Guiana, Guyana, Amazonian Brazil.

Remarks. *T. cunhai* was described by Cascon *et al.* (1991) but its validity has been questioned (Wilkinson and Nussbaum 1997). Based on recent examination (Wilkinson, unpublished) we consider the holotype to be indistinguishable from *T. compressicauda* and thus place *T. cunhai* in the synonymy of *T. compressicauda* (Dumeril and Bibron, 1841). Wilkinson (1991; 1996c) provided formal synonymies for several species recognised by Taylor (1968) but not included in our previous treatment (Nussbaum and Wilkinson 1989). Wilkinson and Nussbaum (1999) identified nine uniquely derived characters supporting monophyly of the genus.

2.4 ACKNOWLEDGEMENTS

We cannot individually acknowledge all of our many colleagues: field workers, bench workers, researchers, curators, students and technicians who have facilitated our work in one way or another, but we wish to express our sincere thanks for these contributions without which most of our work would not be possible. We thank David Gower, Hendrik Müller and Samantha Mohun for reviewing the manuscript and Alexander Kupfer, Daniel Boone, Peter Stafford, and John Measey for providing photographs. MW gratefully acknowledges funding from University of Glasgow New Initiatives Fund, the Museum and Zoology Research Funds of the Natural History Museum, London, the Percy Sladen Memorial Trust and the NERC

(GST/02/832 and GR9/02881). RAN received grants from the National Geographic Society and the U.S. National Science Foundation in support of his field research with caecilians.

2.5 LITERATURE CITED

Barbour, T. 1924. A new *Gymnophis* [sic] from Colombia. Proceedings of the Biological Society of Washington 37: 125-126.

Beddome, R. H. 1870. Descriptions of new reptiles from the Madras Presidency. Madras Monthly Journal of Medical Science 2: 169-176.

Berthold, A. A. 1859. Einige neue Reptilien des Akademie Zoologisches Museums zu Göttingen. Nachricten von der Gesellschaften der Wissenschaftlicen zu Göttingen 1859(1-20): 179-181.

Bhatta, G. and Prasanth, P. 2004. *Gegeneophis nadkarnii* — a caecilian (Amphibia: Gymnophiona: Caeciliidae) from Bondla Wildlife Sanctuary, Western Ghats. Current Science 87: 388-392.

Bhatta, G. and Srinivasa, R. R. 2004. A new species of *Gegeneophis* Peters (Amphibia: Gymnophiona: Caeciliidae) from the surroundings of Mookambika Wildlife Sanctuary, Karnataka, India. Zootaxa 644: 1-8.

Bossuyt, F., Meegaskambura, M., Beenaerts, N., Gower, D. J., Pethiyagoda, R., Roelants, K., Mannaert, A., Wilkinson, M., Bahir, M. M., Manamendra-Arachchi, K., Ng, P. K. L. and Schnieider, C. J., Oommen, O. V. and Milinkovitch, M. C. 2004. Local endemism within the Western Ghats-Sri Lanka biodiversity hotspot. Science 306: 471-489.

Boulenger, G. A. 1882. Catalogue of the Batrachia Gradientia s. Caudata and Batrachia Apoda in the Collection of the British Museum. Taylor and Francis, London. i-viii + 127 pp.

Boulenger, G. A. 1883a. Description of the new genus of Coeciliae. Annals and Magazine of Natural History, series 5, 11: 202-203.

Boulenger, G. A. 1883b. Description of the new genus of Coeciliae. Annals and Magazine of Natural History, ser. 5, 11: 48.

Boulenger, G. A. 1894. Third report on additions to the batrachian collection in the Natural-History Museum. Proceedings of the Zoological Society of London 1894(4): 640-646.

Boulenger, G. A. 1909. A list of the freshwater fishes, batrachians, and reptiles obtained by Mr. J. Stanley Gardiner's expedition to the Indian Ocean. Transactions of the Linnaean Society of London, ser. 2, 12: 291-300.

Cannatella, D. C. and Hillis, D. M. 1993. Amphibian relationships: phylogenetic analysis of morphology and molecules. Herpetological Monographs 7: 1-7.

Cascon, P. and Lima-Verde J. S. 1994. Uma nova especie de *Chthonerpeton* do nordeste Brasileiro (Amphibia, Gymnophiona, Typhlonectidae). Revista-Brasileira-de-Biologia. novembro 54(4): 549-553

Cascon, P., Lima-Verde, J. -S. and Benevides-Marques, R. 1991. Uma nova especie de *Typhlonectes* da Amazonia brasileira (Amphibia, Gymnophiona, Typhlonectidae). Boletim do Museu Paraense Emilio Goeldi Serie Zoologia 7(1): 95-100.

Cuvier, G. 1829. Le Règne Animal. 2nd ed. Déterville and Crochard, Paris.

da Silva, H. R., de Britto-Pereira, M. C. and Caramaschi, U. 2003. A new species of *Chthonerpeton* (Amphibia: Gymnophiona: Typhlonectidae) from Bahia, Brazil. Zootaxa 381: 1–11.

Donoghue, M. J. and Alverson, W. S. 2000. A new age of discovery. Annals of the Missouri Botanical Gardens 87: 110-126.
Duellman, W. E. and Trueb, L. 1986. Biology of Amphibians. McGraw-Hill Book Co., New York. 670 pp.
Duméril, A. M. C. 1859. Reptiles et poissons de l'Afrique occidentale. Étude précédée de considérations générales sur leur distribution géographique. Archives du Muséum d'Histoire Naturelle, Paris 10: 137-268.
Duméril, A. M. C. and Bibron, G. 1841. Erpétologie Générale ou Histoire Naturelle Complète des Reptiles, Vol. 8, Librairie Encyclopédique de Roret, Paris.
Dunn, E. R. 1942. The American caecilians. Bulletin of the Museum of Comparative Zoology, Harvard 91: 339-540.
Dunn, E. R. 1945. A new caecilian of the genus *Gymnopis* from Brazil. American Museum Novitates 1278: 1.
Evans S. E. and Sigogneau-Russell, D. 2001. A stem-group caecilian (Lissamphibia: Gymnophiona) from the Lower Cretaceous of North Africa. Paleontology 44: 259-273.
Feller, A. and Hedges, S. B. 1998. Molecular evidence for the early history of living amphibians. Molecular Phylogeny and Evolution 9: 503-516.
Fitzinger, L. J. 1826. Neue Classification der Reptilien nach ihren natürlichen Verwand-tschaften nebst einer Verwandschaftstafel und einem Verzeichnisse der Reptiliensammlungen des K. K. zoologischen Museums zu Wien. J. G. Heubner, Wien.
Giri, V., Gower, D. J. and Wilkinson, M. 2003. A new species of *Gegeneophis* Peters (Amphibia: Gymnophiona: Caeciliidae) from southern Maharashtra, India, with a key to the species of the genus. Zootaxa 351: 1-10.
Giri, V., Gower, D. J. and Wilkinson, M. 2004. A new species of *Indotyphlus* Taylor (Amphibia: Gymnophiona: Caeciliidae) from the Western Ghats, India. Zootaxa 739: 1-19.
Gower, D. J. and Wilkinson, M. 2002. Phallus morphology in caecilians and its systematic utility. Bulletin of the Natural History Museum (Zoology) 68: 143-154.
Gower D. J. and Wilkinson, M. 2005. The conservation biology of caecilians. Conservation Biology 19(1): 45-55.
Gower, D. J., Bhatta, G., Giri, V., Oommen, O. V., Ravichandran, M. S. and Wilkinson, M. 2004. Biodiversity in the Western Ghats: the discovery of new species of caecilian amphibians. Current Science 87: 739-740.
Gower, D. J., Bahir, M. M., Mapatuna, Y., Pethiyagoda, R., Raheem, D. and Wilkinson, M. 2005. Molecular phylogenetics of Sri Lankan *Ichthyophis* (Amphibia: Gymnopohiona: Ichthyophiidae), with discovery of a cryptic species. Raffles Bulletin of Zoology, in press.
Gower D. J., Kupfer, A., Oommen, O. V., Himstedt, W., Nussbaum, R. A., Loader, S. P., Presswell, B., Müller, H., Krishna, S. B., Boistel, R. and Wilkinson, M. 2002. A molecular phylogeny of ichthyophiid caecilians (Amphibia: Gymnophiona: Ichthyophiidae): out of India or out of South East Asia? Proceedings of the Royal Society (London) B 269: 1563-1569.
Guérin-Méneville, M. F. E. 1829-1844. Iconographie du Règne Animal de G. Cuvier où représentation d'après nature de l'unedes espèces le plus remarquables et souvent non encore figurées, de chaque genre d'animaux. Avec un texte descriptif mis au courant de la science. Ouvrage pouvant servir d'atlas a tous les traités de zoologie. Tome I, Reptiles. J. B. Baillière, Paris.
Haworth, A. H. 1809. Lepidoptera Britannica. Part 2. London. J. Murray. 690 pp.
Hedges, S. B., Nussbaum, R. A. and Maxson, L. R. 1993. Caecilian phylogeny and biogeography inferred from mitochondrial DNA sequences of the 12S rRNA and

16S rRNA genes (Amphibia: Gymnophiona). Herpetological Monographs 7: 64-76.
Hillis, D. M. 1991. The phylogeny of amphibians: current knowledge and the role of cytogenetics. Pp. 7-31. In S. K. Sessions and D. M. Green (eds), *Amphibian Cytogenetics and Evolution*. Academic Press, San Diego.
Jenkins, F. A. and Walsh, D. M. 1993. *An Early Jurassic Caecilian with Limbs*. Nature 365: 246-250.
Kupfer, A. and Müller, H. 2004. On the taxonomy of ichthyophiid caecilians from southern Thailand: A reevaluation of the holotype of *Ichthyophis supachaii* Taylor 1960 (Amphibia: Gymnophiona: Ichthyophiidae). Amphibia-Reptilia 25: 87-97.
Lahanas, P. N. and Savage, J. M. 1992. A new species of caecilian from the Peninsula de Osa of Costa Rica. Copeia 1992: 703-708.
Laurent, R. F. 1986. Ordre des Gymnophiones. Pp. 595-608. In P. -P. Grassé and M. Delsol (eds), *Traité de Zoologie*, Vol. 14, Amphibiens, Masson, Paris, France.
Lawson, D. P. 2000. A new caecilian from Cameroon, Africa (Amphibia: Gymnophiona: Scolecomorphidae). Herpetologica 56(1): 77-80.
Lescure, J., Renous, S. and Gasc, J.-P. 1986. Proposition d'une nouvelle classification des amphibiens gymnophiones.Mémoires de la Société Zoologique de France 43: 145-177.
Linnaeus, C. 1758. Systema Naturae, Ed. 10. Laurentii Salva, Holmiae.
Lütken, C. 1852. *Siphonops brasiliensis*, en ny Art af Ormpaddernes (Caeciliernes) familia. Videnskabelige Meddelelser Dansk Naturhistorisk Forening Kjobenhaun 1851(1852): 52-54.
Lynch, J. D. 1999. Una aproximacion a las culebras ciegas de Colombia (Amphibia: Gymnophiona). Revista de la Academia Colombiana de Ciencias 23 (Supl. Esp.): 317-337.
Mikan, J. C. 1820. Delectus florae et faunae Brasiliensis. Title page, dedication, preface, and 24 plates all unnumbered. J. C. Mikan, Vindobonae.
Milner, A. R. 1993. The Paleozoic relatives of lissamphibians. Herpetological Monographs 7: 8-27.
Moore, T. E., Nussbaum, R. A. and Mockford, E. L. 1984. Caeciliidae in Amphibia and Insecta (Psocoptera): Proposals to remove the homonymy. Z. N. (S.) 2333. Bulletin of Zoological Nomenclature 40: 124-128.
Naylor, B. G. and Nussbaum, R. A. 1980. The trunk musculature of caecilians (Amphibia: Gymnophiona). Journal of Morphology 166: 259-273.
Noble, G. K. 1924. Contributions to the herpetology of the Belgian Congo based on the collection of the American Museum Congo Expedition 1909-1915. Part III. Amphibia: with abstracts from the field notes of Herbert Lang and James P. Chapin. Bulletin of the American Museum of Natural History 49: 147-347.
Nussbaum, R. A. 1977. Rhinatrematidae: A new family of caecilians (Amphibia: Gymnophiona). Occasional Papers of the Museum of Zoology, University of Michigan 682: 1-30.
Nussbaum, R. A. 1979. The taxonomic status of the caecilian genus *Uraeotyphlus* Peters. Occasional Papers of the Museum of Zoology, University of Michigan 687: 1-20.
Nussbaum, R. A. 1981. *Scolecomorphus lamottei*, a new caecilian from West Africa (Amphibia: Gymnophiona: Scolecomorphidae). Copeia 1981(2): 265-269.
Nussbaum, R. A. 1983. The evolution of a unique dual jaw-closing mechanism in caecilians (Amphibia: Gymnophiona) and its bearing on caecilian ancestry. Journal of Zoology, London 199: 545-554.

Nussbaum, R. A. 1984. Amphibians of the Seychelles. Pp. 379-415. In D. R. Stoddard (ed), *Biogeography and Ecology of the Seychelles Islands*. Dr. W. Junk Publishers, The Hague, Boston, Lancaster.

Nussbaum, R. A. 1985. Systematics of caecilians (Amphibia: Gymnophiona) of the family Scolecomorphidae. Occasional Papers of the Museum of Zoology, University of Michigan 713: 1-49.

Nussbaum, R. A. 1986. The taxonomic status of *Lutkenotyphlus* (sic) *brasiliensis* (Lütken) and *Siphonops confusionis* Taylor (Gymnophiona: Caeciliidae). Journal of Herpetology 20(3): 441-444.

Nussbaum, R. A. 1988. On the status of *Copeotyphlinus syntremus, Gymnopis oligozona,* and *Minascaecilia sartoria* (Gymnophiona: Caeciliidae): A comedy of errors. Copeia 1988: 921-928.

Nussbaum, R. A. 1991. Cytotaxonomy of caecilians. Pp. 33-66. In S. K. Sessions and D. M. Green (eds), *Amphibian Cytogenetics and Evolution*. Academic Press, San Diego.

Nussbaum, R. A. and Ducey, P. K. 1988. Cytological evidence for monophyly of the caecilians (Amphibia: Gymnophiona) of the Seychelles Archipelago. Herpetologica 44: 290-296.

Nussbaum, R. A. and Gans, C. 1980. On the *Ichthyophis* (Amphibia: Gymnophiona) of Sri Lanka. Spolia Zeylanica 35 (I and II): 137-154.

Nussbaum, R. A. and Hinkel, H. 1994. Revision of East African caecilians of the genera *Afrocaecilia* Taylor and *Boulengerula* Tornier (Amphibia: Gymnophiona: Caeciliaidae). Copeia 1994(3): 750-760.

Nussbaum, R. A. and Hoogmoed, M. S. 1979. Surinam caecilians, with notes on *Rhinatrema bivittatum* and the description of a new species of *Microcaecilia* (Amphibia, Gymnophiona). Zoologische Mededelingen Rijksmuseum van Natuulijke Historie Leiden 54(14): 217-235.

Nussbaum, R. A. and Naylor, B. G. 1982. Variation in the trunk musculature of caecilians (Amphibia: Gymnophiona). Journal of Zoology, London 198: 383-393.

Nussbaum, R. A. and Pfrender, M. E. 1998. Revision of the African caecilian genus *Schistometopum* Parker (Amphibia: Gymnophiona: Caeciliidae). Miscellaneous Publications of the Museum of Zoology, University of Michigan No. 187, i-iv + 32 pp.

Nussbaum, R. A. and Wilkinson, M. 1987. Two new species of *Chthonerpeton* (Amphibia: Gymnophiona: Typhlonectidae) from Brazil. Occasional Papers of the Museum of Zoology, University of Michigan 716: 1-15.

Nussbaum, R. A. and Wilkinson, M. 1989. On the classification and phylogeny of caecilians (Amphibia: Gymnophiona), a critical review. Herpetological Monographs 3: 1-42.

Nussbaum, R. A. and Wilkinson, M. 1995. A new genus of lungless tetrapod: a radically divergent caecilian (Amphibia: Gymnophiona). Proceeding of the Royal Society of London B, 261: 331-335.

O'Reilly, J. C., Nussbaum, R. A. and Boone, D. 1996. Vertebrate with protrusible eyes. Nature 382: 33.

Parker, H. W. 1936. The amphibians of the Mamfe Division, Cameroons (1) Zoogeography and systematics. Proceedings of the Zoological Society of London Pt. 1: 64-163.

Parker, H. W. 1941. The caecilians of the Seychelles. Annals and Magazine of Natural History, ser. 11, 7: 1-17.

Parker, H. W. 1958. Caecilians of the Seychelles Islands with description of a new species. Copeia 1958: 71-17.

Parsons, T. S. and Williams, E. E. 1963. The relationships of the modern amphibia: a re-examination. Quarterly Review of Biology 38: 26-53.
Peters, W. 1874. Über neue Amphibien (*Gymnopis, Siphonops, Polypedates, Rhacophorus, Hyla, Cyclodus, Euprepes, Clemmys*). Monatsbericht der Deutschen Akademie der Wissenschaften zu Berlin 1874: 616-624.
Peters, W. 1879. Über die Eintheilung der Caecilien und insbesondere über die Gattungen *Rhinatrema* und *Gymnopis* Monatsbericht der Deutschen Akademie der Wissenschaften zu Berlin 1879: 924-943.
Peters, W. 1880. Über Schädel von zwei Coecilien *Hypogeophis rostratus* und *Hypogeophis seraphini*. Sitzung Bericht Geschichte Naturforschung Frankfurt 1880: 53-55.
Pillai, R. S. 1986. Amphibian fauna of Silent Valley, Kerala, S. India. Records of the Zoological Survey of India 84(1-4): 229-242.
Pillai, R. S. and Ravichandran, M. S. 1999. Gymnophiona (Amphibia) of India a taxonomic study. Records of the Zoological Survey of India Occasional Papers 172: 1-117.
Rafinesque, C. S. 1814. Fine del Prodromo d'Erpetologia Siciliana. Specchio delle Scienze o Giornale Enciclopedico di Sicilia, Palermo. Vol. 2: 102-104.
Ravichandran, M. S., Gower, D. J. and Wilkinson, W. 2003. A new species of *Gegeneophis* Peters (Amphibia: Gymnophiona: Caeciliidae) from Maharashtra, India. Zootaxa 350: 1-8.
Reinhardt, J. T. and Lütken, C. F. 1861. Bidrag til kundskab om Brasiliens Padder og Krybdyr. Videnskabelige Meddelelser Dansk Naturhistorisk Forening Kjobenhaun 1861(10-15): 143-242.
Roze, J. 1963. Una nueva especie de cecilidos (Amphibia: Gymnophiona) de Venezuela, con notas sobre los generos *Chthonerpeton* and *Typhlonectes*. Acta Biologica Venezuelica. 3: 279-282.
San Mauro, D., Gower, D. J., Oommen, O. V., Wilkinson, M. and Zardoya, R. 2004. Phylogeny of caecilian amphibians (Gymnophiona) based on complete mitochondrial genomes and nuclear RAG 1. Molecular Phylogenetics and Evolution 33: 413-427.
Savage, J. M. and Wake, M. H. 1972. Geographic variation and systematics of the Middle American caecilians, genera *Dermophis* and *Gymnopis*. Copeia 1972: 680-695.
Savage, J. M. and Wake, M. H. 2001. Reevaluation of the status of taxa of Central American Caecilians (Amphibia: Gymnophiona), with comments on their origin and evolution. Copeia 2001: 52-64.
Scheltinga, D. M., Wilkinson, M., Jamieson, B. G. M. and Oommen, O. V. 2003. Ultrastructure of the mature caecilian spermatozoa (Amphibia: Gymnophiona). Journal of Morphology 258: 179-192.
Schoch, R. R. and Milner, R. A. 2004. Structure and implications of theories on the origin of lissamphibians. Pp. 345- 3777. In: G. Arratia, M. V. H. Wilson and R. Cloutier (eds). Recent Advances in the Origin and Early Radiation of Vertebrates. München, Germany: Verlag Dr. Friedrich Pfeil. 703 pp.
Sheps, J. A., Wilkinson, M. and Orr, K. 1997. Novel cranial muscles and actions in a caecilian. Journal of Morphology 232: 322.
Stejneger, L. 1893. On some collections of reptiles and batrachians from East Africa and the adjacent islands, recently received from Dr. W. L. Abbott and Mr. William Astor Chanler, with descriptions of new species. Proceedings of the United States National Museum 16: 711-741.
Stutchbury, I. 1834. Description of a new species of the genus *Chamaeleon*. Transactions of the Linnaean Society of London, ser 1, 17: 362.

Summers A. P. and Wake, M. H. 2001. Clarification regarding the holotype of *Caecilia volcani* (Amphibia: Gymnophiona). Copeia: 2001: 561–562.

Taylor, E. H. 1960a. A new caecilian genus in India. University of Kansas Science Bulletin 40: 31-36.

Taylor, E. H. 1960b. On the caecilian species *Ichthyophis monochrous* and *Ichthyophis glutinosus* and related species. University of Kansas Science Bulletin 40(4): 37-120.

Taylor, E. H. 1968. The caecilians of the World. University of Kansas Press, Lawrence, Kansas. 848 pp.

Taylor, E. H. 1969. A new family of African Gymnophiona. University of Kansas Science Bulletin 48: 297-305.

Taylor, E. H. 1970. A new caecilian from Ethiopia. University of Kansas Science Bulletin 48: 849-854.

Tornier, G. 1897. Die Kriechtiere Deutsch-Ost-Afrikas. Beiträge zur Systematik und Descendenzlehre, Reimer, Berlin.

Trueb, L. and Cloutier, R. 1991. A phylogenetic investigation of the inter- and intrarelationships of the Lissamphibia (Amphibia: Temnospondyli). Pp. 223–313. In H. P. Schultze and L. Trueb (eds), *Origins of the Major Groups of Tetrapods: Controversies and Consensus*. Cornell University Press, Ithaca, NY.

Wagler, J. 1828. Auszuge aus seinem Systema Amphibiorum. Isis, von Oken 21: 740-744.

Wake, M. H. 1987. A new genus of African caecilian (Amphibia: Gymnophiona). Journal of Herpetology 21: 6-15.

Wake, M. H. 1998. Cartilage in the cloaca: phallodeal spicules in caecilians (Amphibia: Gymnophiona). Journal of Morphology 237 (2): 177-186.

Wake, M. H. 2003. Osteology of caecilians. Pp. 1809-1876. In: H. Heatwole and M. Davies (eds) *Biology of Amphibians* vol. 5, *Osteology*. Surrey Beatty and Sons, Chipping Norton, Australia. 574 pp.

Wake, M. H. and Campbell, J. A. 1983. A new genus and species of caecilian from the Sierra de las Minas of Guatemala. Copeia 1983: 857-863.

Werner, F. 1899. Ueber Reptilien und Batrachier aus Togoland, Kamerun, und Deutsch-Neu-Guinea, Grössenteils aus dem K. Museum für Naturkunde in Berlin. Verhandlungen Zoologische-Botanische Gesellschaft in Wien 49: 132-157.

Wilkinson, M. 1989. On the status of *Nectocaecilia fasciata* Taylor, with a discussion of the phylogeny of the Typhlonectidae (Amphibia: Gymnophiona). Herpetologica 45: 23-36.

Wilkinson, M. 1991. Adult tooth crown morphology in the Typhlonectidae (Amphibia: Gymnophiona): a reinterpretation of variation and its significance. Zeitschrift fur zoologische Systematische und Evolutions-forschung 29: 304-311.

Wilkinson, M. 1992a. The phylogenetic position of the Rhinatrematidae (Amphibia: Gymnophiona): evidence from the larval lateral line system. Amphibia-Reptilia 13: 74-79.

Wilkinson, M. 1992b. Novel modification of the tetrapod cardiovascular system in the West African Caecilian *Herpele squalostoma* (Amphibia: Gymnophiona: Caeciliaidae). Journal of Zoology 228: 277-286.

Wilkinson, M. 1996a. The heart and aortic arches of rhinatrematid caecilians (Amphibia: Gymnophiona). The Zoological Journal of the Linnean Society 118: 135-150.

Wilkinson, M. 1996b. Resolution of the taxonomic status of *Nectocaecilia haydee* (Roze) and a key to the genera of the Typhlonectidae (Amphibia: Gymnophiona). Journal of Herpetology 30: 413-415.

Wilkinson, M. 1996c. The taxonomic status of *Typhlonectes venezuelense* Fuhrmann (Amphibia: Gymnophiona: Typhlonectidae). The Herpetological Journal 6: 30-31

Wilkinson, M. 1997. Characters, congruence and quality: a study of neuroanatomical and traditional data in caecilian phylogeny. Biological Reviews 72: 423-470.

Wilkinson, M. and Nussbaum, R. A. 1992. Taxonomic status of *Pseudosiphonops ptychodermis* Taylor and *Mimosiphonops vermiculatus* Taylor (Amphibia: Gymnophiona: Caeciliaidae), with description of a new species. Journal of Natural History 26: 675-688.

Wilkinson, M. and Nussbaum, R. A. 1996. On the phylogenetic position of the Uraeotyphlidae (Amphibia: Gymnophiona). Copeia 1996: 550-562.

Wilkinson, M. and Nussbaum, R. A. 1997. Comparative morphology and evolution of the lungless caecilian *Atretochoana eiselti* (Taylor) (Amphibia: Gymnophiona: Typhlonectidae). The Biological Journal of the Linnean Society 62: 39-109.

Wilkinson, M. and Nussbaum, R. A. 1999. Evolutionary relationships of the lungless caecilian *Atretochoana eiselti* (Amphibia: Gymnophiona: Typhlonectidae). Zoological Journal of the Linnean Society 126: 191-123.

Wilkinson, M., Müller, H. and Gower, D. J. 2003a. On *Herpele multiplicata* (Amphibia: Gymnophiona: Caeciliidae). African Journal of Herpetology 52: 119-122.

Wilkinson, M., Loader, S. P., Müller, H. and Gower, D. J. 2004. Taxonomic status and phylogenetic relationships of *Boulengerula denhardti* Nieden 1912 (Amphibia: Gymnophiona: Caeciliidae). Mitteilungen aus dem Museum für Naturekunde in Berlin, Zoologische Reihe 80: 41-51.

Wilkinson, M., Loader, S. P., Gower, D. J., Sheps, J. A. and Cohen, B. L. 2003b. Phylogenetic relationships of African caecilians (Amphibia: Gymnophiona): insights from mitochondrial rRNA gene sequences. African Journal of Herpetology 52: 83-92.

Wilkinson, M., Oommen, O. V., Sheps, J. A. and Cohen, B. L. 2002. Phylogenetic relationships of Indian caecilians (Amphibia: Gymnophiona) inferred from mitochondrial rRNA gene sequences. Molecular Phylogenetics and Evolution 23: 401-407.

Wilkinson, M., Sebben, A., Schwartz, E. N. F. and Schwartz, C. 1998. The largest lungless tetrapod: report on a second specimen of the lungless caecilian *Atretochoana eiselti* (Amphibia: Gymnophiona: Typhlonectidae) from Brazil. Journal of Natural History 32: 617-627.

CHAPTER 3

Anatomy with Particular Reference to the Reproductive System

Jean-Marie Exbrayat and Jeanne Estabel

3.1 MAIN ANATOMICAL CHARACTERISTICS OF GYMNOPHIONA

3.1.1 General Characteristics

The general anatomical characteristics of Gymnophiona have been covered by several authors (Sarasin and Sarasin 1887-1890; Taylor 1968; Delsol et al. 1980; Wake 1986; Himstedt 1996; Jared et al. 1999; Exbrayat 2000). The Gymnophiona are elongate animals (Fig. 3.1). The size of the adults is variable. The length is usually between 30 and 50 cm but several species are smaller (*Microcaecilia unicolor* is not more than 20 cm in length) and other species are bigger (*Caecilia thompsoni* can be more than 130 cm in length). The body of Gymnophiona often presents annuli as in *Ichthyophis* or *Siphonops* (Fig. 3.2A). In *Typhlonectes*, there are no annuli but primary, secondary and sometimes tertiary folds of the skin (Fig. 3.2B). A true tail is characterised by several vertebrae that are posterior to the cloacal vent, as in the family Ichthyophiidae. In other species, there are no vertebrae posterior to the cloacal vent and no tail can be seen as in species belonging to the Typhlonectidae. In some Caeciliidae and other families, it exists only as a small appendix posterior to the cloaca, the shield (after Taylor 1968) (Fig. 3.2C).

The cloacal aperture (vent) can be rounded, slit or V shaped. In several Gymnophiona, it is not possible to determine the sex without dissection. Yet, in *Typhlonectes compressicauda*, the shape of cloaca is related to the sex. In males, the vent is situated at the bottom of a depression, giving to the cloaca the shape of a funnel. In females, the vent forms a disc with a central

---

Laboratoire de Biologie générale, Université Catholique de Lyon and Laboratoire de Reproduction et Développement des Vertébrés, Ecole Pratique des Hautes Etudes, 25, rue du Plat, F-69288 Lyon Cedex

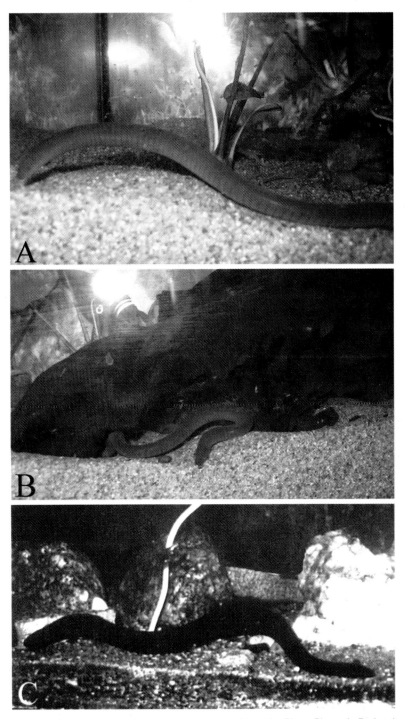

Fig. 3.1. Several Gymnophiona. **A**. *Typhlonectes natans* (photo by Pierre Bleyzac). **B**. A pair of *T. natans* (photo by Pierre Bleyzac). **C**. *T. compressicauda* (photo by Jean-Marie Exbrayat).

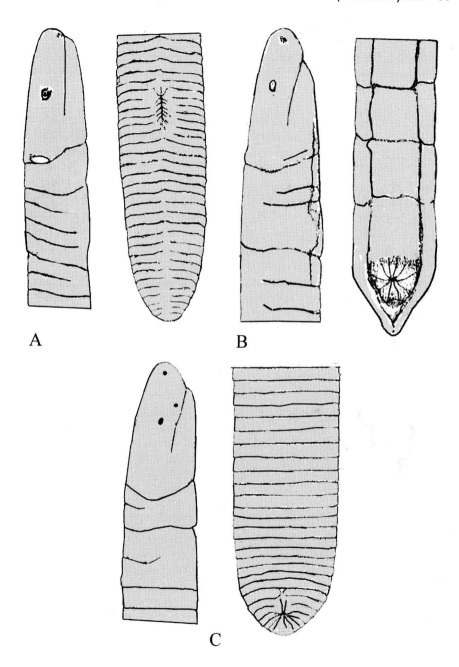

Fig. 3.2. Characteristics of head (on the left) and tail (on the right) in several Gymnophiona. **A.** *Ichthyophis glutinosus* with a true tail. **B.** *Typhlonectes compressicauda* without a true tail. **C.** *Dermophis mexicanus* with a reduced tail. Modified after Taylor, E. H. 1968. *The Caecilians of the World. A taxonomic review*. University of Kansas Press. Lawrence, Kansas, U.S.A., 848 pp, Fig. 24, 242.

bud (Gonçalves 1977). In several species, sexual dimorphism can affect body size (Deletre and Measey 2004; Malonza and Measey 2005).

The eyes are generally difficult to see. They may be superficial but in some species are covered by skin that is more or less thick and transparent (Fig. 3.6E, F). In larval *Scolecomorphus* and in other adult species, the eyes are covered by the bones of the skull.

All the Gymnophiona possess a pair of tentacles that are sense organs, situated on each side of the snout, between nostril and eye (Fig. 3.7A).

By dissection, one can observe the particular characteristics of internal anatomy: the skin is particularly thick. In certain species, it is covered by scales that can be more or less developed in size and in number. The gut is straight with very few loops. The liver is elongate, situated on the right side of the body. It consists of 20 to 40 elements giving it a segmented appearance. The pancreas and the spleen are found on the posterior part of the liver. The respiratory system consists of a trachea and one or two more or less developed lungs according to the species (Fig. 3.3A, B).

Kidneys are two narrow organs extending from the posterior part of the liver to the cloaca. Segmented interrenal glands are situated between anterior parts of kidneys.

Testes are described in chapter **6** and ovaries in chapters **8** and **9** of this volume.

3.1.2 Skeleton and Muscles

The skeleton of Gymnophiona is composed only of the skull and vertebrae (Fig. 3.4A). The structure and development of vertebrae have been studied by several authors (Peters 1874 a, b; Wiedersheim 1879; Marcus and Blume 1926; Marcus 1934, 1937; Mookerjee 1942; Lawson 1963, 1966a; Wake D.B. 1970; Estes and Wake 1972; Taylor 1977; Wake M.H. 1980d; Renous and Gasc 1989). They are of a primitive type and the disposition of the contact faces permits only a horizontal worm-like displacement (Gaymer 1971; Renous and Gasc 1986a, b). The centrum is intermediate between that of Urodela and Anura.

The skull has been studied in *Dermophis mexicanus* (Wake and Hanken 1982) and *Typhlonectes compressicauda* (Wake *et al.* 1985). It consists of numerous bones (Fig. 3.4C, D). The lower jaw consists of two symmetrical halves. Each half-jaw, constituted by two bones, is linked to the other by connective tissue. A retroarticular process is observed. It permits the insertion of powerful muscles that are involved in both mastication and maintaining closure of the mouth when the prey has been caught. A complex system of muscles divided into six types (Bemis *et al.* 1983) is associated with jaws.

Although there are neither girdles nor limbs in Gymnophiona, during embryonic development of *Typhlonectes compressicauda*, limb buds begin to develop in the anterior region from the 3rd to 6th somite, i.e. at the place where limbs develop in other amphibians, but, at metamorphosis, these

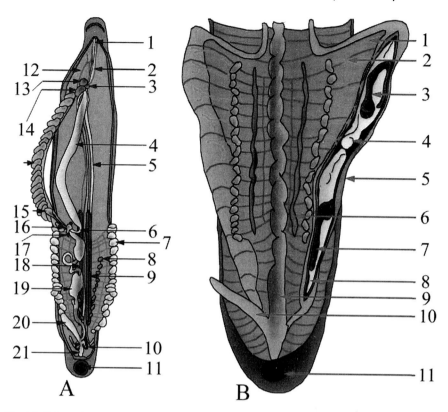

Fig. 3.3. Schematic representation of a *Typhlonectes compressicauda* dissection. **A**. Male. 1, trachea; 2, esophagus; 3, left auricle; 4, stomach; 5, left lung; 6, duodenum; 7, left fat body; 8, left testis; 9, left kidney; 10, left Mullerian duct; 11, cloacal opening; 12, arterial bulb; 13, right auricle; 14, ventricle; 15, gall bladder; 16, spleen; 17, pancreas; 18, right kidney; 19, intestine; 20, urinary bladder; 21, cloaca. **B**. Posterior part of a pregnant female. 1, mesorchium; 2, blood vessel; 3, intra-uterine embryo; 4, gill of an intrauterine embryo; 5, uterus; 6, ovary; 7, intra-uterine embryo; 8, gill of intra-uterine embryo; 9, intestine; 10, gall bladder; 11, opening of cloaca. **A**, from Delsol, M., Exbrayat, J.-M., Flatin, J. and Lescure, J. 1980. Bulletin mensuel de la Société Linnéenne de Lyon 49: 370-379, Fig. 1. **B**, from Exbrayat, J.-M., 1986a. Unpublished D. Sci. Thesis, Université Paris VI, France, Fig. 52.

buds degenerate (Fig. 3.4B). However, in *Ichthyophis glutinosus*, a primitive species, traces of degenerate girdles are observed (Renous *et al.* 1997).

Muscles can be divided into axial, episomatic and infra-vertebral muscles.

3.1.3 Skin

The skin has been studied in adult and larval Gymnophiona by means of light and electron microscopy (Sarasin and Sarasin 1887-1890; Datz 1923; Ochotorena 1932; Lawson 1963; Gabe 1971a; Welsch and Storch 1973a; Fox and Whitear 1978; Fox 1983, 1986, 1987; Zylberberg *et al.* 1980; Riberon and Exbrayat 1996) (Fig. 3.5). The epidermis consists of six to seven layers of cells that become keratinised on the external surface of the skin. Goblet cells

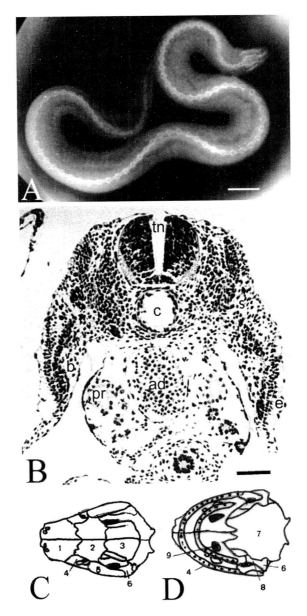

Fig. 3.4. A. Radiography of *Typhlonectes compressicauda*. The skeleton consists of vertebrae and brain case. Scale bar = 20 mm. **B.** section of stage 25 embryo in *T. compressicauda* showing a limb blastema (on the left). b, limb blastema; c, centrum; e, epidermis; tn, neural tube. Scale bar = 60 µm. **C.** dorsal view of brain case in a Gymnophiona. **D.** Ventral view of brain case in a Gymnophiona. 1, nasal bone; 2, frontal bone; 3, parietal bone; 4, maxillary bone; 5, squamosal; 6, quadrate; 7, parasphenoidal; 8, pterigoid; 9, internal nostril. **A.** Original. **B**, from Renous, S., Exbrayat, J.-M. and Estabel, J. 1997. Annales des Sciences Naturelles 18: 11-26, Pl. II, Fig. 3. **C, D**, modified after Porter, K. R. 1972. *Herpetology* W. B. Saunders Company, Philadelphia, Eastbourne, Toronto, U.S.A., England, Canada 524 pp., Figs. 2.12 D, 2.13 D.

are observed at the surface. Merckel cells are situated on or among the cells of the basal layer. These cells are supposed to have a mechanoceptor function. Iridophores are cells with guanine crystals giving an iridescent appearance to the skin.

Scales are present in the dermis of adult Gymnophiona. They have been well studied in several species (Cockerell 1912; Casey and Lawson 1979, 1981; Zylberberg et al. 1980; Zylberberg and Wake 1990). The scales always lie in the dermis, covered by the epidermis. Several scales with a bony structure are contained in a connective capsule that is generally associated with a large granulary gland. Scales of Gymnophiona are homoplastic to those of the teleostean fishes (Zylberberg et al. 1980; Zylberberg and Wake 1990) (Fig. 3.5A, C, E).

The shape, size and number of scales are different according to the species (Wake and Nygren 1987; Riberon and Exbrayat 1996). They are small in small species, and larger in larger species. In *Dermophis mexicanus*, it has been shown that their number increases with growth of the animal. In primitive species, scales cover the entire body. In other species such as *Hypogeophis rostratus*, they cover only the posterior half and they are absent in Typhlonectidae (Fig. 3.5F), though a few degenerate scales have been observed in the caudal region of *Typhlonectes compressicauda* (Wake 1975).

In the dermis, two types of glands are present (Fig. 3.5B-F). The mucous glands open at the surface of the skin by a ductule. More voluminous granular poisonous glands are present in the upper part of the dermis. The dermis also contains melanophores and laminophores (Fox 1983) the function of which is unknown. Several studies have been devoted to the cutaneous glands (Phisalix 1910; Toledo and Jared 1995) with special reference to their venomous activity (Sawaya 1940; Ferroni-Schwartz et al. 1998, 1999).

3.1.4 Brain and Nervous System

Very few works have been devoted to the gymnophionan brain and only some species have been studied (Kuhlenbeck 1922, 1969; Olivecrona 1964; Senn and Reber-Leutenegger 1986; Schmidt and Wake 1990, 1997, 1998; Martin-Bouyer et al. 1995; Estabel and Exbrayat 1998). The neuroanatomical characters were also used as non-traditional characters to study phylogenetic relationships in Gymnophiona (Wake 1993a) as was discussed by Wilkinson (1997). In the adult, the brain is similar to that of other Amphibia. Olfactory nerves are particularly massive and two accessory olfactory lobes associated with vomeronasal organs are also observed on the anterior part of the cerebral hemispheres. A strong flexion of the brain is observed between the diencephalon and the mesencephalon. The thalamus is also very well differentiated from the telencephalon (Fig. 3.6A, B). Such a situation is not found in Anura and Urodela in which all these structures are difficult to distinguish, but it is also observed in reptiles. Cells that line the adult neural tube are disposed as in embryos: they do not look like a true epithelium but seem to be included in a matrix.

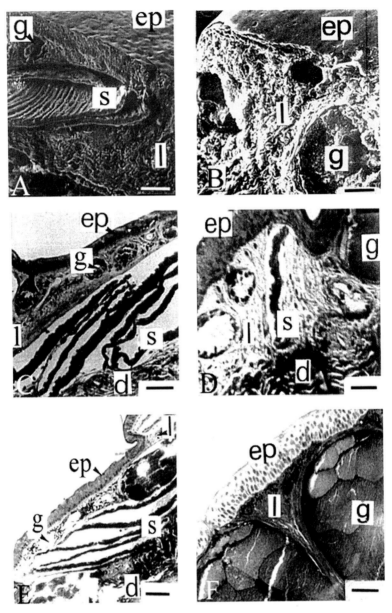

Fig. 3.5. Integument in several Gymnophiona. **A**. SEM view of integument in *Ichthyophis kohtaoensis*. Scale bar = 100 µm. **B**. SEM view of integument in *Siphonops annulatus*. Scale bar = 100 µm. **C**. LM view of integument in *Ichthyophis kohtaoensis*. Scale bar = 120 µm. **D**. LM view of integument in the median zone of *Geotrypetes seraphinii* body. Scales are few numerous and reduced in size. Scale bar = 60 µm. **E**. LM view of integument in the posterior zone of *Geotrypetes seraphinii* body. Scales are numerous and with a normal size. Scale bar = 120 µm. **F**. LM view of integument in *Typhlonectes compressicauda*. Scale bar = 120 µm. d, dense dermis; ep, epidermis; g, gland; l, loose dermis; s, scale. From Riberon, A. and Exbrayat, J.-M. 1996. Bulletin de la Société Herpétologique de France 79: 43-56, Figs. 1-6.

Embryonic development of the brain has been studied in several species (Straub 1985, 1986) and more particularly in *Typhlonectes compressicauda* (Estabel and Exbrayat 1998). In this species, the development of the brain is marked by three main phases that are similar to those of the other vertebrates. The first is a phase of proliferation during which the volume of neural epithelium increases, the second is a phase of differentiation during which the neurons develop extensions, and the last is a phase of maturation. The rhombencephalic flexion appears after hatching (intra-uterine in *Typhlonectes*) (Fig. 3.6C, D). In *Ichthyophis glutinosus* (Olivecrona 1964) and *I. beddomei* (Estabel et al. 1999), two oviparous species, similar features have been observed. Development of the tectum has been studied in several caecilian species (Schmidt and Wake 1997, 1998). Radial cells have been described in *Typhlonectes compressicauda* (Estabel and Exbrayat 1999).

Neurophysiological aspects have also been studied in Gymnophiona, and more particularly in recent works (Welsch et al. 1976; Clairambault et al. 1994; Gonzales and Smeets 1994, 1997; Pinelli et al. 1997, 1999; Hilscher-Conklin et al. 1998; Rastogi et al. 1998). The localisation of gonadotropin-releasing hormone has been described for the brain of *Typhlonectes natans* (Ebersole and Boyd 2000), *T. compressicauda* (Rastogi et al. 1998) and *Ichthyophis beddomei* (Pinelli et al. 1997). In this last species, distribution of Gn-RH-like substance in terminal nerves and in the brain is primitive. Presence of several molecules has been shown in Gymnophiona brain: GFAP (Naujoks-Manteuffel and Meyer 1996), FMR amide-like molecule (Pinelli et al. 1999), vasotocin-like and mesotocin-like hormones (Gonzales and Smeets 1997), AVT (Hilsher-Conklin et al. 1998).

3.1.5 Sensory Organs

Several works have been devoted to the sensory organs in Gymnophiona. Chemical communication has been described in *Typhlonectes natans* (Warbeck et al. 1996).

3.1.5.1 Eyes

Several works have been devoted to the eyes and vision in Gymnophiona (Leydig 1868; Wiedersheim 1879; Sarasin and Sarasin 1887-1890; Norris 1917; Norris and Hughes 1918; Arnold 1935; Ramaswami 1941; Storch and Welsch 1973; Wake 1980e; Himstedt 1989; Himstedt and Manteuffel 1985; O'Reilly et al. 1996) but few have been devoted to visual projections into nervous centers (Clairambault et al. 1980; Himstedt and Manteuffel 1985). As previously said, eyes of Gymnophiona are small, they are covered by skin that may be more or less thin with or without mucous and granular glands and sometimes are covered by skull bones (Fig. 3.6E). The lens usually is not totally lamellar; numerous lens cells contain a nucleus and they are not transparent. The retina is composed only of rods. Optic nerves are always reduced. Ocular muscles are combined with tentacles and no longer with eyes.

88 Reproductive Biology and Phylogeny of Gymnophiona

Nevertheless it has been shown that in *Ichthyophis kohtaoensis*, the retina possesses the same functions as in other Amphibia. The animal exhibits a negative phototropism. Visual stimuli seem to be weak.

In *Typhlonectes compressicauda*, both optic nerves are very poorly developed but a small optic chiasma persists under the diencephalon. In *Ichthyophis*, optic nerves are also very few developed but, in contrast, no

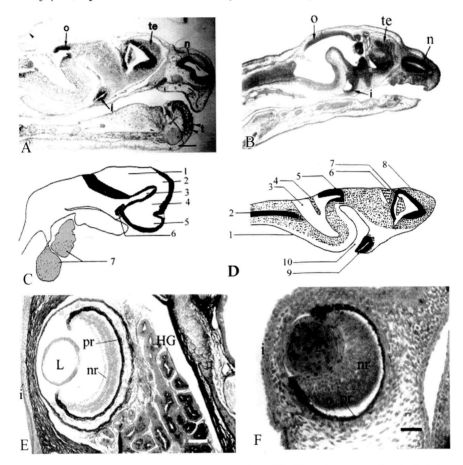

Fig. 3.6. A. Sagittal section of the head in a stage 32 embryo of *Typhlonectes compressicauda*. Scale bar = 320 μm. **B.** sagittal section of the head in a 25 mm embryo of *Ichthyophis beddomei*. Scale bar = 300 μm. n, nostril; o, optical tectum; te, telencephalon; t, tentacle. **C.** Schematic representation of the brain of *Typhlonectes compressicauda* embryo (stage 24, just before hatching). 1, choroidal plexus; 2, mesencephalon; 3, rhombo-mesencephalic flexion; 4, diencephalon; 5, choroidal plexus; 6, hypophysis; 7, heart. **D.** Schematic representation of the brain of *Typhlonectes compressicauda* fetus (stage 31, just before metamorphosis). 1, *medulla oblongata*; 2, spinal cord; 3, ventricle; 4, choroidal plexus; 5, optical tectum; 6, choroidal plexus; 7, choroidal plexus expansions; 8, cerebral hemisphere; 9, infundibulum; 10, rhombomesencephalic flexure. **E.** Eye in adult *T. compressicauda*. Lens is acellular. Scale bar = 150 μm. **F.** Eye in 25 mm embryo of *Ichthyophis beddomei*. Lens is cellular. Scale bar = 150 μm. HG, Harderian glands; i, integument; L, lens; nr, nervous retina; pr, pigmented retina. **A, B, E, F,** original. **C, D**, from Estabel, J. and Exbrayat, J.-M. 1998. Journal of Herpetology 32: 1-10, Figs. 1, 6.

optic chiasma has been observed. The reduction of oculomotor muscles is correlated with a reduction of the corresponding cephalic area in all Gymnophiona. The development of eyes has been the object of very few studies (Wake 1985; Estabel and Exbrayat 2000a) (Fig. 3.6F). Several studies have shown the organisation and development of the serotonergic and dopaminergic systems in the retina (Dünker 1997, 1999; Dünker and Himstedt 1997), the coexistence of tyrosin hydroxilase and GABA in retinal neurons (Dünker 1998a, b).

3.1.5.2 Tentacles

All Gymnophiona possess a pair of tentacles that are situated at the tip of the snout, behind the nostrils. These organs have been described by several authors (Leydig 1868; Wiedersheim 1879; Sarasin and Sarasin 1887-1890; Norris and Hughes 1918; Engelhardt 1924; Laubmann 1927; Marcus 1930; de Villiers 1938; Ramaswami 1941; Jurgens 1971; Badenhorst 1978; Billo 1982, 1986; Fox 1985; Billo and Wake 1987) (Fig. 3.7A).

In *Dermophis mexicanus*, each tentacle is contained in a sac in which the tentacle and corresponding eye are enveloped. This sac is open at the tip of the snout. Each sac is bordered by an epithelium surrounded by connective tissue and a well vascularised spongy tissue. Harderian glands empty into the posterior end of this sac. In this cavity, the wall is folded to form the tentacle itself that can be more or less extruded, according to the species, by means of muscles that are normally devoted to eye activity. The tentacle epidermis is highly innervated. Mucous glands that are situated in the dermis, produce a substance to lubricate the tentacles during their movements. Vomeronasal organs are linked to the tentacles by ductules.

Tentacles are continuously being extruded and withdrawn. By these movements, they can catch chemical substances from the environment that are dissolved in the fluid produced by Harderian gland. Then these substances are injected to the vomeronasal organ where they are analysed.

The protrusible eyes of *Scolecomorphus* (Scolecomorphidae) are discussed by O'Reilly *et al.* (1996; see also chapter 2 of this volume).

3.1.5.3 Olfactory and vomeronasal organs

Some authors have studied olfactory and vomeronasal organs in Gymnophiona (Wiedersheim 1879; Sarasin and Sarasin 1887-1890; Seydel 1895; Laubmann 1927; Marcus 1930; Ramaswami 1941, 1943; Badenhorst 1978; Schmidt and Wake 1990).

Two types of nasal cavities have been described (Schmidt and Wake 1990). Type A is formed by a chamber that is partially covered by respiratory epithelium. In type B, lateral expansions of the main chamber are observed.

Each vomeronasal organ is separated from the corresponding nasal cavity by a narrow cavity. As already mentioned, vomeronasal organs are connected to the tentacles. Nerve fibers arising from the nasal cavities reach the main olfactory bulb and vomeronasal nerve fibers reach a secondary olfactory bulb both situated in the anterior part of the brain.

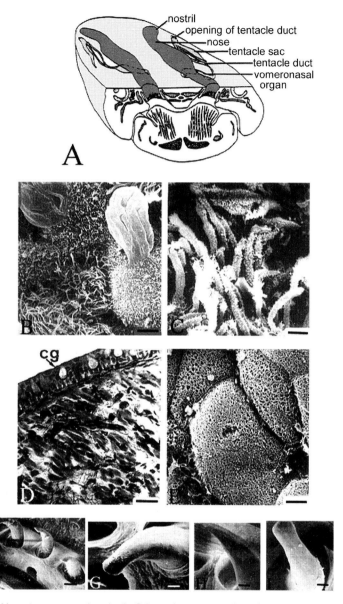

Fig. 3.7. *Typhlonectes compressicauda*. **A**. Schematic representation of a part of a gymnophionan head. **B**. SEM aspect of the dorsal surface of the tongue in stage 29 fetus. Scale bar = 4 µm. **C**. SEM aspect of the dorsal surface of the tongue in stage 32 larva. Scale bar = 4 µm. **D**. LM view of the section of the tongue in new-born. Cg, gland. Scale bar = 65 µm. **E**. SEM aspect of the dorsal surface of the tongue in new-born Sale bar = 4 µm. **F**. Vomerine teeth. Scale bar = 50 µm. **G**. Premaxillary tooth. Scale bar = 50 µm. **H**. Maxillary tooth. Scale bar = 50 µm. **I**. Vomerine tooth. Scale bar = 50 µm. **A**, modified after Billo, R. 1986, Mémoire de la Société Zoologique de France 43: 71-75, Fig. 1a. **B, C, D, E**, from Hraoui-Bloquet, S. and Exbrayat, J.-M. 1997. Bulletin de la Société Herpétologique de France 82-83 : 39-46, Fig. 1-6. **F, G, H, I**, from Greven, H. 1986. Mémoire de la Société Zoologique de France 43: 85-86, Figs. 1.8.1, 1.8.2, 1.8.3, 1.8.4.

3.1.5.4 Ear

The ear of Gymnophiona has been little studied (Wiedersheim 1879; Retzius 1881; Sarasin and Sarasin 1887-1890; de Jaeger 1947; Wever 1975; Wever and Gans 1976; Fritzsch and Wake 1988). Otic capsules are situated at the top of the anterior part of the skull, on each side of the brain. A stapes is observed ventrolaterally in each capsule, except in Scolecomorphidae. A sac extends, in a median situation, into the cephalic cavity. Endolymphatic ducts leave the otic capsule on dorsomedial part and end at a blind sac in the median part of the brain. An extension of the stapes forms a columella that plays an important role in sound perception.

The ventral part of each labyrinth is constituted by a large semicircular vesicle that may have a caudomedial expansion with lagena in which sense cells that are grouped in papilla basilia may be present. There are no sensory cells, however, in *Typhlonectes* nor in several Caeciliidae. In this sac, a single otolith is present. The periotic system is constituted by semicircular ducts. The upper and lower parts of the labyrinth are separated by the utriculosaccular foramen near which we have found papilla neglecta and papilla amphibia that are sensory epithelial zones.

The periotic labyrinth possesses a periotic cisterna, under the stapes, near the fenestra ovalis. The periotic duct passes around the dorsal part of the saccula, and makes contact with the papilla amphibia, papilla basilia and lagena. It ends in the periotic sac. Eight zones with sensory terminals have been observed (Wever 1975; Wever and Gans 1976). Nervous projections of the ear have been studied in *Dermophis mexicanus* (Fritzsch and Wake 1988).

Reception of sound has been studied (Wever 1975). The skin of the posterolateral region of the head is the reception area for sounds. Vibrations applied to the skin are passed to the stapes, and to the fluid contained in the otic capsules and then to sensory terminals that are in contact with these fluids. Gymnophiona are particularly sensitive to low frequencies.

3.1.5.5 Tongue and taste

The tongue is a muscular structure on the dorsal surface of which glands with serous and mucous secretions are found (Zylberberg 1972, 1986). These glands may be scattered throughout the tongue surface, as in *Dermophis mexicanus* (Bemis *et al.* 1983) or are localised in the median region, as in *Hypogeophis rostratus* (Marcus 1932). Taste buds have been studied in several species (Teipel 1932; Wake and Schwenk 1986). In *Typhlonectes compressicauda*, taste buds are situated in the lining of the mouth, between teeth of the lower jaw. In *Ichthyophis* larvae, they are situated on the palate and on the lower jaw, near the teeth. It appears that taste buds only exist in aquatic larvae and adults (Wake and Schwenk 1986). The structure of taste buds in Gymnophiona resembles that of fishes. The tongue and hypoglossal musculature innervation have been investigated (Schmidt *et al.* 1996)

92 Reproductive Biology and Phylogeny of Gymnophiona

3.1.5.6 Lateral line organs
Lateral line organs are characteristic of aquatic animals (Flock 1965, 1967). They occur in aquatic gymnophionan larvae. They are lost in intra-uterine larvae. They also can be found in adults of some aquatic species (Heterington and Wake 1979; Duellman and Trueb 1986). Ampullary organs have been signaled in adult *Typhlonectes compressicauda* (Fritzsch and Wake 1986). The lateral line system has been studied in *Ichthyophis glutinosus* (Sarasin and Sarasin 1887-1890), *Hypogeophis rostratus* (Brauer 1897a, b, 1899), *Grandisonia seychellensis* (Parker 1958), and in larvae of several species (Taylor 1960; Largen *et al.* 1972; Heterington and Wake 1979; Wilkinson 1992). The development of lateral line organs with neuromasts and ampullary organs has been precisely described in *Ichthyophis kohtaoensis* (Dünker *et al.* 2000; see also chapter 12 of this volume). In *Ichthyophis*, two organ types has been found: free neuromasts and ampullary organs (Fig. 12.4G, chapter 12 of this volume, for *Ichthyophis beddomei*). Neuromasts are situated on each side of the head and all along the body. Three cell types constitute one neuromast. One cell type is restricted to the external layer, a second type constitutes a sustaining layer, and the third type is constituted by the sensory cells that are situated at the apex of the organ. The organ is situated on a basal membrane. Nerve fibers arise from the basal part of the neuromast. There is always a blood vessel associated with the neuromast.

Ampullary organs are disposed on the head and more particularly at the upper surface of the snout. They are never found in the posterior part of the animal. They are situated at the base of the epidermis and consist of an external cell layer, with sensory cells, as in free neuromasts. A small duct links each ampullary organ to the surface of the epidermis. Numerous cells and more particularly sensory ones, border the ductule wall. In intrauterine larvae, when a lateral line exists, only ampullary organs are found.

Lateral line organs in Gymnophiona are functional; *Ichthyophis kohtaoensis* larvae react to electric impulses (Himstedt and Fritzsch 1990).

3.1.6 Digestive Tract
The digestive tract consists of mouth, esophagus, stomach, intestine and associated glands such as liver and pancreas.

3.1.6.1 Mouth
The oral cavity contains the tongue, salivary glands and teeth (Bemis *et al.* 1983; Junqueira *et al.* 1999). The tongue has already been described in this chapter in adults. Its development has been studied by several authors (Welsch and Storch 1973a; Hraoui-Bloquet 1995; Hraoui-Bloquet and Exbrayat 1997a, b). During development of *Typhlonectes compressicauda*, the tongue epithelium undergoes several structural variations (Hraoui-Bloquet 1995; Hraoui-Bloquet and Exbrayat 1997a, b) (Fig. 3.7B, C, D, E).

Salivary glands have been studied in *Ichthyophis glutinosus*, *I. kohtaoensis*, *Typhlonectes compressicauda* (Zylberberg 1972, 1986). Tongue epithelium always possesses two cell types: mucous gland cells and ciliated cells that are covered by mucous. Well developed tubular salivary glands have also been shown under the surface of the tongue in *Ichthyophis kohtaoensis* (Fahrenholz 1937; Zylberberg 1986). In *Typhlonectes compressicauda*, these glands are very little developed by comparison with other species. The cells of this gland type synthesize a neutral mucous. In addition, a third glandular cell type has been found in *T. compressicauda*, by means of TEM (Zylberberg 1986). In *T. compressicauda*, reduction of salivary glands can be attributed to its aquatic life.

Teeth of adult Gymnophiona are small. They are found on the upper jaw with a row on premaxillary and maxillary bones, and on the lower jaw. Teeth are also found on prevomer, palatine and parasphenoid bones, as in other Amphibia. Replacement and development of teeth have been studied in *Hypogeophis rostratus* (Lawson 1965 a, b; Wake and Wurst 1979; Casey and Lawson 1981; Greven 1984, 1986).

A fetal dentition that is very different from that of adults is observed in all the viviparous species during their intra-uterine life. They have been studied in several species (Parker and Dunn 1964; Wake 1976, 1978, 1980a) and in *Typhlonectes compressicauda* (Haroui-Bloquet 1995; Hraoui-Bloquet and Exbrayat 1996).

The teeth of Gymnophiona are characteristic of the Amphibia. They consist of a basal pedicel with an external crown (or dentine cone) separated from each other by fibrous connective tissue (Fig. 10.4F, chapter **10** of this volume).

The shape of tooth is characteristic of the species (Fig. 3.7F, G, H, I). In certain Gymnophiona and more particularly the most primitive ones, teeth are monocuspid, in other they are bicuspid or are spoon-shaped (Taylor 1968; Wake 1978; Wake and Wurst 1979; Greven 1984, 1986). Usually all the teeth are of the same type. However, in *Gegenophis ramaswamii*, there are monocuspid and bicuspid types (Greven 1984). The shape of teeth can also be related to alimentary habits. Terrestrial species eat long prey that are caught with the mouth and must be strongly gripped. Teeth that are directed to the back prevent escape. When teeth are sharp, they can be also used for mastication. In *Typhlonectes obesus*, the spoon-shaped teeth aid the animal in finding insect larvae that live in decomposed wood.

3.1.6.2 Gut

The digestive tract has been studied in several species (Rathke 1852; Wiedersheim 1879; Sarasin and Sarasin 1887-1890; Chatterjee 1936; Oyama 1952; Gabe 1969, 1971b; Bons 1986; Delsol *et al.* 1995; Junqueira *et al.* 1999). In Gymnophiona, the alimentary canal is lengthened and straight (Fig. 3.3A, B). It is difficult to externally differentiate esophagus and stomach. The duodenum can be convoluted. The intestine may describe some loops but these are not strongly developed. It ends into the cloaca. The gut wall from the inside out consists of an epithelial layer, a basal membrane, a layer of connective tissue, two or three muscular layers and a connective tissue sheath. Blood vessels, lymphatic spaces and nerves are found in the connective tissue.

In the anterior region of the esophagus, the epithelium is composed of several cell layers. In the posterior region, there is a single cell layer. The apices of these cells are ciliated. Several goblet cells are present and become more numerous posteriorly. Mucous synthesized by these cells is PAS-positive and stains with alcian blue and eosin indicating an acidic glycoproteic nature. No glands have been found in the esophagus (Fig. 3.8A).

In the stomach, gastric glands are observed in the connective tissue. They secrete an acidic mucus, PAS positive and stained with alcian blue; cells situated at the base of these glands are stained by eosin. The wall of stomach lumen is formed by numerous villi that are bordered by a common gastric epithelium with cells containing a neutral mucus (PAS positive) (Fig. 3.8B, C).

In the pyloric sphincter, the epithelium resembles that of the stomach. Glands can or cannot be observed in the connective tissue (Gabe 1971 b; Bons 1986). Muscular layers are particularly thick.

In the intestine, the epithelium consists of several cell layers (Fig. 3.8D). Numerous villosities project into the gut lumen. In the duodenum, the epithelium is constituted by typical enterocytes. Cells of the apical layer possess microvilli, with a characteristic strong alkaline phosphatase activity. Goblet cells with an acidic mucous stained by alcian blue and PAS positive are intercalated between enterocytes. Basal cells generate new cells. Brünner glands that are disposed under the connective tissue can also be observed.

In the posterior region of the gut, villi are lesser numerous than in the duodenum. Then number of enterocytes decreases, in contrast, the number of goblet cells with an acidic mucous increases. Gabe (1972) described the endocrine cells of the gut.

3.1.6.3 Cloaca
The cloaca will be described with genital organs (chapter 5 of this volume).

3.1.6.4 Liver and gall bladder
The liver has been little studied (Welsch and Storch 1971, 1972; Storch et al. 1986). It is surrounded by a thick layer of connective tissue. A peripheral hematopoietic layer is also found in larvae and adults (Hraoui-Bloquet and Exbrayat 1992; Delsol et al. 1995). The main part of the organ is constituted by hepatocytes, blood vessels, bile ductules, pigment cells and Küppfer cells (Fig. 11.5H, chapter 11 of this volume). General microscopic anatomy of the liver is similar to that of the other amphibians. Hepatic blades are constituted by two cell layers. Bile ductules that are found between hepatocytes, without any proper wall, connect to a bile duct. All the bile ducts connect in a single liver duct that joins the duct arising from gall bladder. The chyle canal goes crosses the pancreas and connects to the duodenum.

3.1.6.5 Pancreas
Very few works have been devoted to the pancreas in Gymnophiona (Weysse 1895; Brauer 1897a, b; Siwe 1926; Gabe 1968; Delsol et al. 1995). It resembles

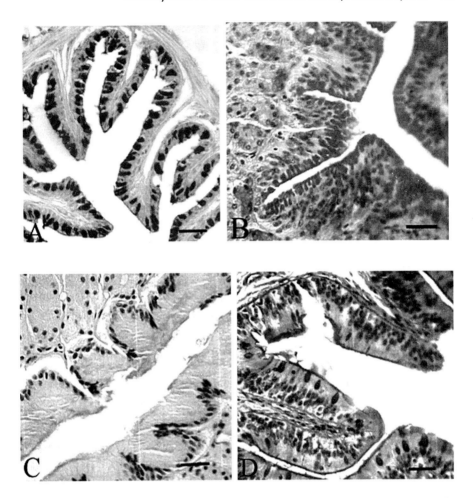

Fig. 3.8. Some aspects of the digestive tract in several Gymnophiona. **A**. Section of esophagus in *Microcaecilia unicolor*. Scale bar = 60 μm. **B**. Section of stomach in *M. unicolor*. Scale bar = 60 μm. **C**. Section of stomach in *Hypogeophis rostratus*. Scale bar = 60 μm. **D**. Section of intestine in *M. unicolor*. Scale bar = 60 μm. **A, C, D**, from Exbrayat, J.-M. 2003. Revue Française d'Histotechnologie 16: 19-29, Figs. 1, 4, 4. **B**, original.

the pancreas of other Amphibia. Exocrine cells are grouped in acini that are limited by a connective tissue. Several acini are grouped in lobules. Connective tissue that separates lobules contains evacuation ductules. Each islet of Langerhans is situated near a blood vessel. These islets contain three cell types. Some of them can also be observed scattered among exocrine cells and in the wall of the ductules.

3.1.6.6 Development of digestive tract

Development of the digestive tract has been studied in few species (Brauer 1897a, b; Hraoui-Bloquet and Exbrayat 1992; Hraoui-Bloquet 1995). First stages of development are described for *Hypogeophis rostratus* (Brauer 1897a, b). In *Typhlonectes compressicauda*, development of the digestive tract has been described in relation to the metamorphosis of this intra-uterine species (Hraoui-Bloquet and Exbrayat 1992; Hraoui-Bloquet 1995) (chapter **11** of this volume).

3.1.7 Respiratory System

The anatomy of the respiratory system in adult Gymnophiona has been studied in several species (Marcus 1927; Baer 1937; Wake 1974; Pattle *et al.* 1977; Welsch 1981; Goniakowska-Witalinska 1995; Kühne and Junqueira 2000).

In adults, the respiratory system is constituted by a trachea that generally extends from the posterior region of the mouth to the anterior part of the liver and a pair of lungs that are variable in size according to the species. In *Chthonerpeton indistinctum*, lungs are equal in length and they reach the posterior third of the body. In *Typhlonectes compressicauda*, they are equal in size and are particularly elongate, reaching the posterior end of body where they describe a loop orientated to the anterior end of the animal (Fig. 3.3A). In numerous other species, lungs are unequal in size, the left being shorter than the right. In *Siphonops annulatus*, for instance, the left lung is approximately 10% of the right one (Kühne and Junqueira 2000). The species *Atretochoana eiselti* does not possess lungs (Nussbaum and Wilkinson 1995; Wilkinson and Nussbaum 1997; Wilkinson *et al.* 1998). In *Ichthyophis orthoplicatus*, *I. paucisulcus* and *Chthonerpeton indistinctum*, the lungs have a tubular structure. The internal wall is divided by connective tissue rows (Pattle *et al.* 1977; Welsch 1981). Cartilaginous rings sustain tissues of the trachea and lungs. The lungs are divided by septa that delimit areas, the "lungenbronchus" (Marcus 1937) that have been well described in *Siphonops annulatus* (Kühne and Junqueira 2000). In these areas, ciliated cells occur according to the species (Pattle *et al.* 1977; Kühne and Junqueira 2000). In connective septa, blood vessels and smooth muscle cells can be observed. Alveolae are disposed in a deep situation between septa. A single pneumocyte type is found. Pneumocytes are columnar cells with apical microvilli that are closely apposed to each other. These cells also contain a superfactant appearing as osmiophilic bodies in their cytoplasm.

The ultrastructure of pneumocytes has been described in *Chthonerpeton indistinctum* (Welsch 1981) and *Ichthyophis orthoplicatus* (Pattle *et al.* 1977). These studies confirm the presence of a single type of pneumocyte, and the presence of superfactant into the cells which resembles lamellar bodies. In *Ichthyophis paucisulcus*, certain pneumocytes have small expansions that penetrate into the connective tissue to reach the walls of blood vessels, constituting the air-blood barrier.

In *Ichthyophis paucisulcus* larvae, the structure of the lungs is similar to that of the adults (Welsch 1981). By electron microscopy, only small differences have been observed in the structure of larval and adult pneumocytes. These concern the presence of abundant glycogen in the larvae and the different appearance of lamellar bodies.

Physical properties of the superfactant have been studied in *Ichthyophis orthoplicatus* (Pattle *et al.* 1977). It seems that the fastest gaseous exchanges occur at very low temperatures, related to the fact that this species burrows deeply, at a relatively low temperature.

It is known that adult amphibians use several respiratory mechanisms: buccopharyngeal, pulmonary and also cutaneous. Several studies have shown that Gymnophiona mainly use lungs for respiration. The skin is very little used, attributed to its structure and, more particularly, its thickness. Buccopharyngeal respiration can be utilized but very few studies have been devoted to this mechanism (Elkan 1958; Bennett and Wake 1974; Gonçalves and Sawaya 1978). In *Atretochoana eiselti*, a species without lungs, buccopharyngeal respiration is mainly used.

In embryos, fetuses and larvae, gills are also observed. In oviparous and viviparous species except *Typhlonectes*, gills are paired triradiate structures on each side of the posterior part of the head (Sarasin and Sarasin 1887-1890; Ramaswami 1954; Breckenridge and Jayasinghe 1979; Breckenridge *et al.* 1987; Bhatta and Exbrayat 1998; Exbrayat *et al.* 1998). Gills persist over the free larval life in oviparous species. In viviparous Gymnophiona, they also persist when embryos are developing in the uterus (Parker and Dunn 1964; Wake 1967, 1969, 1977a). They may degenerate before birth as in *Gymnopis multiplicata* (Wake 1967, 1969). In *Typhlonectes compressicauda*, gills are transformed into exchange organs that constitute a placenta-like structure with the uterine wall (Delsol *et al.* 1981; Exbrayat *et al.* 1981; Sammouri *et al.* 1990; Hraoui-Bloquet 1995, chapter **11** of this volume).

3.1.8 Heart and Circulatory System

Heart anatomy has been studied in *Hypogeophis rostratus* (Marcus 1935; Schilling 1935; Lawson 1966b) and several Asiatic species (Ramaswami 1944).

In *Hypogeophis rostratus*, the heart is composed of an arterial cone, a ventricle, two auricles and a venous sinus (Fig. 3.9A, B). The two auricles are separated by a muscular septum that has been described as complete (Marcus 1935) or fenestrated (Ramaswami 1944; Lawson 1966b). Only one ventricle is found but a large muscular pillar is found in its center. Coronary blood vessels have been described in several species (Schilling 1935, Sawaya 1941; Lawson 1959, 1966b). Coronary arteries originate from the ventricle cavity, traverse the myocardiac muscle and are observed on the surface of the heart.

In *Hypogeophis rostratus*, a single systemico-pulmonary arch arises from the right auricle and gives rise to the dorsal aorta. When it leaves the heart, this systemico-pulmonary arch also gives rise to a pulmonary artery that divides into two branches, one for each lung. Then, the systemic arch gives three other branches, the first two go in a cranial direction and irrigate the esophagus. In *Hypogeophis rostratus*, the systemic arch is connected to the

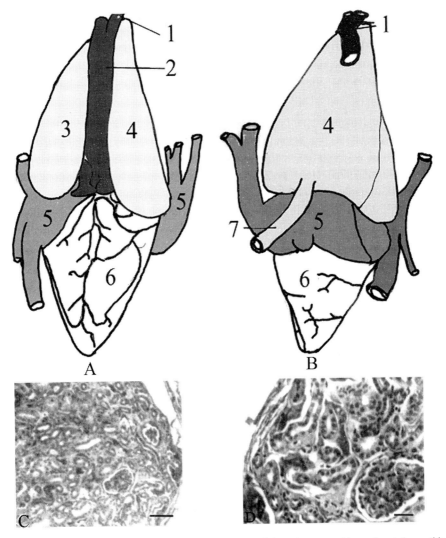

Fig. 3.9. A. Ventral view of a gymnophionan heart. **B.** Dorsal view of a gymnophionan heart. 1, carotida; 2, arterial trunck; 3, right auricle; 4, left auricle; 5, veinous sinuse; 6, ventricle; 7, pulmonary vein. **C.** Section of *Typhlonectes compressicauda* kidney. Scale bar = 200 μm. **D.** Detail of glomerula in the kidney of *T. compressicauda*. Scale bar = 200 μm. Original.

left carotid artery (Marcus 1935; Schilling 1935) but this situation has not been found by all authors in this species (Lawson 1966b).

3.1.9 Blood Cells and Immune System

Blood. The blood of Gymnophiona contains all the cell types that are found in other vertebrates (Cooper and Garcia-Herrera 1966; Charlemagne 1990; Paillot et al. 1997a, b; Jared et al. 1999). Erythrocytes are large cells with a central nucleus. Hemoglobin concentrations in *Typhlonectes compressicauda*,

Boulengerula taitanus and *Siphonops annulatus* are higher than in other amphibians (Toews and McIntyre 1977; Sano-Martins *et al.* 1990). Among other blood cells, 36% of them are thrombocytes, 12% are lymphocytes with a dense nucleus and a very small cytoplast, 25% are monocytes resembling cells with several nuclei. Acidophilic and basophilic cells are also abundant.

Liver. The liver is surrounded by an hematopoietic layer in larvae as well as in adults (Fig. 11.5H, chapter **11** of this volume). The thickness of this layer increases with age. These cells infiltrate the liver to form nodules (Oyama 1952; Cooper and Garcia-Herrera 1968; Storch *et al.* 1986; Paillot *et al.* 1997a, b).

Spleen. The spleen is constituted by a large and a small lobe. It is associated with the pancreas into which it sends infiltrations of lymphoid tissue (Foxon 1964; Welsch and Storch 1982; Welsch and Starck 1986; Paillot *et al.* 1997a; Bleyzac and Exbrayat 2003, 2005; Bleyzac *et al.* 2005). It contains a white pulp with nodules containing cells resembling lymphocytes, and a more abundant red pulp composed of sinusoids limited by endothelial cells. It is in this part of the spleen that senescent erythrocytes are destroyed by macrophages.

Thymus. The thymus is situated under the dermis, behind the otic vesicles (Welsch 1982; Paillot *et al.* 1997a, b). It is composed of three lobules that are separated by connective tissue. Each lobule consists of a medulla and a cortex in which lymphocytes can be observed. In the medulla, lymphocyte-like cells, macrophages, apoptotic lymphocytes and reticulocytes are present. Hassal's corpuscles have been found in embryonic and adult *Typhlonectes compressicauda* (Bleyzac and Exbrayat 2003, 2005; Bleyzac *et al.* 2005).

Lymphatic System. The lymphatic system has been observed in *Hypogeophis rostratus* (Lawson 1963), *Typhlonectes compressicauda* and *T. natans* (Paillot *et al.* 1997a, b). It is composed of two ganglia at the level of the two first annuli behind the head (the collar). Lymphatic vessels arise from these ganglia.

Bone Marrow. Bone marrow has not been observed in most examined species (Lawson 1963; Welsch and Starck 1986; Paillot *et al.* 1997a, b) but has been described in *Ichthyophis glutinosus* (Gabe 1971b).

Immune Reactions. Immune reactions are little known in Gymnophiona but they have been demonstrated in some cases (Cooper and Garcia-Herrera 1968). Protection of embryos against rejection in pregnant females has been described in *Typhlonectes* (Exbrayat *et al.* 1995; Haraoui-Bloquet 1995).

3.1.10 Kidneys and Bladder

Kidneys of Gymnophiona have been described by several authors over many years (Spengel 1876; Wiedersheim 1879; Semon 1892; Chatterjee 1936; Lawson 1959; Garg and Prasad 1962; Wake 1970a; Welsch and Storch 1973b; Sakai *et al.* 1986, 1988a, b; Carvalho and Junqueira 1999). Development of the kidneys has been studied by Brauer (1900).

In Gymnophiona, kidneys are paired elongate structures that extend from the heart to the posterior end of the body (Fig. 3.3A, B). A segmentation of kidneys can be more or less observed (Wake 1970a; Sakai et al. 1986; Carvalho and Junqueira 1999). In males, ductules of the rete testis connect to the kidneys.

Wolffian ducts are closely applied to the external border of each kidney. Each Wolffian duct empties into the cloaca. In males, each Wolffian duct is also connected to the corresponding Mullerian duct before it reaches the cloaca. In females, the Wolffian ducts remain independent of the oviducts. Wolffian ducts receive collector ducts from the kidneys (Sakai et al. 1986).

Each nephron is constituted by a glomerulus, a capsule, a ductule and a collector duct of which anatomical structure and ultrastructure have been studied in *Typhlonectes compressicauda* (Sakai et al. 1986, 1988a, b) (Fig. 3.9C, D). Histology of the kidney has been described in *Siphonops annulatus* (Carvalho and Junqueira 1999). Distal tubules are found in the median part of the kidney. In the dorso-lateral zone, other parts of the Malpighian corpuscle are found in the peripheral zone. The structure of the renal corpuscle does not present any difference from that of other vertebrates, with a glomerulus constituted by a network of capillaries and a Bowman's capsule with podocytes that are applied against capillaries. The basal corpuscle opens into a renal ductule. In *Typhlonectes compressicauda*, the size of renal corpuscle is particularly large, attributable to the aquatic habits of the animal. In *Siphonops annulatus*, lymphocyte-like cells are found in the urinary space of the capsule (Carvalho and Junqueira 1999). Renal ductules are divided into several segments: a collar just after the glomerulus, a proximal ductule, an intermediary segment, a distal ductule and a connection with collector ducts. Nephrostomes have also been observed in *Typhlonectes compressicauda* (Sakai et al. 1986) as well as in *Siphonops annulatus* (Carvalho and Junqueira 1999). These structures appear as invaginations of the kidney surface. They are formed by cubic epithelial cells with long cilia that are directed towards the interior of a proximal renal tubule. These nephrostomes are deduced to convey liquid from the peritoneal cavity. In *T. compressicauda*, proximal tubules possess several lateral expansions that connect nephrostomes, as in *Siphonops annulatus*. If they are not connected to such a structure, they discharge into connective tissue with a virtual lumen (Sakai et al. 1986).

Each Wolffian duct is an elongated tube that lies laterally, parallel to kidney (Wake 1970a; Sakai et al. 1986, 1988a, b; Cavalho and Junqueira 1999). In *Typhlonectes compressicauda*, Wolffian duct is limited by a pseudostratified columnar epithelium, surrounded by a dense connective tissue. Three cell types have been described (Sakai et al. 1986, 1988a, b). The Wolffian cells proper are columnar, the intercalated cells are also columnar but they possess a darker cytoplasm with numerous mitochondria, the basal cells, simpler in structure, are intercalated among the two other types and they do not reach the lumen. Sperm and urine leave the kidney using the unmodified

Wolffian duct that is a more primitive condition than that of other amphibians (Wake 1970a) (Fig. 3.3A, B).

Embryonic development of kidneys has been studied in several genera of Gymnophiona including *Hypogeophis* (Brauer 1902; Wake 1970a, b), *Ichthyophis* (Semon 1892) and *Gymnopis* (Wake 1970a, b). Gymnophiona kidneys have often been considered as the ancestral type of kidneys for vertebrates.

The urinary bladder is an expansion of the cloaca, and is not directly connected to the Wolffian ducts.

The physiology of excretion has not been widely studied in the Gymnophiona. Yet, several works essentially based on histophysiology suggest that excretion in Gymnophiona is similar to that in other vertebrates (Sakai *et al.* 1988a, b; Stiffler *et al.* 1990). As indicated previously, the size of the glomerula is a function of the living habits of the animal (Wake 1970a). Physiological studies on osmoregulation in a burrowing terrestrial species (*Ichthyophis kohtaoensis*) and an aquatic one (*Typhlonectes compressicauda*) reveal differences in glomerular filtration and resorption of water. The terrestrial species can adapt its own osmoregulation characteristics to an aquatic environment; this suggests that mechanisms of osmoregulation are still primitive in these animals (Stiffler *et al.* 1990).

3.1.11 Endocrine Organs

Endocrine organs have been studied in several gymnophionan species. Some of them, as interrenal glands, hypophysis and endocrine gonads will be described and studied in this volume (chapter 5 of this volume). Other endocrine organs are also known in Gymnophiona.

3.1.11.1 Epiphysis

This gland has been studied in *Typhlonectes compressicauda* (Leclercq 1995; Leclercq *et al.* 1995). Leclercq (1995) provided a comprehensive description by LM and TEM. The epiphysis is a pea-shaped organ formed by an expansion of the roof of diencephalon. A parapineal vesicle or a frontal organ was never detected. The pineal organ was completely separated from the diencephalic structures by a basal lamina except for a short pedicle extension that joined the pineal organ with the subcommissural organ at the top of the diencephalon. This pedicle is mainly composed of supportive cells and a few nerve fibres. Rounded cells with a voluminous nucleus border the epiphysis lumen (Fig. 3.10C). By TEM, a majority of these cells resemble pineal photoreceptors (Fig. 3.10D). Their cytoplasm is organised into an outer, an inner segment and a basal segment. The outer segment is composed of a ciliary structure with 3 or 4 cilia containing a 9+1 fibril pattern. The cilia length is estimated at 15 µm and the diameter at 0.8 µm. Various membrane extensions are observed on one side of the cilia. These extensions are not present along the entire length of the cilia and various vesicles with a clear aspect are present. Each cilium had a basal apparatus at the top of the inner segment that is divided into apical and nuclear regions both joined by a thin

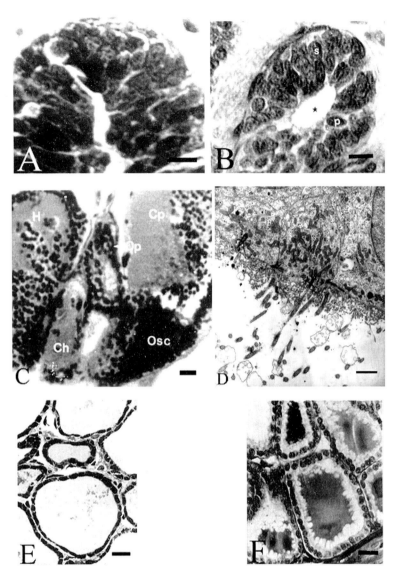

Fig. 3.10. A. Section of *Typhlonectes compressicauda* epiphysis in stage 25 embryo (hatching). Scale bar = 3 μm. **B.** Section of *T. compressicauda* epiphysis in stage 28 embryo. P, photoreceptor structure; star, pineal lumen. Scale bar = 5 μm. **C.** Section of *T. compressicauda* epiphysis in adult. Ch, habenular commissure; Cp, posterior commissure; Op, pineal organ; Osc, under commissural organ. Scale bar = 25 μm. **D.** TEM view of epiphysis in adult *T. compressicauda*. The lumen of the gland is on the lower part of the picture bordered with ciliated epithelial cells and cells with rounded vesicles; Scale bar = 4 μm. **E.** Thyroid in adult *Ichthyophis kohtaoensis* with a low activity. Scale bar = 100 μm. **F.** Thyroid in adult *I. kohtaoensis* with a high activity. Scale bar = 65 μm. **A, B, C**, from Leclercq, B. 1995. Unpublished Diploma of Ecole Pratique des Hautes Etudes, Paris. Figs. 3.6a, b, f. **D**, original by Leclercq. **E, F**, from Welsch, U., Schubert, C.and Storch, V. 1974. Cell and Tissue Research 155: 245-268, Figs. 2a, b.

cytoplasmic portion. The cell membrane had many microvilli lining the lumen and is attached to the surrounding cells by a tight junction. This region, called the ellipsoid, is rich in dense mitochondria lying in the longitudinal axis of the cell and SER, glycogen granules and fine filaments. Ribosome, RER and Golgi apparatus are observed besides the ellipsoid. Striated rootlets are observed extending basically from the centriolar apparatus into the inner segment. They cover the mitochondria. Numerous spherical granules (0.5 to 1.5 µm in diameter) themselves composed of fine dense granules probably containing melanin, are observed within the inner segment. The basal region of the cell is a process terminating in an intricate zone. The contact between photoreceptor-like and round ganglion cells with a clear cytoplasm is made by a synapse in a neuropil area. In the pineal pedicle connecting the diencephalic roof, a large number of fibers that can be surrounded by a myelin sheath are observed.

Supportive cells that are relatively undifferentiated from the ependymal cells surround the photoreceptor-like cells. They are attached to the photoreceptor-like cells by tight junctions. Interdigitations are observed between the neighboring supportive cells. These cells contain a Golgi apparatus, SER, RER, pigment granules (melanin) but they have lost mitochondria. Their nucleus is round. Their bases are divided into many foot processes. Collagen fibers and the basement membrane of the capillary endothelium surround supportive cells.

The embryonic development of the epiphysis has been described in *Typhlonectes compressicauda* (Leclercq *et al.* 1995) (Fig. 3.10A, B). The epiphysis has been observed from the stage of hatching. Before metamorphosis, two cell types can be observed bordering the lumen. Some cells with apical expansions resemble photoreceptors, and other cells do not present any peculiarities. At metamorphosis, the photoreceptor-like cells regress and, finally, only one cell type remains in the gland.

3.1.11.2 Thyroid gland
This gland has been little studied (Maurer 1887; Sarasin and Sarasin 1887-1890; Marcus 1908; Klumpp and Eggert 1935; Welsch *et al.* 1974, 1976; Estabel and Exbrayat 2000b). The adult thyroid gland is similar to that of other Amphibia (Fig. 3.10E, F). Some data concerning embryonic development have been given for two *Ichthyophis* species (Welsch *et al.* 1974) and *Typhlonectes compressicauda* (Estabel and Exbrayat 2000b). In *Ichthyophis*, thyroid activity is greater in larvae than in adults (Welsch *et al.* 1974). In *Chthonerpeton indistinctum*, an increase of follicular cells is observed after an increase of temperature (Welsch *et al.* 1976); in *Typhlonectes compressicauda*, development of the follicles is closely linked to metamorphosis (Estabel and Exbrayat 2000b).

3.1.11.3 Parathyroid glands and ultimobranchial bodies
The ultimobranchial and parathyroid gland are derived from the pharynx. These structures are involved in calcium and phosphate metabolism in other Amphibia. In Gymnophiona, they have been described in *Ichthyophis*

glutinosus (Klumpp and Eggert 1935) and *Chthonerpeton indistinctum* (Welsch and Schubert 1975). Their ablation has no effect in aquatic Gymnophiona (Sasayama *et al.* 1996).

Parathyroid Gland. In Anura and Urodela, the parathyroid gland is composed of clusters in which light and dark cells have been described. By TEM, inactive light cells and active dark cells are distinguished on the basis of their content in organelles and glycogen (Rogers 1965; Coleman 1969; Setoguti *et al.* 1970). In *Ichthyophis glutinosus* the parathyroid gland is not very different from those of the other amphibians, but with only a single type of cell (Klumpp and Eggert 1935). In *Chthonerpeton indistinctum*, the parathyroid gland is a small rounded or oval organ. It is composed of numerous elongated cells. A single type of cell may be observed; each cell contains a pale cytoplasm containing a few dark granules, and an elongated nucleus with two or three nucleoli (Welsch and Schubert 1975). Nerves have not been observed within the gland. A blood vessel occurs on the surface of this organ. By TEM, the surface of the nucleus has a smooth appearance. The nucleus contains heterochromatin. Organelles are not numerous. The mitochondria are scattered in the cytoplasm, they are pale with few cristae, RER and Golgi apparatus are poorly developed, lysosomes are rare. Some dark rounded granules resembling secretory granules of endocrine cells, glycogen particles and lipid inclusions are also observed.

Ultimobranchial Body. The ultimobranchial body of *Chthonerpeton indistinctum* is bigger than the parathyroid gland. It is composed of small follicles limited by one or several layers of cells. These follicles are separated by a connective tissue containing blood vessels. Degenerative cells or granules fill the lumen of the follicles. Two cell types are observed: small dark cells and pale large cells. Pale cells may also be united in clusters. By TEM, the cell clusters appear to be surrounded by a basal lamina. Each cell possesses an indented nucleus with heterochromatin and several nucleoli. In the cytoplasm, numerous electron-dense and larger pale granules are observed, RER and Golgi apparatus are well developed; free ribosomes are numerous, glycogen and lipid droplets are occasional. Numerous lysosomes are observed. Small vesicles are disposed in several rows under the basal plasma membrane.

Larger follicles are surrounded by a basal lamina. They contain two cell types: the first type resembles to cells observed in clusters, and the second type is composed of relatively flat cells without secretory granules. Both cells are poor in organelles. They are interconnected by desmosomes and zonula occludens. The apical cells bear few microvilli. Basally, they contain rows of vesicular elements. Sometimes a third type of cell may be observed; it is located in the basal part of the follicle; its cytoplasm is pale with few organelles. These cells are considered to be replacement cells.

The ultimobranchial body of *Chthonerpeton indistinctum* is innervated by two kinds of nerve fibers: purinergic and cholinergic fibers.

3.2 ANATOMY OF THE MALE GENITAL TRACT

3.2.1 General Anatomy

Genital anatomy of male Gymnophiona has been described by many authors for several species (Spengel 1876; Sarasin and Sarasin 1887-1890; Semon 1892; Tonutti 1931; Chatterjee 1936; Seshachar 1936, 1937, 1939, 1942a, b, c, 1943, 1945; Wake 1968a, b, 1970a, b, 1972, 1981, 1995; Exbrayat and Sentis 1982; Exbrayat 1984a, 1985, 1986a, b, 1992a, 1993; Exbrayat and Delsol 1985; Exbrayat *et al.* 1986, 1998; Exbrayat and Dansard 1992, 1993, 1994; Anjubault and Exbrayat 1998, 2000a, 2004b, Pujol and Exbrayat 2000).

In Gymnophiona, testes are paired lengthened and multilobed organs that discharge into the kidney via the rete testis (Fig. 3.3A). The ducts of the rete testis connect to the kidneys within a glomerulus (Fig. 3.11A, B). Sperms are evacuated via the kidney ductules and Wolffian ducts to the cloaca.

In contrast with other vertebrates, Mullerian ducts are persistent in adult males. They are paired structures observed in the posterior third of the body. They are differentiated as functional glands (Wake 1981; Exbrayat 1985; George *et al.* 2004a, b, 2005a. Mixing of sperm and Mullerian secretions occur in the cloaca.

The male cloaca possesses an erectile part, the phallodeum, that is extruded during reproduction for internal fertilization (Fig. 3.16A).

A pair of fat bodies lies parallel to the testes (Fig. 3.3A).

3.2.2 Structure of the Testes

3.2.2.1 Gross anatomy

Each testis is constituted by several lobes that are variable in number and size according to the species. These lobes are attached to a thin longitudinal duct (Fig. 3.11A, B). The testes are situated in the posterior part of the body and lie parallel to the gut and kidneys. Each testis is linked on one side to the corresponding kidney by a sheet of connective tissue. The arrangement of the lobes has been well studied in numerous species (Wake 1968b). In certain species, the number and the size of the lobes are variable intraspecifically: *Ichthyophis glutinosus*, *Schistometopum gregorii*, *Herpele squalostoma*, *Boulengerula boulengeri*. In a species of *Rhinatrema*, the lobes are equal. Certain genera possess few lobes: *Caecilia*, *Gymnopis* and *Dermophis*. In *Uraeotyphlus oxyurus*, the central lobe is the biggest. In *Geotrypetes seraphinii*, *Siphonops brasiliensis*, *Chthonerpeton viviparus* and *Typhlonectes compressicauda*, the anterior or posterior lobe is the biggest. A partial fusion of lobes has been observed in *Scolecomorphus kirkii*, *S. uluguruensis* and *S. vittatus*. In other species, the anterior lobe is very well developed but the total number of lobes is reduced (*Caecilia ochrocephala*, *Idiocranium russeli*). In *Dermophis mexicanus*, each testis is constituted by a single lengthened lobe, with constrictions showing the fusion of the units (Wake 1968b).

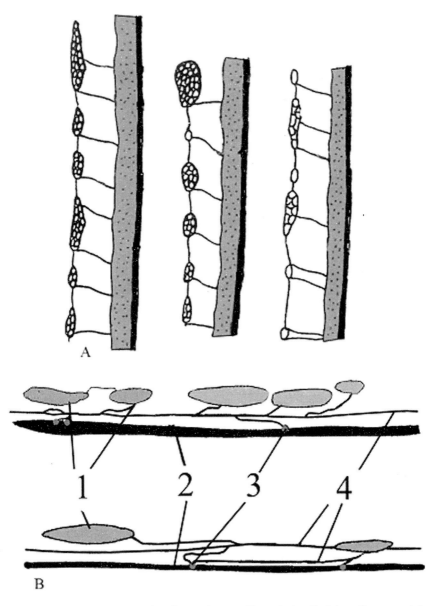

Fig. 3.11. A. Schematic representation of several gymnophionan testes. **B.** Schematic representation of the structure of testis lobes in *T. compressicauda*. 1, lobes; 2, kidneys; 3, glomerula; 4, *rete testis*. **A**, modified after Wake 1968b, Journal of Morphology 126: 291-332. Figs. 2, 3, 4. **B**, from Exbrayat, J.-M. 1986a. Unpublished D. Sci. Thesis, Université Paris VI, France. Fig. 10.

Usually, the number of lobes is about 20 to 30 on each side. In *Typhlonectes compressicauda*, each testis possesses about 12 lobes with major variations between the right and the left testis, and also between individuals: 5 to 22 lobes per testis have been counted in this species (Exbrayat 1986a, b). These variations are not correlated to the size or age of the animal. In *Typhlonectes compressicauda*, the cumulative length of the lobes is constant for animals with a similar size, that are caught at the same period of the year, whatever the number and the arrangement of lobes.

3.2.2.2 Sertoli cells

Each lobe of the testis is divided into lobules that are the seminiferous ductules. In *Typhlonectes compressicauda*, each lobule is 1 to 2 mm^3 in volume, with variations that are narrowly correlated to the yearly sexual cycle. Lobules are separated from each others by connective tissue in which blood vessels, nerves, islets of interstitial (Leydig) cells are observed (Anjubault and Exbrayat 2000a, chapter 5 of this volume). Ductules arising from each tubule are also observed across this connective tissue (Fig. 3.12A, B).

Each lobule contains giant Sertoli cells and germ cells. The structure of gymnophionan testes has been subject to different interpretations. For a long time, the lobules of gymnophionan testes were described as units containing a filamentous matrix with fat droplets, in which isogenic groups of germ cells floated (Sarasin and Sarasin 1887-1890; Seshachar 1936, 1942a; Wake 1968b; De Sa and Berois 1986). Ovoid isolated cell bodies, described as Sertoli cells (Seshachar 1942a, b, 1948; Seshachar and Srinath 1946) were observed at regular interval against the wall of lobule. Rounded cell bodies, deeply stained by nuclear dyes, were observed, scattered throughout the matrix. Other rounded cell bodies were observed among spermatids and spermatozoa. A classical interpretation of this structure which has long been accepted was given by Seshachar (1936, 1942a). Using transmission and scanning electron microscopy, we give a different interpretation (Exbrayat and Dansard 1994; Exbrayat 1986a, b, 2000; Bhatta *et al.* 2001). More recently, the structure of the testis has been precisely described in Indian species (Smita *et al.* 2003, 2004a, b, chapter 6 of this volume).

In the classical interpretation, based exclusively on LM, each lobe is filled by a matrix in which groups of germ cells are floating without any relation to the matrix. As in other Amphibia (and more especially Anura), there are Sertoli cells that originate from follicular cells of the cystic membrane that surrounds spermatogonia. When spermatogonia multiply, follicular cells migrate to the lobule wall, accompanying germ cells, and become Sertoli cells. Then, the nucleus of each follicular cell becomes indented and more voluminous than previously. In contrast, the cytoplasm becomes less abundant and no morphological link is observed between Sertoli cells and the matrix. The deeply stained cell bodies that are observed in the central zone of the lobule are considered to be degenerate nuclei, surrounded by an abundant cytoplasm with osmiophilic granules (fat droplets). These cell bodies were considered to be hypertrophied degenerate spermatocytes or spermatogonia (Seshachar 1936). No interpretation of the origin of the matrix was given.

Use of TEM and SEM electron microscopy (Exbrayat 1986a, b, 2000; Exbrayat and Dansard 1994) resulted in a new interpretation of the structure of the testicular lobules in Gymnophiona.

In *Typhlonectes compressicauda* and in other species, after fracture and observation with a scanning electron microscope, each lobule has the appearance of a cup that is filled by a filamentous structure, except in the center of the lobule that communicates with the evacuation ductule. In certain zones, the filaments separate themselves and groups of germ cells are observed in the spaces (Fig. 3.12C).

Peripheral Sertoli cells formerly described, and the central cells considered previously to be degenerative germ cells (Seshachar 1936) are in fact true Sertoli cells. Cell bodies that were observed in light microscopy and that are deeply stained by nuclear dyes, are in fact nuclei of Sertoli cells. By SEM, these cells are giant cells with a spongy appearance; the hypertrophied cytoplasm forms a filamentous matrix, probably due to fixation by Bouin's fluid or formalin. By use of an iso-osmotic buffered glutaraldehyde and paraformaldehyde solution, we could observe a more homogeneous aspect of Sertoli cells in semithin sections in *Ichthyophis beddomei* (Bhatta et al. 2001). Nuclei are included in a region of the cell that is arranged between the cytoplasmic extensions (Figs. 3.12D, 3.13E). This part of the cell containing the nucleus is linked to the extensions by one of its poles. By TEM, the matrix is seen to be a true cytoplasm in which classical organelles are observed as well as voluminous vacuoles and osmiophilic inclusions (fat droplets) (Fig. 3.13).

These observations permit recognition of the stages in development of Sertoli cells in Gymnophiona. They originate from follicular cells that surround primary spermatogonia. When the spermatogonia divide, the follicular cells increase in size. Their nuclei are first rounded, become indented and polygonal and their cytoplasm develops. These Sertoli cells then leave the periphery of the lobule when they are displaced by a new generation of growing follicular cells. Their nuclei then decrease in volume and become rounded. Their cytoplasm develops progressively and the cells always surround germ cells that increase and differentiate. At the end of their development, Sertoli cells degenerate, their volume reduces and they pass into the efferent ductule, together with spermatozoa, this explains the presence of rounded isolated cells in groups of spermatids or spermatozoa. Observations of growing testes belonging to young animals corroborate this sequence (Exbrayat and Dansard 1994, see also chapter **9** of this volume).

In adult *Typhlonectes compressicauda*, Sertoli cells present histochemical variations during the breeding cycle. The contents of the cytoplasm ("matrix" filaments) are generally of a glycoproteic nature: they are stained by basic fuchsin, PAS positive and more or less stained by alcian blue and metachromatic. They are also stained by dyes used to visualise proteins such as Coomassie blue, and give a positive Hartig and Zacharias reaction. During the breeding period (January to May-June in this species), the cytoplasm of Sertoli cells is particularly acidic. At this period, all types of

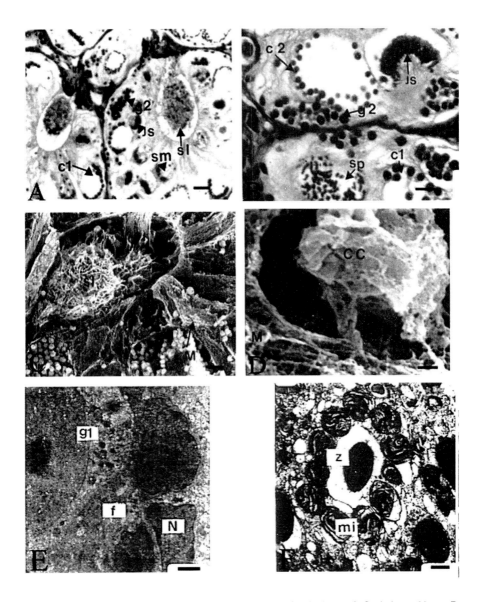

Fig. 3.12. *Typhlonectes compressicauda.* **A.** Section of testis (adult, January). Scale bar = 90 μm. **B.** Section of testis (adult, August). Scale bar = 30 μm. **C.** SEM view of testis. Scale bar = 20 μm. **D.** SEM of the nucleus of a Sertoli cell. Scale bar = 1 μm. **E.** TEM view of primary spermatogonium surrounded with follicle (Sertoli) cells. Scale bar = 1 μm. **F.** TEM view of a spermatozoon surrounded with Sertoli cells. Scale bar = 1 μm. cc, cellular body; c1, primary spermatocytes; f, follicle (Sertoli) cell; g1, primary spermatogonia; g2, secondary spermatogonia; Js, young spermatozoa; M, matrix (Sertoli cells) mi, mitochondrion; N, nucleus; sl, free spermatozoa; sm, matured spermatozoa; sp, old spermatids; z, spermatozoon. **A, B, C, D**, from Exbrayat, J.-M. 1986a. Unpublished D. Sci. thesis, Université Paris VI, France Pl. III, Figs. 3, 5, 7, Pl. IV Figs. 8. **E, F**, from Exbrayat, J.-M. and Dansard, C. 1994. Revue Française d'Histotechnologie 7: 19-26, Figs. 8, 9.

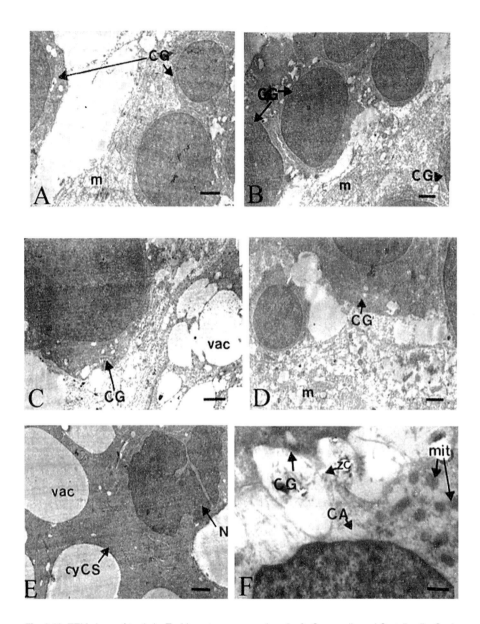

Fig. 3.13. TEM views of testis in *Typhlonectes compressicauda*. **A**. Germ cells and Sertoli cells. Scale bar = 1.5 µm. **B**. Germ cells and Sertoli cells. Scale bar = 1.5 µm. **C**. Germ cells and Sertoli cells. Scale bar = 2 µm. **D**. Germ cells and Sertoli cells. Scale bar = 1.5 µm. **E**. Sertoli cell. Scale bar = 2 µm. **F**. Germ cell associated to a Sertoli cell. Scale bar = 0.35 µm. CA, Sertoli cell associated to a germ cell; CG, germ cell; cyCS, cytoplasm of Sertoli cell; m, matrix (Sertoli cell); mit, mitochondria; N, nucleus of a Sertoli cell; vac, vacuola; zc, contact zone. From Exbrayat, J.-M. 1986a. Unpublished D. Sc. Thesis, Université Paris VI. Pl. V, Figs. 1-4.

germ cells are observed in the lobules. At the end of the breeding period, when spermatogonia and spermatocytes only are observed in the lobules, the cytoplasm is less acidic than previously. From July (last phase of spermatogenesis) to October, Sertoli cells again become acidic. Therefore, Sertoli cells undergo a cycle that is closely linked to that of germ cells (chapter 5 of this volume).

Smita *et al.* (2003, 2004a, b, chapter 6 of this volume) give a more precise interpretation of the "matrix" in several species of Asiatic Gymnophiona. Although several works have been published, the evolution of the interpretations shows that the structure of the testis in Gymnophiona begins to be totally understood though further studies are needed to give a definitive description.

3.2.2.3 Germ cells

Primary spermatogonia are always located at the opening of the efferent ductule. They are surrounded by follicular cells that will give rise to Sertoli cells (Fig. 3.12E). Germ cell differentiation is observed from the periphery of the lobule where primary and secondary spermatogonia are observed, to the center of the lobule, where spermatogonia are evacuated. Germ cells are grouped in isogenic series that can be compact (spermatogonia) (Fig. 3.13A, B, Fig. 3.14A) or arranged to form a hollow sphere. Spermatids and young spermatozoa seem to be embedded in Sertoli cells (Figs. 3.12F, 3.14B, C, D) and mature spermatozoa are free and enter the efferent ductule (Fig. 3.14E, F). Spermatogenesis, spermateleosis (spermiogenesis), structure and ultrastructure of germinal cells have been the object of several works (Seshachar 1936, 1937, 1940, 1942a, 1943, 1945; Wake 1968a, b; Exbrayat and Sentis 1982; De Sa and Berois 1986; Exbrayat 1986a, b, 1993, 2000; Smita *et al.* 2003, 2004a, b, and chapter **6**, Scheltinga *et al.* 2003, Scheltinga and Jamieson in chapter **7** of this volume). Spermatogenesis will be described in chapter 6 of this volume. The quantity of germ cells categories varies throughout the breeding cycle, as will be described in chapter 5.

Sexual cycles have been described in *Ichthyophis* (Sarasin and Sarasin 1887-1890; Seshachar 1936, 1937, 1943), *I. beddomei* (Bhatta 1987), *Gymnopis multiplicata* (Wake 1968b), *Dermophis mexicanus* (Wake 1980b, 1995), *Typhlonectes compressicauda* (Exbrayat and Sentis 1982; Exbrayat and Delsol 1985; Exbrayat 1986b) *Boulengerula taitanus* (Malonza and Measey 2005). Reviews have been given (Exbrayat and Flatin 1985; Exbrayat 1992a; Exbrayat *et al.* 1998, chapter 5 of this volume). Male Gymnophiona have a discontinuous yearly cycle that is characterized in all the species that have been studied by a phase of breeding and a phase of sexual inactivity. These two phases are closely linked to seasons: breeding occurs during the wet or rainy season, sexual quiescence during the dry season. In male *T. compressicauda*, the sexual cycle is under genetic control (Exbrayat and Laurent 1986). In this species, the distribution of germ cells is homogenous in the testes (Exbrayat and Sentis 1982; Pujol and Exbrayat 2000).

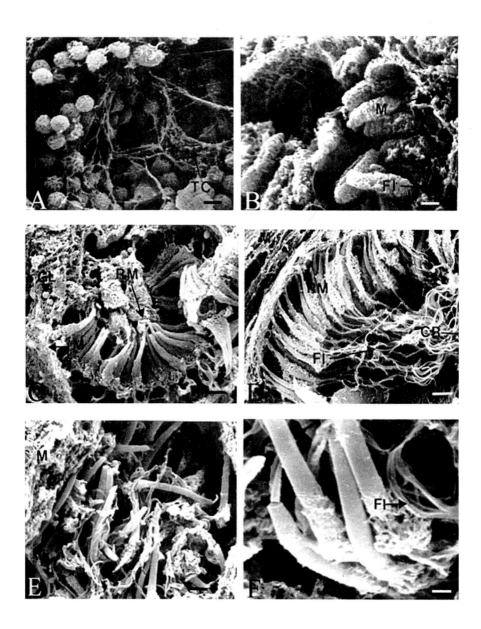

Fig. 3.14. SEM views of testis in *Typhlonectes compressicauda*. **A**. Primary spermatocytes. Scale bar = 7 µm. **B**. Old spermatids. Scale bar = 3 µm. **C**. Spermatids in relation with Sertoli cells. Scale bar = 6 µm. **D**. Spermatozoa. Scale bar = 3 µm. **E**. Free spermatozoa. Scale bar = 3 µm. **F**. Free spermatozoa, details. Scale bar = 1 µm. CR, cell in relation with germ cells (amoeboid cell after Smita et al., chapter 6 of this volume). Fl, flagellum; GL, inclusion; M, matrix (Sertoli cell); RM, relation between germ cell and Sertoli cell. TC, cytoplasm. From Exbrayat J.-M. 1986a. Unpublished D. Sci. thesis, Université Paris VI. Pl. IV, Figs. 1-6.

3.2.2.4 Growth of testes

Several works described embryonic development of gonads in Gymnophiona (Spengel 1876; Brauer 1902; Tonutti 1931; Seshachar 1936; Wake 1968a, b; Anjubault and Exbrayat 2000b, 2004a, and chapter **9** of this volume). Concerning more particularly the growth of the testes, in *Ichthyophis*, each lobe develops from proliferation of germ cells that are situated in a thin longitudinal duct. Fat globules develop in the "matrix" (Sertoli cells), from the periphery and they reach the center of the lobule when this develops (Seshachar 1936). In chapter **9** of this volume, we give data on the embryonic development and sexual differentiation of *Typhlonectes compressicauda*.

In *Dermophis mexicanus*, new-born gonads are dorsal strips of pigmented tissue that extend from the posterior end of the liver to the cloaca. One year after birth, spermatogenesis begins in 90% of males. In two-year-old animals, the testes appear to be active (Wake 1980b).

Testes growth has been particularly studied in *Typhlonectes compressicauda* (Exbrayat 1986a, b, 1993; Exbrayat and Dansard 1994, chapter **5** of this volume). At birth, when the animal is 120 to 180 mm in length, each testis as the appearance of a longitudinal cord on which lobes develop. Testes are now islets of cells that are grouped together at the end of a very small ductule. These lobules are enclosed in connective tissue and are separated from each other by a well defined basal lamina. In each lobule, several primary spermatogonia are surrounded by small somatic cells. Some of spermatogonia are dividing, others possess a single polylobed nucleus; the most developed spermatogonia are surrounded by typical follicular cells. This aspect has been found in animals caught over several months.

At birth, certain animals that are 170 to 180 mm in length already possess relatively developed lobules. The cytoplasm of somatic cells is developed, fibrous-like in LM, and resembles that of Sertoli cells of older animals. The nuclei of these cells are disposed against the lobule wall. A funnel-like aperture that will serve to evacuate spermatozoa is observed. Even secondary spermatogonia have been found in certain new-born.

In older animals (12 to 15 months old, 204 to 245 mm in length), lobules develop further and the follicular cells elongate, their nuclei move away from spermatogonia, their growing cytoplasm lengthens and their bases contact the germ cells. By LM, the cytoplasm shows the filamentous structures and fat globules. The follicular cells can now be considered as Sertoli cells. Secondary spermatogonia begin to develop. In certain animals (caught in July and August), the general structure of the testes remains unchanged and resembles that of the adult but, in addition, primary and secondary spermatocytes and even spermatids and spermatozoa can be observed. At this period, spermatogenesis occurs in adult animals.

During the first year (July to July), germ cells are exclusively primary and secondary spermatogonia. During the second year, the surface of lobule sections increases. Biometric data are characterised by variations that are similar to those found in adult testes. Other germ cell categories are observed from January or February. In May-June (18-month-old animals), spermatids and spermatozoa are eliminated. A proliferation of spermatogonia and spermatocytes is also observed. In August, the quantitative composition of germ cells is similar to that will be observed in adult animals.

3.2.3. Mullerian Ducts

3.2.3.1 In adults
In male Gymnophiona, persistent Mullerian ducts have been observed since the older literature (Spengel 1876; Wiedersheim 1879; Sarasin and Sarasin 1887-1890; Semon 1892). The glandular structure of these organs was first demonstrated by Tonutti (1931). Marcus (1930) considered Mullerian ducts to be auxiliary testes. For some authors, these organs remain rudimentary (Oyama 1952; Romer 1955; Lawson 1959). More recent works have also been devoted to Mullerian ducts in several species (Wake 1970a, 1977 b, 1981; Exbrayat 1985, 1986 a; George *et al.* 2004a, b, 2005, see also chapters 4 and 5 of this volume).

Mullerian ducts are paired organs that are parallel to the kidneys to which they are attached by connective tissue (Fig. 3.3A). At their posterior end, they describe a loop directed anteriorly before connecting to the cloaca. Two morphological types have been described (Wake 1970 a), according to the species and individuals. In the first type, the posterior part of each duct is thick, and the anterior three quarters remain narrow. Mullerian glands of this type have been observed in *Boulengerula boulengeri, B. taitanus, Caecilia* sp., *C. tentaculata, Oscaecilia ochrocephala, Geotrypetes seraphinii, Gymnopis multiplicata, Herpele squalostoma, Hypogeophis rostratus, Ichthyophis glutinosus, I. peninsularis Schistometopum gregorii, Scolecomorphus vittatus, S. uluguruensis, Siphonops brasiliensis, Uraeotyphlus oxyurus*. More recently George *et al.* (2004a, b, 2005) described with precision such a Mullerian gland in *Uraeotyphlus narayani*. In the second type, Mullerian ducts are thick throughout their length, only the anterior end is narrow. This situation has been found in *Chthonerpeton viviparus, Dermophis mexicanus, Schistometopum thomensis, Typhlonectes compressicauda* and some *Gymnopis multiplicata*. These two morphological differences can be either species-specific or linked to the period of the sexual cycle. After histological studies, it is observed that glands are present only in the posterior part of the first type. These glands that are situated all around a central duct correspond to "camera septa" (Tonutti 1931). Cells that contain granular secretions constitute them. Ciliated cells border the gland lumen and central duct. In the second type glands are present throughout the length of the Mullerian duct.

In *Typhlonectes compressicauda*, each Mullerian duct resembles a thin tube that is very narrow in its anterior part. This part extends as a still thinner structure that seems to be mixed with the connective tissue. The thickest part extends from the posterior end of the liver to the cloaca. Each Mullerian duct varies in diameter, length and histological structure during the year. Its diameter is maximal from January until April then it decreases in May. It will again increase in November. The anterior part is differentiated from November to May, but it is not distinguishable from June to October.

As noted in chapters 4 and 5 of this volume, the number, the size and the aspect of gland ductules vary considerably according to the period of sexual cycle.

In January, during the breeding period, each glandular ductule is bordered by a pseudostratified epithelium with ciliated cells (50 × 10 µm) containing an apical nucleus, and glandular cells (20 × 100 µm) containing a basal nucleus. These gland cells contain abundant secretions (chapter 5, Fig. 5.2A, B). In the anterior zone of the secretory region, the cells of the external ductules situated against the wall and the cells at the bottom of certain ductules contain granules stained by acidic fuchsin and azocarmine. They are never PAS positive but they are metachromatic and stained with alcian blue at pH 2.5. Hartig-Zacharias staining gives a pale blue. These properties reveal that these granulations are of a proteic nature, sometimes with an acidic glucidic part. The other ductules of this same region and all the ductule of the other parts of Mullerian duct contain globules that are 1 to 3 µm in diameter. These globules are very evident by SEM. They are PAS positive but neither metachromatic nor stained by alcian blue. They are deeply stained in blue by Hartig-Zacharias reaction. Therefore, these secretions are of neutral glycoproteic nature.

In February, cells belonging to the posterior and the median regions of the Mullerian duct are empty. In April, both types of secretions are always observed in the anterior part. In the median region, the cells are shorter (10 × 70 µm) and contain some globules with the same staining (chapter 5, Fig. 5.2C). The cells situated in the posterior part are empty. The central duct is always bordered by ciliated cells and filled by granules and cytoplasm fragments.

In May, granular secretions are always observed in the cells but the globules continue to be evacuated. The sizes of the secretions remaining in the cells have decreased (10 × 40 µm). They stain very weakly. The ductules have lost their cilia. The wall of the central lumen is in desquamation.

In June, the size of Mullerian duct has greatly decreased. The central duct is not ciliated and the ductules are reduced to small extensions with only one cell type without any cilia or secretion. At this period, the connective tissue is relatively abundant. Fragments of cilia, cytoplasm and PAS positive secretions are observed in the ductule.

In July and August, the cytoplasm of gland cells is stains pink after fuchsin staining or PAS reaction. Several granules can be observed. The cytoplasm is blue stained after Hartig-Zacharias reaction (chapter 5, Fig. 5.2D).

From October, the central ducts and ductules are again ciliated. Two cell types are observed. The gland cells are 15 × 50 µm. The cytoplasm is PAS positive and, in certain of them, small granulations are stained by azocarmine, fuchsin and alcian blue, but they are not stained by Hartig-

Zacharias reaction. Therefore, these secretions are of an acidic glucidic nature.

After October, microanatomical aspect of Mullerian ducts is characteristic of the breeding period.

The Mullerian gland of *Uraeotyphlus narayani* has been studied by George et al. (2004a, b, 2005; see also chapter 4 of this volume) by means of LM and TEM. In this species, the anterior part of the duct extends as a connective strand. The posterior one-third is the glandular part of the organ that consists of a large number of tubular glands arranged around the central duct. These glands are separated by a connective tissue containing blood vessels and amoeboid cells.

During the active spermatogenic phase (July to February), the epithelium of the column of a tubular gland is formed of a single layer of secretory and ciliated cells. Secretory cells contain granules that are PAS positive and stain with alcian blue and mercuric bromophenol blue, showing their glycoproteic nature. The apical cytoplasm is covered with microvilli forming a brush border. The cytoplasm contains granules of varying sizes. The nucleus is in a basal position. Perinuclear and basal cytoplasm is sometimes vacuolated, containing membrane-bound bodies resembling lysosomes andmitochondria. The cytoplasm also contains RER, SER and Golgi apparatus (Fig. 3.15E F). The ciliated cells are broad toward the lumen of the gland. The nucleus, mitochondria, endoplasmic reticulum are situated in the apical part of the cell. Golgi apparatus is not abundant. Large dense bodies are observed in the narrow basal part of the ciliated cells.

At the base of tubular glands, are secretory cells, ciliated cells and amoeboid cells. In these secretory cells, the nucleus is basal; the cytoplasm has a higher electron density than that of columnar secretory cells. In the cytoplasm, secretory material is condensing and granules are elaborated. These cells contain numerous mitochondria, abundant RER, SER and Golgi apparatus. The basal ciliated cells resemble the ciliated cells of the columnar part of the gland. Amoeboid cells are observed between all the other cells.

The duct of tubular gland is lined by a layer of columnar cells, with ciliated dark cells and light cells with microvilli. This duct opens into the central duct. It is composed of ciliated dark cells and light cells with microvilli, sometimes accompanied by cilia.

During sexual quiescence (March until June), the Mullerian gland is regressed. Gland cells are empty. Amoeboid cells at the base of the column are abundant and lymphocytes are observed in the glandular epithelium.

The glandular secretions in *Dermophis mexicanus* and *Typhlonectes compressicauda* have been analysed by Wake (1981). These secretions contain mucopolysaccharides, fructose and acidic phosphatases. Their acidic pH is similar to that of sperm. The Mullerian ducts of Gymnophiona are equivalent to the mammalian prostate (see also chapter 4 of this volume). Prostates possess an epithelium with an embryonic Mullerian origin and also elaborate several elements of the semen of which the chemical nature resembles that of secretions of gymnophionan Mullerian glands.

In Gymnophiona, the Mullerian ducts are active glandular structures that elaborate substances contained in the semen. The secretions of Mullerian ducts and spermatozoa are certainly mixed in the cloaca, but no study indicates such a mixture. In fact, passing of sperm through the cloaca must be very rapid.

3.2.3.2 In young animals

The growth of Mullerian ducts has been studied in *Typhlonectes compressicauda* (Exbrayat 1986a). The lumen of young Mullerian duct is bordered by an epithelium that is one-cell thick (Fig. 3.15A). More or less developed extensions are connected to the lumen. A single layer of cells also limits them. The epithelium is surrounded by a basal lamina that separates it from the connective tissue. Externally, a layer of connective tissue limits each Mullerian duct. The volume of these ducts and more particularly of the extensions varies according to the age of the animal and the season at which the animal has been caught.

At birth (that always occurs from June since September), each Mullerian duct possesses a central lumen with a rounded section, bordered by a columnar epithelium composed of a single layer of undifferentiated cells (8 × 20 µm in diameter) (Fig. 3.15B). These cells have a voluminous well-stained nucleus occupying almost all the cell. The connective tissue is relatively well supplied with blood vessels. In 12 months old animals (June to August), the connective tissue seems to be less abundant (Fig. 3.15C). The diameter of the central duct has increased and in some animals gland extensions begin to be observed. The lumen of the glands is bordered with an epithelium in which tall cells with a basal nucleus and cytoplasm with granulations occur. Cells with short cilia and an apical nucleus are also observed. The central lumen of the Mullerian duct is sometimes ciliated. At this period, all intermediary forms are observed between an undifferentiated duct, without glands and a duct with numerous developed glands.

In 18-month-old animals (January, February), the glands are developed. The high glandular cells are filled with spherical secretions (3 µm in diameter) stained by azocarmine or acidic fuchsin and PAS positive. These secretions are not stained by alcian blue at different pH and are not metachromatic, but they react positively with the Hartig-Zacharias reaction. These histochemical characteristics are those of neutral mucopolysaccharides.

The lumen of each tubular gland is bordered by cilia. The ductules of several neighboring glands unit themselves to empty into the central lumen of the Mullerian duct. This lumen is bordered with numerous cilia.

In the next months, the gland ductule decreases (Fig. 3.15D). The secretory cells are less and less developed. The gland ductule and the central lumen of the Mullerian duct lose their cilia. Numerous fragments of

118 Reproductive Biology and Phylogeny of Gymnophiona

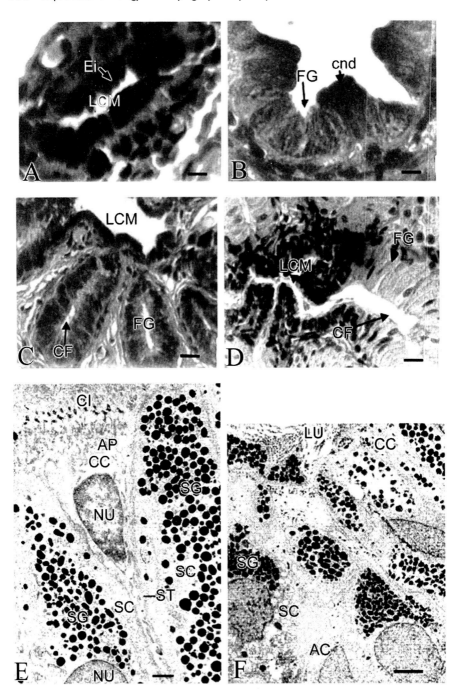

Fig. 3.15. Mullerian glands in *Typhlonectes compressicauda*. **A**. Mullerian gland in a new-born. Scale bar = 15 µm. **B**. Mullerian gland in a new-born. Scale bar = 15 µm. **C**. Mullerian gland in 12 months old

Fig. 3.15 contd

cells are evacuated. By April (19 months), a small animal (232 mm) with well developed glands but without cilia and without secretions in gland cells has been observed. In the testes of this animal, spermatids were present (Exbrayat 1986a). In May (20 months), the connective tissue is again very abundant. The gland cells have greatly decreased. In June (21 months), some animals possess glandular structures that are relatively voluminous with two secretion types, as in the adults. One type is intensely stained with paraldehyde fuchsin and it is situated on the periphery of the gland; the other type is observed into the most central cells. All the gland ducts and the central lumen are ciliated. At this same period, other animals have typical regressive Mullerian ducts.

In two-year-old animals (August), Mullerian ducts are regressive. But just after this period, a new increase of these organs is observed. The animal is now beginning its first adult cycle.

Morphological studies of the total surface of transverse sections, the surface of gland sections or the surface of one single gland section show similar variations with a maximal value in 18-month-old animals (January, February), then a decrease in next May, a new increase in June and a final decrease in July. From August, a new increase of morphometric values is observed and continues to next January, by which the animals are adults. The variations of percentage of the section surface occupied by the gland ductules and the average of the number of the gland ductules observed in each transverse section, show similar variations. These results, added to the histological observations, demonstrate that the increase of Mullerian duct during the two first years of life is due to the increase of the glandular structure.

The variations seem sometimes to be independent of the size of the animal. For instance, in May, a large animal possesses a great quantity of secretions. Conversely, in June, the Mullerian ducts are the most developed in an animal with the smallest size.

Growth of Mullerian ducts takes two years. From the birth to the first year, the ducts enlarge. In one-year-old animals, the gland ductules begin to develop from the central lumen. In the earliest animals, the secretions are already observed in the glands. From 18 to 22 months, during the

---

Fig. 3.15 contd

male. Scale bar = 30 µm. **D**. Transverse section of Mullerian gland in a 19 months old male. Scale bar = 30 µm. LCM, lumen of Mullerian gland; FG, tubular gland; cnd: undifferentiated cell; CF, evacuation duct of a tubular gland; secr, secretion. **E**. TEM view of the upper portion of a tubular gland. Scale bar = 3 µm. **F**. TEM view of the base of a tubular gland. Scale bar = 3 µm. AC, amoeboid cell; AP, apical portion; CC, ciliated cell; CI, cilia; Ei, undifferentiated epithelium; LU, lumen; NU, nucleus; SC, secretory cell; ST, narrow stalk of a ciliated cell; **A, B, C, D**, from Exbrayat, J.-M. 1986a. Unpublished D. Sci. thesis, Université Paris VI. Pl. VII, Figs 1-4. **E, F**, from George, J.M., Smita, M., Oommen, V.O. and Akbarsha, M.A. 2004a. Journal of Morphology 260: 33-56, Figs. 11, 18.

theoretical period of breeding (January, February), the earliest animals possess abundant glycoproteic secretions, but not the others.

It is probable that the animals of which birth occurred at the beginning of the birth period (June-July) develop before the animals of which birth occurs at the end of the birth period (August-September). That may explain why, at different periods of the year, some animals have developed glands in Mullerian ducts and why others have still undifferentiated Mullerian structures. In some animals, numerous cell fragments and small secretions have been observed in the central lumen of the glands or in the duct, that could be the result of the evacuation of small secretions and degenerate cells in an animal that is still not ready for breeding (see chapters 4 and 5 of this volume) (Exbrayat 1986a).

Thus, during these two years, a sexual cycle seems to be sketched out; growth of the Mullerian duct is the main physiological phenomenon during this period.

3.2.4 Structure of the Cloaca

Internal fertilisation is a general rule in Gymnophiona. Accordingly, the male cloaca possesses an intromittent organ, the phallodeum (Fig. 3.16A). Several studies have been devoted to the description of the cloaca in *Hypogeophis rostratus* (Tonutti 1931, 1932, 1933). Comparisons of cloaca morphology in numerous genera and species have been given in an important work (Wake 1972). Cloacal anatomy and its morphological variations during growth and the breeding cycle have been described in *Typhlonectes compressicauda* (Bons 1986; Exbrayat 1991, 1996). The external structure of the phallodeum has also been studied in several other species, especially as a taxonomical character (Taylor 1968; Wake 1972, 1977b, 1981). Spicules have been described in the phallodeum of several species of *Scolecomorphus* genus (Wake 1998).

3.2.4.1 Anatomy

The cloaca is linked to the dorsal part of body by connective tissue that also enwraps the bladder. The cloaca receives the intestine from which it is separated by a sphincter. In its anterior part, it also receives the Wolffian and Mullerian ducts after their fusion. In *Chthonerpeton* and *Typhlonectes*, these ducts emerge at the tips of papillae. The bladder empties ventrally into the cloaca. The posterior part of the male cloaca is the phallodeum. A ventral muscle, the musculus retractor cloacae, links the cloaca to the median part of the body: it enables the return of the phallodeum to its resting position after the sexual act.

The cloaca wall possesses ridges and bulbs arranged in a more or less complex manner according to the species. A pair of blind sacs has also been observed (Wake 1972). These blind sacs are more or less developed and they consist of folds of the dorsal wall. Beneath the blind sacs, several longitudinal gutters are also observed. Dorsal and ventral depressions are situated between these structures. It seems that the most primitive species possess the most complex cloaca structures (Wake 1972). Presence of expansions that form blind sacs is not systematic (Wake 1972; Exbrayat 1991) (Fig. 3.16D, E, F).

In primitive genera and species such as *Ichthyophis kohtaoensis* and *Uraeotyphlus*, or the less primitive *Herpele squalostoma*, these expansions form voluminous blind sacs. In the advanced species *Typhlonectes compressicauda*, such chambers are also found, but they are smaller than in the other species though perfectly differentiated and functional. In *Boulengerula boulengeri*, these expansions are reduced structures. In *Siphonops annulatus* they have been lost and the cloaca is straight. *Microcaecilia unicolor* and *Geotrypetes seraphinii* possess a pair of lateral expansions that are widely open into the central duct of the cloaca. These structures could be the equivalent of blind sacs. In *Scolecomorphus*, several types of differentiation of lateral blind sacs have been observed (Wake 1972). In *Scolecomorphus uluguruensis*, two blind sacs have been found; in *S. kirkii*, only small grooves have been observed and in *S. vittatus*, a pair of small blind sacs that are very well differentiated have been observed.

The cloaca of male Gymnophiona can be divided into three, very well differentiated parts. The most anterior part is the cloacal ampulla. This receives the intestine, bladder and common trunks of Wolffian and Mullerian ducts. The median part, narrow in section, possesses, when they exist, the pair of blind sacs and a central duct. The posterior part is the phallodeum, i.e. the copulatory organ. This organ is contained in a connective tissue capsule from which it is separated by a periphallodeal space. The phallodeum slides into this capsule during its devagination. The three parts of the cloaca exhibit significant histological and morphological variations during the sexual cycle that will be described with some details in chapter 5 of this volume.

The wall of the cloaca always consists of several tissue layers. The outward sequence is: an epithelium the nature of which varies according to the level of the organ and the physiological state; a vascularized connective tissue delimiting numerous longitudinal folds; a longitudinal layer of striated muscles; and a peripheral connective tissue. The different layers are continuous with histological layers of the intestine or of other organs emptying into the cloaca.

In *Hypogeophis rostratus*, certain cloacal zones have been described lacking epithelium (Tonutti 1931, 1932, 1933). In *Boulengerula*, a stratified glandular epithelium has been found bordering each blind sac lumen. Posteriorly in this region, the epithelium is stratified and ciliated but is not glandular; it surrounds the wall of phallodeum. In *Gymnopis*, these two types of epithelium are observed but the transition between them seems to be more progressive: the glandular part is observed also in the phallodeum in a mucous groove, alternating with ciliated zones (Wake 1972). In *Typhlonectes compressicauda*, the anterior part of the cloaca is bordered by a very glandular epithelium that will become ciliated during reproduction; in the phallodeum, the cloacal lumen is bordered by a low ciliated epithelium that becomes progressively and locally thickened to give regular spines that will protrude when the cloaca is evaginated (Exbrayat 1991, 1996).

In Scolecomorphidae, spicules are included in the cloaca wall (Noble 1931; Taylor 1968, 1969; Nussbaum 1985; Wake 1998). The number, size and disposition of the spicules vary according to the species and these variations do not seem to be linked to size, growth or sexual maturity (Wake 1998). Spicules are projecting structures that originate from a plate of tissue forming a cartilaginous structure. These spicules pierce the epithelium bordering the wall of cloaca. It is not still known if these structures are or not involved in reproduction (Wake 1998).

3.2.4.2 Extrusion of phallodeum

The mechanism of extrusion of the erectile phallodeum has been well studied in *Hypogeophis rostratus* (Tonutti 1931) and it seems possible to apply it to all Gymnophiona species from known anatomical data (Wake 1972).

The phallodeum of *Hypogeophis* is a tube bordered by spinous structures that are more or less developed according to the period of sexual cycle (Fig. 5.2E, F, chapter 5 of this volume). In its anterior part, it is in continuity with the median zone of the cloaca. Posteriorly, the internal walls of the phallodeum are connected with the epidermis and dermis. The phallodeum is contained in a connective capsule. The muscular wall contracts and pushes on cloaca so that it slides into the capsule. Its posterior end is extruded. The cloacal tube is inverted and the spinous internal epithelium is now situated on the external face of the phallodeum, enabling copulation. The muscular retractor cloacae will be used for return of the cloaca into its connective tissue (Fig. 3.16B, C).

3.2.4.3 Growth of cloaca

At birth, in male *Typhlonectes compressicauda*, the cloaca is a duct with two rudimentary blind sacs. The anterior region is bordered by a mucous epithelium with a carboxilic acidic secretion. The median region is bordered by an epithelium with undifferentiated cells, some mucous cells and cells with a large vacuole. The posterior region, the phallodeum, is bordered by the same type of epithelium with fewer mucous cells and with other cells that have a squamous apical part. In 12-month-old animals, the structure is the same. In 18-month-old animals, in the normal breeding period, the median part of the cloaca is bordered by a stratified epithelium covered by an acidic mucous secreted by goblet cells of blind sacs. The phallodeum is covered by a squamous epithelium with still several mucous cells.

3.2.5 Fat Bodies

All Amphibians possess a pair of fat bodies associated with the gonads. Variations of size and biochemical composition have been studied in some Anura (*Bufo bufo*, Jorgensen et al. 1979; *B. arenarum*, Penhos 1953; *Rana esculenta*, Roca et al. 1970; Schlaghecke and Blum 1978). These organs have a cycle correlated with periods of breeding.

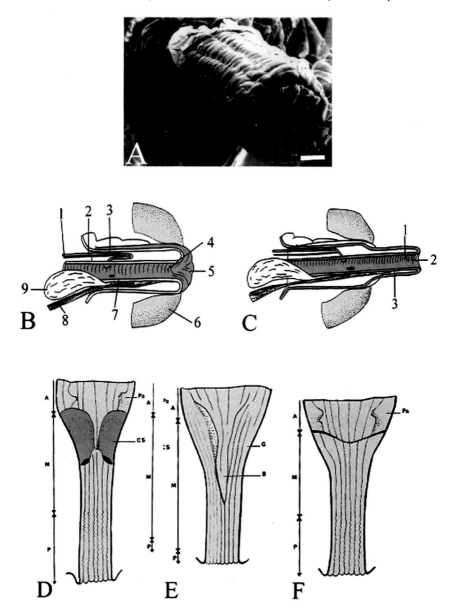

Fig. 3.16. A. SEM view of a devagined phallodeum in *Typhlonectes compressicauda*. Scale bar = 600 µm. **B**. Cloaca during sexual quiescence. 1, Mullerian duct; 2, Wolffian duct; 3, case; 4, phallodeum epithelium; 5, opening of cloaca; 6, posterior part of the body; 7, phallodeum; 8, *musculus retractor cloacae*; 9, urinary bladder. **C**. Extrusion of the phallodeum during sexual activity. 1, phallodeum epithelium; 2, opening of cloaca; 3, phallodeum. **D**: cloaca with a pair of blind sacs. **E**: cloaca with lateral expansions. **F**: cloaca without expansions or blind sacs. A, anterior part; B, central bud; CS, blind sac; G, lateral expansion; M, median part; P, posterior part (phallodeum); Pa, papilla (opening of common duct to Wolffian duct and Mullerian glands). **A**, original. **B, C, D, E, F**, modified after Exbrayat, J.-M. 1991. Bulletin de la Société Herpétologique de France 58: 30–42, Figs. 5a, B, 2a, b, c.

In Gymnophiona, fat bodies have been described by several authors (Sarasin and Sarasin 1887-1890; Fuhrmann 1914; Tonutti 1931; Wake 1968a, b). In these animals, fat bodies are paired segmented organs with 8 to 30 lobes. They extend from the posterior end of the liver to the cloaca. The most voluminous are pale yellow, and the smallest are red orange. Their size is variable and correlated to the size of the animal. In *Siphonops*, the fat bodies are composed of only a few voluminous lobes. In *Typhlonectes compressicauda*, a study has shown variations related to the sexual cycle (Exbrayat 1986a, 1988b, chapter **11** of this volume).

Each fat body is limited by a thin layer of connective tissue. Fat bodies are almost totally composed of adipose cells. The cytoplasm of each cell is filled with voluminous fat vacuoles. Blood vessels are also observed among the cells. In adult *Typhlonectes compressicauda*, the weight varies very little but values are smaller in February and July. Adipose cells generally are 70×80 µm. From birth to adult, fat bodies progressively increase to reach the adult characteristic in 24-month-old animals.

3.3 ANATOMY OF FEMALE GENITAL TRACT

3.3.1 General Anatomy

The first works concerning female genital tract of female Gymnophiona were anatomical with only some indications about their size (Müller 1832; Rathke 1852). The ovaries of *Caecilia gracilis*, *Ichthyophis glutinosus* and *Siphonops annulatus* were the first to be described (Spengel 1876). Morphology and histological structure were studied in *Ichthyophis glutinosus* (Sarasin and Sarasin 1887-1890) and relationships between ovaries and fat bodies in *Hypogeophis rostratus* (Tonutti 1931). The ovaries of *Uraeotyphlus menoni* (Chatterjee 1936) and *Uraeotyphlus oxyurus* (Garg and Prasad 1962) have also been described.

More recently, comparisons of urogenital ducts of numerous species and genera of Gymnophiona have been published (Wake 1968b, 1970a, b, 1972, 1977b). These species were *Caecilia (Oscaecilia) ochrocephala*, *Gegeneophis ramaswamii*, *Geotrypetes seraphinii*, *Icthyophis glutinosus*, *Rhinatrema bicolor*, *Schistometopum gregorii*, *Scolecomorphus kirkii*, *S. vittatus*, *S. uluguruensis*, *Siphonops annulatus*, *Uraeotyphlus menoni*, *Typhlonectes compressicauda*, *Idiocranium russeli*, *Boulengerula taitanus (Afrocaecilia taitana)* and *B. uluguruensis (Afrocaecilia uluguruensis)*. The female genital tract of *Dermophis mexicanus* (Wake 1980 b), *Typhlonectes compressicauda* (Exbrayat 1983, 1986a), *Ichthyophis beddomei* (Masood-Parveez 1987, Masood-Parveez and Nadkarni 1993a, b) and *Chthonerpeton indistintum* (Berois and De Sa 1988) have been more particularly studied.

The genital tract of Gymnophiona is composed by a pair of elongated ovaries, a pair of oviducts, each differentiated as a uterus in viviparous species, and a cloaca that is particularly adapted to internal fertilisation (Fig. 3.17A, B). A pair of fat bodies is associated with the gonads. These fat bodies are subject, like the other parts of female genital tract, to variations linked to the sexual cycle and, in some species, to maintenance of embryos in utero.

3.3.2 Ovaries

Gymnophionan ovaries are paired elongate sac-like structures, parallel to the kidneys and digestive tract. They contain follicles and oocytes at different stages of growth (Figs. 8.2 and 8.4 in chapter 8 of this volume). Each ovary is linked to the corresponding kidney by a sheet of connective tissue, the mesovarium. Another sheet of connective tissue links each to the corresponding fat body. Transverse blood vessels connect these organs to each other. As in other vertebrates, the ovaries are not directly linked to the oviducts. Certain ovaries (length 50 to 90 mm) contain 50 to 100 small cream-colored oocytes that are less than 1 mm in diameter. Other ovaries are characterised by 3 to 12 mature oocytes that are yellow-colored and are about 2 mm in diameter. The larger oocytes are scattered among smaller ones. Certain ovaries are characterised by 20 to 30 oocytes that are yellow or orange colored and 2 to 5 mm in diameter that are often joined together in the anterior part of the gonad.

By histological study, one can observe that each ovary is composed of two to four cell layers among which oocytes develop. Each ovary is surrounded by connective tissue. In old works, the part of the ovary situated between this connective tissue and the oocytes and follicles was described as an "ovarian duct" (Tonutti 1931). This "duct" was supposedly used by oocytes to reach the cloaca and, consequently, the oviduct proper would only have a secondary function. However this hypothesis has long since been abandoned. In fact, the connective tissue has enveloped the ovary, suspending it in the general cavity (Wake 1968b).

After expulsion of the oocyte, a corpus luteum can be observed in oviparous as in viviparous species. Only *Rhinatrema* does not present this structure (Wake 1968b). Corpora lutea are generally ellipsoid, filled by a fat mass and may contain fibrous connective tissue. They are particularly well developed in *Caecilia ochrocephala* (Wake 1968b). They develop according to development of the embryos in *Dermophis mexicanus* (Wake 1980b) and in *Typhlonectes compressicauda* uterus (Exbrayat 1983, 1986a, 1992b; Exbrayat and Collenot 1983) (Fig. 5.8A, B, C, chapter 5 of this volume), that are both viviparous species. These corpora lutea are implicated in endocrine regulation of reproduction. Growth of ovary, aspects of oogenesis and folliculogenesis are described in chapter 8 of this volume. The endocrine aspects are developed in chapter 5 of this volume.

3.3.3 Oviduct and Uterus

The different modes of reproduction in Gymnophiona are the cause of important variations in size, morphology and microscopic anatomy of the genital ducts. Oviducts have been studied in oviparous species: *Ichthyophis glutinosus* (Sarasin and Sarasin 1887-1890), the genera *Ichthyophis* and

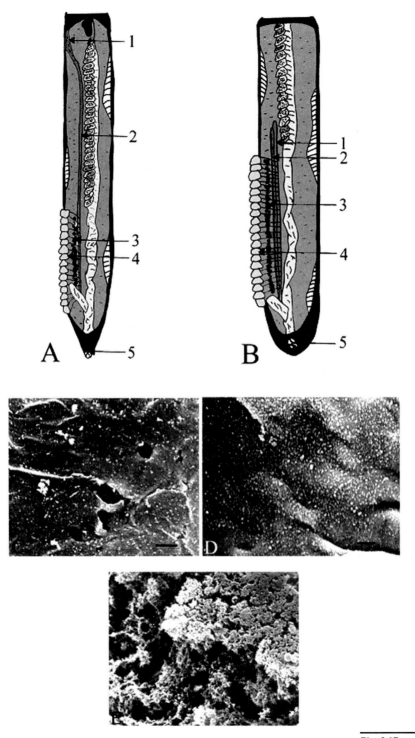

Fig. 3.17 contd

Hypogeophis (Marcus 1939), *Idiocranium* and *Rhinatrema* (Parker 1934, 1936), in *Uraeotyphlus oxyurus* (Garg and Prasad 1962), *Ichthyophis beddomei* (Masood Parveez 1987; Masood Parveez and Nadkarni 1991) and *I. kohtaoensis* (Exbrayat 1989a). The first observation of a viviparous gymnophionan was for *Caecilia (Typhlonectes) compressicauda* (Peters 1874a, b, 1875). Other viviparous Gymnophiona genera have been studied: *Chthonerpeton* (Parker and Wettstein 1929; Parker and Dunn 1964), *Geotrypetes* (Parker 1956), *Schistometopum* (Parker 1956), *Dermophis* (Parker 1956, Wake 1980b), and *Gymnopis* (Dunn 1928; Taylor 1955; Wake 1967, 1977a, b). Several works have been devoted to *Typhlonectes compressicauda* (Exbrayat 1984b, 1986a, 1988a; Hraoui-Bloquet 1995). Study of oviducts in *Microcaecilia unicolor* suggested that this species is viviparous (Exbrayat 1989a). In several reviews, the structure of the oviduct of Gymnophiona has been compared with that of Anura and Urodela (Wake 1993b; Wake and Dickie 1998).

3.3.3.1 The oviducal funnel

The ostium or oviducal funnel has been little studied. Several works indicate that oocytes are extruded by ovaries into the general body cavity where they are grasped by the ostium to be driven into the oviduct (Chatterjee 1936; Oyama 1952; Garg and Prasad 1962; Wake 1970a; Masood Parveez 1987).

In *Ichthyophis kohtaoensis*, *Typhlonectes compressicauda* and *Microcaecilia unicolor*, a comparative study of the funnel has been published (Exbrayat 1989a).

In *Ichthyophis kohtaoensis*, an oviparous species, oviducts open anteriorly by an ostium, at a great distance from the ovaries (Fig. 3.17A). This funnel is directed toward the head and opens into the body cavity. In female sexual quiescence (oocytes less than 1 mm in diameter), the funnel is bordered by a pavement epithelium with a single layer of undifferentiated cells and small pores. In a female with 7 mm diameter oocytes, the funnel is bordered by a single layer of cells with a striated brush-border. In a female bearing large oocytes, the body cavity is covered by a secretion originating from these cells.

In *Typhlonectes compressicauda*, each funnel is a longitudinal gutter elongated parallel to the oviduct to which it is bound (Fig. 3.17B). Communication between funnel and oviduct occurs anteriorly. During

Fig. 3.17 contd

Fig. 3.17. A. Schematic representation of the right female urogenital tract in *Ichthyophis kohtaoensis*. **B**. Schematic representation of the right female urogenital tract in *Typhlonectes compressicauda*. 1, ostium; 2, oviduct; 3, ovary; 4, fat body; 5, cloacal opening. **C**, coelomic wall during sexual quiescence in *I. kohtaoensis*. No secretions can be seen. Scale bar = 2 μm. **D**: coelomic wall during preparation of reproduction in *Ichthyophis kohtaoensis*. Scattered secretions are observed. Scale bar = 2 μm. **E**: coelomic wall during breeding in *I. kohtaoensis*. Abundant secretions are observed. Scale bar = 5 μm.
A, B, modified after Exbrayat, J.-M. 1989. Bulletin de la Société Herptologique de France 52: 34-44.
Figs. 1, 2. **C, D, E**, original.

sexual quiescence, the funnel is poorly developed and is bordered by a single layer of undifferentiated cells. During the breeding period, the funnel develops and the gutter inserts into the parallel ovary. Oocytes are directly emitted into the funnel, without passing through the body cavity. The funnel is bordered by a ciliated epithelium and glandular cells.

In *Microcaecilia unicolor*, all ovaries have few voluminous follicles, but certain of them have oocytes 300 µm in diameter with small yolk platelets separated from developed granular cells by a thin vitelline membrane. In these animals, compact corpora lutea resembling those of *Typhlonectes compressicauda* at the end or just after pregnancy (Exbrayat and Collenot 1983) have been observed. In these animals, the funnel also resembles that of *Typhlonectes compressicauda*: it is an elongated gutter parallel to the oviduct to which it is bound by connective tissue; its wall is covered by glandular cells secreting a PAS positive mucus.

These observations indicate that in oviparous species, the oviduct shows a primitive connection, situated far from the ovary (Fig. 3.17A). In the breeding period, the body cavity is covered by secretions (but no cilia have been observed) facilitating movement of oocytes to the funnel (Fig. 3.17C-E). In *Typhlonectes compressicauda*, a viviparous species, the funnel is more functional, inserting into the ovary (Fig. 3.17B), the case of *Microcaecilia unicolor* is more difficult to interpret for its reproductive mode is not known: neither eggs nor intrauterine embryos has been observed. The form of the funnel suggests that this species is viviparous (Exbrayat 1989a) but this is still only an hypothesis.

During the sexual cycle, the funnel undergoes significant variations that have been particularly studied in *Typhlonectes compressicauda* (Exbrayat 1984b, 1986a, 1988a, chapter 5 of this volume).

3.3.3.2 Oviduct of oviparous species

The oviducts are paired elongated structures, parallel to the ovaries and kidneys; they have been studied in several oviparous species: *Ichthyophis glutinosus* (Sarasin and Sarasin 1887-1890), *Ichthyophis* and *Hypogeophis* genera (Marcus 1939), *Idiocranium* and *Rhinatrema* (Parker 1934, 1936), *Uraeotyphlus oxyurus* (Garg and Prasad 1962), *Ichthyophis beddomei* (Masood Parveez 1987, Masood Parveez and Nadkarni 1991) and *I. kohtaoensis* (Exbrayat 1989a). A comparative study of numerous oviducts belonging to several oviparous species has been made (Wake 1970b). In oviparous species, each oviduct is divided into three regions: (1) an anterior pars recta bordered by gland cells with proteic secretions and acidic mucous glands, (2) a pars convoluta bordered for half of its circumference by glands secreting an acidic mucous, and proteic substances, and for the other half by cells with cilia during the breeding period, (3) the posterior part, a pars utera, is bordered by cells that are ciliated during breeding period and glandular cells with a neutral mucous substance. These structures are subject to a sexual cycle that has been well studied in *Ichthyophis beddomei* (Masood Parveez 1987; Masood Parveez and Nadkarni 1991) (Fig. 3.18A-F).

In *Uraeotyphlus oxyurus*, the posterior two-thirds of the oviducts are superficially fused but each possesses its own aperture into the cloaca (Garg and Prasad 1962). In *Ichthyophis glutinosus*, Spengel (1876) observed that the oviducts are fused with a single common lumen but this observation was not confirmed by later works (Wake and Dickie 1998).

3.3.3.3 Oviduct of viviparous species

It is generally agreed that 50% of gymnophionan species are viviparous (Wake 1977a, 1989, 1992, 1993), but this number is contested by certain authors who estimate that viviparity occurred in 24 to 30% of genera and only in 15 to 17% of species (Wilkinson and Nussbaum 1998). Scolecomorphidae and Typhlonectidae are obligatory viviparous families (Wake 1993b).

Very few studies concerning the structure of the oviduct in viviparous species have been published. In *Scolecomorphus kirkii* without embryos, the oviduct is limited by a connective tissue surrounding a thick layer of muscular cells that are disposed in a fibrous and elastic connective tissue. Numerous vascularized villosities project towards the lumen. The wall of the duct is bordered by a single layer of ciliated cells with a central nucleus. Between the villosities, cells at the bottom are secretory. Each cell possesses a basal nucleus with an abundant apical cytoplasm. The walls of the oviducts are expanded by unfertilized oocytes and eggs. At these places, the oviduct diameter is greater and the wall is flattened (Wake 1970a). In several species, when females bear embryos, the diameter of the oviduct increases as the embryos grow. Blood vessels become very abundant. In *Gymnopis multiplicata*, the wall of the oviduct remains thin at the level of the embryo; connective crests with numerous ramifications are observed in the other zones, epithelial cells are not ciliated and glandular cells are well developed. As the size of the embryos increases, the wall of oviduct is further modified; a thin layer of elastic tissue covers a fibrous zone, no epithelium borders the lumen which is directly in contact with connective tissue, glandular zones are reduced, and certain muscular parts are degraded, perhaps consequently to the action of fetal teeth (Wake 1980a). In *Dermophis mexicanus* and *Gymnopis multiplicata*, at the beginning of gestation, the uterine epithelium is not well developed. It contains free amino-acids and carbohydrates; connective tissue is also not well developed. Subsequently, the connective tissue proliferates with crests projecting into the lumen. The epithelium also proliferates and cells are filled by numerous constituents except lipids. Later, lipids appear in epithelial secretions. At the end of pregnancy, mostly lipids are secreted. After parturition, the epithelium regenerates. In non pregnant animals, the epithelium remains thin with no secretion and connective tissue is not very developed (Wake 1993b; Wake and Dickie 1998). Similar observations have been made (Parker 1956) in *Schistometopum thomensi*.

In pregnant *Chthonerpeton indistinctum*, the duct is composed of a serosa, a muscularis and a mucosa with villosities. The epithelium consists of a single layer of cuboidal or columnar cells with a large irregular nucleus and

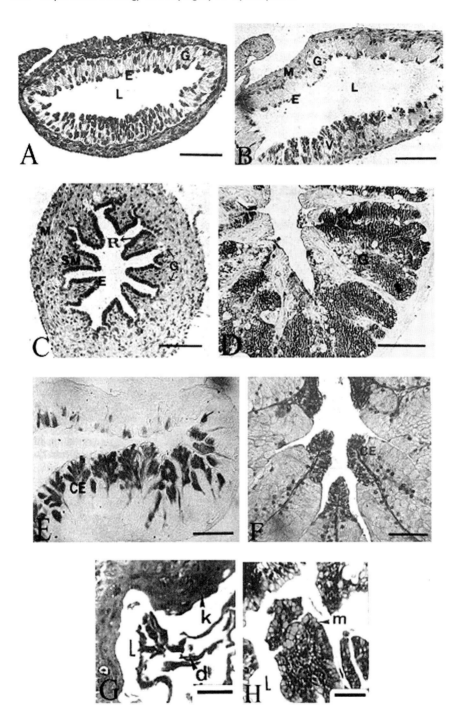

Fig. 3.18 contd

the apical membrane is covered by microvillosities. These cells contain mitochondria, RER, microfilaments and a small Golgi apparatus. They also contain many lipid droplets and glycogen and have apical secretory granules. Under the epithelium, capillaries are numerous. At the beginning of pregnancy, secretions of the epithelium are rich in free amino-acids and carbohydrate. At the end of pregnancy, they contain lipids. In non-pregnant females, the epithelium is flat, muscles are compact and capillaries are less extensive (Wake 1977a, b).

In *Typhlonectes compressicauda*, variations of the oviduct have been observed according to the sexual cycle and gestation (Exbrayat 1984b, 1986a, 1988a; Hraoui-Bloquet et al. 1994; Hraoui-Bloquet 1995, chapters 5 and 11 of this volume). In young animals, before maturation, the oviducts form thin ribbons. By LM, an anterior region with a circular lumen, and a posterior uterus with an elongate lumen can be distinguished. Throughout the length of the oviduct, connective tissue contains many cells with voluminous nuclei. These cells are disposed in several layers surrounding an undifferentiated epithelium. The epithelial cells are columnar (5 × 25 µm) with a nucleus containing a dense chromatin and a narrow layer of cytoplasm. In 18-month-old animals, caught in January-February (breeding period for adults), the epithelium begins differentiate. Tall cells appear in the anterior part and small crypts can be observed. The epithelium is constituted by cells with a pale nucleus and certain of them are ciliated. The uterine region remains unchanged. Differentiation of the anterior region continues in April-May with sometimes a differentiation of crests and crypts in the uterine region. From July to October (24- to 28-month-old animals), all the oviducts again take on an undifferentiated appearance.

In adults, the anterior and uterine regions are strongly differentiated. Each of these regions is subject to significant variations during the biennial

---

Fig. 3.18 contd

Fig. 3.18. A-F. *Ichthyophis beddomei*. **A**. Transverse section of the oviduct: pars recta, August (pre-breeding phase of the reproductive cycle). Scale bar = 10 µm. **B**. Transverse section of the oviduct: pars convoluta, August (pre-breeding phase of the reproductive cycle). Scale bar = 10 µm. **C**. Transverse section of the oviduct: pars uterina, July (end of post-breeding phase of the reproductive cycle). Scale bar = 10 µm. **D**. Transverse section of the oviduct: pars uterina, January (breeding phase of the reproductive cycle). Secretions are PAS positive. Scale bar = 10 µm. **E**. Transverse section of the oviduct: pars convoluta, January (breeding phase of the reproductive cycle). Ciliated cells of epithelium contain secretions stained with alcian blue. Scale bar = 10 µm. **F**. Transverse section of the oviduct: pars uterina, January (breeding phase of the reproductive cycle). Ciliated cells of epithelium are stained with mercuric bromophenol blue. Scale bar = 10 µm. CE, ciliated epithelium; E, cuboidal epithelium; G, glandular cells; L, lumen; M, outer muscular layer; R, rugae-like projections; SM, sub-mucosa; V, villus-like projections. **G**. Transverse section of the posterior part of cloaca in a female *Typhlonectes compressicauda*, period of sexual quiescence. Scale bar = 30 µm. **H**. Transverse section of the posterior part of cloaca in a female *T. compressicauda*, period of breeding. Scale bar = 36 µm. d, squamous cell; k, keratinized cell; L, lumen; m, mucous cell. **A, B, C, D, E, F**: from Masood-Parveez, U. and Nadkarni, B. 1991. Journal of Herpetology 25: 234-237. Fig. 1a, b, c, d, e, f. **G, H**: from Exbrayat, J.-M. 1996. Bulletin de la Société Zoologique de France 121: 93-98. Figs. 7, 8.

breeding cycle of *Typhlonectes* (chapters **5** and **11** of this volume). In January-February of the first year, the oviduct becomes ciliated, with numerous glands. Several eggs and unfertilized oocytes are observed. From April to August-September, i.e. during pregnancy, the anterior region of the oviduct regresses, showing undifferentiated cells without connective crests. In October, the cells begin to differentiate. From January to April of the second year, the anterior region has a differentiated appearance, as in the first year and, during the rest of the year devoted to sexual inactivity, the anterior oviduct remains quiescent. The uterine region is also subject to variations. In January to April of the first year, connective tissue develops and the epithelial cells become secretory. During pregnancy, the wall is submitted to variations that are closely linked to embryonic growth. After parturition (August to October of the first year), the uterus becomes undifferentiated. At the beginning of the second year, a new differentiation occurs but, in April, it lapses and the uterine part of the oviduct becomes undifferentiated to the end of this year of sexual inactivity.

In *Microcaecilia unicolor*, a little known species, the length of the oviducts is 30 to 40% of the total length of the animal. The oviduct is divided into three regions. The wall of anterior region has connective crests with undifferentiated cells. Between the crests, gland cells produce an acidic mucus. The median region possesses a transversly lengthened lumen bordered, in one half by a very straight and smooth wall and in the other half by a wall with weakly developed crests. As previously said, the mode of reproduction is not known in this species, but it could be viviparous. The modifications of general structure in the uterine wall are closely linked to embryonic development (see chapter **11** of this volume). When embryos possess fetal teeth, at the end of their development, these are hypothesized to stimulate secretions by abrading the oviductal epithelium (Wake 1977a, 1980a, 1993b; Wake and Dickie 1998).

3.3.3.4 Fertilization

Fertilization occurs anteriorly in the oviduct. Eggs and the first stages of embryogenesis are found in this region (Delsol *et al.* 1981; Exbrayat *et al.* 1981; Hraoui-Bloquet and Exbrayat 1993; Hraoui-Bloquet 1995; Wake and Dickie 1998, chapter **12** of this volume). Within the oviducts, each egg is embedded into a gelatinous capsule. *Typhlonectes compressicauda* copulation has been studied by Billo *et al.* (1985). Fertilization has been particularly studied in *Typhlonectes compressicauda* (Hraoui-Bloquet and Exbrayat 1993; Hraoui-Bloquet 1995).

3.3.4 Conclusions

In oviparous Gymnophiona, the oviduct is divisible into three regions, each possessing several glands which elaborate the egg envelopes. In *Ichthyophis*, these glands elaborate a proteic mucus. The structure seems to be the same in all oviparous species studied. The glandular system appears to be less complex than in Anura and Urodela (Salthe and Mecham 1974) in which eggs are surrounded by several layers (up to five) with different biochemical characteristics, according to the number of glandular zones (Salthe 1963; see also volume 2, Anura, in this series). The first membrane that surrounds the egg, the vitelline membrane, is proteic and is elaborated in the ovary. The other envelopes consist of more or less acidic mucoplolysaccharides, or are mucoproteic, and they are elaborated in the oviduct. Generally, glands of the oviduct in Urodela and Anura are complex (Vilter 1966, 1967, 1968a, b; Vilter and Thorn 1967; Pereda 1969, 1970; Boisseau and Jego 1972; Boisseau and Joly 1972; Boisseau 1973a, b; Boisseau et al. 1974; Jego 1974; Pujol and Exbrayat 1987). In Gymnophiona, the glandular system seems to be lesser complex and very constant from species to species. This glandular system is also implied in the elaboration of the envelope of eggs, and has been particularly studied in *Ichthyophis glutinosus* (Breckenridge and de Silva 1973; Breckenridge and Jayasinghe 1979).

In viviparous species, there is an adaptation to intrauterine growth of embryos. Thus, in *Typhlonectes compressicauda* and other species, the oviduct is divided into two regions. The anterior region is short, flexuous, with glands involved in the synthesis of all the egg envelopes. These thin envelopes will be resorbed during development (Parker 1956). The posterior region of the oviduct is longer than the first, and it is differentiated as a uterus, the wall of which will be greatly modified according to embryonic growth. This wall permits exchanges between embryo or fetus and its mother. In other Amphibia, intrauterine development of embryos is not often observed (Exbrayat 1989b; see also volume 1, Urodela, and volume 2, Anura, in this series). Among Urodela, *Salamandra atra* can bear embryos over four or even five years. In this species, only the first egg develops to give an animal which subsists on the yolk mass, other oocytes and secretions of the uterine wall with epithelial cells (Vilter and Vilter 1962; Vilter 1967, 1968b, 1986). In *Salamandra salamandra*, embryos spend a more or less long time in the oviduct according to altitude. Genital ducts do not vary notably during the pregnancy (Joly and Boisseau 1973; Greven et al. 1975; Lostanlen et al. 1976; Greven 1980 a, b; Greven and Robeneck 1980a, b; Greven and Rutherbories 1984; Joly 1986). In this species, the epithelium of oviduct is implicated in regulation of the medium in which the embryos develop. Other Urodela, such as *Mertensiella*, are also viviparous.

In Anura, only the genus *Nectophrynoides* possesses both oviparous and viviparous species (Wake 1980 c; Xavier 1986). *Nectophrynoides occidentalis* is the only known viviparous species in this genus. Embryonic development is sustained by a maternal nutritive substance. Embryos live in a uterine fluid that is absorbed by the mouth (Lamotte and Tuchmann-Duplessis 1948; Vilter and Lamotte 1956; Vilter and Lugand 1959; Vilter 1967; Xavier 1971, 1977, 1986).

Viviparity in *Ascaphus* (Ascaphidae) and other Anura is described in detail in chapter 7 of volume 2 in this series.

Gymnophiona have a particular place among viviparous Amphibia, as it is in gymnophionans that viviparity appears to be the most frequent, affecting 50% of species (Wake 1977 a, b). Genital ducts are highly adapted to intrauterine growth and the life of the embryo. In *Typhlonectes compressicauda*, in addition, there is a true placentation involving embryonic gills and the uterine wall.

As these conclusions are based on very few observations, it is possible that other structural types will be found in other species, as already suggested by *Microcaecilia unicolor* that seems to have an intermediate structure between the oviparous and viviparous oviduct.

3.3.5 Cloaca

3.3.5.1 Adults
The cloaca in female Gymnophiona has been poorly studied (Wake 1972; Exbrayat 1991, 1996). In female *Typhlonectes compressicauda*, the anterior region of the cloaca is connected to the digestive tract, bladder, Wolffian ducts and oviducts. The posterior region is homologous with the median region and phallodeum in males. This posterior region is bordered by a wall consisting of a stratified epithelium. During sexual quiescence, this epithelium consists of very large cells that are highly keratinised. A squamous layer is observed at the apical surface of the epithelium (Fig. 3.18G). At commencement of breeding, the epithelium develops. Certain apical cells produce a carboxylic acidic mucus (stained by alcian blue at pH 2.5). During pregnancy, the height decreases and mucus is less abundant. Connective tissue is particularly lacunar during the breeding period (January to May) and pregnancy (April to September) (Exbrayat 1996) (Fig. 3.18H). A musculus retractor cloacae has been observed associated with female cloaca in some species (Wilkinson 1990).

3.3.5.2 Growth
At birth, the female cloaca of *Typhlonectes compressicauda* is a small duct, the anterior part of which is bordered by a mucous epithelium with carboxylic secretions, as in males. The posterior part is bordered by an epithelium composed of undifferentiated cells, several mucous cells with carboxylic acidic secretions and cells with a large vacuole. In 18-month-old animals, the posterior part is bordered by a thicker stratified epithelium producing abundant carboxylic acidic mucus. At the end of the normal breeding period (24 months), posterior part is bordered by a keratinised epithelium resembling that of adult female in sexual activity (Exbrayat 1991).

3.3.6 Fat Bodies
The general appearance of the fat bodies is similar to that of males (Sarasin and Sarasin 1887-1890; Fuhrmann 1914; Tonutti 1931; Wake 1968a). In the viviparous *Typhlonectes compressicauda*, several variations have been noticed according to pregnancy. During sexual quiescence (July to December), the

density of fat bodies is relatively constant. Yet an increase has been observed in August (Exbrayat 1986a, 1988b). During the breeding period and at the beginning of pregnancy, fat bodies are well developed, and then they decrease during gestation. At the end of pregnancy, these organs become very small. It seems that in this species, fat bodies function as nutritional reserves involved in development of embryos (Fig. 11.12, chapter 11 of this volume).

3.4 CONCLUSIONS

The Gymnophiona is a particularly interesting zoological group. Characters of all the species have a great homogeneity. The majority of their anatomical features are an amphibian type but they also possess discrete characters reminiscent fishes or amniotes. The skin is naked but also glandular and even in primitive species is covered with scales. Fat bodies are associated with the gonads as in other Amphibia. A metamorphosis occurs during development (see chapter 12 of this volume). Caecilians are tetrapods even their girdles and limbs do not develop and show rudimentary development in some species.

Yet, certain Amphibian features are not present: they lack green pigment in the retina and an operculum (a piece of bone found in Anura and Urodela behind each stapes) is absent (Rage 1985). Certain features of embryonic development are very different from those is observed in other Amphibia. The development of choanes, for instance, resembles amniotes (Rage 1985).

The reproductive system is also peculiar. Testes and ovaries have a segmented appearance, as do other organs such as kidneys, fat bodies and liver. Males have persistent Mullerian ducts that become functional glands the activity of which is related to the breeding cycle. Sertoli cells are different from those of other Vertebrates, they are giant with many lipoid droplets though their ultrastructural anatomy is still debatable (see also Smita *et al.*, chapter 6 of this volume). Development of gonads is very late, well after metamorphosis, at least in *Typhlonectes compressicauda* (Anjubault and Exbrayat 2000b, c), possibly related to presence of Mullerian ducts in adults. An intromittent organ is present in males and permits internal fertilization. The female genital system is adapted to oviparity or viviparity.

Although new studies devoted to Gymnophiona have been published in recent years, this order is still very little known. Studies generally concern few individuals belonging to few species, and studies in depth, including monographs, are rare.

In the future, there is little doubt that new theories on the origin of Gymnophiona, and new data on their development will be published and provide a better understanding of the biology and phylogeny of this strange group. This knowledge is essential for conservation of these animals. It is to be hoped that such knowledge of caecilians will be developed before disappearance of the group!

3.5 ACKNOWLEDGMENTS

The authors thank Catholic University of Lyon (U.C.L.) and Ecole Pratique des Hautes Etudes (E.P.H.E.) that allowed the freedom in organizing their research fields and supported the works on Gymnophiona. They also thank Singer-Polignac Foundation that supported the missions necessary to collect the material of study. The authors also thank all the colleagues, searchers, technical assistants, students who contribute to this work. And more especially, Michel Delsol, Honorary Professor at he U.C.L. and at E.P.H.E. who is at the origin of these works in Gymnophiona, Marie-Thérèse Laurent, who made thousands of sections, Elisabeth Anjubault, Pierre Bleyzac, Jean-Lou Dorne, Geneviève Escudié, Janine Flatin, Madeleine Gueydan, Souad Hraoui-Bloquet, Bertrand Leclercq, Romain Paillot, Jean-Pierre Parent, Paulette Pujol, Jean-Lescure, C.N.R.S. and Museum National d'Histoire Naturelle de Paris, Sabine Renous, also C.N.R.S. and Museum National d'Histoire Naturelle de Paris, Gérard Morel, C.N.R.S. They also thank colleagues met at different occasions: Marvalee H. Wake, Gopalakrishna Bhatta. They also thank Barrie Jamieson, who invited them to contribute to this collection devoted to reproduction and phylogeny of Amphibia and also for his patience in reviewing the manuscript. They thank Jean-François Exbrayat for his help in preparing the illustrations.

3.6 LITERATURE CITED

Anjubault, E. and Exbrayat, J.-M. 1998. Yearly cycle of Leydig-like cells in testes of *Typhlonectes compressicaudus* (Amphibia, Gymnophiona) Pp. 53-58. In C. Miaud and R. Guyetant (eds), *Current Studies in Herpetology*, Proceedings of the 9th General Meeting of the Societas Europaea Herpetologica, Le Bourget du Lac, France.

Anjubault, E. and Exbrayat, J.-M. 2000a. Cycle annuel des cellules du tissu interstitiel des testicules chez *Typhlonectes compressicaudus*, Amphibien Gymnophione. Bulletin de la Société Zoologique de France 125: 133.

Anjubault, E. and Exbrayat J.-M. 2000b. Development of gonads in *Typhlonectes compressicauda* (Amphibia, Gymnophiona). XVIIIth International Congress of Zoology, Athens, August-September 2000: 51.

Anjubault, E. and Exbrayat J.-M. 2004a. Contribution à la connaissance de l'appareil génital de *Typhlonectes compressicauda* (Duméril and Bibron, 1841), Amphibien Gymnophione. I. Gonadogenèse. Bulletin Mensuel de la Société Linnéenne de Lyon, 73: 379-392.

Anjubault, E. and Exbrayat J.-M. 2004b. Contribution à la connaissance de l'appareil génital de *Typhlonectes compressicauda* (Duméril et Bibron, 1841), Amphibien Gymnophione. II. Croissance des gonades et maturité sexuelle des mâles. Bulletin Mensuel de la Société Linnéenne de Lyon, 73: 393-405.

Arnold, W. 1935. Das Auge von *Hypogeophis*, Beitrag zur Kenntnis der Gymnophionen XXVII. Morphologisches Jahrbuch 76: 589-625.

Badenhorst, A. 1978. The development and the phylogeny of the organ of Jacobson and the tentacular apparatus of *Ichthyophis glutinosus* (Linn.). Annals of University of Stellenbosch 1: 1-26.

Baer, J. G. 1937. L'appareil respiratoire des Gymnophiones. Revue Suisse de Zoologie 44: 353-377.

Bemis, W., Schwenk, K. and Wake, M. H. 1983. Morphology and function of the feeding apparatus in *Dermophis mexicanus* (Amphibian : Gymnophiona). Zoological Journal of Linnean Society 77: 75-96.
Bennett, A. F. and Wake, M. H. 1974. Metabolic correlates of activity in the Caecilian *Geotrypetes seraphini*. Copeia 1974: 764-769.
Berois, N. and De Sa, R. 1988. Histology of the ovaries and fat bodies of *Chthonerpeton indistinctum*. Journal of Herpetology 22: 146-151.
Bhatta, G. K. 1987. Some aspects of reproduction in the apodan amphibian *Ichthyophis*. PhD. Dissertation, Karnatak University, Dharwad, India.
Bhatta, G. K. and Exbrayat, J.-M. 1998. Premières observations sur le développement embryonnaire d'*Ichthyophis beddomei*, Amphibien Gymnophione ovipare. Bulletin de la Société Zoologique de France. 124: 117-118.
Bhatta, G. K., Anjubault, E. and Exbrayat, J.-M. 2001. Structure et ultrastructure des testicules d'*Ichthyophis beddomei* (Peters, 1879), Amphibien Gymnophione. Annales du Muséum du Havre 67: 11-12.
Billo, R. 1982. Zur Anatomie der tentakelregion von *Ichthyophis kohtaoensis* (Ichthyophiidae, Gymnophiona). Diplomarbeit, Basel.
Billo, R. 1986. Tentacle apparatus of Caecilians. Mémoire de la Société Zoologique de France 43: 71-75.
Billo, R. and Wake, M. H. 1987. Tentacle development in *Dermophis mexicanus* (Amphibian: Gymnophiona) with an hypothesis of tentacle origin. Journal of Morphology 192: 101-111.
Billo, R. R., Straub, J. O. and Senn, D. G. 1985. Vivipare Apoda (Amphibia : Gymnophiona) *Typhlonectes compressicaudus* (Duméril und Bibron, 1841): Kopulation, Tragzeit und Geburt. Amphibia Reptilia 6: 1-9.
Bleyzac, P. and Exbrayat, J.-M. 2003. First observations on the ontogeny of immune organs in *Typhlonectes compressicauda* (Duméril and Bibron, 1842), Amphibia, Gymnophiona. 12th Congress SEH, St Petersburg, August 2003, Abstract.
Bleyzac, P. and Exbrayat, J.-M. 2005. Some aspects of the ontogenesis of the immune system organs of *Typhlonectes compressicauda* (Duméril and Bibron, 1841), Amphibia, Gymnophiona. Pp. 120-123. In N. Ananjeva and O. Tsinenko (eds), *Herpetologia Petropolitana*. Proceedings of the 12th Ordinary General Meeting of Societas Europaea Herpetologica. St. Petersburg. Russian Journal of Herpetology 12 (Supplement).
Bleyzac, P., Cordier, G. and Exbrayat, J.-M. 2005. A morphological description of embryonic development of immune organs in *Typhlonectes compressicauda* (Amphibia, Gymnophiona). Journal of Herpetology 39 : 57-65.
Boisseau, C. 1973a. Etude ultrastructurale de l'oviducte du triton *Pleurodeles waltlii* Michah. I. Ultrastructure des cellules épithéliales de l'oviducte moyen différencié. Journal de Microscopie 18: 341-358.
Boisseau, C. 1973b. Etude ultrastructurale de l'oviducte du triton *Pleurodeles waltlii* Michah. II. - Morphogenèse des glandes et différenciation des cellules épithéliales de l'oviducte moyen. Journal de Microscopie 8: 359-382.
Boisseau, C. and Jego, P. 1972. Caractérisation histochimique des mucosubstances de l'oviducte de *Pleurodeles waltlii* Michah (Amphibien, Urodèle, Salamandridé). Comptes Rendus de la Société de Biologie 166: 1774-1779.
Boisseau, C. and Joly, J. 1972. Données sur l'histologie de l'oviducte de *Pleurodeles waltlii* Michah (Amphibien Urodèle, Salamandridé). Comptes Rendus de la Société de Biologie 166: 1770-1773.
Boisseau, C., Jego, P., Joly, J. and Picheral, B. 1974. Organisation et caractérisation histochimique des gangues ovulaires sécrétées par l'oviducte de *Pleurodeles waltlii*

Michah. (Amphibien, Urodèle, Salamandridé). Comptes Rendus de la Société de Biologie 168: 1102-1107.
Bons, J. 1986. Données histologiques sur le tube digestif de *Typhlonectes compressicaudus* (Duméril et Bibron, 1841) (Amphibien Apode*)*. Mémoires de la Société Zoologique de France 43: 87-90.
Brauer, A. 1897a. Beitrage zur kenntniss der Entwicklungsgeschichte und der Anatomie der Gymnophionen. Zoologisches Jahrbuch für Anatomie 10: 277-279.
Brauer, A. 1897b. Beitrage zur kenntniss der Entwicklungsgeschichte und der Anatomie der Gymnophionen. Zoologisches Jahrbuch für Anatomie 10: 389-472.
Brauer, A. 1899. Beitrage zur kenntniss der Entwicklung und Anatomie der Gymnophionen. II. Die Entwicklung der aüssern Form. Zoologisches Jahrbuch für Anatomie 12: 477-508.
Brauer, A. 1900. Zur kenntniss der Entwicklung der Excretionsorgane der Gymnophionen. Zoologischer Anzeiger 23: 353-358.
Brauer, A. 1902. Beitrage zur kenntniss der Entwicklung und Anatomie der Gymnophionen. III. Die Entwicklung der Excretionsorgane. Zoologisches Jahrbuch für Anatomie 16: 1-176.
Breckenridge, W. R. and de Silva, G. I. S. 1973. The egg case of *Ichthyophis glutinosus* (Amphibia, Gymnophiona). Proceedings of Ceylon Association for Advancement of Sciences 29: 107.
Breckenridge, W. R. and Jayasinghe, S. 1979. Observations on the eggs and larvae of *Ichthyophis glutinosus*. Ceylon Journal of Sciences (Biological Sciences) 13: 187-202.
Breckenridge, W. R., Shirani, N. and Pereira, L. 1987. Some aspects of the biology and development of *Ichthyophis glutinosus* (Amphibia, Gymnophiona). Journal of Zoology 211: 437-449.
Carvalho, E. T. C. and Junqueira, L. C. U. 1999. Histology of the kidney and urinary bladder of *Siphonops annulatus* (Amphibia, Gymnophiona). Archives of Histology Cytology 62: 39-45.
Casey, J. and Lawson, R. 1979. Amphibians with scales: the structure of the scale in *Hypogeophis rostratus*. British Journal of Herpetology 5: 831-833.
Casey, J. and Lawson, R. 1981. A histological and scanning electron microscope study of the teeth of Caecilian Amphibians. Archives of Oral Biology 26: 48-58.
Charlemagne, J. 1990. Immunologie des poïkilothermes. Pp. 447-457. In P.-P. Pastoret, A. Govaerts and H. Bazin (eds), *Immunologie Animale*, Médecine et Sciences., Flammarion, Paris.
Chatterjee, B. K. 1936. The anatomy of *Uraeotyphlus menoni* Annandale. Part I: digestive, circulatory, respiratory and urogenital systems. Anatomischer Anzeiger 81: 393-414.
Clairambault, P., Cordier-Picouet, M. J. and Pairault, C. 1980. Premières données sur les projections visuelles primaires d'un Amphibien Apode (*Typhlonectes compressicaudus*). Comptes Rendus des Séances de l'Académie des Sciences de Paris, série D, 291: 283-286.
Clairambault, P., Christophe, N., Pairault, C., Herbin, M., Ward, R. and Reperant, J. 1994. Organisation of the serotoninergic system in the brain of two amphibian species, *Ambystoma mexicanum* (Urodela) and *Typhlonectes compressicaudus* (Gymnophiona). Anatomy and Embryology 190: 87-99.
Cockerell, T. D. A. 1912. The scales of *Dermophis*. Science 36: 681.
Coleman, R. 1969. The fine structure of ultimobranchial secretory cells in the anuran: *Rana temporaria* L. and *Bufo bufo* L. Zeitschrift für Zellforschung und mikroskopische Anatomie 100: 201-214.

Cooper, E. L. and Garcia-Herrera, F. 1966. Peripheral blood cells in the Caecilian, *Typhlonectes compressicauda*. American Zoologist 6: 352.
Cooper, E. L. and Garcia-Herrera, F. 1968. Chronic skin allograft rejection in the Apodan *Typhlonectes compressicauda*. Copeia 1968: 224-229.
Datz, E. 1923. Die Haut von *Ichthyophis glutinosus*. Jena Zeitschrift Naturwissenschaften 59: 311-342.
De Jaeger E. F. J. 1947. Some points in the development of the stapes of *Ichthyophis glutinosus*. Anatomische Anzeiger 96: 203-210.
De Sa, R. and Berois, N. 1986. Spermatogenesis and histology of the testes of the Caecilian *Chthonerpeton indistinctum*. Journal of Herpetology, 20: 510-514.
De Villiers, C. G. S. 1938. A comparison of some cranial features of the East African Gymnophiones *Boulengerula boulengeri* Tornier and *Scolecomorphus uluguruensis* Boulenger. Anatomische Anzeiger 86: 1-26.
Deletre M. and Measey G. J. 2004. Sexual selection vs ecological causation in a sexually dimorphic caecilian, *Schistometopum thomense* (Amphibia Gymnophiona Cae ciliidae). Ethology Ecology and Evolution 16 (3): 243-253.
Delsol, M., Flatin, J. and Exbrayat, J.-M. 1995. Le tube digestif des Amphibiens adultes. Pp. 497-508. In P.P. Grassé and M. Delsol (eds), *Traité de Zoologie*, tome XIV, fasc.I A. Masson, Paris.
Delsol, M., Exbrayat, J.-M., Flatin, J. and Lescure, J. 1980. Particularités du groupe des Batraciens Apodes. Bulletin mensuel de la Société Linnéenne de Lyon 49: 370-379.
Delsol, M., Flatin, J., Exbrayat, J.-M. and Bons, J. 1981. Développement de *Typhlonectes compressicaudus* Amphibien Apode vivipare. Hypothèse sur sa nutrition embryonnaire et larvaire par un ectotrophoblaste. Comptes Rendus des Séances de l'Académie des Sciences de Paris, série III 293: 281-285.
Duellman, W. E. and Trueb, L. 1986. *Biology of Amphibians*. McGraw-Hill Inc., U.S.A. 670 pp.
Dünker, N. 1997. Development and organization of the retinal dopaminergic system of *Ichthyophis kohtaoensis* (Amphibia; Gymnophiona). Cell and Tissue Research 289: 265-274.
Dünker, N. 1998a. A double-label analysis demonstrating the partial coexistence of tyrosine hydroxylase and GABA in retinal neurons of *Ichthyophis kohtaoensis* (Amphibia: Gymnophiona). Cell and Tissue Research 294: 387-30.
Dünker, N. 1998b. Colocalization of serotonin and GABA in retinal neurons of *Ichthyophis kohtaoensis* (Amphibia: Gymnophiona). Anatomy Embryology 197: 69-75.
Dünker, N. 1999. Serotonergic neurons and processes in the adult and developing retina of *Ichthyophis kohtaoensis*. Anatomy Embryology 199: 35-43.
Dünker, N. and Himstedt, W. 1997. Organization and development of the serotonergic system in the retina of *Ichthyophis kohtaoensis* (Amphibia: Gymnophiona) in From membrane to mind. Proceedings of the 25th Gottingen Neurobiology Conference, p. 231.
Dünker, N., Wake, M. H. and Olson, W. M. 2000. Embryonic and larval development in the Caecilian *Ichthyophis kohtaoensis*. Journal of Morphology 243: 3-34.
Dunn, E. R. 1928. Notes on central American Caecilians. Proceedings of New England Zoological Club 10: 71-76.
Ebersole, T. J. and Boyd, S. 2000. Immunocytochemical localization of gonadotropin-releasing hormones in the brain of a viviparous Caecilian Amphibian, *Typhlonectes natans* (Amphibia: Gymnophiona). Brain Behavior Evolution 55: 14-25.

Elkan, E. 1958. Further contributions to the buccal and pharyngeal mucous membranes in urodeles. Proceedings of Zoological Society of London 131: 335-355.
Engelhardt, F. 1924. Tentakelapparat und Auge von *Ichthyophis*. Jena Zeitschrift Naturwissenschaften 60: 241-305.
Estabel, J. and Exbrayat, J.-M. 1998. Brain development of *Typhlonectes compressicaudus* (Dumeril and Bibron, 1841). Journal of Herpetology 32: 1-10.
Estabel, J. and Exbrayat, J.-M. 1999. Mise en évidence des cellules radiales chez *Typhlonectes compressicaudus* (Amphibien, Gymnophione) par immunofluorescence et immunoenzymologie. Revue Française d'Histotechnologie 12: 49-54.
Estabel, J. and Exbrayat, J.-M. 2000a. Structure and development of eyes in *Typhlonectes compressicauda* (Amphibia, Gymnophiona) Abstract. XVIIIth International Congress of Zoology, Athens, August-September 2000: 53.
Estabel, J. and Exbrayat, J.-M. 2000b. Development of thyroid gland in *Typhlonectes compressicauda* (Amphibia, Gymnophiona). Abstract. XVIIIth International Congress of Zoology, Athens, August-September 2000: 52.
Estabel, J., Bhatta, G. K. and Exbrayat, J.-M. 1999. Comparison of brain development in two Caecilians *Typhlonectes compressicaudus* and *Ichthyophis beddomei*. Pp. 105-111. In C. Miaud and R. Guyetant (eds) *Current Studies in Herpetology*, Proceedings of the 9th General Meeting of the Societas Europaea Herpetologica, Le Bourget du Lac, France.
Estes, R. and and Wake, M. H. 1972. The first fossil record of Caecilian Amphibians. Nature 239: 228-231.
Exbrayat, J.-M. 1983. Premières observations sur le cycle annuel de l'ovaire de *Typhlonectes compressicaudus* (Duméril et Bibron, 1841), Batracien Apode vivipare. Comptes Rendus des Séances de l'Académie des Sciences de Paris 296: 493-498.
Exbrayat, J.-M. 1984a. Cycle sexuel et reproduction chez un Amphibien Apode: *Typhlonectes compressicaudus* (Duméril et Bibron, 1841). Bulletin de la Société Herpétologique de France 32: 31-35.
Exbrayat, J.-M. 1984b. Quelques observations sur l'évolution des voies génitales femelles de *Typhlonectes compressicaudus* (Duméril et Bibron, 1841), Amphibien Apode vivipare, au cours du cycle de reproduction., Comptes Rendus des Séances de l'Académie des Sciences de Paris 298: 13-18.
Exbrayat, J.-M. 1985. Cycle des canaux de Müller chez le mâle adulte de *Typhlonectes compressicaudus* (Duméril et Bibron, 1841), Amphibien Apode. Comptes Rendus des Séances de l'Académie des Sciences de Paris 301: 507-512.
Exbrayat, J.-M. 1986a. Quelques aspects de la biologie de la reproduction chez *Typhlonectes compressicaudus* (Duméril et Bibron, 1841), Amphibien Apode. Doctorat ès Sciences Naturelles, Université Paris VI, France.
Exbrayat, J.-M. 1986b. Le testicule de *Typhlonectes compressicaudus*; structure, ultrastructure, croissance et cycle de reproduction. Mémoires de la Société Zoologique de France 43: 121-132.
Exbrayat, J.-M. 1988a. Croissance et cycle des voies génitales femelles chez *Typhlonectes compressicaudus* (Duméril et Bibron, 1841), Amphibien Apode vivipare. Amphibia Reptilia 9: 117-137.
Exbrayat, J.-M. 1988b. Variations pondérales des organes de réserve (corps adipeux et foie) chez *Typhlonectes compressicaudus*, Amphibien Apode vivipare au cours des alternances saisonnières et des cycles de reproduction. Annales des Sciences Naturelles, Zoologie., Paris, 13éme série 9: 45-53.
Exbrayat, J.-M. 1989a. Quelques observations sur les appareils génitaux de trois Gymnophiones; hypothèses sur le mode de reproduction de *Microcaecilia unicolor*. Bulletin de la Société Herpétologique de France 52: 34-44.

Exbrayat, J.-M. 1989b. Quelques aspects de l'évolution de la viviparité chez les Vertébrés. Pp. 143-169. In J. Bons et M. Delsol (eds), *Evolution Biologique. Quelques données actuelles*, Boubée, AAA, Paris, France.

Exbrayat, J.-M. 1991 Anatomie du cloaque chez quelques Gymnophiones. Bulletin de la Société Herpétologique de France 58: 30-42.

Exbrayat, J.-M. 1992a. Appareils génitaux et reproduction chez les Amphibiens Gymnophiones. Bulletin de la Société Zoologique de France 117: 291-296.

Exbrayat, J.-M. 1992b. Reproduction et organes endocrines chez les femelles d'un Amphibien Gymnophione vivipare, *Typhlonectes compressicaudus* Bulletin de la Société Herpétologique de France 64: 37-50

Exbrayat, J.-M. 1993. Quelques aspects de la reproduction chez *Typhlonectes compressicaudus* (Duméril et Bibron, 1841), Amphibien Gymnophione. Cahiers de l'Institut Catholique de Lyon, Série Sciences 7: 1-263.

Exbrayat, J.-M. 1996. Croissance et cycle du cloaque chez *Typhlonectes compressicaudus* (Duméril et Bibron, 1841), Amphibien Gymnophione. Bulletin de la Société Zoologique de France 121: 99-104.

Exbrayat, J.-M. 2000. *Les Gymnophiones, ces curieux Amphibiens*. Boubée, Paris, France 443 pp.

Exbrayat, J.-M. 2003. Apport de l'histologie à l'étude de quelques organs et de quelques aspects endocrines de la reproduction d'animaux rares issus de collecctions. Revue Française d'Histotechnologie 16: 19-29.

Exbrayat, J.-M. and Collenot, G. 1983. Quelques aspects de l'évolution de l'ovaire de *Typhlonectes compressicaudus* (Duméril et Bibron, 1841), Batracien Apode vivipare. Etude quantitative et histochimique des corps jaunes. Reproduction Nutrition Développement 23: 889-898.

Exbrayat, J.-M. and Dansard, C. 1992. Ultrastructure des cellules de Sertoli chez *Typhlonectes compressicaudus*, Amphibien Gymnophione. Bulletin de la Société Zoologique de France 117: 166-167.

Exbrayat, J.-M. and Dansard, C. 1993. An ultratructural study of the evolution of Sertoli cells in a Gymnophionan Amphibia. Biology of the Cell 79: 90.

Exbrayat, J.-M. and Dansard, C. 1994. Apports de techniques complémentaires à la connaissance de l'histologie du testicule d'un Amphibien Gymnophione. Bulletin de l'Association Française d'Histotechnologie 7: 19-26.

Exbrayat, J.-M. and Delsol, M. 1985. Reproduction and growth of *Typhlonectes compressicaudus*, a viviparous Gymnophione. Copeia 1985: 950-955.

Exbrayat, J.-M. and Flatin, J. 1985. Les cycles de reproduction chez les Amphibiens Apodes. Influence des variations saisonnières. Bulletin de la Société Zoologique de France 110: 301-305.

Exbrayat, J.-M. and Laurent, M.-T. 1986. Quelques observations sur la reproduction en élevage de deux Amphibiens Apodes : *Typhlonectes compressicaudus* et un *Ichthyophis*. Possibilité de rythmes endogènes. Bulletin de la Société Herpétologique de France 40: 52-62.

Exbrayat, J.-M. and Sentis, P. 1982. Homogénéité du testicule et cycle annuel chez *Typhlonectes compressicaudus* (Duméril et Bibron, 1841), Amphibien Apode vivipare. Comptes Rendus des Séances de l'Académie des Sciences de Paris 294: 757-762.

Exbrayat, J.-M., Delsol, M. and Flatin. J. 1981. Premières remarques sur la gestation chez *Typhlonectes compressicaudus* (Duméril et Bibron, 1841) Amphibien Apode vivipare. Comptes Rendus des Séances de l'Académie des Sciences de Paris 292: 417-420.

Exbrayat, J.-M., Delsol, M. and Flatin. J. 1986. *Typhlonectes compressicaudus*, Amphibien Apode vivipare de Guyane. Pp. 119-124. In Sepanguy-Sepanrit (ed),

Le littoral guyanais, Actes du colloque Le Littoral Guyanais, Fragilité de l'Environnement, Cayenne, Guyane Française.

Exbrayat, J.-M., Pujol, P. and Hraoui-Bloquet, S. 1995. First observations on the immunological materno-foetal relationships in *Typhlonectes compressicaudus,* a viparous Gymnophionan Amphibia. Pp. 271-273. In G. A. Llorente, A. Montori, X. Santos and M. A. Carretero (eds), *Scientia Herpetologica,* Proceedings of the 7th General Meeting of the Societas Europaea Herpetologica, Barcelona, Spain.

Exbrayat, J.-M., Pujol, P. and Leclercq, B. 1998. Quelques aspects des cycles sexuels et nycthéméraux chez les Amphibiens. Bulletin de la Société Zoologiqe de France 123: 113-124.

Exbrayat, J.-M., Bhatta, G. K., Estabel, J. and Paillot, R. 1998. First observations on embryonic development of *Ichthyophis beddomei,* an oviparous Gymnophionan Amphibia. Pp. 113-120. In C. Miaud and R. Guyetant (eds), *Current Studies in Herpetology,* Proceedings of the 9th General Meeting of the Societas Europaea Herpetologica, Le Bourget du Lac, France.

Farenholz, C. 1937. Drüsen der mundhöle. Pp. 115-210. In L. Bolk, E. Goppert, E. Kallius und W. Lubosch (eds), *Handbuch Verlag Anatomie der Wirbeltiere.*

Ferroni-Schwartz, E. N., Schwartz, C. A. and Sebben, A. 1998. Occurrence of hemolytic activity in the skin secretion of the caecilian *Siphonops paulensis.* Natural Toxins 6: 179-182.

Ferroni-Schwartz, E. N., Schwartz, C. A., Sebben, A., Largura, S. W. R. and Mendes, E. A. 1999. Indirect cadiotoxic activity of the caecilian *Siphonops paulensis* (Gymnophiona, Amphibia) skin secretion. Toxicon 37: 47-54.

Flock, A. 1965. Electron microscope and electrophysiological studies on the lateral line canal organ. Acta oto-laryngologica supplementum 199: 1-90.

Flock, A. 1967. Ultrastructure and function in the lateral line organs. Pp. 163-197. In P. Cahn (ed), *Lateral Line Detectors,* Bloomington, Indiana University Press, U.S.A.

Fox, H. 1983. The skin of *Ichthyophis* (Amphibia: Caecilia): an ultrastructural study. Journal of Zoology 199: 223-248.

Fox; H. 1985. The tentacles of *Ichthyophis* (Amphibia: Caecilia) with special reference to the skin. Journal of Zoology 205 : 223-234.

Fox, H. 1986. Early development of caecilian skin with special reference to the epidermis. Journal of Herpetology 20: 154-157.

Fox, H. 1987. On the fine structure of the skin of larval juvenile and adult *Ichthyophis* (Amphibia: Caecilia). Zoomorphology 107: 67-76.

Fox, H. and Whitear, M. 1978. Observations on Merckel cells in amphibians. Biology of the Cell 32: 223-232.

Foxon, G. E. H. 1964. Blood and respiration. Pp. 151-210. In Moore J. A. (ed), *The Physiology of the Amphibia,* Academic Press, New York, U.S.A.

Fritzsch, B. and Wake, M. H. 1986. The distribution of ampullary organs in Gymnophiona. Journal of Herpetology 20: 90-93.

Fritzsch, B. and Wake, M. H. 1988. The inner ear of gymnophione amphibians and its nerve supply: a comparative study of regressive events in a complex sensory system (Amphibia, Gymnophiona). Zoomorphology 108: 201-217.

Fuhrmann, O. 1914. Le genre *Typhlonectes.* Mémoires de la Société Neuchâteloise de Sciences Naturelles 5: 112-138.

Gabe, M. 1968. Données histologiques sur le pancréas endocrine d'*Ichthyophis glutinosus* L. Batracien gymnophione. Archives d'Anatomie, d'Histologie et d'Embryologie Normale et Expérimentale 51: 231-246.

Gabe, M. 1969. Emplacement des cellules à gastrine dans l'estomac de quelques Sauropsidés et Batraciens. Comptes Rendus des Séances de l'Académie des Sciences de Paris 268: 3088-3090.

Gabe, M. 1971a. Données histologiques sur le tégument d'*Ichthyophis glutinosus* L. Annales de Sciences Naturelles 13: 573-608.

Gabe, M. 1971b. Apport de l'histologie à l'étude des relations phylétiques des Gymnophiones. Bulletin Biologique 105: 125-127.

Gabe; M. 1972. Données histologiques sur les cellules endocrines gastro-duodénales des Amphibiens. Archivium Histologicum Japonicum 35: 51-81.

Garg, B. L. and Prasad, J. 1962. Observations of the female urogenital organs of limbless amphibian *Uraeotyphlus oxyurus*. Journal of Animal Morphology and Physiology 9: 154-156.

Gaymer, R. 1971. New method of locomotion in limbless terrestrial vertebrates. Nature 234: 150-151.

George, J. M., Smita, M., Oommen, V. O. and Akbarsha, M. A. 2004a. Histology and ultrastructure of male Mullerian gland of *Uraeotyphlus narayani* (Amphibia: Gymnophiona). Journal of Morphology 260: 33-56.

George J. M., Smita, M., Kadalmani, B., Girija, R., Oommen, O. V. and Akbarsha, A. 2004b. Secretory and basal cells of the epithelium of the tubular glands in the male Mullerian gland of the Caecilian *Uraeotyophlus narayani* (Amphibia: Gymnophiona). Journal of Morphology 262: 760-769.

George J. M., Smita, M., Kadalmani, B., Girija, R., Oommen, O. V. and Akbarsha, A. 2005. Contribution of the secretory material of Caecilian (Amphibia: Gymnophiona) male Mullerian gland to motility of sperm: a study in *Uraeotyphlus narayani*. Journal of Morphology 263: 227-237.

Gonçalves, A. A. 1977. Dimorfismo sexual de *Typhlonectes compressicaudus* (Amphibia Apoda). Boletim de Fisiologia animal—Universidad de Sao Paulo 1: 141-142.

Gonçalves, A. A. and Sawaya, P. 1978. Oxygen uptake by *Typhlonectes compressicaudus* related to the body weight. Comparative Biochemistry Physiology 61A: 141-143.

Goniakowska-Witalinska, L. 1995. The histology and ultrastructure of the amphibian. In L. M. Pastor (ed). *Histology, Ultrastructure and Immunohistochemistry of the Respiratory Organs in Non-mammalian Vertebrates*, Servicio de Publicationes, Universidade de Murcia, Murcia

Gonzales, A. and Smeets, W. J. A. J. 1994. Distribution of tyrosine hydroxylase immunoreactivity in the brain of *Typhlonectes compressicauda* (Amphibia, Gymnophiona): further assessment of primitive and derived traits of amphibian catecholamine systems. Journal of Chemical Neuroanatomy 8: 19-32.

Gonzales, A. and Smeets, W. J. A. J. 1997. Distribution of vasotocin- and mesotocin-like immunoreactivities in the brain of *Typhlonectes compressicauda* (Amphibia, Gymnophiona): further assessment of primitive and derived traits of amphibian neuropeptidergic systems. Cell and Tissue Research 287: 305-314.

Greven, H. 1980a. Ultrastructural investigations of the epidermis and the gill epithelium in the intrauterine larvae of *Salamandra salamandra* (L.) (Amphibia, Urodela). Zeitschift für mikroscopische Anatomie 94: 196-208.

Greven, H. 1980b. Ultrahistochemical and autoradiographic evidence for epithelial transport in the uterus of the ovoviviparous salamander, *Salamandra salamandra* (L.) (Amphibia, Urodela). Cell and Tissue Research 212: 147-162.

Greven, H. 1984. The dentition of *Gegenophis ramaswamii* Taylor 1964 (Amphibia, Gymnophiona), with comments on monocuspid teeth in the Amphibia. Zeitschrift für Zoologie Systematisch Evolution Forschungen, 22: 342-348.

Greven, H. 1986. On the diversity of tooth crowns in Gymnophiona. Mémoires de la Société Zoologique de France 43: 85-86.

Greven, H. and Robenek, H. 1980a. On the lumenal plasmalemma and associated structures of the uterine epithelium in pregnant and non-pregnant females of *Salamandra salamandra* (L.) (Amphibia, Urodela). EUREM 80, The Hague 2: 34-35.

Greven, H. and Robenek, H. 1980b. Intercellular junctions in the uterine epithelium of *Salamandra salamandra* (L.) (Amphibia, Urodela). A freeze-fracture study. Cell and Tissue Research 212: 163-172.

Greven, H. and Ruterbories, H. J. 1984. Scanning electron microscopy of the oviduct of *Salamandra salamandra* (L.) (Amphibia, Urodela). Zeitschift für mikroscopische Anatomie 98: 49-62.

Greven, H., Kuhlmann, D. and Reineck, U. 1975. Anatomie und Histologie des Oviductes von *Salamandra salamandra* (Amphibia, Urodela). Zoologische Beitrag 2: 325-345.

Heterington, T. E. and Wake, M. H. 1979. The lateral line system in larval *Ichthyophis* (Amphibian: Gymnophiona). Zoomorphology 93: 209-225.

Hilscher-Conklin, C., Conlon, M. and Boyd, S. 1998. Identification and localization of neurohypophysial peptides in the brain of a caecilian amphibian, *Typhlonectes natans*. The Journal of Comparative Neurology 394: 139-151.

Himstedt, W. 1989. The caecilian *Ichthyophis kohtaoensis* is not blind. First World Congress of Herpetology, Canterbury, U.K.: R7.

Himstedt, W. 1996. *Die Blindwühlen*. Westarp Wissenschaften, Magdeburg. 159 pp.

Himstedt, W. and Fritzch, B. 1990. Behavioral evidence for electroception in larvae of the Caecilian *Ichthyophis kohtaoensis* (Amphibia, Gymnophiona). Zoologische Jahrbuch, Physiologie 94: 484-492.

Himstedt, W. and Manteuffel, G. 1985. Retinal projections in the caecilian *Ichthyophis kohtaoensis* (Amphibia, Gymnophiona). Cell and Tissue Research 239: 689-692.

Hraoui-Bloquet, S. 1995. Nutrition embryonnaire et relations materno-foetales chez *Typhlonectes compressicaudus* (Duméril et Bibron, 1841), Amphibien Gymnophione vivipare. Thèse de Doctorat E.P.H.E., Lyon, France.

Hraoui-Bloquet, S. and Exbrayat, J.-M. 1992. Développement embryonnaire du tube digestif chez *Typhlonectes compressicaudus* (Duméril et Bibron, 1841), Amphibien Gymnophione vivipare. Annales de Sciences Naturelles, Zoologie, Paris, 13éme série 13: 11-23.

Hraoui-Bloquet, S. and Exbrayat, J.-M. 1993. La fécondation chez *Typhlonectes compressicaudus* (Duméril et Bibron, 1841), Amphibien Gymnophione. Bulletin de la Société Zoologique de France 118: 356-357.

Hraoui-Bloquet, S. and Exbrayat, J.-M. 1996. Les dents de *Typhlonectes compressicaudus* (Amphibia, Gymnophiona) au cours du développement. Annales de Sciences Naturelles, Zoologie, Paris, 13éme série 17: 11-23.

Hraoui-Bloquet, S. and Exbrayat, J.-M. 1997a. Développement embryonnaire de la langue de *Typhlonectes compressicaudus*, Amphibien Gymnophione. Bulletin de la Société Zoologique de France 122: 452.

Hraoui-Bloquet, S. and Exbrayat, J.-M. 1997a. Développement de la langue de *Typhlonectes compressicaudus* (Duméril et Bibron, 141), Amphibien Gymnophione vivipare. Bulletin de la Société Herpétologique de France 82-83: 39-46.

Hraoui-Bloquet, S., Escudié, G. and Exbrayat, J.-M. 1994. Aspects ultrastructuraux de l'évolution de la muqueuse utérine au cours de la phase de nutrition orale des embryons chez *Typhlonectes compressicaudus*, Amphibien Gymnophione vivipare. Bulletin de la Société Zoologique de France 119: 237-242.

Jared, C., Navas, C. A. and Toledo, R. C. 1999. An appreciation of the physiology and morphology of the Caecilians (Amphibia: Gymnophiona). Comparative Biochemistry Physiology Part A, 123: 313-328.

Jego, P. 1974. Composition en glucides des différents segments de l'oviducte et des gangues ovulaires chez *Pleurodeles waltlii* Michah (Amphibien Urodèle). Comparative Biochemistry Physiology 48B: 435-446.

Joly 1986. La reproducion de la salamandre terrestre. Pp. 471-486. In P.P. Grassé et M. Delsol (eds), *Traité de Zoologie*, tome XIV, fasc.I A. Masson, Paris

Joly, J. and Boisseau, C. 1973. Localisation des spermatozoïdes dans l'oviducte de la Salamandre terrestre, *Salamandra salamandra* L. (Amphibien Urodèle) au moment de la fécondation. Comptes Rendus des Séances de l'Académie des Sciences de Paris 277: 2537-2540.

Jorgensen, C. B., Larsen, L. O. and Lofts, B. 1979. Annual cycle of fat bodies and gonads in the toad *Bufo bufo* (L.) compared with cycles in other temperature zone anurans. Kongelike danske Videnskabernes Selskabs Biologiske Skrifter 22: 1-37.

Junqueira, L. C. U., Jared, C. and Antoniazzi, M. M. 1999. Structure of the Caecilian *Siphonops annulatus* (Amphibia, Gymnophiona): general aspect of the body, disposition of the organs and structure of the mouth, esophagus and stomach. Acta Zoologica 80: 75-84.

Jurgens, J. D. 1971. The morphology of the nasal region of Amphibia and its bearing on the phylogeny of the group. Annals of University of Stellenbosch 46: 1-46.

Klumpp, W. and Eggert, B. 1935. Die Schilddrüse und die branchiogenen Organe in *Ichthyophis glutinosus*. L. Zeitschrift für wissen der Zoologie 146: 329-381.

Kuhlenbeck, H. 1922. Zur Morphologie des Gymnophionengehirns. Jena Zeitschrift Naturwissenschaften 58: 453-484.

Kuhlenbeck, H. 1969. Observations on the rhombencephalon in the Gymnophion *Siphonops annulatus*. Anatomical Research 163: 311.

Kühne, B. and Junqueira, L. C. U. 2000. Histology of the trachea and lung of *Siphonops annulatus* (Amphibia, Gymnophiona). Revista Brazileira de Biologia 60: 167-172.

Lamotte, M. and Tuchman-Duplessis, H. 1948. Structure et transformations gravidiques du tractus génital femelle chez un Anoure vivipare (*Nectophryoïdes occidentalis* Angel). Comptes Rendus des Séances de l'Académie des Sciences de Paris 226: 597-599.

Largen, M. J., Morris, P. A. and Yalden, D. W. 1972. Observations on the Caecilian *Geotrypetes grandisonae* Taylor (Amphibia Gymnophiona) from Ethiopia. Monitore Zoologico Italiano 8: 185-205.

Laubmann, W. 1927. Uber die Morphogenese von Gehern und Gereuchsorgan der Gymnophionen. Beitrag zur Kenntniss der Gymnophionen. Zeitschrift für Anatomie und Entwicklung 84: 597.

Lawson, R. 1959. The anatomy of *Hypogeophis rostratus* Cuvier. Amphibia: Apoda or Gymnophiona. PhD. Dissertation University of Durham, King's College.

Lawson, R. 1963. The anatomy of *Hypogeophis rostratus* Cuvier (Amphibia : Apoda or Gymnophiona). I. The skin and skeleton. Proceedings of the University of Durham Philosophical Society Series A, 13: 254-273.

Lawson, R. 1965a. The development and replacement of teeth in *Hypogeophis rostratus* (Amphibia, Apoda). Journal of Zoology 147: 352-362.

Lawson, R. 1965b. The teeth of *Hypogeophis rostratus* (Amphibia, Apoda) and tooth structure in the amphibia. Proceedings of Zoological Society of London 145: 321-325.

Lawson, R. 1966a. The development of the centrum of *Hypogeophis rostratus* (Amphibia, Apoda) with special reference to the notochordal (intra-vertebral) cartilage. Journal of Morphology 118: 137-148.

Lawson, R. 1966b. The anatomy of the heart of *Hypogeophis rostratus* (Amphibia, Apoda) and its possible mode of action. Journal of Zoology 149: 320-336.

Leclercq, B. 1995. Contribution à l'étude du complexe pinéal des Amphibiens actuels. Diplôme de l'E.P.H.E., Lille, France.

Leclercq, B., Martin-Bouyer, L. and Exbrayat, J. M. 1995. Embryonic development of pineal organ in *Typhlonectes compressicaudus* (Dumeril and Bibron, 1841), a viviparous Gymnophionan Amphibia. Pp. 107-111. In G. A. Llorente, A. Montori, X. Santos and M. A. Carretero (eds), *Scientia Herpetologica*, Proceedings of the 7th General Meeting of the Societas Europaea Herpetologica, Barcelona, Spain.

Leydig, F. 1868. Uber die Schleichenlurche. Zeitschrift für wissen der Zoologie 18: 283-286.

Lostanlen, D., Boisseau, C. and Joly, J. 1976. Données ultrastructurales et physiologiques sur l'utérus d'un Amphibien ovovivipare *Salamandra salamandra* L. Annales de Sciences Naturelles, Zoologie, 12ème série 18: 113-144.

Malonza, P.K. and Measey, G.J. 2005. Life history of an African caecilian: *Boulengerula taitanus* Loveridge 1935 (Caeciilidae: Amphibia: Gymnophiona). Tropical Zoology 18:49-66.

Marcus, H. 1908. Beitrage zur Kenntniss der Gymnophionen. I. über das Schlundspaltengebiet. Archiv für Mikroskopie und Anatomie 71: 695-774.

Marcus, H. 1927. Lungenstudien. Morphologisches Jahrbuch 58: 100-127.

Marcus, H. 1930. Beitrag zur Kenntniss der Gymnophionen. XIII. Uber die Bildung von Geruchsorgan, Tentakel und Choanen bei *Hypogeophis*, nebst Vergleisch mit Dipnoen und *Polypterus*. Zeitschrift für Anatomie und Entwicklung 91: 657-691.

Marcus, H. 1932. Weitere Versuche und Beobachtungen über die Vorderdarmentwicklung bei den Amphibien. Zoologische Jahrbuch, Anatomie 55: 581-602.

Marcus, H. 1934. Das Integument. Beitrag zur Kenntniss der Gymnophionen. XXI. Zeitschrift für Anatomie und Entwicklung 103: 189-234.

Marcus, H. 1935. Zur stammensgeschichte der Herzens. Morphologisches Jahrbuch 76: 92-103.

Marcus, H. 1937. Uber Myotome, Horizontalseptum und Rippen bei *Hypogeophis* und Urodelen. Beitrag zur Kenntniss der Gymnophionen. XXX. Zeitschrift für Anatomie und Entwicklung 107: 531-532.

Marcus, H. 1939. Beitrag zur kenntnis der Gymnophionen. Ueber keimbahn, keimdruusen, Fettkörper und Urogenitalverbindung bei *Hypogeophis*. Biomorphosis 1: 360-384.

Marcus N. and Blume, W. 1926. Uber Wirbel und Rippen bei *Hypogeophis rostratus* nebst Bemerkungen uber *Torpedo*. Zeitschrift für Anatomie und Entwicklung 80: 1-78.

Martin-Bouyer, L., Godard, J. P. and Mairie, J.-M. 1995. 3-D reconstitution of the brain of *Typhlonectes compressicaudus* (Dumeril and Bibron, 1841) (Amphibia, Gymnophiona) Pp. 98-100. In G. A. Llorente, A. Montori, X. Santos and M. A. Carretero (eds), *Scientia Herpetologica*, Proceedings of the 7th General Meeting of the Societas Europaea Herpetologica, Barcelona, Spain.

Masood-Parveez, U. 1987. Some aspects of reproduction in the female Apodan Amphibian *Ichthyophis*. PhD. Dissertation Karnatak University, Dharwad, India.

Masood-Parveez, U. and Nadkarni, B. 1991. Morphological, histological, histochemical and annual cycle of the oviduct in *Ichthyophis beddomei* (Amphibia: Gymnophiona). Journal of Herpetology 25: 234-237.

Masood-Parveez, U. and Nadkarni, B. 1993a. The ovarian cycle in an oviparous gymnophione amphibian, *Ichthyophis beddomei* (Peters). Journal of Herpetology 27: 59-63;

Masood-Parveez, U. and Nadkarni, B. 1993b. Morphological, histological and histochemical studies of the ovary of an oviparous Caecilian, *Ichthyophis beddomei* (Peters). Journal of Herpetology 27: 63-69.

Maurer, F. 1887. Schilddrüse, Thymus und Kiemenreste der Amphibien. Morphologische Jahrbuch 13: 296-382.

Mookerjee, H. K. 1942. Development of the vertebral column in Gymnophiona. Proceedings of 29th Indian Scientific Congress 15: 256-268.
Müller, J. 1832. Beitrag zur Anatomie und Natursgeschichte der Amphibien. Zeitschrift Jahrbur für Physiologie 4: 195-275.
Naujoks-Manteuffel, C. and Meyer, D. L. 1996. Glial fibrillary acidic protein in the brain of the caecilian *Typhonectes natans* (Amphibia, Gymnophiona) an immunocytochemical study. Cell and Tissue Research 283: 51-58.
Noble G. K. 1931. *The Biology of the Amphibia*. McGraw-Hill, NewYork. 577 pp.
Norris, H. W. 1917. The eyeball and associated structures in the blindworms. Iowa Academy of Sciences Proceedings 24: 299-300.
Norris, H. W. and Hughes, S. P. 1918. The cranial and anterior spinal nerves of the caecilian amphibians. Journal of Morphology 31: 490-557.
Nussbaum, R. A. 1985. Systematics of Caecilians (Amphibia: Gymnophiona) of the family Scolecomorphidae. Occasional Papers of Museum of Zoology, University of Michigan 713: 1-49.
Nussbaum, R. A. and Wilkinson, M. 1995. A new genus of lungless tetrapod: a radically divergent caecilian (Amphibia: Gymnophiona). Proceedings of Royal Society of London 26: 331-335.
Ochotorena, I. 1932. Nota acerca la histologie de la piel de *Dermophis mexicanus* Dumeril y Bibron. Anales del Instituto de Biologia Mexico 3: 363-370.
Olivecrona, H. 1964. Notes on forebrain morphology in the Gymnophion (*Ichthyophis glutinosus*). Acta Morphologica 6: 45-53.
O'Reilly, J. C., Nussbaum, R. A. and Boone, D. 1996. Vertebrate with protrusible eyes. Nature 382: 33.
Oyama, J. 1952. Microscopical study of the visceral organs of Gymnophiona, *Hypogeophis rostratus*. Kumamoto Journal of Sciences 1B: 117-125.
Paillot, R., Estabel, J. and Exbrayat, J.-M. 1997a. Organes hématopoiétiques et cellules sanguines chez *Typhlonectes compressicaudus* et *Typhlonectes natans*. Bulletin mensuel de la Société Linnéenne de Lyon 66: 124-134.
Paillot, R., Estabel, J. and Exbrayat, J.-M. 1997b. Liver, bone marrow substitut in adult Gymnophionan. 3rd World Congress Herp., Prague, Czech Republic: 157.
Paillot, R., Estabel, J., Keller, E. and Exbrayat, J.-M. 1997c. Le foie, organe hématopoiétique chez les Gymnophiones. Bulletin de la Société Zoologique de France 122: 455.
Parker, H. W. 1934. Reptiles and Amphibians from Southern Ecuador. Annals and Magazine of Natural History 10 : 264-273.
Parker, H. W. 1936. The Caecilians of the Mamfe Division, Cameroons. Proceedings of Zoological Society of London 1936: 135-163.
Parker, H. W. 1956. Viviparous caecilians and amphibian phylogeny. Nature 178: 250-252.
Parker, H. W. 1958. Caecilians of the Seychelles, a description of a new species. Copeia 12: 71-76.
Parker, H. W. and Dunn, E. R. 1964. Dentitional metamorphosis in the Amphibia. Copeia 1964: 75-86.
Parker, H. W. and Wettstein, O. 1929. A new caecilian from southern Brazil. Annals and Magazine of Natural History 10 : 594-596.
Pattle, R. E., Schock, C., Creasey, J. M. and Hughes, G. M. 1977. Surpelling film, lung surfactant, and their origin in newt, caecilian, and frog. Journal of Zoology 182: 125-136.
Penhos, J. C. 1953. Rôle des corps adipeux du *Bufo arenarum*. Boletin de la Sociedad argentina de Biologia 28: 1095-1096.

Pereda, J. 1969. Histochimie des mucopolysaccharides de l'oviducte et des gangues muqueuses de l'ovocyte de *Pleuroderma bibroni*. Annales d'Histochimie 14: 55-66.
Pereda, J. 1970. Etude histochimique de la distribution des sialomucines dans l'oviducte et les gangues muqueuses des ovocytes de *Rana pipiens*. Comportement dans l'eau des différentes gangues. Journal d'Embryologie expérimentale et de Morphologie 24: 1-12.
Peters, W. 1874a. Observations sur le développement du *Caecilia compressicauda*. Annales des Sciences Naturelles, Zoologie, série 5: article 13.
Peters, W. 1874b. Derselbe las ferner über die Entwicklung der Caecilien und besonders der *Caecilia compressicauda*. Monatsberichte der Deutschen Akademie der wissenschaften zu Berlin 1874: 45-49.
Peters, W. 1875. Uber die Entwicklung der Caecilien. Monatsberichte der Deutschen Akademie der wissenschaften zu Berlin 1875: 483-486.
Phisalix, M. 1910. Répartition et spécificité des glandes curtanées chez les Batraciens. Annales des Sciences Naturelles, Zoologie 12: 183-201.
Pinelli, C., d'Aniello, B., Fiorentino, M., Bhat, G. K., Saidapur, S. K. and Rastogi, R. K. 1997. Distribution of gonadotropin-releasing hormone immunoreactivity in the brain of *Ichthyophis beddomei* (Amphibia: Gymnophiona). The Journal of Comparative Neurology 384: 283-292.
Pinelli C., d'Aniello, B., Fiorentino, M., Calace, P., di Meglio, M., Iela, L., Meyer, D. L., Sacnara, J. and Rastogi, R. 1999. Distribution of FMRFamide-like immunoreactivity in the amphibian brain: comparative analysis. The Journal of Comparative Analysis 414: 275-305.
Porter, K. R. 1972. *Herpetology*. W. B. Saunders Company, Philadelphia, Eastbourne, Toronto, U. S. A., England, Canada 524 pp.
Pujol, P. and Exbrayat, J.-M. 1987. Observations préliminaires sur la structure et les sécrétions mucipares de l'oviducte de *Bufo regularis* (Reuss, 1834). Amphibien Anoure tropical. Bulletin de la Société Herpétologique de France 44: 8-15.
Pujol, P. and Exbrayat, J.-M. 2000. Mise en évidence de l'homogénéité des testicules multilobés de deux Amphibiens par des méthodes morphométriques. Bulletin de la Société Herpétologique de France 95: 53-66.
Rage, J.-C. 1985. Origine et phylogénie des Amphibiens. Bulletin de de la Société Herpétologique de France 34 : 1-19.
Ramaswami, L. S. 1941. Some aspects of the cranial morphology of *Uraeotyphlus narayani* Seshachar (Apoda). Record of Indian Museum of Calcutta 43: 143-207.
Ramaswami, L. S. 1943. An account of the head morphology of *Gegenophis carnosus* (Beddome), Apoda. Journal of Mysore University 3: 205-220.
Ramaswami, L. S. 1944. An account of the heart and associated vessels in some genera of Apoda. Proceedings of Zoological Society, London 114: 117-138.
Ramaswami, L. S. 1954. The external gills of *Gegenophis* embryos. Anatomische Anzeichen 101: 120-122.
Rastogi, R. K., Meyer, D. L., Pinelli, C., Fiorentino, M. and d'Aniello, B. 1998. Comparative analysis of GnRH neuronal systems in the amphibian brain. General and Comparative Endocrinology 112: 330-345.
Rathke, H. 1852. Bemerkungen uber mehrere Korpertheile der *Coecilia annnulata*. Mullers Arkiv 1852: 334-350.
Renous, S. and Gasc, J.-P. 1986a. Le fouissage des Gymnophiones (Amphibia). Hypothèse morphofonctionnelle fondée sur la comparaison avec d'autres Vertébrés tétrapodes. Zoologische Jahrbuch, Anatomie 114: 95-130.
Renous, S. and Gasc, J.-P. 1986b. Hypothèse d'étude concernant la locomotion des Gymnophiones. Mémoires de la Société Zoologique de France 43: 133-143.

Renous, S. and Gasc, J.-P. 1989. Réflexion sur les règles générales de l'organisation des vertébrés terrestres à corps allongé: exemple des Amphibiens Gymnophiones. Zoologische Jahrbuch, Anatomie 118: 231-249.

Renous, S., Exbrayat, J.-M. and Estabel, J. 1997. Recherche d'indices de membres chez les Amphibiens Gymnophiones. Annales des Sciences Naturelles 18: 11-26.

Retzius, G. 1881. Das Gehörorgan der Wirbelthiere. I. Das Ghörorgan der Fishes und Amphibien. Samson und Wallin, Stockholm.

Riberon, A. and Exbrayat, J.-M. 1996. Quelques aspects de la structure du tégument des Amphibiens Gymnophiones. Bulletin de la Société Herpétologique de France 79: 43-56.

Roca, A., Maurice, B., Baraud, J., Mauget, C. and Cambar, R. 1970. Etude des lipides du corps adipeux de *Rana esculenta*. Comptes Rendus des Séances de l'Académie des Sciences de Paris 270: 1278-1281.

Rogers, D. C. 1965. An electron microscope study of the parathyroid gland of the frog (*Rana clamitans*). Journal of Ultrastructural Research 13: 478-499.

Romer, A. S. 1955. *The Vertebrate Body*. Second edition. W. B. Sanders Co., Philadelphia-London, 644 pp.

Sakai, T., Billo, R. and Kriz, W. 1986. The structural organization of the kidney of *Typhlonectes compressicaudus* (Amphibia, Gymnophiona). Anatomy and Embryology 174: 243-252.

Sakai, T., Billo, R. and Kriz, W. 1988a. Ultrastructure of the kidney of a south american Caecilian, *Typhlonectes compressicaudus* (Amphibia, Gymnophiona). II: distal tubule, connecting tubule, collecting duct and Wolffian duct. Cell and Tissue Research 252: 601-610.

Sakai, T., Billo, R., Nobiling, R., Gorgas, K. and Kriz, W. 1988b. Ultrastructure of the kidney of a South American Caecilian, *Typhlonectes compressicaudus* (Amphibia, Gymnophiona) I: renal corpuscle, neck segment, proximal tubule and intermediate segment. Cell and Tissue Research 252: 589-600.

Salthe, N. S. 1963. The egg capsules in the Amphibia. Journal of Morphology 113: 161-171.

Salthe, N. S and Mecham, J. S. 1974. Reproductive and courtship patterns. Pp. 309-521. In B. Lofts (ed), *Physiology of Amphibia*, vol. 2. Academic Press, NewYork.

Sammouri, R., Renous, S., Exbrayat, J.-M. and Lescure, J. 1990. Développement embryonnaire de *Typhlonectes compressicaudus* (Amphibia Gymnophiona). Annales des Sciences Naturelles, Zoologie, 13ème série 11: 135-163.

Sano-Martins, I. S., Jared, C. and Argolo, A. J. S. 1990. Estudo Hematologico de *Siphonops annulatus* (Amphibia, Appoda, Caeciliidae). Londrina, Brasil: Resumos do XVIII Congresso Brasileiro de Zoologia: 405.

Sarasin, P. and Sarasin, F. 1887-1890. Ergebnisse Naturwissenschaftlicher Forschungen auf Ceylon. Zur Entwicklungsgeschichte und Anatomie der Ceylonischen Blindwuhle *Ichthyophis glutinosus*. C.W. Kreidel's Verlag, Wiesbaden.

Sasayama, Y., Taniguchi, K., Suzuki, N. and Srivastav, A. J. 1996. The non-effect of parathyroidectomy in the aquatic limbless newt (Apoda, Amphibia). Okajimas Folia Anatomica Japanensis 72: 329-332.

Sawaya, P. 1940. Sobre o veneno das glandlas cutaneas, a secreçao e o craçao de *Siphonops annulatus*. Boletins da Faculdade de Filosofia, Ciencias e Lebras Universita. S. Paulo 4: 207-270.

Sawaya, P. 1941. Contribuitao para o estudo da fisiologica do sistema circulatorio do amfibio *Siphonops annulatus* (Mikan). Boletins da Faculdade de Filosofia, Ciencias e Lebras Universita S. Paulo 5: 209-233.

Schilling, C. 1935. Das Herzs von *Hypogeophis* und seine Entwicklung. Morphologisches Jahrbuch 76: 52-91.
Schlaghecke, R. and Blüm, R. 1978. Seasonal variations in fat body metabolism of the green frog *Rana esculenta* (L.). Experientia 34: 1019-1020.
Schmidt, A. and Wake, M. H. 1990. Olfactory and vomeronasal system of caecilians (Amphibia: Gymnophiona). Journal of Morphology 205: 255-268.
Schmidt, A. and Wake, M. H. 1997. Cellular migration and morphological complexity in the caecilian brain. Journal of Morphology 231: 11-27.
Schmidt, A. and Wake, M. H. 1998. Development of the tectum in Gymnophiones, with comparison to other Amphibians. Journal of Morphology 236: 233-245.
Schmidt, A., Wake, D. B. and Wake, M. H. 1996. Motor nuclei of nerves innervating the tongue and hypoglossal musculature in a Caecilian (Amphibia: Gymnophiona), as revealed by HRP transport. The Journal of Comparative Neurology 370: 342-349.
Semon, R. 1892. Studien über den Bauplan des Urogenitalsystem der Wirbeltiere. Dargelegt an der Entwicklung dieses organysystems bei *Ichthyophis glutinosus*. Jena Zeitschrift. Naturwissenschaften 26: 89-203.
Senn, D. G. and Reber-Leutenegger, S. 1986. Notes on the brain of Gymnophiona. Mémoires de la Société Zoologique de France 43: 65-66.
Seshachar, B. R. 1936. The spermatogenesis of *Ichthyophis glutinosus* (Linn.) I. The spermatogonia and their division. Zeitschrift für Zellforschung und mikroskopische Anatomie 24: 662-706.
Seshachar, B. R. 1937. The spermatogenesis of *Ichthyophis glutinosus* (Linn.). II. The meiotic divisions. Zeitschrift für Zellforschung und mikroskopische Anatomie 27: 133-158.
Seshachar, B. R. 1939. Testicular ova in *Uraeotyphlus narayani* Seshachar. Proceedings of Indian Academy of Sciences 10: 213-217.
Seshachar, B. R. 1940. The apodan sperme. Current Science 9: 464-465.
Seshachar, B. R. 1942a. Stages in the spermatogenesis of *Siphonops annulatus* Mikan. and *Dermophis gregorii* (Blgr) (Amphibia: Apoda). Proceedings of Indian Academy of Sciences 15: 263-277.
Seshachar, B. R. 1942b. The Sertoli cells in Apoda. Journal of Mysore University 3: 65-71.
Seshachar, B. R. 1942c. Origin of intralocular oocytes in male Apode. Proceedings of Indian Academy of Sciences 15: 278-279.
Seshachar, B. R. 1943. The spermatogenesis of *Ichthyophis glutinosus* (Linn.), III. Spermateleosis. Proceedings of National Institute of Sciences of India 9: 271-285.
Seshachar, B. R. 1945. Spermateleosis in *Uraeotyphlus narayani* Seshachar and *Gegenophis carnosus* Beddome (Apoda). Proceedings of National Institute of Sciences of India 11: 336-340.
Seshachar, B. R. 1948. The nucleolus of the apodan Sertoli cell. Nature 161: 558-559.
Seshachar, B. R. and Srinath, K. V. 1946. Studies on the nucleolus. I. The nucleolus of the Apodan Sertoli cell. Proceedings of Indian Scientific Congress 3: 96.
Setoguti, T., Isono, H. and Sakurai, S. 1970. Electron microscopic study on the parathyroid gland of the newt *Triturus pyrrhogaster* (Boie) in natural hibernation. Journal of Ultrastructural Research 31: 46-60.
Seydel, C. 1895. Uber die Nasenhohle und das Jacobson'sche Organ der Amphibien. Morphologisches Jahrbuch 23: 453-543.
Siwe, S. A. 1926. Pankreasstudien. Morphologisches Jahrbuch 5: 84-307.
Smita, M., Oommen, O. V., Jancy, M. G. and Akbarsha, M. A. 2003. Sertoli cells in the testis of Caecilians, *Ichthyophis tricolor* and *Uraeotyphlus* cf. *narayani* (Am-

phibia: Gymnophiona): Light and electron microscopic perspectives. Journal of Morphology 258: 317-326.
Smita, M., Oommen, O. V., Jancy, M. G. and Akbarsha, M. A. 2004a. Stages in spermatogenesis of two species of caecilians *Ichthyophis tricolor* and *Uraeotyphlus* cf. *narayani* (Amphibia: Gymnophiona): Light and electron microscopic study. Journal of Morphology 261: 92-104.
Smita, M., Jancy, M. G., Girija, R., Akbarsha, M. A. and Oommen, O. V. 2004b. Spermiogenesis in Caecilians *Ichthyophis tricolor* and *Uraeotyphlus* cf. *narayani* (Amphibia: Gymnophiona): Analysis by Light and Transmission Electron Microscopy. Journal of Morphology (in press).
Spengel, J. W. 1876. Das Urogenitalsystem der Amphibien. I. Theil. Der Anatomische Bau des Urogenitalsystem. Arbeitenaus demm Zoologzootom Institute Wurzburg 3: 51-114.
Stiffler, D. F., Deruyter, M. L. and Talbot, C. R. 1990. Osmotic and ionic regulation in the aquatic caecilian *Typhlonectes compressicaudus* and the terrestrial caecilian *Ichthyophis kohtaoensis*. Physiological Zoology 63: 649-668.
Storch, V. and Welsch, U. 1973. ZurUltrastruktur von Pigmentepithel und Photoreceptoren der Seitenaugen von *Ichthyophis kohtaoensis* (Gymnophiona, Amphiba). Zoologische Jahrbuch Anatomie 90: 160-173.
Storch, V., Prosi, F., Gorgas, K., Hacker, H. J., Rafael, J. and Vsiansky, P. 1986. The liver of *Ichthyophis glutinosus* Linn. 1758 (Gymnophiona). Mémoires de la Société Zoologique de France 43: 91-106.
Straub, J. O. 1985. Contributions of the cranial anatomy of the genus *Grandisonia* Taylor 1968 (Amphibia : Gymnophiona). PhD. Dissertation, Basel.
Straub, J. O. 1986. Aspects of the cranial anatomy of *Grandisonia diminutiva* Taylor. Mémoires de la Société Zoologique de France 43: 55-63.
Taylor, E. H. 1955. Additions to the known herpetological fauna of Costa Rica with comments on other species. University of Kansas Science Bulletin 37: 499-575.
Taylor, E. H. 1960. On the caecilian species *Ichthyophis monochrous* and *Ichthyophis glutinosus* with description of related species. University of Kansas Science Bulletin 40: 37-120.
Taylor, E. H. 1968. *The Caecilians of the World. A Taxonomic Review*. University of Kansas Press, Lawrence, Kansas, U.S.A., 848 pp.
Taylor, E. H. 1969. A new family of African Gymnophiona. University of Kansas Science Bulletin 48: 297-305.
Taylor, E. H. 1970. The lateral-line sensory system in the Caecilian family Ichthyophiidae (Amphibia: Gymnophiona). University of Kansas Science Bulletin 48: 861-868.
Taylor, E. H. 1977. Comparative anatomy of Caecilian anterior vertebrae. Kansas university Bulletin 51: 219-231.
Teipel, H. 1932.- Beitrag zue Kenntnis der Gymnophionen. XVI. Die Zunge.Zeitschrift für Anatomie und Entwicklung: 726-746.
Toews, D. and Macintyre, D. 1977. Blood respiratory properties of a viviparous Amphibian. Nature, 266: 464-465.
Toledo, R. C. and Jared, C. 1995. Cutaneous granular glands and amphibian venoms. Comparative Biochemistry and Physiology 111: 1-29.
Tonutti, E. 1931. Beitrag zur Kenntnis der Gymnophionen. XV. Das Genital-system. Morphologisches Jahrbuch 68: 151-292.
Tonutti, E. 1932. Vergleichende morphologische Studen uber Eddarm und Kopulations-organe. Morphologisches Jahrbuch 70: 101-130.
Tonutti, E. 1933. Beitrag zur Kenntnis der Gymnophionen. XIX. Kopulations-organe bei Weiteren Gymnophionenarten. Morphologisches Jahrbuch 72: 155-211.

Vilter, A. and Vilter, V. 1962. Rôle de l'oviducte dans l'uniparité utérine de la salamandre noire des Alpes orientales (*Salamandra atra*). Comptes Rendus de la Société de Biologie 156: 49-51.
Vilter, V. 1966. Histochimie de l'oviducte mature du *Triturus alpestris* L. de montagne. Comptes Rendus de la Société de Biologie 160: 2245-2250.
Vilter, V. 1967. Histologie de l'oviducte chez *Salamandra atra* mature, Urodèle totalement vivipare de haute montagne. Comptes Rendus de la Société de Biologie 161: 260-264.
Vilter, V. 1968a. Histochimie quantitative des sécrétions glandulaires de l'oviducte de *Triturus alpestris* Laur. sexuellement mature. Archives d'Anatomie, Histologie, Embryologie 51: 751-764.
Vilter, V. 1968b. Histochimie quantitative des sécrétions mucipares acides de l'oviducte mature de *Salamandra atra* des Alpes vaudoises. Comptes Rendus de la Société de Biologie 162: 76-80.
Vilter, V. 1986. La reproduction de la salamandre noire. Pp. 487-495. In P.P. Grassé et M. Delsol (eds), *Traité de Zoologie*, tome XIV, fasc.I B, Masson, Paris, France.
Vilter, V. and Lamotte, M. 1956. Evolution post-gravidique de l'utérus chez *Nectophrynoïdes occidentalis* Ang., Crapaud totalement vivipare de la Haute-Guinée. Comptes Rendus de la Société de Biologie 150: 2109-2113.
Vilter, V. and Lugand, A. 1959. Trophisme intra-utérin et croissance embryonnaire chez *Nectophrynoides occidentalis* Ang., Crapaud totalement vivipare du mont Nimba (Haute-Guinée). Comptes Rendus de la Société de Biologie 153: 29-32.
Vilter, V. and Thorn, R. 1967. Histologie de l'oviducte et mode de reproduction d'un Urodèle cavernicole d'Europe: *Hydromantes genei* (Temminck et Schlegel). Comptes Rendus de la Société de Biologie 161 : 1222-1227.
Wake, D. B. 1970. Aspects of vertebral evolution in the modern Amphibia. Forma et Functio, 3: 33-40.
Wake, M. H. 1967. Gill structure in the Caecilian genus *Gymnopis*. Bulletin of South California Academy of Sciences 66: 109-116.
Wake, M. H. 1968a. The comparative morphology and evolutionary relationships of the urogenital system of Caecilians. Ph.D. Dissertation, University of California.
Wake, M. H. 1968b. Evolutionary morphology of the Caecilian urogenital system. Part I: the gonads and fat bodies. Journal of Morphology 126: 291-332.
Wake, M. H. 1969. Gill ontogeny in embryos of *Gymnopis* (Amphibia : Gymnophiona). Copeia 1969: 183-184.
Wake, M. H. 1970a. Evolutionary morphology of the caecilian urogenital system. Part II: the kidneys and urogenital ducts. Acta Anatomica 75: 321-358.
Wake, M. H. 1970b. Evolutionary morphology of the caecilian urogenital system. Part III: the bladder. Herpetologica 26: 120-128.
Wake, M. H. 1972. Evolutionary morphology of the caecilian urogenital system. Part IV: the cloaca. Journal of Morphology 136: 353-366.
Wake, M. H. 1974. The comparative morphology of the caecilian lung. Anatomical Record 178: 483.
Wake, M. H. 1975. Another scaled caecilian (Gymnophiona, Typhlonectidae). Herpetologica 31: 134-136.
Wake, M. H. 1976. The development and replacement of teeth in viviparous Caecilians. Journal of Morphology 148: 33-63.
Wake, M. H. 1977a. Fetal maintenance and its evolutionary significance in the Amphibia : Gymnophiona. Journal of Herpetology 11: 379-386.
Wake, M. H. 1977b. The reproductive biology of Caecilians. An evolutionary perspective. Pp. 73-100. In D.H. Taylor and S. I. Guttman (eds), *The Reproductive Biology of Amphibians*, Miami University, Oxford, Ohio.

Wake, M. H. 1978. Comments on the ontogeny of *Typhlonectes obesus* particulary its dentition and feeding. Papeis avulsos de Zooogia 32: 1-13.

Wake, M. H. 1980a. Fetal tooth development and adult replacement in *Dermophis mexicanus* (Amphibia : Gymnophiona): Fields versus clones. Journal of Morphology 166 : 203-216.

Wake, M. H. 1980b. Reproduction, growth and population structure of the Central American Caecilian *Dermophis mexicanus*. Herpetologica 36: 244-256.

Wake, M. H. 1980c. The reproductive biology of *Nectophrynoides malcolmi* (Amphibia: Bufonidae) with comments on the evaluation of reproductive modes in the genus *Nectophrynoides*. Copeia 1980: 194-209.

Wake, M. H. 1980d. Morphometrics of the skeleton of *Dermophis mexicanus* (Amphibia: Gymnophiona). Part I. The vertebrae, with comparison to other species. Journal of Morphology 165: 117-130.

Wake, M. H. 1980e. Morphological information on caecilian eye function. American Zoologist 20: 785.

Wake, M. H. 1981. Structure and function of the male Mullerian gland in Caecilians (Amphibia: Gymnophiona), with comments on ilts evolutionary significance. Journal of Herpetology 15: 17-22.

Wake, M. H. 1986. A perspective on the systematics and morphology of the Gymnophiona (Amphibia). Mémoires de la Société Zoologique de France 43: 21-38.

Wake, M. H. 1985. The comparative morphology and evolution of the eyes of Caecilians (Amphibia, Gymnophiona). Zoomorphology 105: 277-295.

Wake, M. H. 1989. Metamorphosis of the hyobranchial apparatus in *Epicrionops* (Amphibia, Gymnophiona, Rhinatrematidae): replacement of bone by cartilage. Annales des Sciences naturelles, Zoologie, 13ème série, Paris 10: 171-182.

Wake, M. H. 1992. Patterns of peripheral innervation of the tongue and hyobranchial apparatus in Caeclians (Amphibia: Gymnophiona). Journal of Morphology 212: 3753.

Wake, M.H. 1993a. Non-traditional characters in the assessment of caecilian phylogenetic relationships. Herpetological Monographs 7: 42-55.

Wake, M. H. 1993b. Evolution of oviductal gestation in Amphibians. The Journal of Experimental Zoology 266: 394-413.

Wake, M. H. 1995. The spermatogenic cycle of *Dermophis mexicanus* (Amphibia: Gymnophiona). Journal of Herpetology 29:119-122.

Wake, M. H. 1998. Cartilage in the cloaca: phallodeal spicules in caecilians (Amphibia: Gymnophiona). Journal of Morphology 237: 177-186.

Wake, M. H. and Dickie, R. 1998. Oviduct structure and function and reproductive modes in Amphibians. The Journal of Experimental Zoology 282: 477-506.

Wake, M. H. and Hanken, J. 1982. Development of the skull of *Dermophis mexicanus* (Amphibia: Gymnophiona) with comments on skull kinesis and amphibian relationships. Journal of Morphology 173: 203-223.

Wake, M. H. and Nygren, K. M. 1987. Variation in scales in *Dermophis mexicanus*: are Gymnophione scales of systematic utility? Fieldiana 36: 1-8.

Wake, M. H. and Schwenk, K. 1986. A preliminary report on the morphology and distribution of taste buds in Gymnophiones, with comparison to other Amphibians. Journal of Herpetology 20: 254-256.

Wake, M. H. and Wurst, G. Z. 1979. Tooth crown morphology in Caecilians (Amphibia : Gymnophiona). Journal of Morphology 159: 331-341.

Wake, M. H., Exbrayat, J.-M. and Delsol, M. 1985. The development of the chondrocranium of *Typhlonectes compressicaudus* (Gymnophiona), with comparison to other species. Journal of Herpetology 19: 568-577.

Warbeck, A., Breiter, I. and Parzefall, J. 1996. Evidence for chemical communication in the aquatic Caecilian *Typhlonectes natans* (Typhlonectidae, Gymnophiona). Mémoires de Biospéléologie 23: 37-41.
Welsch, U. 1981. Fine structural and enzyme histochemical observations on the respiratory epithelium on the Caecilian lungs and gills. A contribution to the understanding of the evolution of the vertebrate respiratory epithelium. Archivium Histologicum Japonicum 14: 117-133.
Welsch, U. 1982. Morphologische Beobachtungen am thymus larvaler und adulter Gymnophionen. Zoologische Jahrbuch, Anatomie 107: 288-305.
Welsch, U. and Schubert, C. 1975. Observations on the fine structure, enzyme histochemistry and innervation of parathyroid gland and ultimobranchial body of *Chthonerpeton indistinctum* (Gymnophiona, Amphibia). Cell and Tissue Research 164: 105-119.
Welsch, U. and Starck, M. 1986. Morphological observations on blood cells and blood cell forming tissues of Gymnophiona. Mémoires de la Société Zoologique de France 43: 107-115.
Welsch, U. and Storch, V. 1971. Fine structural and enzyme histochemical observations on the notochord of *Ichthyophis glutinosus* and *Ichthyophis kohtaoensis* (Gymnophiona, Amphibia). Zeitschrift für Zellforschung und mikroskopische Anatomie 117: 443-450.
Welsch, U. and Storch, V. 1972. Elektrononmikroskopische Untersuchungen an den Leber von *Ichthyophis kohtaoensis* (Gymnophiona). Zoologische Jahrbuch Anatomie 89: 621-635
Welsch, U. and Storch, V. 1973a. Die Feinstruktur verhornter und nichtverhornter ektodermaler Epithelien und der Hautdrüsen, embryonaler und adulter Gymnophionen. Zoologische Jahrbuch Anatomie 90: 323-342.
Welsch, U. and Storch, V. 1973b. Elektronenmikroskopische Beobachtungen am Nephron adulter Gymnophionen (*Ichthyophis kohtaoensis* Taylor). Zoologische Jahrbuch Anatomie 90: 311-322.
Welsch, U. and Storch, V. 1982. Light and electron microscopical observations on the Caecilian spleen. A contribution to the evolution of lymphatic organs. Developmental and Comparative Immunology 6: 293-302.
Welsch, U., Schubert, C. and Storch, V. 1974. Investigation on the thyroid gland of embryonic, larval and adult *Ichthyophis glutinosus* and *Ichthyophis kohtaoensis* (Gymnophiona, Amphibia). Histology, fine structure and studies with radioactive iodide (^{131}I). Cell and Tissue Research 155: 245-268.
Welsch, U., Schubert, C. and Siak Hauw Tan 1976. Histological and histochemical observations on the neurosecretory cells in the diencephalon of *Chthonerpeton indistinctum* and *Ichthyophis paucisulcus* (Gymnophiona, Amphibia). Cell and Tissue Research 175: 137-145.
Welsch, U., Schubert, C., Kirmse, L. and Storch, V. 1976. The influence of temperature on the thyroid gland of *Chthonerpeton indistinctum* (Gymnophiona, Amphibia). Cell and Tissue Research 165: 455-465.
Wever, E. G. 1975. The Caecilian ear. Journal of Experimental Zoology 191: 63-71.
Wever, E. G. and Gans, C. 1976. The Caecilian ear. Further observations. Proceedings of National Academy of Sciences, U. S. A. 73: 3744-3746.
Weysse, A. W. 1895. Uber die ersten Anlagen der Hauptanhangsorgane des Darmkanals deim Frosch. Archiv für Mikroskopie und Anatomie 46: 632-645.
Wiedersheim, R. 1879. *Die Anatomie der Gymnophionen*. Jena, Gustav Fisher, 101 pp.
Wilkinson, M. 1990. The presence of a Musculus retractor cloacae in female caecilians. Amphibia Reptilia 11: 300-304.

Wilkinson, M. 1992. The phylogeneic position of the Rhinatrematidae (Amphibia: Gymnophiona): evidence from the larval lateral line system. Amphibia Reptilia 13: 74-79.
Wilkinson, M. 1997. Characters, congruence and quality: a study of neuroanatomical and traditional data in caecilian phylogeny. Biological Review 72: 423-470.
Wilkinson, M. and Nussbaum, R. 1997. Comparative morphology and evolution of the lungless caecilian *Atretochoana eiselti* (Amphibia: Gymnophiona). Biological Journal of the Linnean Society 6: 39-109.
Wilkinson, M. and Nussbaum, R. 1998. Caecilian viviparity and amniote origins. Journal of Natural History 32: 1403-1409.
Wilkinson, M., Sebben, A., Schwartz, E. N. F. and Schwartz, C. A. 1998. The largest lungless tetrapod: report on a second specimen of *Atretochoana eiseltii* (Amphibia: Gymnophiona: Typhlonectidae) from Brazil. Journal of Natural History 32: 617-627.
Xavier, F. 1971. Recherches sur l'endocrinologie sexuelle de la femelle de *Nectophrynoides occidentalis* Angel (Amphibien Anoure vivipare). Doctorat ès Sciences Naturelles, Université Paris VI.
Xavier, F. 1977. An exceptional reproductive strategy in Anura: *Nectophrynoides occidentalis* Angel (Bufonidae), an example of adaptation to terrestrial life by viviparity. Pp. 545-552. In M. K. Hecht, P. C. Googy and B. M. Hecht (eds), *Major Patterns in Vertebrate Evolution,* Plenum Press, New York, U.S.A.
Xavier, F. 1986. La reproduction des Nectophrynoïdes. Pp. 497-513. In P.-P. Grassé et M. Delsol (eds), *Traité de Zoologie,* Tome XIV, fasc. I-B, Masson (Paris).
Zylberberg, L. 1972. Données histologiques sur les glandes linguales d'*Ichthyophis glutinosus* (L.) Batracien gymnophione. Archives d'Anatomie Microscopique, Morphologie Expérimentale 61: 227-242.
Zylberberg, L. 1986. L'épithélium lingual de deux Amphibiens Gymnophiones: *Typhlonectes compressicaudus* et *Ichthyophis kohtaoensis*. Mémoires de la Société Zoologique de France 43: 83-84.
Zylberberg, L. and Wake, M. H. 1990. Structure of the scales of *Dermophis* and *Microcaecilia* (Amphibia: Gymnophiona), and a comparison to dermal ossifications to other Vertebrates. Journal of Morphology 206: 25-43.
Zylberberg, L., Castanet, J. and de Ricqles, A. 1980. Structure of the dermal scales in Gymnophiona (Amphibia). Journal of Morphology 165: 41-54.

CHAPTER 4

Caecilian Male Mullerian Gland, with Special Reference to *Uraeotyphlus narayani*

Mohammad A. Akbarsha[1], George M. Jancy [1],
Mathew Smita[2] and Oommen V. Oommen[2]

4.1 INTRODUCTION

The caecilians are a unique group of vertebrates in that the Mullerian duct is retained in the adult males as a functional glandular structure, and this feature was first reported by Spengel (1876). This is an enigmatic aspect of caecilian biology. How and why this happens is important in understanding the evolutionary significance of the caecilians and, more importantly, their terrestrialization to different grades and their internal fertilization. In all other vertebrate males, the Mullerian duct regresses in the embryonic stage itself in response to the Mullerian inhibiting hormone (MIH) secreted by the Sertoli cells of the embryonic testis, and any further development of the duct is aborted. Therefore, the retention of Mullerian duct in the adult caecilians raises the question as to whether the Sertoli cells of the embryonic testis in Gymnophiona do not, in fact, secrete MIH.

In the very few anurans in which internal fertilization occurs, the mechanism of sperm transfer is mostly known and secretion of a fluid medium to suspend the spermatozoa has not been reported but Sever *et al.* (2002) have demonstrated insemination via a penis in *Ascaphus* together with addition of cloacal secretions to the semen. A few urodeles in which internal fertilization is the practice, the spermatozoa are fabricated into spermatophores and are acquired by the females. Therefore, the males do not need to secrete a fluid medium to suspend the sperm for ejaculation

---

[1] Department of Animal Science, Bharathidasan University, Tiruchirappalli 620 024, Tamil Nadu, India
[2] Department of Zoology, University of Kerala, Kariavattom 695 581, Thiruvananthapuram, Kerala, India

(Wake and Dickie 1998; Onitake *et al*. 2000). The mechanism of sperm transfer in the context of internal fertilization in the caecilians is different from that of urodeles. The spermatozoa are directly ejaculated into the female tract making use of the eversible phallodeum as the phallus (Wake 1977). Thus, a medium to suspend the spermatozoa, as in the amniotic vertebrates, is imminent. But the organs in the higher vertebrates concerned with the secretion of the seminal plasma, viz., the sexual segment of the kidney (Bishop 1959; Sarkar and Sivanandappa 1989), epididymis (Depeiges and Dufaure 1977; Robaire and Hermo 1998) and ampulla ductus deferentis (Akbarsha and Meeran 1995, Daisy *et al*. 2000) as in the case of reptiles, and the prostate gland, seminal vesicles (Luke and Coffey 1994) and ampulla ductus deferentis (Aumuller and Seitz 1990; Setchell *et al*. 1994) as in the case of mammals are absent in the caecilians. The Mullerian gland is the only glandular structure anatomically connected with the urinogenital system and, therefore, hypothetically, it would be the candidate organ to contribute the seminal plasma.

Tonutti (1931) suggested that the Mullerian duct might be a functional analogue of the prostate gland of mammals and would secrete fluid for sperm transport. Wake (1970, 1977, 1981) suggested that the secretion of the Mullerian gland could be a vehicle for sperm transport and, perhaps, nutrition. She also attempted to trace the evolution of a part of the mammalian prostate gland from the Mullerian duct. Exbrayat (1985, 1986) suggested that the secretions of the Mullerian gland are evacuated during copulation, and probably mixed with spermatozoa in the cloaca so as to process the sperm towards fertilization. Thus, a role for the Mullerian gland of caecilians has remained only speculative. However, recent publications of the authors of this chapter (George *et al*. 2004a, b, 2005), and some more findings due for publication, based on the light microscopic, transmission electron microscopic, biochemical and sperm physiological studies in *Uraeotyphlus narayani*, have thrown considerable light on the functional and evolutionary significance of the caecilian male Mullerian gland. The present review is an attempt to summarize these findings and discuss them in relation to the existing literature. The morphology, types, anatomy and histology of Mullerian gland are reviewed in adequate detail in chapters **3** and **5** of this volume and, hence, repetition is avoided as far as possible.

4.2 ANATOMY OF THE MULLERIAN DUCT

The Mullerian duct of *Uraeotyphlus narayani*, on each side, is an elongated organ, white in colour and extends up to the level of the transverse septum alongside the kidney, parallel to the testis lobes. The posterior one-third of the duct, from beyond the level of the last testis lobe, is a cylindrical glandular organ, the Mullerian gland, and anteriorly extends up to the liver as a connective tissue strand (Fig. 4.1A). Thus, it belongs to the first type described by Exbrayat and Estabel in this volume (Chapter **3**). In animals measuring 18–22 cm long, the Mullerian gland measures 5–7 cm long, and the diameter at the widest region is 1.6–2.2 mm. At the posterior end the Mullerian gland is continued as a duct parallel to the urinogenital duct (George *et al.* 2004a). In *Gymnopis multiplicata* (Wake 1977) and *Typhlonectes compressicauda* (Exbrayat 1985), the posterior ends of the Mullerian duct and the urinogenital duct open separately into the cloaca. Wake (1970) found in her specimen of *Uraeotyphlus* the two ducts to enter the cloaca at the same place. However, Chatterjee as early as 1936, found that in *Uraeotyphlus menoni* the Mullerian duct narrows at the posterior end and joins the urinogenital duct to form a common duct that opens into the cloaca, which is the case in *Uraeotyphlus narayani* also (George *et al.* 2004a) (Fig. 4.1A). While this would indicate differences in the urinogenital anatomy between species, it also warrants reinvestigation of the anatomy of this particular portion of the male urinogenital system in the species studied earlier. The joining of the two ducts before opening into the cloaca, at least in these two species of caecilians, is a clear indication that the secretion of the male Mullerian gland would potentially mix with sperm before arrival at the cloaca for ejaculation, strengthening the concept of Tonutti (1931), Wake (1977, 1981) and Exbrayat (1985, 1986) that the male Mullerian gland of caecilians would provide a medium in which spermatozoa are suspended during ejaculation and would even provide nutritional support to them subsequent to ejaculation.

4.3   INTERNAL STRUCTURE OF THE MULLERIAN GLAND

The Mullerian gland has an outer pleuroperitoneum followed by a layer of smooth muscle bundles, a layer of thick but diffuse connective tissue, individual tubular glands ("camera septa" of Tonutti 1931), ducts of the tubular glands and the central duct, all occurring in six distinct zones (Fig. 4.1B). The glandular zone consists of a large number of tubular glands arranged in a centripetal manner around the circumference of the central duct. In a single section, the number of glands may vary from 35 to 55. Each gland is tall and cuboidal, broad at the base and narrow towards the central duct. The glands are separated by loose connective tissue, which is continuous with the outer connective tissue boundary of the whole organ. There is a similar connective tissue in the zone that includes the ducts of the glands, i.e., between the tubular glands and the central duct. The connective tissue contains capillaries and amoeboid cells, differing in fine detail among the different zones. The tubular glands connect to the central longitudinal duct through the ducts of the tubular glands. The central duct continues at the posterior end, beyond the Mullerian gland, to join the urinogenital duct. Between March and June the testis lobes as well as the constituent lobules regress and the interlobular tissue increases (Smita *et al.* 2003, 2004). The Mullerian gland also regresses in respect of diameter of the whole gland, height of the tubular glands, diameter of the tubular glands at different levels and diameter of the ducts of the tubular glands and the central duct. On the other hand, the density of the tissues around

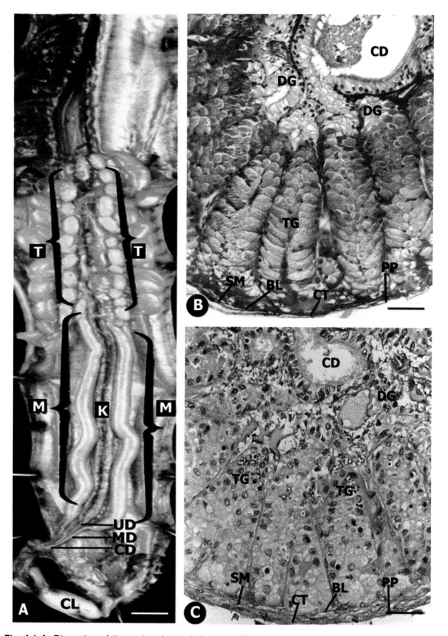

Fig. 4.1 A. Dissection of the male urinogenital system of *Uraeotyphlus narayani* showing testis lobes (T), kidney (K), Mullerian gland (M), urinogenital duct (UD), Mullerian duct (MD), common duct (CD) and cloaca (CL). **B, C**. Transverse section of a Mullerian gland of *U. narayani* during showing central duct (CD), duct of the gland (DG), tubular glands (TG), basement membrane (BL), smooth muscle bundles (SM), connective tissue (CT) and pleuroperitonium (PP). **B.** During the period of active spermatogenesis. **C.** During the period of spermatogenic quiescence. LM, haematoxylin and eosin stained paraffin section. Scale bar: A = 0.5 mm; B, C = 0.120 μm. From George, J.M., Smita, M., Oommen, O.V. and Akbarsha, M.A. 2004a. Journal of Morphology 260: 33-56, Figs. 1, 4, 45.

the entire gland, between the tubular glands and between the tubular glands and the central duct increases. The lumen of the tubular glands is almost fully regressed. The capillaries in the connective tissue between the tubular glands and the central duct are more profuse (Fig. 4.1C) (George et al. 2004a).

4.4 LIGHT MICROSCOPIC HISTOLOGY OF TUBULAR GLANDS

The tubular gland is differentiated based on the epithelial lining and the basal lamina into two portions, the tubular and basal portions (the 'column' and the 'base', according to George et al. 2004a) (Fig. 4.1B, C). The epithelium along the tubular portion is tall columnar and formed of a single layer of cells. The nuclei of these cells are located at two different levels, one in which nuclei are basally located and the other in which the nuclei are apical (Fig. 4.2A). The epithelium of the basal portion (i.e., closer to the outer boundary) is taller than at the tubular portion and the intercellular associations are such that a longitudinal array of intercellular spaces is discernible (Fig. 4.2B). Such intercellular spaces are absent between the cells lining the tubular portion (Fig. 4.2A, B). The epithelial cells of the tubular portion line the lumen of the tubular glands all around, whereas the cells at the basal portion line the basal aspect of the lumen (Fig. 4.2B). The profile of the lumen at the junction between the tubular and basal portions is such that the lumen of the tubular portion, on reaching the basal portion, spreads out on the top of the epithelial cells of the basal portion. Thus, the lumen at this part of the tubular glands, in a mid-sagittal section, has the shape of "Y" (Fig. 4.2C). Therefore, the epithelial cells of the basal portion decrease in height from the mid-point towards the periphery. A third cell type, pyramid-shaped and darkly stained, absent in the epithelium of the tubular portion, is abundant among the tall columnar cells of the basal portion of the tubular glands (Fig. 4.2B). The lumen of the tubular gland is either empty or contains ovoid granules or a flocculent material (Fig. 4.2E). The basal lamina of the tubular portion is thick and compact (Fig. 4.2A) whereas that of the basal portion is thin and diffuse (Fig. 4.2B, C).

The point of transition of the tubular gland into its duct is abrupt (Fig. 4.2D). The duct of the tubular gland is lined by a single row of tall columnar cells with clear cytoplasm and the nuclei located at different heights. The ducts take a tortuous course through the connective tissue between the tubular glands and the central duct (Figs. 4.1B, C, 4.2C) to open into the latter. Occasionally, the ducts of two tubular glands join into a common duct (Fig. 4.2E). The loose connective tissue surrounding the ducts of the tubular glands contains capillaries and amoeboid cells (Fig. 4.2E). At the point of transition of tubular glands into the ducts (i.e., at the neck), the lining epithelium is taller than at the duct proper (Fig. 4.2D). The central duct is large and its epithelial lining is tall and columnar with the cells differing in the intensity of staining (Fig. 4.2F).

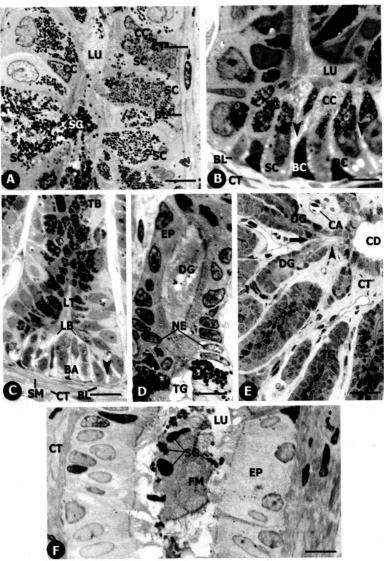

Fig. 4.2 A. Histology of tubular portion of tubular gland. **B.** Histology of basal portion of tubular gland. **C.** Nature of lumen of tubular and basal portions of tubular gland. (In **B** and **C**, the peculiar intercellular spaces between cells at the basal portion are indicated by arrowheads). **D.** Difference in the epithelium between the duct of tubular gland and the neck. **E.** Ducts of two tubular glands merging into a common duct (arrow), which in turn leads towards the central duct (arrowhead). **F.** Central duct of Mullerian gland. BA, basal portion of tubular gland; BC, basal cell; BL, basement membrane; CA, capillary; CC, ciliated cell; CD, central duct; CT, connective tissue; DG, duct of tubular gland; EP, epithelium; FM, flocculent material; LB, lumen at the basal portion; LT, lumen at the tubular portion; LU, lumen; NE, neck; NU, nucleus; SC, secretory cell; SG, secretion granule; SM, smooth muscle bundles; TB, tubular portion of tubular gland; TG, tubular gland. LM, TBO stained semithin section. Scale bar: A, B, D, F = 15 μm; C = 30 μm; E = 60 μm. **A.** Original. **B-F.** From George, J.M., Smita, M., Oommen, O.V. and Akbarsha, M.A. 2004a. Journal of Morphology 260: 33-56, Figs. 6, 7, 28, 29, 36.

4.5 ULTRASTRUCTURAL ORGANIZATION OF EPITHELIAL CELLS OF TUBULAR GLANDS

Transmission electron microscopic observations revealed that the cells with basal nuclei of both the tubular and basal portions of the tubular glands are non-ciliated secretory cells (hereafter referred to as secretory cells) and those with apical nuclei are ciliated non-secretory cells (hereafter referred to as ciliated cells). The pyramid-shaped darkly stained cells of the basal portion are basal cells.

4.5.1 Secretory Cells of Tubular Portion

These cells are tall and cuboidal in shape, broad at the base and narrow towards the lumen. They are always larger than the ciliated cells (Fig. 4.3A). The apical cytoplasm possesses a large number of microvilli, which form the brush border (Fig. 4.3B). The nucleus is quite large and occupies a basal position (Fig. 4.3C). The supranuclear cytoplasm contains secretory granules of varying sizes and densities. The granules have a diameter of 0.5 – 3.0 μm. These cells, by and large, match in ultrastructural organization the other glandular epithelial cells concerned with synthesis of secretory material and formation of secretion granules. The prominent Golgi apparatus is supranuclear and presents as two or more vertical or horizontal stacks. The Golgi cisternae are fenestrated, and the Golgi vesicles are prominent (Fig. 4.3D). The secretory material first appears in the Golgi area as an amorphous material in small to large membrane-bound vesicles (Fig. 4.3D). The vesicles increase in size, and their contents undergo condensation gradually, resulting in the secretion granules (Fig. 4.3E-H). The granules present even in the most apical cytoplasm are in different phases of condensation (Fig. 4.3B). The secretion granules, since their first appearance in the Golgi area, are invariably confined to the medullary region of the cytoplasm (Fig. 4.3A, B). This is in contravention to the observation that generally in the secretory cells most of the secretion granules are dispersed to the cortex during the first three hours after biogenesis (Rudolf *et al.* 2001). The significance of confinement of secretory granules to the medullary cytoplasm in the secretory cells of the tubular portion of the Mullerian gland is worth further investigation. The infranuclear cytoplasm is invariably free from such granules but rich in mitochondria and endoplasmic reticulum (Fig. 4.3C). In fact, mitochondria are confined to the perinuclear and basal cytoplasm.

4.5.2 Secretory Cells of Basal Portion

The secretory cells of the basal portion differ among themselves in abundance and electron density of the secretion granules and electron density of the cytoplasm (Fig. 4.4A). In these cells the Golgi apparatus, more prominent than in the secretory cells of the tubular portion, is located in the cytoplasm basal to the nucleus. This is in a position slightly eccentric and where the nucleus is either flat or slightly indented (Fig. 4.4B). The manifestation of the

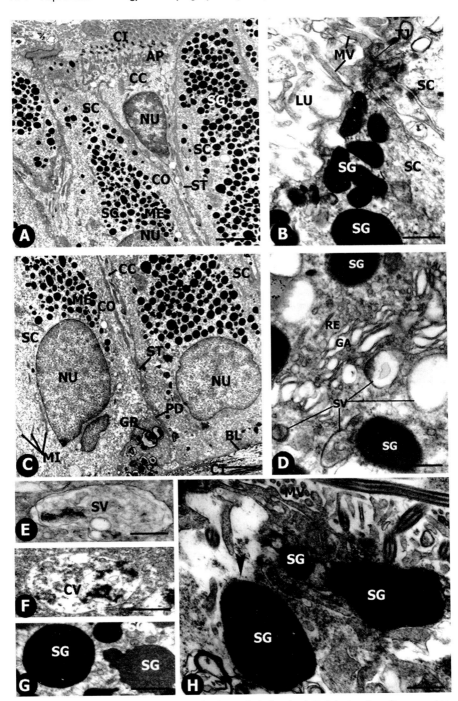

Fig. 4.3. A. Upper portion of the epithelium of column of tubular gland. **B.** Apical portion of two secretory cells. **C.** Lower portion of epithelium of tubular portion of tubular gland. **D-H.** Secretory cell of the tubular

Fig. 4.3 contd

secretion granules in these cells is as in any other secretory cell with one difference, viz., the mitochondria accumulate as dense aggregations around the secretory vesicles (Fig. 4.4B-F). An intimate association between the mitochondria and condensing vacuoles is also indicated (Fig. 4.4F). The secretion granules, unlike in the secretory cells of the tubular portion, are not limited to the medullary cytoplasm (Fig. 4.4A), and measure 0.2–2.0 µm in diameter (George *et al.* 2004b). The secretory material, after release into the lumen, invariably retains its identity as dense granules (Fig. 4.4G).

The microvilli lining the luminal surface of the secretory cells of both the tubular and basal portions may be involved in absorption and/or endocytosis of materials from the lumen.

4.5.3 Release of Secretion Granules from Secretory Cells

The mature granules are released into the lumen of the tubular glands, as structured granules, through exocytosis (Fig. 4.3H). Such granules are traceable to the central duct of the Mullerian gland. Such a mechanism of release of the secretion granules is an interesting observation. In general, the glandular epithelial cells fabricate the secretory material as discrete granules, but the release of the material contained therein into the lumen is in a solublized form in an apocrine mechanism (Burgoyne and Morgan 2003). However, in several reptiles the epididymal epithelial principal cells fabricate the secretory material as dense granules and release them as structured granules through exocytosis such that the granules maintain their identity even after their release into the lumen (Depeiges and Dufaure 1977; Dufaure and Saint Girons 1984, Manimekalai and Akbarsha 1992, Akbarsha and Manimekalai 1999). Similar is the mechanism of release of secretion granules in the mammalian prostate gland (Cohen 2001, Gesase and Satoh 2003). In the reptiles, in the absence of male accessory reproductive

---

Fig. 4.3 contd

portion of tubular glands, showing stages in the first appearance, condensation and release of the secretory material. **D:** Rough endoplasmic reticulum associates with the *cis*-phase of the Golgi apparatus and secretory vesicles appear at the *trans*-phase. Secretion granules are present around the Golgi apparatus. **E.** The content of the secretory vesicle undergoes condensation and becomes a condensing vacuole. **F.** The content of the condensing vacuole is much more condensed. **G.** The secretion granules are formed but in different electron densities. **H.** Highly condensed secretion granules in the most apical cytoplasm are released into the lumen (arrowheads) as dense granules from between the bases of microvilli. AP, apical expanded portion of ciliated cell; BL, basement membrane; CI, cilia; CC, ciliated cell; CO, cortical cytoplasm; CT, connective tissue; CV, condensing vacuole; GA, Golgi apparatus; GR, dark dense granules at the pyramidal basal portion of ciliated cell; LU, lumen; ME, medullary cytoplasm; MI, mitochondria; MV, microvilli; NU, nucleus; PD, basal pyramidal peduncle of ciliated cell; RE, rough endoplasmic reticulum; SC, secretory cell; SG, secretion granules; ST, narrow stalk of ciliated cell; SV, secretory vesicle; TJ, tight junction. TEM. Scale bar: A, C = 3.0 µm; B, D, = 0.5 µm; E-G: 0.3 µm; H = 0.2 µm. **A-C.** From George, J.M., Smita, M., Oommen, O.V. and Akbarsha, M.A. 2004a. Journal of Morphology 260: 33-56 Figs. 11, 13, 12. **D-F, H.** From George, J.M., Smita, M., Kadalmani, B., Girija, R., Oommen, O.V. and Akbarsha M.A. 2004b. Journal of Morphology 262: 760-769, Figs. 2A-C, 2G. G. Original.

166 Reproductive Biology and Phylogeny of Gymnophiona

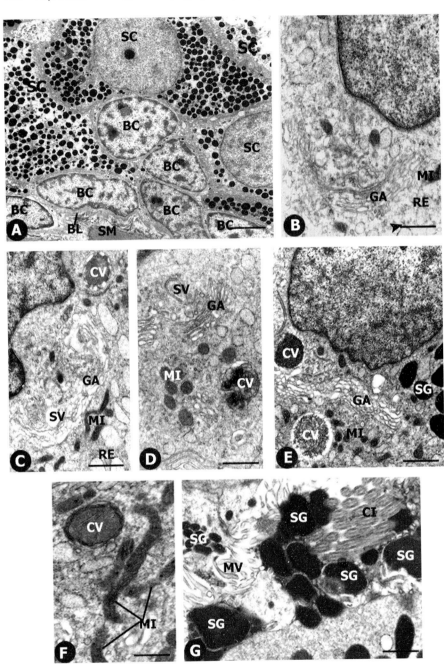

Fig. 4.4. A. Secretory and basal cells of basal portion of tubular gland. **B-F.** Golgi area (GA) of secretory cells of basal portion of tubular gland, showing different stages in the formation of secretion granules. **B**: Minute vesicles are pinched off (arrowheads) from the rough endoplasmic reticulum, which associate with the *cis*-phase of the Golgi apparatus. **C**: There are a few mitochondria, secretory vesicles and

Fig. 4.4 contd

glands comparable to those in mammals, the epididymis contributes several of the proteins of the seminal plasma (Depeiges and Dufaure 1980, 1981). It is also an established fact that in the mammals a major portion of the seminal plasma is contributed by the prostate gland (Luke and Coffey 1994). Perhaps this similarity between the Mullerian gland of the caecilians and the epididymis of reptiles on the one hand and between the Mullerian gland of the caecilians and the prostate gland of mammals on the other would indicate an evolutionary relationship in the mechanisms of secretion of the seminal plasma, the medium for transport of sperm among the tetrapods in the context of terrestrialization and internal fertilization.

4.5.4 Ciliated Cells

The ciliated cells are narrower than the secretory cells (Fig. 4.3A). They are broad towards the luminal end and connect to a cuboidal basal peduncle through a narrow stalk that runs between the secretory cells. The elongated ovoid nucleus lies in the apical expanded portion. The cells are characterized by tall cilia along the luminal border (Fig. 4.5A). The ciliated cells are totally devoid of secretion granules. However, large dense bodies are present in the cytoplasm of the basal peduncle (Fig. 4.3C). One of the prominent features of the ciliated cells is the accumulation of ribbon-like mitochondria at the luminal end, in between the basal bodies of the cilia (Fig. 4.5A). The endoplasmic reticulum is prominent, whereas the Golgi apparatus is only poorly manifested. The supranuclear cytoplasm abounds with vacuoles and membrane-bound vesicles. The ultrastructural organization of the ciliated cells indicates a role in transport of the secretory material from the lumen of the tubular gland towards the duct of the gland.

4.5.5 Basal Cells

The pyramid-shaped darkly stained cells present only in the epithelium of the basal portion of the tubular glands were earlier described as amoeboid cells (George et al. 2004a) but later identified as basal cells (George et al. 2004b). These cells are present mostly at the basal portion of the epithelium (Fig. 4.5B) but, occasionally, deeper also (Fig. 4.5C). The tissue around the basal aspect of the tubular glands contains irregularly shaped cells, which

---

Fig. 4.4 contd
condensing vacuoles in the Golgi area. **D**: Secretory vesicles are produced in the Golgi apparatus. The Golgi area is rich in condensing vacuoles and mitochondria. **E:** The Golgi area is rich in condensing vacuoles and mitochondria. Close by there are secretion granules. **F**: Mitochondria associate with a condensing vacuole. **G**: A portion of the central duct of the Mullerian gland, showing the epithelium, lumen and the dense granules, which arrived from the epithelium. BC, basal cell; BL, basement membrane; CC, ciliated cells; CI, cilia; CV, condensing vacuole; GA, Golgi apparatus; MI, mitochondria; MV, microvilli; RE, rough endoplasmic reticulum; SC, secretory cells; SG, secretion granules; SM, smooth muscle bundles; SV, secretory vesicles. TEM. Scale bars: A= 3 µm; B–E = 0.8 µm; F, G = 0.5 µm. **A-F.** From George, J.M., Smita, M., Kadalmani, B., Girija, R., Oommen, O.V. and Akbarsha M.A. 2004b. Journal of Morphology 262: 760-769, Figs. 6A, 4A-E. **G**. Original.

168 Reproductive Biology and Phylogeny of Gymnophiona

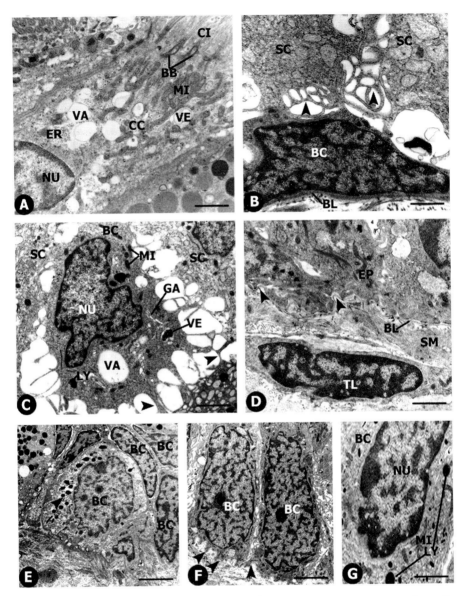

Fig. 4.5. A. Apical portion of a ciliated cell of tubular portion of tubular gland showing nucleus, cilia along the luminal border, basal bodies of the cilia, dense mitochondria in the apical cytoplasm, abundant endoplasmic reticulum in the supranuclear cytoplasm and vacuoles. **B-G.** Basal cells and their establishment in the epithelium. **B**: Epithelium of basal portion of tubular gland showing a basal cell resting on the basement membrane. The basal cell and secretory cell associate through cytoplasmic processes (arrowheads). The basal cell appears deficient in cytoplasmic organelles. **C**: Magnified view of a basal cell possessing heterochromatic nucleus, large vacuoles, membrane-bound vesicles containing an electron-dense material, lysosomes, mitochondria, Golgi apparatus and delicate and

Fig. 4.5 contd

in all probability are peritubular tissue leukocytes, possessing elongated, irregular, and highly heterochromatic nuclei. A few such cells closely approximate the thin basement membrane (Fig. 4.5D). The morphology of the nuclei of the basal cells resident in the epithelium and those of the peritubular tissue leukocytes is almost alike. The peritubular leukocytes appear to penetrate through the basement membrane and arrive at the epithelium (Fig. 4.5D, E). This inference is based purely on histological and ultrastructural evidence. Once in the epithelium, the basal cells are pyramidal in shape, with the nucleus also of the same shape (Fig. 4.5B, F). The basal cells form into a three-dimensional latticework on top of the basement membrane. Wherever the basal cell is adherent to the basement membrane, the membrane of the basal cell projects deep into the basement membrane as shoe-like structures (Fig. 4.5F). The basal cells appear to move into deeper parts of the epithelium occasionally, but confine their distribution closer to the basement membrane. Once established in the epithelium, the basal cells increase the amount of their cytoplasm. A prominent Golgi apparatus, abundant mitochondria and numerous lysosomes appear (Fig. 4.5G). The plasma membrane of the basal cells produces finger-like processes that interdigitate with similar processes produced by the plasma membrane of the secretory cells (Fig. 4.5B, C).

The occurrence of basal cells in the epithelium of the basal portion of the tubular glands, and their total absence in the epithelium of the tubular portion is interesting. The basement membrane surrounding the tubular portion is thick and does not permit penetration of the progenitor basal cells into the epithelium. On the other hand, the basement membrane surrounding the basal portion is fairly thin and, apparently, permits the progenitor basal cells to penetrate it (George et al. 2004b).

In an animal collected during February (i.e., early regression phase of the testis), the histological organization of the epithelium, particularly of the basal portion of the tubular glands, was indicative of an increase in the abundance of basal cells, and an atypical aggregation of secretion granules

---

Fig. 4.5 contd

branching cytoplasmic processes (arrowheads), which associate with the processes of secretory cells. D: Basal portion of the tubular gland showing basement membrane, connective tissue, a tissue leukocyte and smooth muscle bundles. The basement membrane is folded in (arrowheads) providing passageway for the tissue leukocyte to enter the epithelium. E: Sparse cytoplasmic organelles (arrowhead) appear in the basal cell on its arrival at the epithelium. F: Newly established basal cells are pyramid-shaped and produce shoe-like projections (arrowheads) into the basement membrane. G: A magnified picture of an established basal cell showing prominent cytoplasmic organelles. BB, basal body; BC, basal cell; BL, basement membrane; CC, ciliated cell; CI: cilia; CT, connective tissue; EP, epithelium; ER, endoplasmic reticulum; GA, Golgi apparatus; LY, lysosomes; MI, mitochondria; NU, nucleus; SC, secretory cell; SM, smooth muscle bundles; TL, tissue leukocyte; VA, vacuoles; VE, membrane-bound vesicles. TEM. Scale bars: A, C, E, F = 1.5 µm; B = 3.0 µm; D, G = 1 µm. **A,D-F**. Original. **B, C**. From George, J.M., Smita, M., Oommen, O.V. and Akbarsha, M.A. 2004a. Journal of Morphology 260: 33-56, Figs. 23, 24. G. From George, J.M., Smita, M., Kadalmani, B., Girija, R., Oommen, O.V. and Akbarsha M.A. 2004b. Journal of Morphology 262: 760-769, Fig. 5H.

in the cytoplasm of the secretory cells (Fig. 4.6A). The basal cells accumulated granules, acquiring them from the neighbouring secretory cells (Fig. 4.6B). The granules accumulated as large dense aggregations, in which each granule appeared like a multivesicular body (Fig. 4.6C). Heightened lysosomal activity was indicated, as a result of which the granules became diffuse and were losing their identity (Fig. 4.6D). Such granules, and the aggregates they constituted, appeared like the lipofuscin material present in the epididymal epithelial basal cells (Yeung et al. 1994; Akbarsha et al. 2000). The nuclei of the cells were altered in morphology in such a way as to accommodate the large mass (Fig. 4.6E) in the same manner as basal cells of epididymal epithelium (Akbarsha et al. 2000). A rare picture suggested that such basal cells, carrying a large mass of lipofuscin material penetrate the basement membrane to reach the peritubular tissue (Fig. 4.6F).

The epididymal epithelium of amniotes has basal cells (Robaire and Hermo 1988), which match in organization with the basal cells of *Uraeotyphlus narayani* Mullerian gland. Recent studies have shown that the basal cells arrive at the epididymal epithelium from the peritubular tissue and are derivatives of tissue monocytes (Seiler et al. 1999, Holschbach and Cooper 2002). The basal cells of the mammalian epididymal epithelium are, thus, a migratory cell population and express macrophage antigen (Yeung et al. 1994; Seiler et al. 1999, 2000; Holschbach and Cooper 2002). The basal cells acquire the disintegration products from the overlying principal cells and convert them into lipofuscin material through lysosomal activity (Akbarsha et al. 2000). The lipofuscin material accumulates in the cytoplasm of the basal cell to such a large size that the nucleus is indented at that pole of the cell. Subsequently, the basal cells leave the epithelium, carrying the lipofuscin material (Yeung et al. 1994, Seiler et al. 1999, Akbarsha et al. 2000). The situation occurring in the basal cells of the epithelium of the basal portion of the tubular glands of *U. narayani* Mullerian gland appears to be similar to that in the epididymal epithelium of amniotes. The basal cells play a scavenging role in removing the secretory granules from the epithelium of the basal portion of the tubular glands when the Mullerian gland regresses as an aspect of seasonality of reproduction (Smita et al. 2003, 2004, George et al. 2004a, b). The significance of confinement of basal cells to the basal portion, and their absence from the tubular portion, although partly explained by the nature of the basement membrane at the respective portions, is intriguing.

4.6   ULTRASTRUCTURAL ORGANIZATION OF DUCT EPITHELIUM OF TUBULAR GLANDS

The duct of the tubular glands has not been reported in any other species and it may be an aspect species-specific or was missed earlier. This duct is long and tortuous, which is clearly reflected in the sections along the various planes of the same duct in a single transverse or mid-sagittal section of the Mullerian gland. The epithelium of this duct has two kinds of cells,

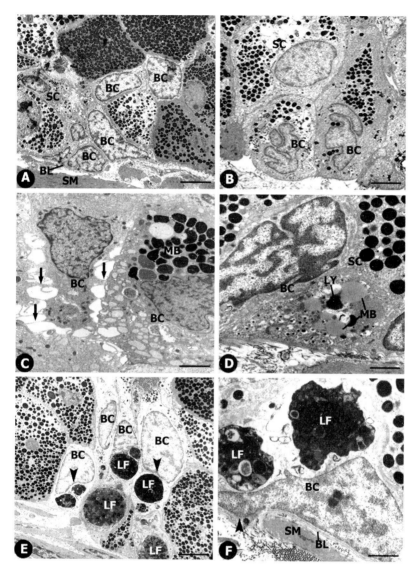

Fig. 4.6. A-F. Different phases in the acquisition of granules by the basal cells in the epithelium of the basal portion of the tubular glands of a regressing Mullerian gland and formation of lipofuscin material. **A**: A few basal cells remain crowded among secretory cells. **B**. The basal cell acquires granules from the secretory cells and accumulates them. **C**. The basal cell contains vesicles, which are comparable to multivescicular bodies. Arrows point to finger-like processes of basal cells. **D**. Association of lysosomes with the multivescicular bodies, resulting in lipofuscin material. **E**. The cells contain dense masses of lipofuscin material. The nucleus is indented to accommodate the lipofuscin material (arrowheads). **F**. A basal cell containing large masses of lipofuscin material appears to move into the peritubular tissue (arrowhead). BC, basal cell; BL, basal lamina; LF, lipofuscin material; LY, lysosome; MB, multivesicular body; SC, secretory cell; SM, smooth muscle bundle. TEM. Scale bars: A = 6 μm; **B-D**, F = 3 μm; E= 4 μm. **D**. From George, J.M., Smita, M., Kadalmani, B., Girija, R., Oommen, O.V. and Akbarsha, M.A. 2004b. Journal of Morphology 262: 760-769, Fig. 6D. **A-C, E, F**. Original.

light and dark (Fig. 4.7A). The dark cells are ciliated, where as the light cells possess microvilli. The light cells possess a large nucleus, occupying almost the entire cell, with limited areas of heterochromatization and one or two nucleoli. The cytoplasm is pale and contains mitochondria, smooth endoplasmic reticulum, polyribosomes, lysosomes, membrane-bound vesicles containing an electron-dense content and vacuoles. This organization of the light cells indicates a role in some kind of processing of the luminal secretory material. A few light cells possess dark dense granules but the latter are different from those in the secretory cells of the tubular glands (Fig. 4.7B). These cells may not be secretory, but endocytic; the granules are likely the product of processing of the endocytosed material. The dark cells possess a centrally or basally located irregular nucleus with dense patches of heterochromatin. The vacuolated cytoplasm is more electron dense than the light cells and abounds with parallel stacks of smooth endoplasmic reticulum, and mitochondria and lysosomes are distributed between these stacks. The apical membrane between the cilia has shallow indentations forming tubular coated pits (Fig. 4.7B). The tall cilia and the abundant mitochondria in the apical cytoplasm of the dark cells indicate a role for these cells in transport of the secretory material from the tubular glands into the central duct.

At the point of transition of the tubular gland into the duct (i.e., at the neck), the lining epithelium is taller than at the duct proper (Fig. 4.7C). The cell types are the same as at the duct proper but both the cell types are taller and possess elongated nuclei. The nuclei of the dark cells are densely heterochromatic whereas those of the light cells are euchromatic. The cytoplasm of the dark cell is thoroughly vacuolated and contains abundant mitochondria around the basal bodies whereas that of the light cell is poorly vacuolated but abounds with mitochondria throughout.

4.7   ULTRASTRUCTURAL ORGANIZATION OF CENTRAL DUCT EPITHELIUM

The cellular organization of the central duct almost matches that of the duct of the tubular gland, except that some of the light cells in the former are also ciliated, and the dark cells possess both cilia and microvilli, which are very unusual attributes (Fig. 4.7D).

4.8   ORGANIZATION OF THE MULLERIAN GLAND DURING THE PERIOD OF TESTICULAR QUIESCENCE

During the period of spermatogenic quiescence, the glandular duct lumen and the central duct are invariably empty (Fig. 4.1C). Although the cellular organization of different parts of the Mullerian gland compares with that of testis during the period of active spermatogenesis, the secretory cells of both the tubular and basal portions lack secretory granules. On the other hand, a few large-sized lightly staining granules and lymphocytes with polymorphic nuclei are present in the epithelium. Thus, in *Uraeotyphlus*

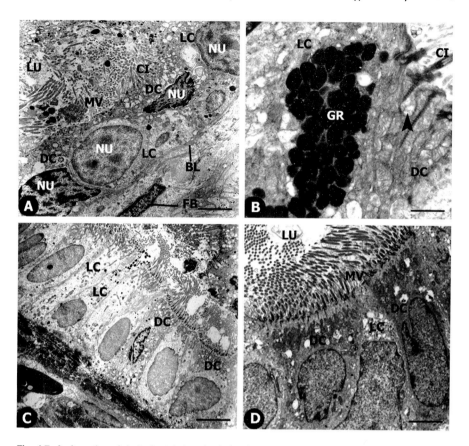

Fig. 4.7. A. A portion of duct of a tubular gland showing light and dark cells. Scale bar = 3.0 μm. **B.** Epithelium of duct of a tubular gland showing a light cell possessing dense granules. On one side is a dark cell with a coated pit (arrowhead) at the base of a cilium. Scale bar = 1.0 μm. **C.** Epithelium of the neck connecting tubular gland and its duct, showing tall epithelium consisting of ciliated dark cells and microvillated light cells. Scale bar = 3.0 μm. **D.** Epithelium of central duct showing dark cells with cilia towards the lumen and light cells with microvilli. BL, basal lamina; CI, cilia; DC, dark cells; FB, fibroblast; GR, dense granules; LC, light cells; LU, lumen; MV, microvilli; NU, nucleus. TEM. Scale bars: A, D = 5 μm; B = 0. 3 μm; C = 2 μm. **A, B, D.** From George, J.M., Smita, M., Oommen, O.V. and Akbarsha, M.A. 2004a. Journal of Morphology 260: 33-56, Figs. 31, 34, 37. **C.** Original.

narayani the structure of the Mullerian gland has a relationship with the structure and spermatogenic activity of the testis, which is also the case in *Typhlonectes compressicauda* (Exbrayat 1985). This relationship suggests a kind of androgen dependence of the Mullerian gland in respect of structure as well as function, an aspect worth investigating.

4.9 BIOCHEMICAL COMPOSITION OF MULLERIAN GLAND SECRETION

Wake (1977, 1981) was the first to attempt to find the biochemical composition of the secretory material of Mullerian gland in *Typhlonectes compressicauda* and *Dermophis mexicanus*. The intracellular secretion granules in these caecilians are rich in monosaccharides, mucopolysaccharides, lipids and acid phosphatase. Exbrayat (1985) reported in *T. compressicauda* two kinds of secretion granules, one which is not PAS⁺ but metachromatic and stained with alcian blue at pH 2.5 indicating it to be protein nature which is sometimes acidic glucosidic, and the other one PAS⁺ but neither metachromatic nor stained by alcian blue indicating it to be a neutral glycoprotein. In *Uraeotyphlus narayani* the secretion granules are strongly PAS⁺, mercuric bromophenol blue⁺ and alcian blue⁺. These reactivities are more prominent in the cells lining the column than the base (Fig. 4.8A). The proteins of the Mullerian gland secretion of *U. narayani* separated out into several fractions of which 14 were prominent (Fig. 4.8B). The fractions ranged in molecular weight from 220 kDa to 3 kDa. Five of these fractions were PAS⁺ and, therefore, glycoproteins (Fig. 4.8C). These observations (due for publication) conclusively prove that i) the Mullerian gland secretion is rich in proteins, and ii) some of them are glycoproteins, attributes fulfilling the requirement as constituents of the seminal plasma (Luke and Coffey 1994).

4.10 HOMOGENEITY OF PROTEINS OF MULLERIAN GLAND SECRETION AND SECRETORY MATERIAL OF LIZARD EPIDIDYMIS AND RAT VENTRAL PROSTATE AND SEMINAL VESICLES

Immunodiffusion, immunoelectrophoretic and immunoblot analyses of the protein extracts of epididymis of the agamid lizard *Calotes versicolor*, and epididymis, ventral prostate and seminal vesicles of Wistar albino rat against *Uraeotyphlus narayani* anti-Mullerian gland serum, raised in rabbit. This revealed that a few protein fractions each of the lizard epididymis, rat ventral prostate and rat seminal vesicles are homogenous to Mullerian gland secretory proteins but not the proteins of rat epididymis (observations due for publication). These results indicate that Mullerian gland secretes proteins that are similar to proteins of the reptilian epididymis (reptiles lack seminal vesicles and prostate gland), and mammalian ventral prostate and seminal vesicles. It adds strength to the hypothesis that the Mullerian gland of the caecilians plays a role equivalent to that of reptilian epididymis and mammalian seminal vesicles and ventral prostate in secreting seminal proteins. The results also support the hypothesis that the Mullerian gland is the progenitor of the prostate gland of mammals.

4.11 CONTRIBUTION OF MULLERIAN GLAND SECRETORY MATERIAL TO MOTILITY OF SPERM

The live spermatozoa of *Uraeotyphlus narayani* released in a physiological solution are pyriform. The sperm has a spatulate acrosome with an acrosome vesicle anteriorly, cylindrical nucleus, slightly expanded midpiece and a long flagellum provided with an undulating membrane on one side. For an account of caecilian sperm, see Scheltinga et al. (2003) and chapter 7 of this volume. There is a lobe of granular cytoplasm, the mitochondrial vesicle (Kouba et al. 2003) or the cytoplasmic droplet (Lee and Jamieson 1992; Scheltinga and Jamieson 2002a, b), occurring anywhere along the length of the sperm from the anterior end of the nucleus to the posterior end of the midpiece, more frequently around the midpiece (Fig. 4.8D). The spermatozoa that were engaged in prolonged motility or ceased motility lacked this structure. The spermatozoa were motile as soon as they were released into the physiological saline solution. Three kinds of motility, slow to rapid forward progression, side-wise lashing of the tail and very rapid corkscrew-like movement of the flagellum, beginning from the midpiece and ending at the tip of the tail, were observed (George et al. 2005). Even spermatozoa from cysts that were not spermiated had flagella that exhibited lashing movement (Fig. 4.8E, F). Therefore, in this species, and perhaps in all caecilians, the spermatozoa do not require to be initiated into motility through a post-testicular mechanism an aspect, which is essential in the mammals (Cooper 1995). Hence, the Mullerian gland secretion may not have any role in post-testicular physiological maturation of sperm. Perhaps, the caecilian sperm do not undergo any post-testicular maturation. However, the observations that the Mullerian gland is profusely secretory as the secretory material is traceable to the central duct of the gland and the latter joins the urinogenital duct in *Uraeotyphlus* sps. (Chatterjee 1936; George et al. 2004a) or the cloaca directly in most of the caecilians (Wake 1977) strongly suggest that the Mullerian gland secretes material concerned with aspects of the semen and sperm physiology.

In order to obtain direct evidence for role of Mullerian gland secretion in the physiology of the sperm, the secretion material of the Mullerian gland was mixed with sperm released from the testis and the sperm motility was analyzed (George et al. 2005). It was found that the structured secretion granules from the central duct of the Mullerian gland when put in a buffer swell to a large size (Fig. 4.8G), and the sperm were enhanced in rate as well as duration of motility when mixed with this preparation. The sperm that were not mixed with Mullerian gland secretion were motile at a rate of 1.82 ± 0.21 μm/sec, for 45.28 ± 3.61 min, where as those mixed with Mullerian gland secretion, the rate was 7.52 ± 0.47 μm/sec and the duration was 600 min and even beyond (George et al. 2005). The data substantiate a role for the secretory material of the Mullerian gland of *Uraeotyphlus narayani* (and perhaps, all caecilians) in motility and longevity of spermatozoa and support the opinions of Tonutti (1931) that caecilian male Mullerian gland secretion would form the medium in which the sperm would be suspended during ejaculation, and Wake (1981) that the secretory material, on being added on to the sperm, would provide nutritional support for the latter for their motility.

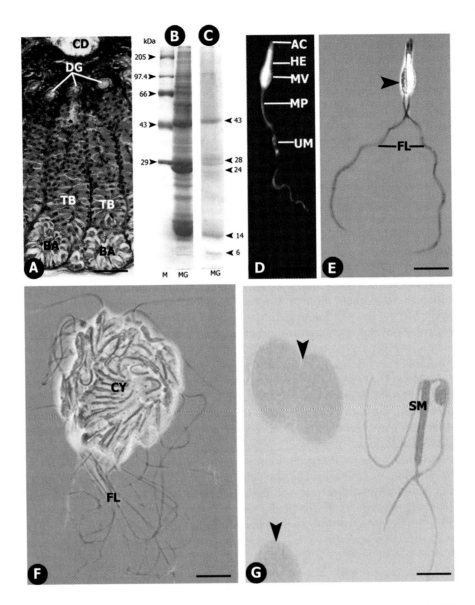

Fig. 4.8. A. Secretory cells of tubular portion (TU) of the tubular gland are strongly reactive to PAS, where as those at the base (BA) are only poorly reactive. Ducts of tubular glands (DG) and central duct (CD) are also shown. PAS and haematoxylin stained paraffin section. Scale bar = 120 µm. **B.** Protein fractions of Mullerian gland secretion in Coomasie brilliant blue stained SDS-PAGE gel. Lane M, standard molecular weight proteins. Lane MG, Mullerian gland proteins. **C.** PAS-stained SDS-PAGE showing the glycoproteic nature of a few of Mullerian gland proteins (MG). **D.** An unstained sperm in dark field illumination showing acrosome (AC) on top of the head (HE), mitochondrial vesicle (MV), midpiece (MP) and undulating membrane of the flagellum (UM). Scale bar = 12 µm. **E.** Unstained sperm

Fig. 4.8 contd

4.12 SUMMARY AND CONCLUSIONS

Thus, in addition to contributing newer knowledge in respect of structure of caecilian Mullerian gland, these series of studies have demonstrated that Mullerian gland secretory material is concerned with ejaculation and sperm physiology. The contributory factors of the male Mullerian gland secretion to motility of sperm could involve the proteins, glycoproteins, monosaccharides, acid phosphatase, etc. (Wake, 1981, George et al. 2005). Each one of these seminal constituents contributes in one way or the other to the sustenance of sperm motility in the amniotes (Luke and Coffey 1994). In mammals the prostate gland and seminal vesicles secrete these various substances and are added on to the sperm only at the time of ejaculation. Therefore, we support the hypothesis of Wake (1981) that the male Mullerian gland of caecilians is an attempt towards secretion of a seminal plasma, a requirement for internal fertilization in the context of a primitive form of terrestrialization, which in the mammals is taken care of by the prostate gland and seminal vesicles. Thus, the Mullerian gland of caecilians is proved to be the male accessory reproductive gland. Tonutti (1931), for the first time, suggested that the Mullerian gland might be a functional analog of the mammalian prostate gland. Wake (1981) attempted to trace the evolution of a part of the mammalian prostate gland from the Mullerian duct. The studies on the protein profile of the Mullerian gland secretion and the antigenic homogeneity of some of these proteins with secretory proteins of male accessory glands of amniotes indicate an evolutionary relationship between caecilian male Mullerian gland and amniote male accessory reproductive glands.

According to George et al. (2005), the medium for sperm transport and its nutritional support for the sperm appeared in the following sequence in the evolution of internal fertilization in vertebrates:

1. The Mullerian gland secreting the substance in the form of discrete granules which partly flocculate after discharge into the lumen and partly remain granular for suspending and supporting the sperm in the caecilians.

Fig. 4.8 contd

in phase-contrast illumination. Maceration resulted in dissociation of sperm held together in pairs or larger aggregations. The photomicrograph is that of a duplex sperm demonstrating lashing of the flagellum (FL). Note the duplex sperm retain the mitochondrial vesicle (arrowhead). Scale bar = 12 μm. **F.** An unstained mature sperm cyst (CY) in phase-contrast illumination showing lashing of the flagellum (FL) of sperm. Scale bar = 12 μm. **G.** Eosin-stained smear of sperm (SM) suspension mixed with the secretory material of the Mullerian gland (arrowheads). Having been prepared long after extraction standing, the granules of the Mullerian gland secretory material have considerably increased in size. Scale bar = 10 μm. **A.** From George, J.M., Smita, M., Oommen, O.V. and Akbarsha, M.A. 2004a. Journal of Morphology 260: 33-56, Fig. 9. **D-G.** From George, J.M., Smita, M., Kadalmani, B, Girija, R., Oommen, O.V. and Akbarsha, M.A. 2005. Journal of Morphology 263: 227-237, Figs. 7B, 6B, 5, 8. B, C. Original.

2. The ductus epididymidis secreting proteins/glycoproteins which are partly solublized and partly as discrete granules. The male accessory glands *viz.*, the ampulla ductus deferentis and renal sex segment also secrete large amounts of mucoid substances and enzymes, all of which play vital roles in the initiation and sustenance of sperm motility in several reptiles.

3. The ductus epididymidis secreting proteins and glycoproteins in solublized form, thereby playing a critical role in the post-testicular physiological maturation of the sperm. They contribute to initiation of sperm motility and acquisition of fertilizing ability. The accessory glands such as the prostate gland and seminal vesicles each secrete a substance with a variety of constituents for forming the medium of transport and for sustenance of sperm motility when in the female tract in the case of mammals.

The packaging of the secretory materials into granules in the caecilian is, perhaps, an adaptation for seasonal reproduction (Smita *et al.* 2003, George *et al.* 2004a, b) If so, it would have a specific purpose of releasing their content in a phased manner commensurate with the requirement of proteins, sugar and enzymes during the storage and/or motility of sperm in the female reproductive tract.

The following intriguing aspects of caecilian Mullerian gland are worth investigating:

1) Whether the gland is active in secretion during the period when the testis is spermatogenetically active and regresses without any trace of secretion during the period of spermatogenic quiescence. As already mentioned, this suggests androgen dependence of the structure and function of the gland. This needs to be established.

2) Whether the Mullerian duct, by its origin and role in the female, is an estrogen-responsive tissue. It would be interesting to find how and when this tissue shifts the responsiveness to androgens in the male caecilians, because this would involve androgen receptors instead of estrogen receptors.

3) It would be pertinent to find the relevance of MIH in the male caecilians.

4) The difference between the histology, histochemistry and ultrastructure of the tubular and basal portions of the tubular glands is also worth examining. This also is true of the secretory cells between the two portions. It would be worth investigating whether the secretory cells of the two portions produce different secretions, with different roles too.

5) Now that the secretory material of the Mullerian gland is established as contributing to sperm transfer as well physiology, it would be worth finding the role of each of the constituents of the secretory material, including the protein fractions, in the physiology of the sperm. The outcome would throw further light on the evolution of the mechanisms

of secretion of seminal proteins in the vertebrates in general and caecilians, in particular.

Currently, the authors are working on clarifying these aspects.

4.13 ACKNOWLEDGEMENTS

We thank John Wiley and Sons, Publishers, Journal of Morphology, for granting permission to reproduce the figures. We acknowledge the ultramicrotome and transmission electron microscope facilities of the Department of Gastrointestinal Sciences, Wellcome Trust Research Laboratory, Christian Medical College and Hospital (CMC&H), Vellore, India. We thank Ms. R. Girija and Mr. B. Kadalmani, Department of Animal Science, Bharathidasan University, Tiruchirappalli, and Dr. M.M. Ibrahim Sha, Department of Zoology, Jamal Mohamed College, Tiruchirappalli, for the technical assistance. The financial assistance from University Grants Commission, New Delhi, India, separately to M.A. Akbarsha and O.V. Oommen, is acknowledged. The FIST grant from DST, Government of India, New Delhi, separately to the laboratories of M.A. Akbarsha and O.V. Oommen, is also acknowledged. We thank Dr. Mark Wilkinson and Dr. David Gower, Natural History Museum, London, and Dr. B.G.M. Jamieson, School of Integrative Biology, University of Queensland, Australia, for academic support.

4.14 LITERATURE CITED

Akbarsha, M. A. and Meeran, M. M. 1995. Occurrence of ampulla in the ductus deferens of *Calotes versicolor* Daudin. Journal of Morphology 225: 261-268.

Akbarsha, M. A and Manimekalai, M. 1999. Histological differentiation along the turtle ductus epididymidis with a note on secretion of seminal proteins as discrete granules. Journal of Endocrinology and Reproduction 3: 36–46.

Akbarsha, M. A., Agnes, V. F. and Girija, R. 2000. Epididymal epithelial basal cell: structure, development and role—a review. Journal of Endocrinology and Reproduction 4: 1–12.

Aumuller, G. and Seitz, J. 1990. Protein secretion and processes in male sex accessory glands. International Review of Cytology 121: 127-231.

Bishop, J. E. 1959. A histological and histochemical study of the kidney tubule of the common garter snake *Thamnophis sirtalis* with special reference to the sexual segment in the male. Journal of Morphology 104: 307-357.

Burgoyne, R. D., Morgan, A. 2003. Secretory granule exocytosis. Physiological Reviews 83: 581–632

Chatterjee, B. K. 1936. The anatomy of *Uraeotyphlus menoni* Annandale. Part I. The digestive, circulatory, respiratory and urinogenital systems. Anatomische Anzeiger 81: 393-414.

Cohen, R. J. 2001. Characterization of cytoplasmic secretory granules (PSG) in prostatic epithelium and their transformation-induced loss in dysplasia and adenocarcinoma. Human Pathology 29: 1488–1494.

Cooper, T. G. 1995. Role of the epididymis in mediating changes in the male gamete during maturation. Advances in Experimental Medicine and Biology 377: 87-101.

Daisy, P., Meeran, M. M. and Akbarsha, M. A. 2000. Histological variation along the vas deferens of the Indian garden lizard *Calotes versicolor* Daudin with special reference to the ampulla ductus deferentis. Journal of Endocrinology and Reproduction 4: 74-88.
Depeiges, A. and Dufaure, J. P. 1977. Secretory activity of the lizard epididymis and its control by testosterone. General and Comparative Endocrinology 33: 473-479.
Depeiges, A. and Dufaure, J. P. 1980. Major proteins secreted by the epididymis of *Lacerta vivipara*. Isolation and characterization by electrophoresis of the central core. Biochemica et Biophysica Acta 628: 109–115.
Depeiges, A. and Dufaure, J. P. 1981. Major proteins secreted by the epididymis of *Lacerta vivipara*. Identification by electrophoresis of soluble proteins. Biochemica et Biophysica Acta 667: 260-266.
Dufaure, J. P. and Saint-Girons, H. 1984. Histologie comparée de épididyme et ses sécrétions chez les reptiles (lézards et serpents). Archives de Microscopie 73: 15-26.
Exbrayat, J.-M. 1985. Cycle des canaux de Müller chez le mâle adulte de *Typhlonectes compressicaudus* (Dumeril et Bibron, 1841), Amphibien apode. Comptes Rendus des Séances de l'Académie des Sciences de Paris 301: 507-512.
Exbrayat, J.-M. 1986. Quelques aspects de la biologie de la reproduction chez *Typhlonectes compressicaudus,* Amphibien Apode. Doctorat ès Sciences Naturelles. Université Paris VI, France.
George, J. M., Smita, M., Oommen, O. V. and Akbarsha, M. A. 2004a. Anatomy, histology and ultrastructure of male Mullerian gland of *Uraeotyphlus narayani* (Amphibia: Gymnophiona). Journal of Morphology 260: 33-56.
George, J. M., Smita, M., Kadalmani, B., Girija, R., Oommen, O. V. and Akbarsha, M. A. 2004b. Secretory and basal cells of the epithelium of the tubular glands in the male Mullerian gland of the caecilian *Uraeotyphlus narayani* (Amphibia: Gymnophiona). Journal of Morphology 262: 760-769.
George, J. M., Smita, M., Kadalmani, B., Girija, R., Oommen, O. V. and Akbarsha, M. A. 2005. Contribution of the secretory material of caecilian (Amphibia: Gymnophiona) male Mullerian gland to motility of sperm: a study in *Uraeotyphlus narayani.* Journal of Morphology 263: 227-237.
Gesase, A. P. and Satoh, Y. 2003. Apocrine secretory mechanism: recent findings and unresolved problems. Histology and Histopathology 18: 597– 608.
Holschbach, C. and Cooper, T. G. 2002. A possible extra-tubular origin of epididymal basal cells in mice. Reproduction 123: 517–525.
Kouba, A. J., Vance, C. K., Frommeyer, M. A. and Roth, T. L. 2003. Structural and functional aspects of *Bufo americanus* spermatozoa: effects of inactivation and reactivation. Journal of Experimental Zoology. Part A. Comparative Experimental Biology 295: 172-182.
Lee, M. S. Y. and Jamieson, B. G. M. 1992. The ultrastructure of the spermatozoa of three species of myobatrachid frogs (Anura, Amphibia) with phylogenetic considerations. Acta Zoologica 73: 213-222.
Luke, C. M. and Coffey, D. S. 1994. The male sex accessory tissue: structure, androgen action and physiology. Pp. 1435-1487. In E. Knobil and J. D. Neill (eds), *The Physiology of Reproduction.* Raven Press Ltd., New York.
Manimekalai, M. and Akbarsha, M. A. 1992. Secretion of glycoprotein granules in the epididymis of the agamid lizard *Calotes versicolor* (Daudin) is region-specific. Biological Structures and Morphogenesis 4: 96–101.
Onitake, K., Takai, H., Ukita M., Mizuno, J., Sasaki, T., and Watanabe, A. 2000. Significance of egg-jelly in the internal fertilization of the newt *Cynops pyrrhogaster.* Comparative Biochemistry and Physiology 126B: 121-128.

Robaire, B. and Hermo, L. 1988. Efferent ducts, epididymis and vas deferens: Structure, functions and their regulation. Pp. 999-1080. In E. Knobil and J. D. Neill (eds), *The Physiology of Reproduction.* Raven Press Ltd., New York.
Rudolf, R., Salm, T., Rustom, A. and Gerdes, H. H. 2001. Dynamics of immature secretory granules: role of cytoskeletal elements during transport, cortical restriction, and F-actin dependent tethering. Molecular Biology of Cell 12: 1353-1365.
Sarkar, H. B. D. and Shivanandappa, T. 1989. Reproductive cycles of reptiles. Pp. 224-271. In S. K. Saidapur (ed.), *Reproductive Cycles of Indian Vertebrates.* Allied Publishers Ltd., New Delhi, India.
Scheltinga, M. D. and Jamieson, B. G. M. 2002a. The mature spermatozoon. Pp. 201-274. In B. G. M. Jamieson (series ed), D. M. Sever (volume ed.). *Reproductive Biology and Phylogeny of Urodela.* Science Publishers, Inc., Enfield, New Hampshire.
Scheltinga, M. D. and Jamieson, B. G. M. 2002b. Spermatogenesis and the mature spermatozoon: form, function and phylogenetic implications. Pp. 192-251. In B. G. M. Jamieson (series and volume ed)., *Reproductive Biology and Phylogeny of Anura.* Science Publishers, Inc., Enfield, New Hampshire.
Scheltinga, D. M., Wilkinson, M., Jamieson, B. G. M. and Oommen, O. V. 2003. Ultrastructure of the mature spermatozoa of caecilians (Amphibia: Gymnophiona). Journal of Morphology 258: 179-192.
Seiler, P., Cooper, T. G., Yeung, C. H. and Nieschlag, E. 1999. Regional variation in macrophage antigen expression by murine epididymal basal cells and their regulation by testicular factors. Journal of Andrology 20: 738–746.
Seiler, P., Cooper, T. G. and Nieschlag, E. 2000. Sperm number and condition affect the number of basal cells and their expression of macrophage antigen in the murine epididymis. International Journal of Andrology 23: 65–76.
Setchell, B. P., Maddocks, S. and Brooks, D. E. 1994. Anatomy, vasculature, innervation and fluids of the male reproductive tract. Pp. 1063-1175. In E. Knobil and J. D. Neill (eds)., *The Physiology of Reproduction.* Raven Press Ltd., New York
Sever, D. M., Hamlett, W. C., Slabach, R., Stephenson, B. and Verrell, P. A. 2002. Internal fertilization in the Anura with special reference to mating and female sperm storage in *Ascaphus.* Pp. 319-341. In B. G. M. Jamieson (series and volume ed.). *Reproductive Biology and Phylogeny of Anura.* Science Publishers, Inc., Enfield, New Hampshire.
Smita, M., Oommen, O. V., George, J. M. and Akbarsha, M. A. 2003. Sertoli cells in the testis of caecilians, *Ichthyophis tricolor* and *Uraeotyphlus* cf. *narayani* (Amphibia: Gymnophiona): Light and electron microscopic perspective. Journal of Morphology 258: 317-326.
Smita, M., Oommen, O. V., Jancy, M. G. and Akbarsha, M. A. 2004. Stages in spermatogenesis in two species of caecilians, *Ichthyophis tricolor* and *Uraeotyphlus* cf. *narayani* (Amphibia: Gymnophiona): Light and electron microscopic study. Journal of Morphology 261: 92-104.
Spengel, J. W. 1876. Das Urogenital system der Amphibien. I. Theil Der annatomische Bau des Urogenitalsystems. Arbeitenaus demm Zoologyzootom. Institute in Wurzburg 3: 51-114.
Tonutti, E. 1931. Beitrag zur Kenntnis der Gymnophionen XV. Das Genital-system. Morphologisches Jahrbuch 68: 151-292.
Wake, M. H. 1970. Evolutionary morphology of the caecilian urogenital system. Part II. The kidneys and urogenital ducts. Acta Anatomica 75: 321-358.
Wake, M. H. 1977. Reproductive biology of caecilians: an evolutionary perspective. Pp. 73-102. In S. Guttman and D. Taylor (eds), *The Reproductive Biology of the Amphibia.* Plenum Press, New York.

Wake, M. H. 1981. Structure and function of the male Mullerian gland in caecilians, with comments on its evolutionary significance. Journal of Herpetology 126: 291-332.

Wake, M. H. and Dickie, R. 1998. Oviductal structure and function and reproductive modes in amphibians. Journal of Experimental Zoology 282: 477-506.

Yeung, C. H., Nashan, D., Sorg, C., Oberpenning, F., Schulze, H., Nieschlag, E. and Cooper, T. G. 1994. Basal cells of the human epididymis: antigenic and ultra-structural similarities to tissue-fixed macrophage. Biology of Reproduction 50: 917–926.

CHAPTER 5

Endocrinology of Reproduction in Gymnophiona

Jean-Marie Exbrayat

5.1 INTRODUCTION

Endocrine organs of Gymnophiona have received little attention. Some works have been devoted to endocrine organs as interrenal glands (Rathke 1852; Semon 1892; Brauer 1902; Dittus 1936; Gabe 1971; Bhatta 1987; Masood Parveez 1987; Dorne and Exbrayat 1996), adenohypophysis (Stendell 1914; Laubmann 1926; Pillay 1957; Pasteel and Herlant 1962; Gabe 1972; Schubert et al. 1977; Schubert and Welsch 1979; Exbrayat 1989b, 1992a, b; Exbrayat and Morel 1990-1991, 1995), pineal gland (Leclercq 1995; Leclercq et al. 1995), thyroid (Maurer 1887; Sarasin and Sarasin 1887-1890; Klump and Eggert 1935; Welsch et al. 1974, 1976), parathyroid glands and ultimobranchial bodies (Klump and Eggert 1935; Welsch and Schubert 1975).

Concerning endocrine regulation of reproduction, some works have been devoted to correlations between male and female sexual cycles and variations of endocrine tissues such as Leydig cells in testes, ovarian follicles in the female and corpora lutea during pregnancy in females of viviparous species. Correlations have also been studied between gonads and other sexual organs and endocrine glands such as the hypophysis or interrenal tissue. More recent works have also shown variations of prolactin receptors in *Typhlonectes compressicauda* sexual tissues (Exbrayat et al. 1996a, b; Exbrayat and Morel 2003).

In this chapter are described sexual cycles in male and female Gymnophiona, with acquisition of sexual maturity; variations of gonadal endocrine tissues in both males and females; and the structure of the hypophysis and interrenal organs correlated to variations observed in sexual organs.

Laboratoire de Biologie générale de l'Université catholique de Lyon and Laboratoire de Reproduction et Développement des Vertébrés, 25 rue du Plat, 69288 Lyon Cedex 02, France.

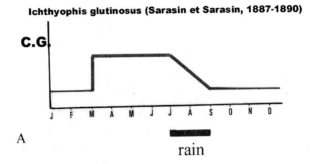

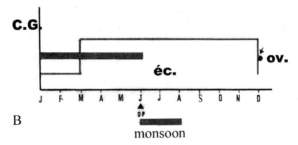

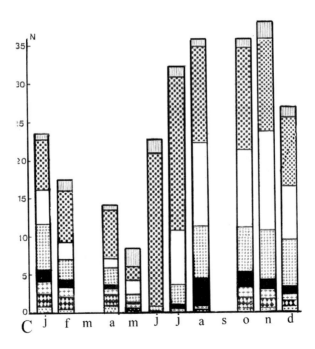

Fig. 5.1 contd

5.2 VARIATIONS OF SEXUAL ORGANS IN MALES

5.2.1 Testes

In Gymnophiona, males exhibit discontinuous sexual cycles not only in Asiatic species (*Ichthyophis*, Sarasin and Sarasin 1887-1890; Seschachar 1936, 1937, 1943; Bhatta 1987) but also in South and Central American species (*Gymnopis multiplicata*, Wake 1968a, b; *Dermophis mexicanus*, Wake 1980, 1995; *Typhlonectes compressicauda*, Exbrayat and Sentis 1982; Exbrayat and Delsol 1985; Exbrayat 1986a, b; *Chthonerpeton indistinctum*, De Sa and Berois 1986; and African species (*Boulengerula taitanus*, Malonza and Measey 2005).

In *Ichthyophis glutinosus*, from Sri Lanka (Fig. 5.1A), only a few germ cells were observed in the testes during winter. In spring, they underwent division and ultimately differentiated into spermatozoa that were evacuated during next summer, in the breeding season. In an *Ichthyophis* (named *I. glutinosus* but certainly *I. beddomei*, after Taylor 1968) from Mysore (India), Seshachar observed spermatogenesis in March; evacuation of spermatozoa occurred in December and lobules remained empty until March when a new spermatogenetic cycle commenced. In the Indian species *Ichthyophis beddomei*, Bhatta (1987) showed a discontinuous spermatogenesis; the breeding season was January-February (Fig. 5.1B).

In *Gymnopis multiplicata*, from Costa Rica, testes contained only several spermatogonia from December to March. Spermatogenesis occurred in June, and by October, all germ cells categories were observed in the lobules, the breeding season being December-January (Wake 1968b). In *Dermophis mexicanus*, from Guatemala, spermatogenesis remained active for most of the year and breeding occurred from March to June. During this period, testes regressed (Wake 1980, 1995).

In *Typhlonectes compressicauda*, from French Guyana, yearly variations were observed (Exbrayat and Sentis, 1982; Exbrayat, 1986a, b; chapter 3 of this volume) (Fig. 5.1C). Lobes and lobules are at maximal size in December-

Fig. 5.1 contd

Fig. 5.1. Schematic representations of breeding cycles in several Gymnophiona. **A.** Male *Ichthyophis glutinosus* from Sri Lanka (from data given in Sarasin and Sarasin 1887-1890); germ cells (GC) fill the testis from March to July, then they are evacuated during breeding period (July until September); testis last empty until next March. cg., germ cells. **B.** Male and female *Ichthyophis beddomei* from Sringeri (India) (from data given in Bhatta 1987) or *Ichthyophis* sp. from Mysore (India) (from data given in Seshachar 1936); in both the species, male germ cells (GC) fill the testis from March to December, period of breeding; in females, vitellogenesis (thick grey line), ovulation (ov) occurs in December, after egg laying (ec) in July, embryos develop during monsoon. **C.** Male sexual cycle in *Typhlonectes compressicauda*; spermatogenesis occurs from June to August; period of sexual quiescence is observed from August until December at which evacuation of germ cells begins; from December to May-June, a spermatogenesis is also observed to reconstitute the stock of post-meiotic germ cells. ▦ primary spermatogonia; ▦ secondary spermatogonia; ☐ primary spermatocytes; ▦ secondary spermatocytes and young spermatides; ■ old spermatides; +++ young spermatozoa; ▦ mature spermatozoa; ▦ free spermatozoa. N: number of germ cells/µm² of section. From Exbrayat, J.-M. 1986a. Unpublished D. Sci. Thesis, Université Paris VI, France, Fig. 30.

January, just at the beginning of the breeding season; from February until May, they decrease in size in conjunction with evacuation of spermatozoa. Then, progressive growth has been observed until December. In this species, spermatogenesis starts in June with major multiplication of secondary spermatogonia and their division to give primary spermatocytes. In July, spermatids and the first spermatozoa appear. In August, a new stock of germ cells develops in the testes and remains constant until November. From December to May, the breeding season, the total number of germ cells progressively decreases; numerous spermatozoa are seen in evacuation ducts. However, the number of postmeiotic cells remains constant from August to February and even April, suggesting that evacuated spermatozoa are replaced by development of sperm from lower categories. In April-May, the last spermatozoa are evacuated and the lowest stock of germ cells is reached. By means of ^{3}H thymidine injection, the male sexual cycle has been elucidated in *T. compressicauda*. The cycle is divisible into three phases. From June to August, spermatogenesis occurs and the stock of germ cells is formed. From August to December, during the period of sexual inactivity, this stock remains stable, then from December until May, is the breeding season, characterised by both evacuation of sperms and spermatogenesis. This cycle is closely linked to seasonal variations: breeding during the rainy season and sexual quiescence during the relatively dry season.

5.2.2 Mullerian Glands

In Gymnophiona, Mullerian ducts persist in adult males. Spengel (1876) was the first to verify their presence in several species. Subsequently, other authors (Wiedersheim 1879; Sarasin and Sarasin 1887-1890; Semon 1892) described them in other Gymnophiona but could not attribute a function to them. For Tonutti 1931, these organs were glandular structures. For Marcus 1939, they were auxiliary testes. Other authors considered them to be rudimentary organs (Oyama 1952; Lawson 1959). The glandular structure of Mullerian ducts has since been studied in several species (Wake 1970, 1977b, 1981; Exbrayat 1985, 1986a; George *et al.* 2004a, b, 2005, chapters 3 and 4 of this volume).

Mullerian ducts (or Mullerian glands) are paired organs that are parallel to kidneys. At their posterior end, they connect to the cloaca. Two main morphologies have been observed in different species by Wake (1970a). In all types, Mullerian ducts are constituted by more or less voluminous glands that connect to a central duct. George *et al.* (2004a, b, 2005) recently published an utrastructural study of these organs in the Asiatic species *Uraeotyphlus narayani* (see chapters 3 and 4 of this volume).

In *Typhlonectes compressicauda*, each Mullerian duct presents important variations throughout the year. Its diameter increases in November, is maximal from January to April and decreases in May. The number and appearance of gland cells varies. In January, each gland is bordered by a stratified ciliated epithelium with an apical nucleus and gland cells with a basal nucleus in which abundant glycoproteic secretions are observed (Fig.

5.2A, B). The central duct of the Mullerian gland is ciliated. In February-April, gland cells of posterior and median parts of the Mullerian duct are empty. The central duct, always ciliated, contains secretions and cell fragments (Fig. 5.2C). In May, it is occluded; gland ducts are no longer ciliated, and glandular structures have regressed, being reduced by June to small expansions with only one cell type. At this period, connective tissue is relatively abundant. From July until October, glandular structures remain poorly developed (Fig. 5.2D) but by October, the central duct and gland ducts become ciliated. Two cell types are differentiated. In certain gland cells, glucidic substances have been observed. In November, the Mullerian ducts present the characteristic appearance of the breeding season.

Wake (1981) has analysed the chemical nature of Mullerian secretions in *Dermophis mexicanus* and *Typhlonectes compressicauda*. These secretions contain mucopolysaccharides, fructose, acid phosphatase. pH is acidic and resembles the pH of sperm. The Mullerian ducts are the equivalents of prostates in mammals (George *et al.* 2004a, b, 2005, chapter 4 of this volume). The cycle of these organs is precisely correlated with that of the testes.

5.2.3 Cloaca

Internal fertilisation occurs in all Gymnophiona. The male cloaca possesses an intromittent organ, the phallodeum. Very few works have been devoted to this organ and its variations during sexual cycle (Tonutti 1931, 1932, 1933; Taylor, 1968; Wake 1972; Bons 1986; Exbrayat 1991, 1996). The cloaca is linked to dorsal organs of the body. It receives the posterior gut and, anteriorly, receives the Wolffian and Mullerian ducts after these have united on each side. The bladder is a ventral expansion of the cloaca. The posterior part of the male cloaca is the phallodeum. The median part, in several species, may put out a pair of "blind sacs" (*Ichthyophis kohtaoensis*, *Uraeotyphlus narayani*, *Herpele squalostoma*, *Typhlonectes compressicauda*). In species such as *Scolecomorphus kirkii*, *Microcaecilia unicolor* and *Geotrypetes seraphinii*, only lateral folds are present on each side of the central duct (Wake 1972; Exbrayat 1991, 1996) (Fig. 3.16D, E, F, chapter 3 of this volume). Cyclic variations of cloacal anatomy have been observed in *Typhlonectes compressicauda* (Exbrayat 1996); the total length of the cloaca (measured from insertion of Wolffian and Mullerian ducts to the posterior part) is maximal during the breeding season and minimal during the period of sexual inactivity. Histological aspects also varied according to the sexual cycle.

During the breeding season, January to April, the phallodeum wall consists of a series of expansions. The epithelium between expansions is only one cell layer thick. The epithelium of the expansions is stratified and may be composed of as many as 14 cell layers; its basal cells resemble the cells situated between the expansions but have an elongate nucleus (Fig. 5.2F). Other cells possess a rounded nucleus. The apical layer consists of keratinised cells with a degenerated nucleus that may be eliminated. In

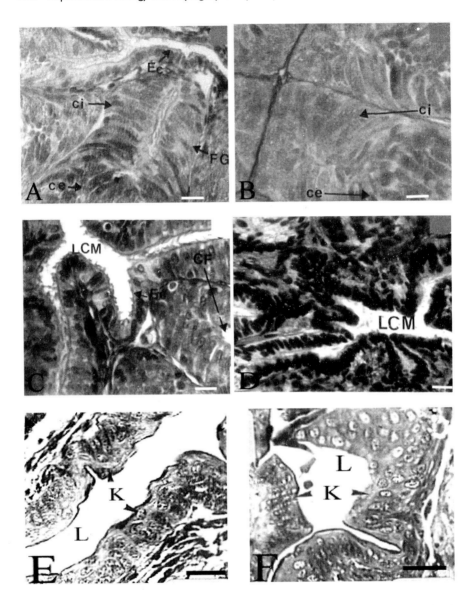

Fig. 5.2. *Typhlonectes compressicauda*, adult male. **A.** Transverse section (TS) of Mullerian duct in January (beginning of breeding period). Scale bar = 30 µm. **B.** detail of a gland tubule in January. Scale bar = 30 µm. **C.** TS of Mullerian gland in April (breeding period). Scale bar = 30 µm. **D.** TS of Mullerian gland in August (resting period). Scale bar = 30 µm. **E.** Posterior part of cloaca in male during sexual quiescence. Scale bar = 30 µm. **F.** Posterior part of cloaca in male during breeding. Scale bar = 30 µm. ce, gland cell on the peripheral (external) part of the Mullerian duct; CF, evacuation ductule; ci, gland cell on the central (internal) part of the Mullerian duct; Ec, ciliated epithelium; FG, gland tubule; K, keratinised cells; L, lumen of cloaca; LCM, central lumen of Mullerian duct. **A, B, C, D:** from Exbrayat, J.-M. 1986a, Unpublished D. Sci. Thesis, Université Paris VI, France. Pl. VII, Figs. 5-7, Pl. VIII, Fig. 4. **E, F,** from Exbrayat, J.-M. 1996. Bulletin de la Société Zoologique de France 121: 99-104, Figs. 2, 3.

May, only two cells layers are observable in the epithelium and keratinised substances have been observed in the central duct in which residues originating from the wall are also present. At this period, the thickness of the epithelial wall is the same throughout the phallodeum and expansions are no longer present (Fig. 5.2).

During sexual inactivity the epithelium retains two to three layers of cells. Until August, small expansions are observed on the wall. Cells of basal layer migrate towards the duct lumen. Connective tissue is very vascular with numerous cells. In October, expansions are more developed with two or three cells layers, apical ones being keratinised. The phallodeum is ready for breeding.

During the breeding season, the median region of the cloaca is covered by a pseudostratified epithelium with ciliated cells and several goblet cells secreting acid mucus. Basal cells are hyaline with an optically empty cytoplasm and a rounded nucleus. Numerous blood vessels are observed in the surrounding connective tissue. In January, at the beginning of the reproductive season, the walls of the blind sacs are composed of folds with ciliated crests. In certain zones, the epithelium is stratified, with an apical layer of narrow ciliated cells with elongate nuclei. The deep layer contains ovoid cells, more hyaline than the others, and inserted between other cells. Small undifferentiated cells originating from connective tissue are present between the other cell types. In February, the ciliated cells of the central duct are replaced by gland cells with abundant acid mucus. The blind sacs become very glandular, cilia are shorter than previously and goblet cells are increasingly numerous. Replacement cells arise from connective tissue. Blood vessels become less abundant in this. At the end of the breeding season, the epithelia covering the blind sacs and the central duct are no longer ciliated and mucous substances are very rare. During sexual quiescence, the epithelium covering the blind sacs and central duct walls becomes undifferentiated. Large cells originating from the wall, and later small mucous and ciliated cells are observable in the lumen. In August, renewed development of the wall occurs.

The anterior part, the ampulla, is glandular and resembles the posterior region of the intestine, with numerous goblet cells with an acid mucus.

Thus, the male genital tract presents major variations during the sexual cycle and the cycle itself is subject to seasonal variations (Exbrayat and Flatin 1985; Exbrayat *et al.* 1998), testes, Mullerian ducts and the cloaca are target organs for hormones implicated in regulation of reproduction.

5.2.4 Interstitial Tissue in the Testes

Interstitial tissue is known to have a steroidogenic activity and to be implicated in regulation of reproduction. In Gymnophiona, this tissue is situated within the connective tissue that separates the lobules of the testes (Fig. 5.3A). Leydig cells are united in groups which resemble those of the Anura. The cytoplasm of these cells is uniformly grey after staining by Masson Goldner trichroma. Biometric variations affect this tissue according to the sexual cycle. In *Gymnopis multiplicata*, Wake (1968b) has observed a well developed interstitial tissue when the lobules contain only primary spermatogonia as germ cells during the breeding season (December to January). The tissue then decreases during spermatogenesis.

Variations of interstitial tissue have been studied in *Typhlonectes compressicauda* (Exbrayat 1986a, b; Anjubault and Exbrayat 1998, 2004). From November to February, the interstitial tissue area increases and reaches, for each month, 22 to 25% of the total surface of the section. Each cell (25 µm in diameter) is polyhedral with a central nucleus (5 µm in diameter). In April, the relative surface of interstitial tissue reaches 27% of the total surface and the cells having reached their maximal size (Fig. 5.3B). Between May and August, the interstitial tissue is minimal. From May to June, the period at which the germ cells consist only of primary and secondary spermatogonia, the interstitial tissue nevertheless occupies 24 to 28% of the total surface, probably linked to a decrease in total size of the lobule section. In July and August, the period of sexual inactivity, the tissue surface represents only 18 to 20% of the entire surface. In May, cells are diminishing, with reduced cytoplasm; in June, cells are flattened with a small nucleus and a cytoplasm poor in lipids. Interstitial tissue decreases until October (Fig. 5.3C).

δ5 3β hydroxysteroid dehydrogenase (δ5 3β HSDH) activity has been detected in several testes sections of animal caught in April (Fig. 5.3D, E). During the breeding season, February, immunocytochemistry using AEC as a chromophore revealed a substantial presence of testosterone (Anjubault and Exbrayat 1998, 2004) in Leydig cells. In April cells with well developed cytoplasm were also labelled. In May and June, the label appeared less strong than previously. At this period, orange staining indicating the presence of anti-testosterone hormone was also found in follicle cells surrounding spermatogonia. In November, labeling was slight in the Leydig cells.

5.2.5 Acquisition of Sexual Maturity in Males

Acquisition of sexual maturity has been poorly studied in Gymnophiona. Several works, some of them old, concern embryonic development of the gonads (Spengel 1876; Brauer 1902; Tonutti 1931; Seshachar 1936; Wake 1968a, b). Development of the male genital tract from birth to sexual maturity has been studied only in *Dermophis mexicanus* (Wake 1980) and *Typhlonectes compressicauda* (Exbrayat 1986a, b; Exbrayat and Dansard, 1994). In both species at birth, each testis resembles a string on which lobes develop. After one year, commencement of spermatogenesis has been observed in 90% of male *Dermophis mexicanus*. After two years, testes are active but lobes still small with only a few lobules in which true spermatogenesis is seen. It seems that *Dermophis mexicanus* is able to breed when three years old. In *Typhlonectes compressicauda*, at birth, only spermatogonia, somatic cells that surround germ cells, and connective tissue are observable, neither Sertoli cells nor Leydig cells being observed.

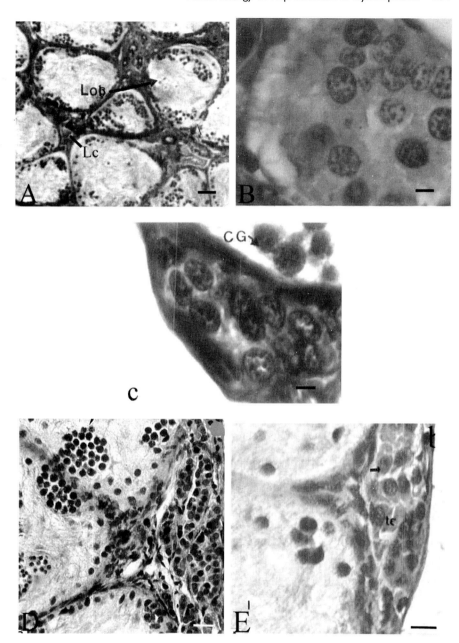

Fig. 5.3. *Typhlonectes compressicauda*, adult male. **A.** TS of a testis (June, sexual quiescence). Scale bar = 90 µm. **B.** Leydig-like cells (April, breeding).Scale bar = 10 µm. **C.** Leydig-like cells (July, sexual quiescence). Scale bar = 10 µm. **D.** development of Leydig-like cells during breeding period. Scale bar = 30 µm. **E.** Visualisation of testosterone (te) in Leydig-like cells (arrow) during breeding period. Scale bar = 15 µm. Lc, Leydig-like cell; Lob, lobule. **A, D, E,** original. **B, C,** from Exbrayat, J.-M. 1986a. Unpublished D. Sci. Thesis, Université Paris VI, France. Pl. VI, Figs 1, 3.

Using an anti-testosterone serum, this hormone could be detected in connective tissue (Anjubault and Exbrayat, chapter 9 of this volume).

In *Typhlonectes compressicauda* males that are one-year-old, lobules are developed with Sertoli and Leydig cells. Germ cells are essentially spermatogonia. In July-August of the same year, it has been possible to observe some primary and secondary spermatocytes and even, in a few individuals, some spermatids and spermatozoa. Males that were two years old seem to be ready for breeding. At this period, Leydig cells contain testosterone as indicated by anti-testosterone serum. In parallel, Mullerian glands and the cloaca are not differentiated during the first year of life. Morphological differentiation resembling that of adults is observed in two years old animals. These data show that sexual maturity is acquired in two-year-old males, i.e. during their third year of life, as in *Dermophis mexicanus*.

5.3 CYCLIC VARIATIONS OF SEXUAL ORGANS IN FEMALES

In Gymnophiona, about 50% of species are viviparous (Wake 1977b). Cycles in oviparous and viviparous species will now be considered.

5.3.1 Ovaries

5.3.1.1 Oviparous species

Female breeding cycle of the oviparous *Ichthyophis beddomei* has been studied by Masood Parveez (1987) and Masood Parveez and Nadkarni (1993a, b). The female sexual cycle is yearly and discontinuous. Oogonia are produced throughout the year but undergo division only in January-February. Small oocytes surrounded by one layer of follicle cells are present throughout the year. The mean diameter of follicles progressively increases to August, when the first vitellogenic oocytes appear. The size of follicles also increases, 37 or 38 times, and reaches 750 µm in diameter. From December till February, the breeding season, numerous follicles are laid in the oviducts. From March until July, the ovary regresses; a few oocytes with yolk platelets are still present but they rapidly degenerate (Fig. 5.4A). Eggs are laid on the ground by females in December and January, just after internal fertilisation. Cycles of others oviparous species resemble that of *Ichthyophis beddomei*. In *I. glutinosus*, in Sri Lanka, copulation and egg-laying have been observed from May to September, during the wet season (Sarasin and Sarasin 1887-1890, Breckenridge and Jayasinghe 1979). In Mysore, breeding of an *Ichthyophis* was observed from December to March during the dry winter; hatching then occurred in May-June, at the first rains (Seshachar 1936).

5.3.1.2 Viviparous species

Some viviparous Gymnophiona have a short pregnancy. In *Chthonerpeton indistinctum*, pregnancy is four months long according to Barrio (1969). Breeding occurs at the end of August and beginning of September. Births of young animals have been observed in January-February. Other viviparous species that have been well studied had a long gestation (*Dermophis mexicanus*, Wake 1980; *Typhlonectes compressicauda*, Exbrayat et al. 1981; Exbrayat 1983, 1986a, 1988a; *Gymnopis multiplicata*, Wake, personal communication). In these females, sexual cycles are biennial.

Female *Dermophis mexicanus* carrying embryos and females with a regressive tract have been observed throughout the second year in a biennial cycle (Wake 1980). In June (commencement of cycle), certain females possess ovaries with oocytes at maximal size, others contain very young embryos. Corpora lutea are observed throughout pregnancy (Wake 1977a, b). Parturition occurs at the end of the first year in May-June when numerous small animals have been observed in the field. The female genital tract then atrophies and ovaries no longer contain mature follicles. The second year of the cycle is a period of sexual quiescence (Fig. 5.4B).

In *Typhlonectes compressicauda*, the female sexual cycle, though again biennial, differs markedly from that of *Dermophis mexicanus*. From October (commencement of cycle) to January, the number of females with vitellogenic oocytes increases. In December-January, 80 to 100% females are vitellogenic. In February, 36% females are pregnant and 5 to 6% are still vitellogenic. In April, 66% females are pregnant, 7% possessing mature follicles. In May-June, 50 to 60% females carry embryos. Other females exhibit a regressive genital tract. From July to October, the ratio of pregnant

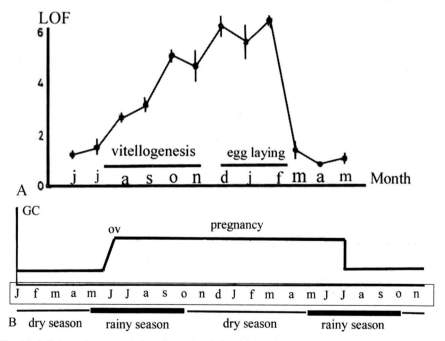

Fig. 5.4. A. Schematic representation of sexual cycle in adult female *Ichthyophis beddomei* (from June to May). LOF: largest ovarian follicle. **B.** Schematic representation of sexual cycle in adult female *Dermophis mexicanus* (from January of the first year to November of the second one). ov, ovulation. **A**, modified after Masood-Parveez, U. 1987, PhD thesis, Karnatak University, Darwad, India. **B**, after data of Wake, M. H. 1980. Herpetologica 36: 244-256.

females decreases. At this stage, numerous small animals have been observed in the field. Thus the first year of the cycle starts in October and vitellogenesis occurs until January. Ovulation and fertilisation occur from February to April. The duration of pregnancy is estimated to be 6 to 7 months (Fig. 8.3, chapter 8 of this volume). After parturition, renewed vitellogenesis occurs from October to January of the second year, but ovulation does not occur and the ovaries diminish, many atretic follicles being observed. Sexual inactivity extends to the following October at which a new vitellogenesis occured, starting a new two years long cycle (Exbrayat 1983, 1986b).

In all observed *Typhlonectes compressicauda* females, dividing oogonia and premeiotic oocytes are observed in germinal cysts. These nests are more particularly developed in February to April, just after ovulation. At this period, the increase in numbers of oocytes is correlated with that of follicles. In all pregnant or non-pregnant females, atretic follicles are observed, suggesting that the stock of oocytes is renewed throughout the year. It seems that there is a continuous evolution of oocytes and follicles throughout the life of the animal. When follicles and oocytes have reached a threshold, they can continue to develop, if it is during breeding period; in contrast, if it is period of sexual inactivity they degenerate, bringing on atretic follicles.

5.3.2 Oviducts

Significant variations in genital ducts have been observed in Gymnophiona during the yearly cycle. Oviducts are paired elongate structures parallel to the kidneys and ovaries. Their anterior end opens as a funnel that receives oocytes provided from the ovaries at ovulation. At the posterior end, each oviduct opens into the cloaca. In oviparous species, egg envelopes are elaborated by oviduct glands. In viviparous species, embryos develop in the uterus, middle and posterior part of the oviduct. Reviews of evolution of oviductal structure in Amphibians related to mode of parity have been published, with special reference to Gymnophiona (Wake 1977b, Wake and Dickie 1998).

5.3.2.1 Oviparous species

Genital ducts of oviparous species have been studied in *Ichthyophis glutinosus* (Sarasin and Sarasin 1887-1890), *Ichthyophis* and *Hypogeophis* (Marcus 1939), *Idiocranium* and *Rhinatrema* (Parker 1934, 1936), and *Uraeotyphlus oxyurus* (Garg and Prasad 1962). Masood Parveez (1987) and Masood Parveez and Nadkarni (1991) described morphology and variations of the female genital tract in *Ichthyophis beddomei* during the sexual cycle. Exbrayat (1989a) described the microanatomy of female oviducts in *Ichthyophis kohtaoensis* and *Microcaecilia unicolor* (but it is possible that this last species is viviparous).

Variations of oviducts in oviparous species have received little attention (Wake 1970; Exbrayat 1989a). Nevertheless, *Ichthyophis beddomei* has been the subject of a more detailed study (Masood Parveez 1987; Masood Parveez and Nadkarni 1991) (Fig. 3.18, chapter 3 of this volume). Each oviduct is divided into three parts: anterior pars recta, middle pars convoluta and a posterior pars utera connecting to the cloaca. The pars recta is always bordered by gland cells containing proteic secretions and also a few acid mucous granules. The pars convoluta wall is bordered, for half of its circumference, by gland cells secreting an acid mucus and proteic substances. Beneath these cells several layers of undifferentiated cells are observed. The other half of transverse sections possesses crests that project into the lumen of the oviduct and are bordered by a single columnar epithelium at the top of which cilia are present during the breeding season (October to February). The wall of the pars utera consists of connective tissue surrounded by a thick muscular layer. Crests project into the lumen, the wall of which is bordered by a single columnar epithelium. Cells of this epithelium are covered by cilia during the breeding season. Between crests, glandular cells secreting a neutral mucous substance are present. All the glands are particularly well developed from December until February (the breeding season). After ovulation (March-April), the glands regress and are at their minimum in June-July. Similar structures are present in other oviparous species (Wake 1970; Exbrayat 1989a). In all species that have been studied, variations of the oviducts were always linked to the sexual cycle and were under endocrine control.

5.3.2.2 Viviparous species

In *Typhlonectes compressicaudus*, the wall of the oviducal funnel is poorly developed during the period of sexual quiescence. Epithelium bordering the wall is composed of a single layer of undifferentiated cells, characterised by a large nucleus surrounded by a thin crown of cytoplasm. In December-January, just before ovulation, epithelial cells become voluminous and rounded and their apical border develops cilia (Fig. 5.5A, B, E). In February, when ovulation occurs, the funnel wall is composed of developed crests between which are crypts containing cells with proteic secretions. In April and in pregnancy or during sexual quiescence, the funnel reverts to an inactive appearance (Exbrayat 1988a) (Fig. 5.5C, D, F).

In viviparous species, several observations have shown variations of the oviduct in relation to the sexual cycle and pregnancy (Wake 1970; Exbrayat 1984, 1988a; Hraoui-Bloquet 1995). The oviduct is divided into two parts: the anterior part and the posterior uterus. Before pregnancy, the anterior part becomes receptive to oocytes. Connective tissue is lacunal and vascularised, with numerous cells. The lumen wall is bordered by a simple columnar epithelium with undifferentiated cells (Fig. 5.6A). The diameter of this region then increases, and the anterior oviduct becomes elongate and flexuous. Connective tissue sends crests into the lumen that are bordered with epithelium in which some ciliated cells are observed. Glandular cells develop between the crests. In December-January, at the beginning of the breeding season, gland cells contain acid mucous secretions. Other mucous cells with sulphated mucus lie between the ciliated cells (Fig. 5.6C). At ovulation, cilia are covered by a thick mucous layer. At the level of oocytes, cilia are lost. At the beginning of pregnancy, this anterior part of the oviduct regresses.

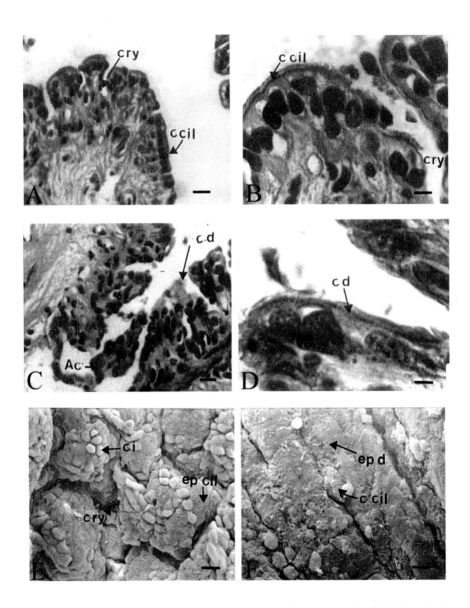

Fig. 5.5. Funnel in adult female *Typhlonectes compressicauda*. **A.** Transverse section (TS) of funnel ready to receive the oocytes (February, period of breeding) Scale bar = 100 µm. **B.** Detail of funnel in February. Scale bar = 110 µm. **C.** TS of degradation of funnel in April. Scale bar = 100 µm. **D.** Detail of funnel in April. Scale bar = 10 µm. **E.** SEM view of funnel in February. Scale bar = 15 µm. **F.** SEM view of funnel in April. Scale bar = 15 µm. ccil, ciliated cells; cd, desquamating cell; ci, undifferentiated cell; cry, crypt; ep cil, ciliated epithelium; epd, desquamating epithelium; AC, group of cells. From Exbrayat, J.-M. 1986a. D. Sc. Unpublished Thesis, Université Paris VI, France, Pl. XI, Figs. 1-4, 7, 8.

Oocytes that have not been fertilised degenerate (Fig. 5.6D), and yolk platelets become free in the lumen where they are ingested by embryos as food. After pregnancy, the oviduct wall is thin, epithelial cells have degenerated and certain cells have been evacuated into the lumen. At the beginning of the second year of the sexual cycle (October), the oviduct again becomes prepared for pregnancy, but it degenerates rapidly at the theoretical period of ovulation.

The uterus is relatively slender during sexual quiescence (2 to 2.5 mm in diameter) but increases significantly in width before ovulation. The wall that was smooth (Fig. 5.6E, F) has become bordered by lacunal and vascularised crests with numerous blood cells. Three types of epithelial cells are observed: gland cells between cells with acid mucous secretions; rounded ciliated cells at the top of crests; and cells with numerous microvilli (Fig. 5.7A). At ovulation, the third type of cell produces long cytoplasm expansions and the ciliated cells are covered by acid mucous produced from the gland cells. Lipid substances also cover all cell types. At the beginning of pregnancy, the appearance remains the same (Fig. 5.7B). In females carrying free embryos in the uterus, the uterine wall is distended, glandular cells have reduced secretions; the cells with cytoplasm expansions are degraded at their contact with the embryos but they are unchanged where not in this contact (Fig. 5.7C, D). At the end of pregnancy, all the epithelia cells have degenerated (Fig. 5.7E) and the connective tissue is exposed at the luminal surface (Fig. 5.6B, 5.7F). At this stage, the epithelium of the gills of the embryo acquires a narrow contact with this connective tissue to give a structure resembling a placenta (Hraoui-Bloquet 1995; chapter 11 of this volume). After parturition, the uterine region becomes narrow. An undifferentiated tissue replaces the different cell types observed previously. At the beginning of the second year of the cycle, a new differentiation of the uterine wall takes place but, in February, at the theoretical period of ovulation, it degenerates. In the following October, a new differentiation marks the beginning of a new biennial cycle (see chapter 11 of this volume).

5.3.3 Female Cloaca

In *Typhlonectes compressicauda* the female cloaca also exhibits variations closely linked to the sexual cycle (Exbrayat 1996). The female cloaca can be divided in an anterior part which receives the intestine, bladder, oviducts and Wolffian ducts, and a posterior part. During sexual inactivity, the posterior part is covered with a stratified epithelium with keratinised cells. When breeding occurs, this epithelium is covered with gland cells with acid mucus. During pregnancy, the epithelium becomes low, with little mucus (Fig. 3.18G, H, chapter 3 of this volume).

5.3.4 The Endocrine Ovary

In vertebrates, it is known that the ovary possesses endocrine functions that are closely linked to the activity of the hypophysis. In Gymnophiona, general

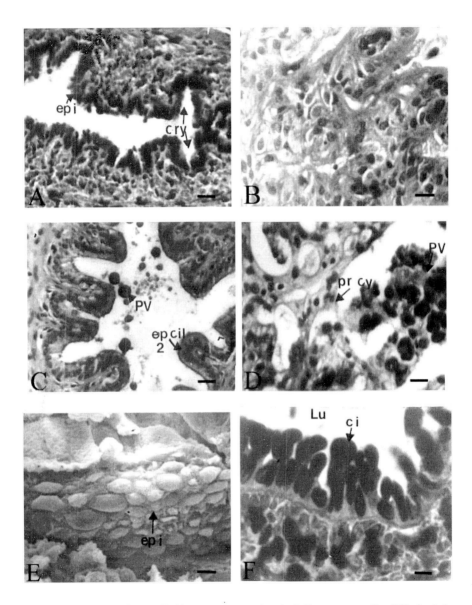

Fig. 5.6. Oviduct in adult female *Typhlonectes compressicauda*. **A.** Transverse section (TS) of anterior oviduct in August (sexual quiescence). Scale bar = 100 µm. **B.** TS of uterus in a female with embryos (middle of pregnancy): connective tissue shows an important proliferation. Scale bar = 100 µm. **C.** TS of anterior differentiated oviduct in February. Scale bar = 100 µm. **D.** TS of anterior oviduct with degraded oocyte. Scale bar = 100 µm. **E.** SEM view of an undifferentiated anterior oviduct. Scale bar = 4 µm. **F.** TS of uterus (August, sexual quiescence). Scale bar = 30 µm. ci, undifferentiated cell. cry, crypt; epi, undifferentiated epithelium; ep cil, ciliated epithelium; Lu, lumen of the uterus; prcy, cytoplasmic proliferation; PV, yolk platelets; From Exbrayat, J.-M. 1986a. Unpublished D. Sci. Thesis, Université Paris VI, France, Pl. XII, Figs. 1, 5, 8, Pl. XII, Figs. 1, 5, Pl. XIV Fig. 3.

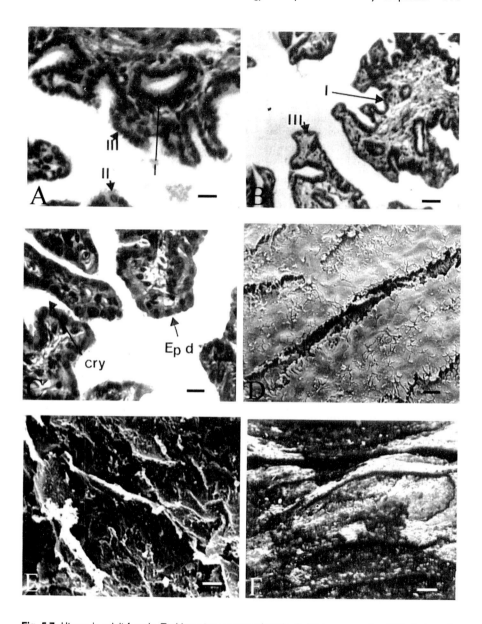

Fig. 5.7. Uterus in adult female *Typhlonectes compressicauda*. **A.** Trasverse section (TS) of uterus in February (just before pregnancy). Scale bar = 100 μm. **B.** TS of uterus in April (at the beginning of pregnancy). Scale bar = 100 μm. **C.** TS of uterus (end of pregnancy). Scale bar = 100 μm. **D.** SEM view of uterine epithelium at the end of pregnancy. Scale bar = 30 μm. **E.** SEM view of a degraded uterine epithelium at the end of pregnancy. Scale bar = 1.5 μm. **F.** SEM view of uterine epithelium before parturition. Scale bar = 80 μm. cry, crypt; Epd, degraded epithelium; I, gland cells between the villi; II, ciliated cells at the top of the epithelium; III, ciliated cells. From Exbrayat, J.-M. 1986a. Unpublished D. Sci. Thesis, Université Paris VI, Pl. XIII, Figs. 6, 7, Pl. XIV, Fig. 1, Pl. XV, Figs. 3, 5, 6.

endocrinology has not been well studied. Some data nevertheless exist for the oviparous *Ichthyophis beddomei* (Masood Parveez 1987) and the viviparous *T. compressicauda* (Exbrayat and Collenot 1983; Exbrayat 1986a, 1992a, b).

5.3.4.1 Oocytes and follicles before vitellogenesis

The only detailed studies on the development of ovaries during reproductive cycles of Gymnophiona are those on *Ichthyophis beddomei* (Masood Parveez 1987; Masood Parveez and Nadkarni 1993a, b) and *Typhlonectes compressicauda* (Exbrayat and Collenot 1983; Exbrayat 1986a, 1992a) though descriptions of ovaries have been given for other species (Berois and de Sa 1988; Wake1968a, b; chapter 8 of this volume).

In *Typhlonectes compressicauda*, oogonia are situated in germ nests which form small masses regularly spaced along each ovary. These oogonia (20 µm diameter) have been observed in the ovaries of all species studied. In *Typhlonectes compressicauda*, the nests are more numerous in February to April i.e. during the breeding season. The smallest primary oocytes (50 µm in diameter) may or may not be surrounded by small follicular cells originating from somatic cells of the ovary. Granulosa cells and their nuclei are very elongate and the cells are closely adpressed to each other. They are separated from the oocytes by a thin vitelline envelope. Growth of the follicle has been studied in detail in *Typhlonectes compressicauda* (Exbrayat 1986a).

The follicles became increasingly voluminous. Oocytes reach 150 to 300 µm in diameter with a central or slightly eccentric nucleus. Individualisation of chromosomes occurs. The appearance of the follicular layer is constant. When the oocytes reach about 600 µm, glucidic inclusions are observed in their cytoplasm. At this stage, connective tissue begins to organise itself around the follicle. When the follicle exceeds 750 µm in diameter, it has entered the previtellogenic phase, the oocyte becomes increasingly charged with glycoproteic platelets and its nucleus is displaced to the periphery. Granulosa cells are always flattened and contiguous with each other. Peripheral connective tissue is organised to form a theca. Finally, follicles reach about 1200 µm in diameter. In *Ichthyophis beddomei*, the development of oocytes and follicles was of the same type.

5.3.4.2 Oocytes and follicles during vitellogenesis

In *Typhlonectes compressicauda*, the biggest oocyte and follicles are 1000 to 2000 µm in diameter. The oocyte cytoplasm is filled with numerous yolk platelets. Its voluminous nucleus (10 µm in diameter) is pushed against the plasma membrane. In *Ichthyophis beddomei*, follicles reach 750 µm in diameter. In this species, it is difficult to differentiate granulosa cells from theca cells (Masood Parveez 1987). In *Typhlonectes compressicauda*, in contrast, follicle cells and vitelline membrane exhibit spectacular morphological variations. The less developed follicles, with a vitellogenic oocyte, are surrounded by a thin vitellin membrane. A striated zone corresponding to microvilli of oocytes underlies its plasma membrane. The follicular cells are much enlarged and contiguous. The granulosa cells became cubical and contiguous. The vitelline membrane becomes progessively thicker but then becomes very loose and the follicular cells seem to float within the follicle. Certain of the follicular cells are disposed in two layers, one against the theca, others against the oocyte.

5.3.4.3 Oocytes and follicles after ovulation

In viviparous species, the follicle becomes a corpus luteum after ovulation. Corpora lutea have been also observed in several oviparous species: *Hypogeophis rostratus* (Tonutti 1931), *Ichthyophis beddomei* (Masood Parveez 1987) and further species (Wake 1968a, b). In *Ichthyophis beddomei*, corpora lutea have been observed in females which have eggs or oocytes in the oviducts. Granulosa cells are hypertrophied and invade the space let free by the oocyte after its expulsion from the follicle. The cytoplasm of follicle cells is voluminous and the nucleus takes on an irregular shape. The connective tissue has also hypertrophied and represents about 50% the total mass of the postovulatory follicle.

In viviparous species, corpora lutea are laid down throughout pregnancy and are modified according to the stages of the embryo in the oviduct. Development of corpora lutea has been studied in *Typhlonectes compressicauda* (Exbrayat and Collenot 1983; Exbrayat 1986a). After ovulation, follicle cells proliferate and invade the central cavity left free. They are disposed in about 10 layers and leave a large central cavity filled by a substance (Fig. 5.8A). These hypertrophied cells are spherical (30 µm in diameter) or elongated (20 × 40 µm). Corpora lutea are 1200 to 2000 µm in diameter. At this stage blood vessels are limited to the peripheral theca. The granulosa cells continue to invade the central cavity. Cytoplasmic bridges have been observed between cells that otherwise have little contact. At the periphery, cells originating from the theca infiltrate between the granulosa cells. Blood vessels, containing blood cells, develop between all the cells (Fig. 5.8B). The corpora lutea then regress and the central cavity fills with cells (Fig. 5.8C). Numerous cells arising from the ganulosa contain vacuoles. At the end of pregnancy, the corpora lutea degenerate. Cells became high or very elongate, with a degenerate nucleus. Numerous blood cells are present. Finally, a degenerating corpus luteum is reduced to a small mass that is rapidly integrated into the connective tissue (Fig. 5.8D).

In *Typhlonectes compressicauda*, δ5 3β hydroxysteroid dehydrogenase (δ5 3β HSDH, an enzyme implied in steroid synthesis) activity has been observed in cells of hollow or compact corpora lutea (Fig. 5.8I), more specifically in cells arising from the granulosa in which numerous lipidic granules have been demonstrated. After ovulation, in degenerative corpora lutea, no δ5 3β HSDH is detectable. The incubation of ovaries of pregnant *Typhlonectes compressicauda* females with radio active pregnenelone, a precursor of progesterone, has shown that the ovary elaborates radio-active progesterone into the incubation bath. These results showed that corpora lutea have an endocrine activity linked to pregnancy.

In *Typhlonectes compressicauda*, a biometric study has shown that about 20 corpora lutea are present in each ovary, immediately after ovulation.

This number remains constant through a great part of the pregnancy. Just before the end of pregnancy, however, the number of corpora lutea decreases, indicating that these structures are resorbed. Just after parturition, about 10 corpora lutea, at the end of their development and which rapidly degenerate, have been observed. Thus the corpora lutea seem to be functional through most of gestation but, at the end, when no longer essential, they degenerate.

5.3.4.4 Atretic follicles

In oviparous and viviparous species, certain oocytes degenerate and the corresponding follicles develop atretic bodies. In *Ichthyophis beddomei*, two types of atretic follicles have been recognized (Masood Parveez 1987) (Fig. 8.4E, F, G, chapter **8** of this volume). The first type involves atresy of growing follicles in which oocytes degenerate *in situ* and are resorbed. In the second type, atresy of follicles with vitellogenic oocytes occurs. Atresy began by hypertrophy of granulosa cells. These cells multiply and invade the oocyte cytoplasm. Granulosa cells become phagocytic and eliminate yolk platelets; numerous blood vessels invade the theca; and granulosa cells fill the space left free by the degenerating oocyte. The cells are then filled with a brown pigment. Thecal cells proliferate and finally the follicle is reduced to a mass that is integrated into the connective tissue. In *Chthonerpeton indistinctum*, Berois and de Sa (1988) observed the presence of atretic follicles (Fig. 8.2F, chapter **8** of this volume).

In *Typhlonectes compressicauda*, the biggest atretic follicles are 1200 to 2000 μm in diameter. In them vitellogenic oocytes degenerate and the cytoplasm of the oocyte becomes vacuolate; granulosa cells are hypertrophied and migrate into the oocyte after lysis of the plasma membrane and phagocytose of yolk platelets (Fig. 5.8E, F). Phagocytic cells contain a lipidic substance that stains with Sudan black (Fig. 5.8G). In 750 μm atretic follicles, cells invading oocytes have the appearance of adipocytes. Finally, atretic follicles degenerate, giving a mass that is, again, incorporated into the connective tissue.

In *Ichthyophis beddomei*, histochemical detection of δ5 3β HSDH and 17β HSDH has given positive results in the granulosa cells and the oocyte cytoplasm in follicles with previtellogenic and vitellogenic oocytes. Label was also positive in hypertrophied cells of atretic follicles. In *Ichthyophis beddomei*, the ovary thus contains enzymes that are necessary for steroid biosynthesis.

In *Typhlonectes compressicauda*, δ5 3β HSDH label was positive in follicles with previtellogenic, vitellogenic and mature oocytes and in cells of the internal theca (Fig. 5.8H). The use of an antibody directed against estrogenic hormone (estriol and 17b estradiol) on sections of *Typhlonectes compressicauda* ovary has provided new information. Indirect immunocytochemistry with peroxidase secondary antibody has shown that labeling is mainly observed in granulosa of follicles with vitellogenic and mature oocytes (Fig. 5.8J). When granulosa cells are still flattened, the slight label is localised against

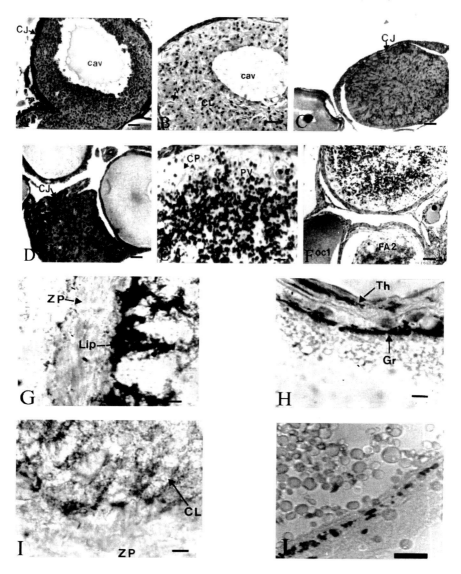

Fig. 5.8. Endocrine ovary in *Typhlonectes compressicauda*. **A.** Corpus luteum at the beginning of pregnancy. Scale bar = 100 μm. **B.** Corpus luteum in the middle of pregnancy. Scale bar = 30 μm. **C.** Corpus luteum at the end of pregnancy. Scale bar = 30 μm. **D.** Regressive corpus luteum after birth. Scale bar = 10 μm. **E.** Atretic follicle. Scale bar = 30 μm. **F.** Several atretic follicles. Scale bar = 100 μm. **G.** Detection of lipids in an atretic follicle. Scale bar = 10 μm. **H.** Detection of $\delta 5\ 3\beta$ HSDH activity in a follicle with a vitellogenic oocyte. Scale bar = 10 μm. **I.** Detection of $\delta 5\ 3\beta$ HSDH activity in a corpus luteum in adult female *T. compressicauda*. Scale bar = 10 μm. **J.** Visualisation of 17β estradiol in a follicle with a vitellogenic follicle in adult female *T. compressicauda*. Scale bar = 10 μm. cav, central cavity; CJ, corpus luteum; CL, luteal cell; cp, phagocytic cell; FA1, atretic follicle of type 1; FA2, atretic follicle of type 2; Gr, granulosa; Lip, lipid; PV, yolk platelet; oc1, oocyte 1; Th, external theca; v, blood vessel; zp, peripheral zone. **A, C, F, H, I**, from Exbrayat, J.-M. and Collenot, G. 1983. Reproduction, Nutrition, Développement 23: 889-898, Pl. I, Figs. 3, 6, Pl. II, Figs. 7, 8, 10. **B, D, E, G**, from Exbrayat, J.-M. 1986a. Unpublished D. Sci. thesis, Université Paris VI, Pl. IX, Figs. 4, 5, 8, Pl. X, Fig. 4. **J**, original.

the vitelline membrane. When follicular cells have become cubical or polyhedral, label is situated in the intercellular spaces. Labelling was particularly evident in the granulosa of the most developed follicles where it was diffusing into the particularly thick vitelline membrane. No label has been observed in atretic follicles.

5.3.5 Acquisition of Sexual Maturity in Females

The sexual maturity in females and the development of the early genital tract in Gymnophiona have been studied only in *Dermophis mexicanus* and *Typhlonectes compressicauda* (Wake 1980; Exbrayat 1986a). In *Dermophis mexicanus*, gonads of the new-born resemble a small ribbon. In one-year-old animals, vitellogenic oocytes are observed in 90% of females. The smallest pregnant females are 300 to 330 mm in length, which is the size of two-year-old animals. In *Geotrypetes seraphinii*, sexual maturation is also reached in two years old animals (Wake 1977a, b). In *Typhlonectes compressicaudus*, ovaries of the new-born also form small ribbons with oogonia and oocytes but without follicular cells. When ovaries are more developed, 150 to 300 µm in diameter, oocytes are surrounded by flattened follicle cells. In more developed ovaries, oogonia are grouped in germinal nests between which more or less developed follicles are seen. In 20-month-old animals, 600 to 750 µm follicles with a thin vitelline membrane are observed. In 25-month-old animals, the first previtellogenic oocytes and atretic follicles are present after which vitellogenesis occurs. Animals became adult in January-February of the third year (24- to 30-month-old animals). This has been confirmed by observation of the smallest pregnant females that were 300 mm length, i.e. with the size of a two-year-old animal. In young *T. compressicauda*, female genital ducts are very slender structures not visibly differentiated into parts. However, during the second year of life, some slight morphological variations indicate the presence of the different parts and the announcement of a sexual cycle (Exbrayat 1988a). Similar observations have also been given for the cloaca (Exbrayat 1996).

5.4 HYPOPHYSIS IN GYMNOPHIONA

The gymnophionan hypophysis has been very little studied. The majority of works have been esssentially devoted to anatomical and histological aspects (Stendell 1914; Laubmann 1926; Pillay 1957; Pasteel and Herlant 1961; Kuhlenbeck 1970; Gabe 1972; Schubert *et al.* 1977; Schubert and Welsch 1979). It is only in *Ichthyophis beddomei* (Bhatta 1987) and *Typhlonectes compressicauda* (Exbrayat 1989b; Exbrayat and Morel 1990-1991, 1995) that cytological variations of some cells types have been correlated to sexual activity and cyclic variations of target organs. On the other hand, hypophysal cytology has been detailed by means of classic histological staining and immunocytochemistry in *Typhlonectes compressicauda* (Zuber-Vogeli and Doerr-Schott 1981; Doerr-Schott and Zuber-Vogeli 1984, 1986;

Exbrayat 1989b; Exbrayat and Morel 1990-1991) and in *Chthonerpeton indistinctum* (Schubert *et al.* 1977).

5.4.1 General Anatomy of the Hypophysis

The gymnophionan hypophysis is always a flattened and more or less elongate organ (Fig. 5.9A). In *Ichthyophis glutinosus* it is particularly long (Gabe 1972) whereas in *Typhlonectes compressicauda* it is ovoid.

In *Gegenophis carnosus*, Pillay (1957) has described a neurohypophysis situated in the anterior part of the organ, a ventral distal lobe (pars distalis) and an intermediate lobe (pars intermedia) situated in the rostral part of the pars tuberalis. In *Schistometopum thomense*, the pars intermedia is situated in the caudal portion of the pars distalis (Kuhlenbeck 1970). In Typhlonectidae (*Chthonerpeton indistinctum*, Schubert *et al.* 1977; and *Typhlonectes compressicauda*, Zuber-Vogeli and Doerr-Schott 1981, Doerr-Schott and Zuber-Vogeli 1984, 1986), there is not a pars intermedia. In *Ichthyophis beddomei* (Bhatta 1987), the hypophysis is a flattened elongate organ. The pars tuberalis is situated in the rostral part of the adenohypophysis and is closely applied to the median eminence. The pars intermedia is situated in the caudal part of pars distalis, beneath the pars nervosa.

In all species, a long infundibular ridge runs dorsally across the hypophysis to join a voluminous posterior nervous lobe. In the pars distalis, five cell types have been found in *Typhlonectes compressicauda* (Zuber-Vogeli and Doerr-Schott 1981) and *Ichthyophis beddomei* (Bhatta 1987). Three cell types resemble cell types found in other amphibians. Two cell types possess proteic granules. They are (1) lactotropic cells, or A1 cells (Kerr 1965; Van Oordt 1974) which are generally scattered through the pars distalis except its rostral region; and (2) somatotropic cells, or A2 cells, situated in the caudodorsal part of distal lobe. Two further cell categories possess glycoproteic inclusions: (3) thyrotropic cells (or B1 cells) also scattered in the pars distalis except its rostral part, and (4) gonadotropic cells (B2 cell) situated throughout the pars distalis except rostrally. The last cell type (5), also with glycoproteic secretions are corticotropic cells, B3 cells, situated in the rostral region of the pars distalis, in contact with the hypophysis secondary capillary system.

5.4.2 Cycle of the Hypophysis in Males

Cell types belonging to the pars distalis have been found to present variations during the year. Four cell types have been studied in Gymnophiona: gonadotropic cells, prolactin cells, somatotropic cells and corticotropic cells. In *Typhlonectes compressicauda*, an important network of capillaries is present throughout the adenohypophysis. The diameter of these blood vessels is relatively narrow through the year except in June, at the end of the breeding season, when the testes are almost empty, when it is large.

5.4.2.1 Gonadotropic cells

Gonadotropic cells react with anti LH serum (Doerr-Schott and Zuber-Vogeli 1984) (Fig. 5.9B). In January, their cytoplasm shows blue Cleveland and Wolf staining, with several pink granules scattered in the cytoplasm and, sometimes, orange-stained globules. From February to April, granules become very numerous and fill the well-developed cytoplasm. Cells reach their maximal size and are particularly abundant in the median part of the gland. In May-June, cell size greatly decreases and granules are much less numerous, but orange-stained globules are abundant. In June, cells without granules are at their minimal size and even the nuclei of some seem reduced in size. From July to October, sexual quiescence, cell size increases to reach a mean yearly value; the cytoplasm still possesses a few granules. From October to February, the cell size again increases, to a maximal value (Fig. 5.10A). Nuclear size is correlated with that of cell.

5.4.2.2 Lactotropic cells

Lactotropic cells react positively with anti PRL-serum (Doerr-Schott and Zuber-Vogeli 1984) (Fig. 5.9C). Irrespective of their size, these cells always contain the same type of voluminous orange-stained globule. The cells are particularly well developed during the rainy season (that is the breeding season) in February to April. Between April and June, the cells decrease but always contain granules. From June to August, the cell size increases to reach a mean value at which they remain to January, after a slight decrease in October. Nucleus size is correlated to that of cell (Fig. 5.10B).

Contrary at the variations of size, a count of lactotropic cells has shown that the number of cells is relatively reduced (270 cells/mm^3, Exbrayat and Morel 1995) during the breeding season. Just after breeding, the number of cells is maximal (about 500 cells/mm^3) and remain maximal to the next period of reproduction. Quantification of mRNAs after visualisation by *in situ* hybridisation shows a constant label during all the periods of the cycle, indicating that total mRNAs are expressed at the same level throughout the year. Despite incomplete data, a schema can be proposed for the activity of lactotropic cells. Just after reproduction and during sexual quiescence, the number of cells remains constant, synthesizing a constant quantity of prolactin to be discharged into the organism. During reproduction the synthesis ratio is always the same but involves fewer cells. Correspondingly, these cells possess an increased quantity of mRNAs that permits an increased synthesis of prolactin. This hormone could be stored in the cells, explaining the increase in their volume. During breeding, prolactin would be discharged into the organism.

A study using *in situ* hybridisation has permitted visualisation of mRNAs coding for prolactin receptors (Exbrayat et al. 1996a, Exbrayat and Morel 2003). These receptors are found in all the organs (Exbrayat et al. 1997) and more particularly in the testes (germ cells, Sertoli cells and Leydig cells). The study of mRNAs expression of prolactin receptors in Mullerian ducts indicates the importance of prolactin in regulation of the physiology of sexual organs during the cycle. mRNAs have been observed in histological sections of Mullerian ducts, localised in connective tissue and mainly in glandular structures (Exbrayat and Morel 2003). During sexual inactivity, visualisation of mRNAs gives a very scattered label (Fig. 5.11A).

Quantification by density analysis also gives minimal values. In contrast, during the breeding season, when glands are very well developed, the signal is increasingly dense (Fig. 5.11B). Quantification analysis of density gave a high value. There is a close correlation between the volume of prolactin cells, variations of testes, spermatogenesis, and Mullerian gland activity.

5.4.2.3 Somatotropic cells

Somatotropic cells react positively to anti-GH serum (Doerr-Schott and Zuber-Vogeli 1984) (Fig. 5.9D). Throughout the year, cytoplasm of these cells is uniformly filled with small pink granules after Cleveland and Wolfe staining. The cytoplasm stains brown with PAS orange G. In August, granules seem more voluminous and abundant than in other months. Biometric variations are relatively insignificant. However, there are maximal values from February to April and also a peak in August (Fig. 5.10C). Nuclear size is again correlated with cell size.

5.4.2.4 Corticotropic cells

Corticotropic cells react with anti ACTH serum (Doerr-Schott and Zuber-Vogeli 1984) (Fig. 5.9E). Staining of cytoplasm is very slight. Some variations on size and cytological aspects of these cells have been observed. In January, the size is minimal. Cytoplasm is purple-stained after Cleveland and Wolfe staining and pink after PAS orange G. They are filled with granules. Staining by aldehyde fuchsin picro indigo carmine gives them a purple coloration. If this staining is done after oxidisation, they appear blue-stained. In February and April, cell size increases and the cytoplasm has a reticulated appearance; the basal region of cells that abut against blood vessels show a particularly intense color. In May, cell volume decreases and the cytoplasm often contains granules. In June, cytoplasm appears very dense, purple-stained by fuchsin aldehyde after oxidisation, brown pink after Cleveland and Wolfe staining and brick pink with PAS orange G. From August to January, cell size progressively decreases and cytoplasm is reticulated or dense (Fig. 5.10D). Nuclear and cellular sizes again correlate.

In *Ichthyophis beddomei*, Bhatta (1987) has reported that gonadotropic cells are filled with granules throughout the year. However, vacuoles appear in the cytoplasm in January and February, during the breeding season. Presence of granules only in March could be correlated with proliferation of spermatogonia. Corticotropic cells are active in August and September and in December-January, correlated with the increase of interrenal activity at these periods (rainy season in July-August and breeding in January-February). In this species, as in *Typhlonectes compressicauda*, a close correlation exists between activity of certain cell types in the hypophysis and activity of the animal.

5.4.3 Cycles of the Hypophysis in Females

In only two species, the oviparous *Ichthyophis beddomei* and the viviparous *Typhlonectes compressicauda* the cycles of cell types have been studied in the

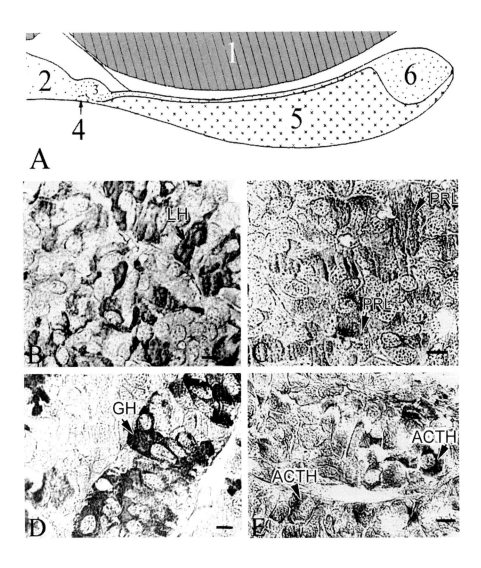

Fig. 5.9. A. Schematic representation of the hypophysis in adult female *Typhlonectes compressicauda*. 1, brain; 2, hypothalamus; 3, median eminence; 4, pars tuberalis; 5, pars distalis; 6, pars nervosa. **B.** Visualisation of gonadotropic cells by immunocytochemistry. Scale bar = 10 µm. **C.** Visualisation of lactotropic cells by immunocytochemistry. Scale bar = 10 µm. **D.** Visualisation of somatotropic cells by immunocytochemistry. Scale bar = 10 µm. **E.** Visualisation of corticototropic cells by immunocytochemistry. Scale bar = 10 µm. ACTH, cell with ACTH; GH, cell with growth hormone; LH, cell with LH hormone; PRL, cell with prolactine. **A**, modified after Zuber-Vogeli, M., Doerr-Schott, J. 1981. Comptes rendus des Séances de l'Académie des Sciences de Paris, série D 292: 503-506, Fig. 1. **B, C, D, E**, modified after Exbrayat, J.-M. and Morel, G. 1990-1991. Biological Structures and Morphogenesis, 3: 129-138, Figs. 2A, B, C, D.

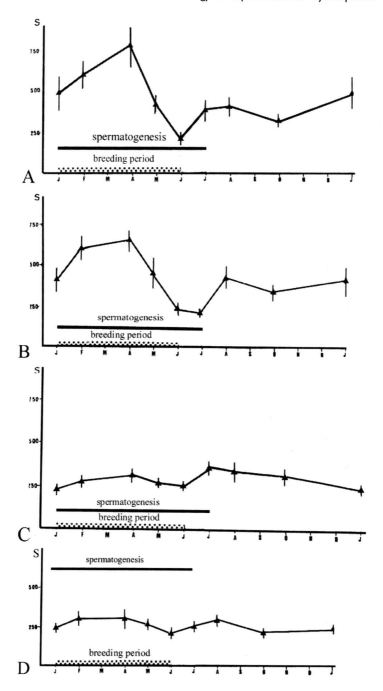

Fig. 5.10. Yearly variations of hypophysis in adult male *Typhlonectes compressicauda*. **A.** Variations of gonadotropic cell surface. **B.** Variations of lactotropic cell surface. **C.** Variations of somatotropic cell surface. **D.** Variations of corticotropic cell surface. Modified after Exbrayat, J.-M. 1989, Biological Structures and Morphogenesis, 2: 117-123, Figs. 1, 2, 4, 3.

adenohypophysis (Masood Parveez 1987; Exbrayat and Morel 1990-1991). Four to five cell types presented major variations through the year or in the sexual cycle. In *I. beddomei*, only gonadotropic and corticotropic cells have been studied.

5.4.3.1 Gonadotropic cells

The gonadotropic cells of *Ichthyophis beddomei* are very large, with cytoplasm that is PAS positive, stained by alcian blue and aldehyde fuchsin. Their nucleus increases from June until November when granules accumulate. The cell diameter is maximal in November. They lose granules in December and January, correlated with the breeding season. From March until May, the diameter of nuclei decreases and the cytoplasm appears empty: this period corresponds to the phase at which the ovary is regressing after breeding. Activity of gonadotropic cells is closely linked to the sexual cycle.

In *Typhlonectes compressicauda*, from October to January, at the beginning of its biennial reproductive cycle, gonadotropic cells that react with an anti-LH serum become more numerous and are progressively filled with granules. After Cleveland and Wolfe staining, the cytoplasm appears blue and the granules pink or purple. During this time, orange-stained globules are also observed in the cells (Fig. 5.11C, D). From February until April, in pregnant females with young embryos, cell size reaches a maximum. Granules fill the cytoplasm in which orange globules are more frequent. Besides these developed cells, smaller gonadotropic cells occur. During pregnancy, from April to October, the number of cells lacking granules increases. The most developed cells decrease in size and cytoplasmic granules are increasingly scattered though globules are more and more numerous and voluminous (Fig. 5.11E). At parturition, these cells have reached their smallest size; the cytoplasm is blue-stained by Cleveland and Wolfe trichroma; it contains few granules and many voluminous globules stained by orange G. Many cells are minute with blue cytoplasm restricted to the vicinity of the nucleus (Fig. 5.11F). From October to February, at the beginning of the second year of the sexual cycle, gonadotropic cells develop again and attain the appearance previously observed in January-February during the first year. In February, at ovulation, when the more voluminous follicles become atresic, gonadotropic cells decrease in size and rapidly lose their granules. The more developed cells have a small cytoplasm that is blue-stained by Cleveland and Wolfe trichroma. Orange-stained globules are very rare. The appearance of the cells changes in August and October at the beginning of a new biennial cycle. Cytoplasm becomes increasingly abundant. Granules appear in October, indicating the beginning of the new cycle (Fig. 5.12A). Nuclear sizes are correlated with the size of cells.

5.4.3.2 Lactotropic cells

In *Typhlonectes compressicauda*, lactotropic cells reacting with anti prolactine serum are uniformly filled with voluminous orange-stained globules after Cleveland and Wolfe trichroma. Only slight volume variations have been observed that concern size of cells and nuclei. From October until January, lactotropic cells reach a maximal volume that is maintained at the beginning of parturition after a slight decrease in February when ovulation occurs. Then cells progressively decrease in size. They have a minimal size at the end of parturition. In October, at the beginning of the second year of the cycle, lactotropic cells again increase in size until January. After the theoretical time of ovulation, these cells decrease. During sexual inactivity, the size of lactotropic cells is very variable in the same hypophysis yet a maximal value is observed in May and August. In October, these cells are again at their maximal size, before commencement of a new cycle (Fig. 5.12B).

Contrary to volumetric variations, the number of cells varies considerably throughout the cycle. In vitellogenic females, from the beginning to the middle of pregnancy, the cell number is low (about 250 cells/mm^3). As in males, cell volume is maximal. At the end of pregnancy, when cell size has decreased, this number is high (425 cells/mm^3). In females caught in October, the number of prolactin cells is progressively higher (375 cells/mm^3). In females that have not achieved gestation, the cell volume decreases but the number of prolactin cells increases to reach a high number in the following October (Fig. 5.12B).

From the beginning to the end of pregnancy, an increase of mRNAs encoding for prolactin has been observed. After parturition, during vitellogenesis and during sexual quiescence, mRNAs ratio remain constant and relatively low (Exbrayat and Morel 1995), suggesting that prolactin synthesis is relatively slight during vitellogenesis as during sexual quiescence. At the beginning of pregnancy, a slight accumulation of prolactin has been observed. The demand for prolactin then increases, and increased secretion is achieved by an augmentation of both cell number and mRNAs encoding for prolactin, and a decrease in the size of cells which synthesise prolactin without storage.

At parturition, cells have a reduced size, and they no longer synthesize much prolactin, as indicated by a decreasing quantity of mRNAs. During the second year of the cycle, an increase in cell numbers has been observed but not of their size or RNAs synthesis.

A study by *in situ* hybridization of mRNAs coding for prolactin receptors has indicated an increase of these receptors in the ovaries (Fig. 5.11G). At the beginning of pregnancy, these RNAs are infrequent in corpora lutea with a cavity. They became very abundant in compact corpora lutea (Fig. 5.11H) and are almost absent in degenerative corpora lutea (Exbrayat *et al.* 1996b; Exbrayat and Morel 2003). On the other hand, mRNAs coding for prolactin receptors are more abundant in the liver (a reserve organ, Exbrayat 1988b) of pregnant than of resting females. These observations reveal the important role of prolactin in regulation of the female sexual cycle and gestation.

5.4.3.3 Somatotropic cells

In *Typhlonectes* compressicauda, somatotropic cells react with anti GH serum. They contain granules that are pink-stained after Cleveland and Wolfe trichroma and brown after PAS orange G. In January and February,

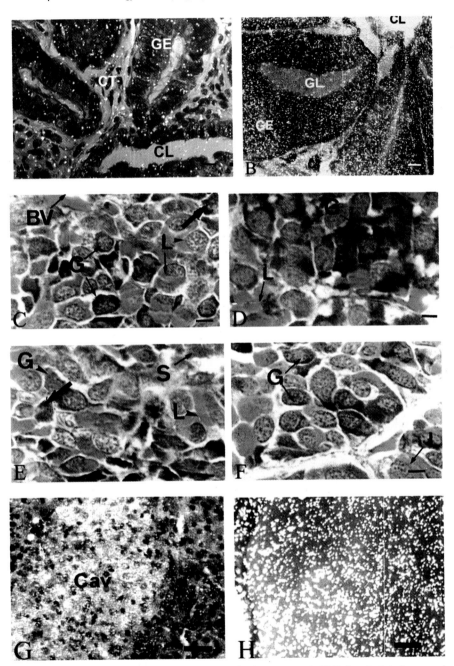

Fig. 5.11. A. Visualisation of mRNAs coding for prolactin receptors in Mullerian duct during sexual quiescence. Scale bar = 30 µm. B. Visualisation of mRNAs coding for prolactin receptors in Mullerian duct during reproduction. Scale bar = 90 µm. C. Cytology of hypophysis in an adult female *Typhlonectes compressicauda* in October (sexual quiescence). Scale bar = 10 µm. D. Cytology of hypophysis in an adult female *T. compressicauda* in January (vitellogenesis). Scale bar = 10 µm. E. Cytology of

Fig. 5.11 contd

somatotropic cells reach their maximal size. During pregnancy cell size does not vary. In October, it is minimal. During the year of sexual quiescence, biometric variations have been observed: cell size is greatest in May and August and smallest in April and June (Fig. 5.12C).

5.4.3.4 Corticotropic cells

Masood Parveez (1987) also studied variations of corticotropic cells. These cells are rounded or slightly elongate. The nuclear diameter increases in June, parallel to the increase of interrenal adrenocortical cells. In *Ichthyophis beddomei*, the cycle of corticotropic cells is linked to interrenal activity, being itself linked to the reproductive cycle and seasonal activity.

In *Typhlonectes compressicauda*, corticotropic cells react with an anti ACTH serum. In January, these cells are poorly developed. Their cytoplasm is filled with minute globules that are pink-stained by Cleveland and Wolfe trichroma, and blue-stained by aldehyde fuchsin and purple when this last staining was done after oxidisation. In February, cytoplasm is pale purple after Cleveland and Wolfe trichroma and is filled with a reticulated and granular substance. In females with embryos at the beginning or middle of development, these cells are particularly numerous. Cytoplasm is always reticulated and size slightly increased. At the end of pregnancy and at parturition, the cytoplasm is very dense and pale purple after Cleveland and Wolfe staining. During sexual quiescence, in April, the cytoplasm is reticulated. It becomes dense, and cells are more voluminous than previously. In June-July, size decreases, though cytoplasms always had the same appearance. In August, cell volume increases again and cytoplasm is granular. In October, corticotropic cells of all females have a small size (Fig. 5.12D).

5.5 INTERRENAL GLANDS

Interrenal glands in Gymnophiona were described for the first time in *Siphonops annulatus* (Rathke 1852). Development of these glands has been described in *Ichthyophis glutinosus* (Semon 1892; Brauer 1902; Dittus 1936), *Hypogeophis rostratus* and *H. alternans* (Brauer 1902). Cytological aspects of interrenal glands have been described for *Ichthyophis glutinosus* (Gabe 1971). Anatomical and histological descriptions of the glands have also been

Fig. 5.11 contd

hypophysis in an adult female *T. compressicauda* in June (with developed embryos in the uterus). Scale bar = 10 μm. **F.** Cytology of hypophysis in an adult female *T. compressicauda* in August (after parturition). Scale bar = 10 μm. **G**: visualisation of mRNAs coding for prolactin receptors in a young corpus luteum. Scale bar = 100 μm. **H**: visualisation of mRNAs coding for prolactin receptors in a corpus luteum in the middle of pregnancy. Scale bar = 100 μm. BV, blood vessel; cav, cavity in the central part of corpus luteum; CL, lumen of Mullerian duct; CT, connective tissue; G, gonadotrophic cells; GE, glandular epithelium; L, lactotrophic cells; S, somatotrophic cells. **A, B, G, H,** from Exbrayat, J.-M. and Morel, G. 2003. Cell and Tissue Research 312: 361-367, Figs. 1a, c, 3a, b. **C, D, E, F,** modified after Exbrayat, J.-M. and Morel, G. 1990-1991. Biological Structures and Morphogenesis, 3: 129-138, Figs. 1b, c, d, e.

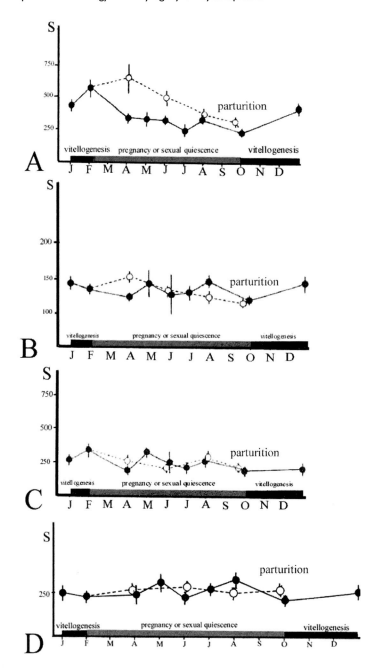

Fig. 5.12. Yearly variations of hypophysis in adult female *Typhlonectes compressicauda*. **A.** Variations of gonadotropic cell surface. **B.** Variations of lactotropic cell surface. **C.** Variations of somatotropic cell surface. **D.** Variations of corticotropic cell surface. Full line, non-pregnant females; interrupted line, pregnant females. Modified after Exbrayat, J.-M. and Morel, G. 1990-1991. Biological Structures and Morphogenesis, 3: 129-138, Figs. 3, 4, 6, 5.

published for *Ichthyophis kohtaoensis, Siphonops annulatus* and *Typhlonectes compressicauda* (Dorne and Exbrayat 1996). Some cyclical variations have been reported in *Ichthyophis beddomei* (Bhatta 1987, Masood Parveez 1987; Masood Parveez *et al.* 1992).

5.5.1 Anatomy and Histology of Interrenal Glands

In *Siphonops annulatus*, each interrenal gland extends from the anterior tip of the kidney to its posterior third. The anterior part of the gland is constituted by a small mass, the posterior part by small nodules (Rathke 1852). In *Ichthyophis* (Dittus 1936; Masood Parveez 1987; Masood Parveez *et al.* 1992; Dorne and Exbrayat 1996), each interrenal gland is situated on the ventral part of the kidney and is composed of four portions. The first portion is located on the anterior part of the kidney. It is composed of small islets near the union of the aortic roots. Posteriorly, a second portion surrounds kidney veins and several islets are scattered on the internal part of kidney. In the third portion, situated near middle region of the kidney, islets of interrenal tissue are scattered on the kidney, all along the caudal vein. The fourth part begins at the posterior end of the caval vein and ends at the posterior part of the kidney (Fig. 5.13A, B, C). Interrenal islets are limited by a capsule constituted with a connective tissue that is thicker on the side of the general cavity than on the side near the kidney. The capsule supports a highly vascularised glandular tissue that is disposed in strings. This tissue contained adrenocortical cells and chromaffin cells (Gabe 1971; Masood Parveez 1987; Masood Parveez *et al.* 1992; Dorne and Exbrayat 1996) (Fig. 5.13D, E, F).

5.5.2 Adrenocortical Cells

Adrenocortical cells constitute the most important part of the interrenal gland of *Ichthyophis beddomei*. The cells are prismatic, cuboidal or polyhedral. Nuclei are situated in their ventral part with two or three nucleoli. The cytoplasm contains abundant lipids that are visualised by Sudan black staining. In *I. kohtaoensis, Siphonops annulatus* and *Typhlonectes compressicauda*, they react positively with anticortisol serum (Dorne and Exbrayat 1996). In *I. beddomei*, δ5 3β HSDH activity has been detected (Bhatta 1987, Masood Parveez 1987, Masood Parveeez *et al.* 1992). Presence of glycogen has been shown in adrenocortical cells of *I. glutinosus* (Gabe 1971).

Cyclical variations of cell and nuclear size have been demonstrated in the female of *Ichthyophis beddomei* (Masood Parveez 1987; Masood Parveeez *et al.* 1992). A first increase in nuclear size is observed in June-July and another in January-February. Maximal cell size is also observed in July. Cell size decreases between August and September, then increases again from October till January. These variations showed a peak of activity in December-January, during the breeding season. The June-July peak seems to be linked to variations of hydromineral balance at this period and not to the breeding status. Variations are closely linked to variations of corticotropic cells in the hypophysis as was shown above. In *Typhlonectes compressicauda*, cyclical

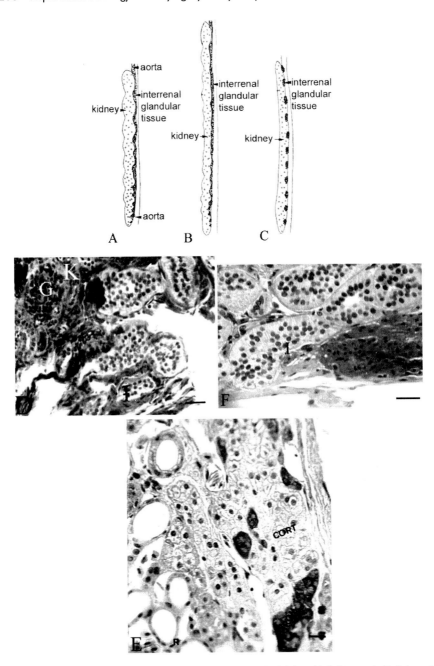

Fig. 5.13. A. Schematic representation of right interrenal gland in *Ichthyophis kohtaoensis*. B. Schematic representation of right interrenal gland in *Siphonops annulatus*. C. Schematic representation of right interrenal gland in *Typhlonectes compressicauda*. D. Section of *Ichthyophis kohtaoensis* interrenal gland. Scale bar = 30 μm. E. section of *Siphonops annulatus* interrenal gland. Scale bar = 30 μm. F. section of *T. compressicauda* interrenal gland. Scale bar = 30 μm; chrom, chromaffin cells; cort, cortical tissue; G, glomerulus; I, interrenal glandular tissue; R, kidney. Original.

variations of corticotropic cell tissue have not been studied but variations of corticotropic cells in the hypophysis suggest that they show variations of the same type as in *I. beddomei*.

5.5.3 Adrenal Cells

Chromaffin cells are scattered between adrenocortical islets (Gabe 1971; Dorne and Exbrayat 1996). Their nuclei are irregularly shaped. The cytoplasm stains with fuchsin and they contain PAS positive granules. They synthesize adrenaline and noradrenaline that is visualised by Hillarp and Höeckfelt staining. Two types of adrenal cells have been visualised in *Ichthyophis glutinosus* (Gabe 1971) and have also been found in other species (Dorne and Exbrayat 1996). Some of them possess numerous granules and resemble noradrenaline cells. In *I. beddomei* females, nuclei size of all the chromaffin cells showed variations with a maximal size in January, the breeding season. However, no precision has been given concerning the nature of cells: with adrenaline or with noradrenaline (Masood Parveez 1987).

Finally, interrenal glands of Gymnophiona resemble these cells in other vertebrates. Cyclical variations have been observed only in *Ichthyophis beddomei* and are linked to cyclical activity and, particularly, breeding.

5.6 PROPOSAL FOR A SCHEME OF ENDOCRINE REGULATION OF SEXUAL CYCLES IN GYMNOPHIONA

5.6.1 In Males

In male Gymnophiona, the interstitial tissue of testes showed variations related to the sexual cycle and to development of sexual organs such as the cloaca and Mullerian glands. Such correlations suggest that Leydig cells are involved in development of sexual characters. Development of interstitial cells during the breeding season is also correlated with reproductive behavior of the animals at this period. Despite the lack of experimental studies in Gymnophiona, it is possible to compare these observations with those for Anura and Urodela in which testosterone has been shown to be implicated in the development of secondary sexual characters (Greenberg 1942; Cei 1944; Blair 1946). Activity in *Rana esculenta* is linked to variations of hypothalamic and hypophyseal hormones and testosterone (Del Rio *et al.* 1980). During the breeding season, increases in steroid hormone blood levels have been shown in *Rana esculenta* (Lupo *et al.* 1988). In *Rana pipiens*, implantation of testosterone cyprionate stimulated the nuptial excrescence on the thumb in castrated males (Lynch and Blackburn 1995). In *Bufo gredosicola*, interstitial tissue is abundant during the period in which thumb pads develop (Fraile *et al.* 1989). Comparing the data for Gymnophiona, we may deduce that endocrine regulation of male breeding is related to testosterone production. It is also influenced by the hypophysis as will now be discussed.

The different cell types in the hypophysis in several species of Gymnophiona have been shown to undergo cyclical variations connected with several events.

Gonadotropic cells reach a maximal development during the breeding season, when interstitial tissue is the most developed. They regress at the end of the reproductive period and this is correlated with regression of the interstitial tissue. These variations are also correlated with variations in Mullerian glands and the cloaca. More particularly, in *Typhlonectes compressicauda*, the increase in gonadotropic cells begins in October, in the middle of the dry season when the biotope has dried up and the animals probably live in burrows in the mud, as has been confirmed in other populations of *Typhlonectes* (Moodie 1978). It is well known that gonadotropic cells become active under the influence of hypothalamic factors, which are subject to external conditions. This phase could be a determining factor in initiating the reproductive cycle.

Lactotropic cells also followed a cycle that is superimposed on that of the genital organs but likewise in accordance with seasonal changes. In addition, mRNAs coding for prolactin receptors have been demonstrated in Mullerian ducts with a great abundance during breeding and a decrease during sexual quiescence. In lower vertebrates, prolactin has a stimulating or inhibiting effect on the testes and on the secondary sexual organs (Mazzi *et al.* 1967; Vellano *et al.* 1967; de Vlaming 1979). A similar phenomenon seems to exist in Gymnophiona. Furthermore, prolactin plays a significant role in the regulation of water-balance (Chadwick 1940; de Vlaming 1979). In the case of the aquatic *Typhlonectes compressicauda*, the succession of periods of inundation and of drought, could perhaps account for the animals being sensitive to variations of osmotic pressure and of the composition of the biotope. It would therefore be indispensable that the water-balance be regulated. The lactotropic cells are certainly involved in this phenomenon. A similar phenomenon could also exist in Asiatic species that are subject to monsoon. This does not, however, exclude a relationship with the reproductive cycle as in other amphibians (Zuber-Vogeli 1968; Van Oordt *et al.* 1968; Rastogi and Chieffi 1970).

Corticotropic cells also have a cycle closely linked to that of interrenal glands and periods of activity of the animals. In *Ichthyophis beddomei*, it has been shown that these cells are more particularly developed during the breeding season and are thus linked to the activity of the animals at this period.

In male Gymnophiona, endocrine regulation has not been studied by experimentation but an examination of correlations between sexual organs, periods of the sexual cycle and major endocrine organs such as Leydig, gonadotropic, lactotropic, corticotropic and interrenal cells indicates a mode of endocrine regulation of reproduction as complex as that of other vertebrates.

5.6.2 In Females

Endocrinology of reproduction in females involves an endocrinal ovary. The endocrine activity of this organ can be divided in two aspects: 1) endocrinology of follicular evolution from sexual quiescence to ovulation in both oviparous and viviparous species, and 2) endocrinology of pregnancy involving corpora lutea, mainly in viviparous species but also oviparous species at the beginning of development when the young embryo is still in the genital tract of female.

In all the species studied, in the granulosa surrounding previtellogenic oocytes, cells possess a steroidogenic activity but, in *Typhlonectes compressicauda*, the only species that has been studied in this way, immunocytochemistry failed to reveal the presence of steroidogenic hormones.

Before the breeding season, vitellogenic oocytes are increasingly numerous and granulosa cells began increasingly voluminous. It has been possible to detect estrogenic hormones in granulosa cells as these become more and more numerous. At the end of this period, ovulation occurs. It is possible that, at this period, presence of estrogens and progesterone reaches a threshold permitting ovulation. Although studies are few and scattered, examination of the number, size and structure of follicles in oviparous or viviparous species leads to similar conclusions. Correspondingly, during this period, growth of the genital tract, oviduct, uterus, and cloaca occurs, as these are prepared for reproduction and pregnancy.

In viviparous species, pregnancy appears to be under the control of the corpora lutea. In *Typhlonectes compressicauda*, corpora lutea first develop cavities during the proliferation of granulosa cells and thecal cells. They then become vascularised, providing a means of driving the hormonal secretions into the general blood circulation. At this time, the ovary is able to synthesize progesterone. At the end of pregnancy, corpora lutea degenerate, in correlation with the birth of offspring. These observations suggest that corpora lutea are active in the maintenance of pregnancy. At the beginning of gestation, the corpora lutea are probably relatively inactive but they become steroidogenic during the greatest part of gestation. At the end, the number of active corpora lutea decreases and hormones become less and less abundant. Then expulsion of embryos (parturition) occurs. This development of corpora lutea is closely correlated with the development of the genital tract to provide embryos to the uterus which, after birth, rapidly returns to a resting state.

These results may now be discussed in the context of what is known about Anura and Urodela. Corpora lutea resembling those of *Typhlonectes* have been observed in several Amphibia, including *Salamandra atra* (Vilter and Vilter 1960, 1962), *S. salamandra* (Joly 1964, 1965, 1971; Joly and Picheral 1972) and the frog *Nectophrynoides occidentalis* (Ozon and Xavier 1968; Lamotte *et al.* 1964; Xavier *et al.* 1970; Xavier and Ozon 1971).

Reproduction in female Gymnophiona is also under pituitary control. Gonadotropic cells present morphological variations correlated to the sexual cycle. A synchronisation between ovulation, gonadotropic cells and vitellogenesis has been observed in Anura and Urodela amphibians whether oviparous or viviparous (Zuber-Vogeli 1968; Lofts 1974; Vilter 1986; Xavier 1986).

The studies of gymnophionan hypophysis cells during pregnancy have been limited to *Typhlonectes compressicauda*. In this species, gonadotropic cells are well developed at the end of the vitellogenic period, just before breeding, as in the oviparous species *Ichthyophis beddomei*. Gonadotropic cells are still well developed at the beginning of gestation, but decrease in size and number at the end of pregnancy. In other viviparous amphibia, results were very different from one species to another. In *Nectophrynoides occidentalis*, gonadotropic cells continue to be very numerous and well developed during pregnancy (Zuber-Vogeli 1968). In *Salamandra atra*, the pituitary gland decreases in volume after ovulation, but remains at an average size until the disappearance of the corpora lutea (Vilter 1986). In contrast, in the two years cycle of *S. salamandra fastuosa* the gonadotropic cells remain inactive during gestation (Joly 1986).

During the second year of the sexual cycle of *Typhlonectes compressicauda*, gonadotropic cells are well developed before the theoretical period of ovulation, then they quickly decrease, in correlation with degeneration of the follicles (Exbrayat 1983, 1986a; Exbrayat and Collenot 1983). These observations are reminiscent of experiments involving removal of the pituitary gland in Amphibia (Lofts 1974) which led to the degeneration of oocytes approaching maturity and to the formation of atretic bodies.

The activity of gonadotropic cells in the hypophysis is correlated with those of the genital tract and can be divided in two phases: 1) in oviparous and viviparous species they are involved in preparation of the tubular region of the oviduct for ovulation and 2) in viviparous species they are involved in preparation of the uterus for pregnancy. If gestation takes place, there is a parallel between the continuation of the development of gonadotropic cells at the beginning of gestation and the continuation followed by the development of the tubular region of the oviduct. During the year of sexual inactivity, in *Typhlonectes compressicauda*, gonadotropic cells regress quickly. The tubular and uterine parts of the oviduct also regress (Exbrayat 1986a, 1988a). In other viviparous amphibians, correlations have also been observed between gonadotropic cells and the tubular region of the oviduct (*Nectophrynoides occidentalis* Zuber-Vogeli 1968, Xavier 1986; *Salamandra atra* Vilter 1966, 1968; *S. salamandra* Joly 1986). In viviparous species, correlation between development of the uterus and that of the pituitay gland is, however, less distinct. In *Nectophrynoides occidentalis*, gonadotropic cells remain developed during pregnancy (Zuber-Vogeli 1968). In *S. salamandra*, gonadotropic variations are not linked to those of genital tract (Joly 1986). In *Salamandra atra*, after ovulation and during gestation, only the tubular region of the oviduct followed the evolution of the pituitary gland, while the uterine part remained unaffected (Vilter 1986). In *Typhlonectes compressicauda*, variations in the weight of reserve organs also follow a cycle parallel to that of the gonadotropic cells (Exbrayat 1988a). In Anura and Urodela, it has been shown that the gonadotropic hormones stimulate the biosynthesis of yolk precursors (Redshaw 1972; Follet and Redshaw 1974). A relationship of the same kind certainly exists in *Typhlonectes compressicauda*.

Prolactin cells also increase in size and number during vitellogenesis and at the beginning of pregnancy in Gymnophiona so far studied. During the year of sexual inactivity in biennial species, these cells diminish in size after vitellogenesis and the theoretical date of ovulation. During pregnancy, these cells diminish progressively in size, following development and regression of corpora lutea. At the same time, mRNAs coding for prolactin receptors are very abundant in corpora lutea in the middle of pregnancy, indicating that corpora lutea are target organs for prolactin. To understand the role of prolactin cells during gestation in viviparous gymnophiona, it is useful to compare data obtained from other amphibia. In *Nectophrynoides occidentalis*, corpora lutea persist under the influence of prolactin cells (Zuber-Vogeli 1968; Zuber-Vogeli and Doerr-Schott 1976; Xavier 1986). It is possible that such a mechanism exists in *Typhlonectes compressicauda* and other viviparous gymophionans.

Corticotropic cells, like interrenal cells, are well developed during the breeding season in all the animals that have been studied, suggesting that these cells are involved in breeding behavior.

5.6.3 Conclusion

Despite a dearth of experimental works, it is possible to understand the endocrine control of both male and female reproductive cycles in Gymnophiona. These cycles are closely correlated with external factors. External signals are translated into hormonal signals that direct the development of genital tracts. In both males and females, it is possible to perceive a role of the endocrinal gonad, of lactotropic, gonadotropic and corticotropic cells, and of interrenal glands, in influencing breeding behaviour, gametogenesis, and development of the genital tracts and of reserve organs.

5.7 ACKNOWLEDGEMENTS

The author thanks Catholic University of Lyon (U.C.L.) and Ecole Pratique des Hautes Etudes (E.P.H.E.) that allowed the freedom in organizing his research fields and supported the works on Gymnophiona. He also thanks Singer-Polignac Foundation that supported the missions necessary to collect the material of study. The author also thanks more especially, Michel Delsol who is at the origin of these works in Gymnophiona, Marie-Thérèse Laurent, who made thousands of sections, Elisabeth Anjubault, Jean-Lou Dorne, Bertrand Leclercq. He thanks Barrie Jamieson, who invited him to contribute to this book series and also for his patience in reviewing the manuscript. He thanks Jean-François Exbrayat for his help in preparing the illustrations.

5.8 LITERATURE CITED

Anjubault, E. and Exbrayat, J.-M. 1998. Yearly cycle of Leydig-like cells in testes of *Typhlonectes compressicaudus* (Amphibia, Gymnophiona). Pp. 53-58. In C. Miaud and R. Guyetant (eds), *Current Studies in Herpetology*. Proceedings of the 9th General Meeting of Societas Europaea Herpetologica. Le Bourget du Lac, France.

Anjubault, E. and Exbrayat J.-M. 2004. Contribution à la connaissance de l'appareil génital de *Typhlonectes compressicauda* (Duméril et Bibron, 1841), Amphibien Gymnophione. I. Gonadogenèse. Bulletin Mensuel de la Société Linnéenne de Lyon 73: 379-392.

Barrio, A. 1969. Observaciones sobre *Chthonerpeton indistinctum* (Gymnophiona, Caecilidae) y su reproduccion. Physis 28: 499-503.

Berois, N. and De Sa, R. 1988. Histology of the ovaries and fat bodies of *Chthonerpeton indistinctum*. Journal of Herpetology 22: 146-151.

Bhatta, G. K. 1987. Some aspects of reproduction in the apodan amphibian *Ichthyophis*. PhD. Dissertation, Karnatak University, Dharwad, India.

Blair, A. P. 1946. The effects of various hormones on primary and secondary sex characters in juvenile *Bufo fowleri*. Journal of Experimental Zoology 103: 365-400.

Bons, J. 1986. Données histologiques sur le tube digestif de *Typhlonectes compressicaudus* (Duméril et Bibron, 1841) (Amphibien Apode). Mémoires de la Société Zoologique de France. 43: 87-90.

Brauer, A. 1902. Beitrage zur kenntniss der Entwicklung und Anatomie der Gymnophionen. III. Die Entwicklung der Excretionsorgane. Zoologisches Jahrbuch für Anatomie 16: 1-176.

Breckenridge, W. R. and Jayasinghe, S. 1979. Observations on the eggs and larvae of *Ichthyophis glutinosus*. Ceylon Journal of Sciences (Biol. Sci.) 13: 187-202.

Cei, G. 1944. Analisi biogeografica e ricerche biologiche e sperimenti sul ciclo sessuale annuo delle rana rosse d'Europa. Monitore Zoologico Italiano N.S. Suppl. 54: 1-117.

Chadwick, C. S. 1940. Identity of prolactine with water drivefactor in *Triturus vridescens*. Proceedings of Society for Experimental Biology and Medicine 45: 334-337.

De Sa, R. and Berois, N. 1986. Spermatogenesis and histology of the testes of the Caecilian *Chthonerpeton indistinctum*. Journal of Herpetology 20: 510-514.

De Vlaming, V. L. 1979. Actions of prolactine among the vertebrates. Pp. 561-642. In E. J. W. Barrington (ed) *Hormones and Evolution*, Academic Press, New York.

Del Rio, G., Citarella, F. and d'Istria, M. 1980. Androgen receptor in the thumb pad of *Rana esculenta*: dynamic aspects. Journal of Endocrinology 85: 279-282.

Dittus, P. 1936. Interrenalsystem und chromaffine Zellen im Lebensablauf von *Ichthyophis glutinosus*. Zeitschrift für wissen der Zoologie 147: 459-512.

Doerr-Schott, J. and Zuber-Vogeli, M. 1984. Immunohistochemical study of the adenohypophysis of *Typhlonectes compressicaudus* (Amphibia, Gymnophiona). Cell and Tissue Research 235: 211-214.

Doerr-Schott, J. and Zuber-Vogeli, M. 1986. Cytologie et immunocytologie de l'hypophyse de *Typhlonectes compressicaudus*. Mémoire de la Société Zoologique de France 43: 77-79.

Dorne, J.-L. and Exbrayat, J.-M. 1996. Quelques aspects de l'anatomie et de l'histologie des glandes interrénales chez trois Gymnophiones. Bulletin de la Société Zoologique de France 121: 146-147.

Exbrayat, J.-M. 1983. Premières observations sur le cycle annuel de l'ovaire de *Typhlonectes compressicaudus* (Duméril et Bibron, 1841), Batracien Apode vivipare. Comptes rendus des Séances de l'Académie des Sciences de Paris 296: 493-498.

Exbrayat, J.-M. 1984. Quelques observations sur l'évolution des voies génitales femelles de *Typhlonectes compressicaudus* (Duméril et Bibron, 1841), Amphibien Apode vivipare, au cours du cycle de reproduction. Comptes rendus des Séances de l'Académie des Sciences de Paris 298: 13-18.

Exbrayat, J.-M. 1985. Cycle des canaux de Müller chez le mâle adulte de *Typhlonectes compressicaudus* (Duméril et Bibron, 1841), Amphibien Apode. Comptes rendus des Séances de l'Académie des Sciences de Paris 301: 507-512.

Exbrayat, J.-M. 1986a. Quelques aspects de la biologie de la reproduction chez *Typhlonectes compressicaudus* (Duméril et Bibron, 1841), Amphibien Apode. Doctorat és Sciences naturelles, Université de Paris VI, France.

Exbrayat, J.-M. 1986b. Le testicule de *Typhlonectes compressicaudus*: structure, ultrastructure, croissance et cycle de reproduction. Mémoires de la Société Zoologique de France 43: 121-132.

Exbrayat, J.-M. 1988a. Croissance et cycle des voies génitales femelles chez *Typhlonectes compressicaudus* (Duméril et Bibron, 1841), Amphibien Apode vivipare. Amphibia Reptilia: 117-137.

Exbrayat, J.-M. 1988b. Variations pondérales des organes de réserve (corps adipeux et foie) chez *Typhlonectes compressicaudus*, Amphibien Apode vivipare au cours des alternances saisonnières et des cycles de reproduction. Annales des Sciences Naturelles, Zoologie, 13éme série 9: 45-53.

Exbrayat, J.-M. 1989a. Quelques observations sur les appareils génitaux de trois Gymnophiones; hypothéses sur le mode de reproduction de *Microcaecilia unicolor*. Bulletin de la Société Herpétologique de France 52: 34-44.

Exbrayat, J.-M. 1989b. The cytological modifications of the distal lobe of the hypophysis in *Typhlonectes compressicaudus* (Duméril and Bibron, 1841), Amphibia Gymnophiona, during the cycles of seasonal activity. I—In adult males. Biological Structures and Morphogenesis 2: 117-123.

Exbrayat, J.-M. 1991 Anatomie du cloaque chez quelques Gymnophiones. Bulletin de la Société Herpétologique de France 58: 30-42.

Exbrayat, J.-M. 1992a. Appareils génitaux et reproduction chez les Amphibiens Gymnophiones. Bulletin de la Société Zoolologique de France 117: 291-296.

Exbrayat, J.-M. 1992b. Reproduction et organes endocrines chez les femelles d'un Amphibien Gymnophione vivipare, *Typhlonectes compressicaudus*, Bulletin de la Société Herpétologique de France 64: 37-50

Exbrayat, J.-M. 1996. Croissance et cycle du cloaque chez *Typhlonectes compressicaudus* (Duméril et Bibron, 1841), Amphibien Gymnophione. Bulletin de la Société Zoologique de France 121: 99-104.

Exbrayat, J.-M. and Collenot, G. 1983. Quelques aspects de l'évolution de l'ovaire de *Typhlonectes compressicaudus* (Duméril et Bibron, 1841), Batracien Apode vivipare. Etude quantitative et histochimique des corps jaunes. Reproduction., Nutrition, Développement 23: 889-898.

Exbrayat, J.-M. and Dansard, C. 1994. Apports de techniques complémentaires à la connaissance de l'histologie du testicule d'un Amphibien Gymnophione. Revue Française d'Histotechnologie 7: 19-26.

Exbrayat, J.-M. and Delsol, M. 1985. Reproduction and growth of *Typhlonectes compressicaudus*, a viviparous Gymnophione. Copeia 1985: 950-955.

Exbrayat, J.-M. and Flatin, J. 1985. Les cycles de reproduction chez les Amphibiens Apodes. Influence des variations saisonnières. Bulletin de la Société Zoologique de France 110: 301-305.

Exbrayat, J.-M. and Morel, G. 1990-1991. The cytological modifications of the distal lobe of the hypophysis of *Typhlonectes compressicaudus* (Duméril and Bibron, 1841), Amphibia Gymnophiona, during the cycles of seasonal activity. II—In adult males. Biological Structures and Morphogenesis, 3: 129-138.

Exbrayat, J.-M. and Morel, G. 1995. Prolactin (PRL)-coding mRNA in *Typhlonectes compressicaudus*, a viviparous gymnophionan Amphibian. An in situ hybridization study. Cell and Tissue Research 280: 133-138.

Exbrayat, J.-M. and Morel, G. 2003. Visualization of gene expression of prolactin-receptor (PRL-R) by in situ hybridization in reproductive organs of *Typhlonectes compressicauda*, a gymnophionan amphibian. Cell and Tissue Research 312: 361-367.

Exbrayat, J.-M. and Sentis, P. 1982. Homogénéité du testicule et cycle annuel chez *Typhlonectes compressicaudus* (Duméril et Bibron, 1841), Amphibien Apode vivipare. Comptes rendus des Séances de l'Académie des Sciences de Paris 294: 757-762.

Exbrayat, J.-M., Delsol, M. and Flatin J. 1981. Premières remarques sur la gestation chez *Typhlonectes compressicaudus* (Duméril et Bibron, 1841) Amphibien Apode vivipare. Comptes rendus des Séances de l'Académie des Sciences de Paris 292: 417-420.

Exbrayat, J.-M., Ouhtit, A. and Morel, G. 1996a. Prolactin (PRL) and prolactin receptor (PRL-R) mRNA expression in *Typhlonectes compressicaudus* (Amphibia, Gymnophiona) female sexual organs. An in situ hybridization study. 18th Conference of European Society of Comparative Endocrorinology, Rouen, France, Septembre 1996. Annales d'Endocrinologie 57: supplt, add.

Exbrayat, J.-M., Ouhtit, A. and Morel, G. 1996b. Prolactin (PRL) and prolactin receptor (PRL-R) mRNA expression in *Typhlonectes compressicaudus* (Amphibia, Gymnophiona) male genital organs. An in situ hybridization study. 18th Conference of European Society of Comparative Endocrinology, Rouen, France, Sept. 1996. Annales d'Endocrinologie 57: supplement, 55.

Exbrayat, J.-M, Ouhtit, A. and Morel, G. 1997. Visualization of gene expression of prolactin receptors (PRL-R) by in situ hybridization, in *Typhlonectes compressicaudus*, a gymnophionan Amphibian. Life Science 61: 1915-1928.

Exbrayat, J.-M., Pujol, P. and Leclerq, B. 1998. Quelques aspects des cycles sexuels et nycthéméraux chez les Amphibiens. Bulletin de la Société Zoologique de France 123: 113-124.

Follett, B. K., Redshaw, M. R. 1974. The physiology of vitellogenesis. Pp. 219-308. In Lofts B. (ed). *Physiology of the Amphibia*. vol. II, Academic Press, New York, London.

Fraile, B., Paniagua, R., Saez, F. J., Rodriguez, M. C. and Lizana, M. 1989. Immunocytochemical and quantitative study of interstitial cells in the high mountains toad *Bufo bufo gredosicola* during the spermatogenetic cycle. Herptological Journal 1: 330-335.

Gabe, M. 1971. Données histologiques sur la glande surrénale d'*Ichthyophis glutinosus*. Archives de Biologie 82: 1-23.

Gabe, M. 1972. Contribution à l'histologie du complexe hypothalamo-hypophysaire d'*Ichthyophis glutinosus* L. (Batracien Apode). Acta Anatomica 81: 253-269.

Garg, B. L. and Prasad, J. 1962. Observations of the female urogenital organs of limbless amphibian *Uraeotyphlus oxyurus*. Journal of Animal Morphology and Physiology 9: 154-156.

George, J. M., Smita, M., Oommen, O. V. and Akbarsha, M. A. 2004a. Anatomy, histology and ultrastructure of male Mullerian gland of *Uraeotyphlus narayani* (Amphibia: Gymnophiona). Journal of Morphology 260: 33-56.

George, J. M., Smita, M., Kadalmani, B., Girija, R., Oommen, O. V. and Akbarsha M. A. 2004b. Secretory and basal cells of the epithelium of the tubular glands in the male Mullerian gland of the caecilian *Uraeotyphlus narayani* (Amphibia: Gymnophiona). Journal of Morphology 262: 760-769.

George, J. M., Smita, M., Kadalmani, B, Girija, R., Oommen, O. V. and Akbarsha, M. A. 2005. Contribution of the secretory material of caecilian (Amphibia: Gymnophiona) male Mullerian gland to motility of sperm: a study in *Uraeotyphlus narayani*. Journal of Morphology 263: 227-237.

Greenberg, B. 1942. Some effects of testosterone on the sexual pigmentation and other sex characters on the cricket frog (*Acris gryllus*). Journal of Experimental Zoology 91: 435-446.

Hraoui-Bloquet, S. 1995. Nutrition embryonnaire et relations materno-foetales chez *Typhlonectes compressicaudus* (Duméril et Bibron, 1841), Amphibien Gymnophione vivipare. Thèse de Doctorat E.P.H.E., Lyon, France.

Joly, J. 1964. Présence de lipides biréfringents dans les corps jaunes post-ovulaires de la salamandre tachetée. Comptes Rendus des Séances de l'Acdadémie des Sciences de Paris 258 : 3563-3565.

Joly, J. 1965. Mise en évidence histochimique d'une d5 3b hydroxystéroïde-déshydrogénase dans l'ovaire de l'Urodèle *Salamandra salamandra* (L.) à différents stades du cycle sexuel. Comptes rendus des Séances de l'Académie des Sciences de Paris 261: 1569-1571.

Joly, J. 1971. Les cycles sexuels de *Salamandra* (L.). I. Cycle sexuel des mâles. Annales des Sciences naturelles, Zoologie, 12ème série 13: 451-503.

Joly, J. 1986. La reproduction de la salamandre terrestre. Pp. 471-486. In P.P. Grassé et M. Delsol (eds), *Traité de Zoologie*, tome XIV, fasc.I B, Masson, Paris.

Joly, J. and Picheral, B. 1972. Ultrastructure, histochimie et physiologie du follicule pré-ovulatoire et du corps jaune de l'Urodèle ovovivipare *Salamandra salamandra* (L.). General and Comparative Endocrinology 18: 235-259.

Kerr, T. 1965. Histology of the distal lobe of the pituitary of *Xenopus laevis* Daudin. General and Comparative Endocrinology: 232-240.

Klumpp, W. and Eggert, B. 1934. Die Schilddrüse und die branchiogenen Organe in *Ichthyophis glutinosus*. L. Zeitschrift für wissen der Zoologie 146: 329-381.

Kuhlenbeck, H. 1970. A note on the morphology of the hypophysis in the Gymnophione *Schistometopum thomense*. Okajimas Folia Anatomica Japonica 46: 307-319.

Lamotte, M., Rey and P. and Vogeli, M. 1964. Recherches sur l'ovaire de *Nectophrynoides occidentalis*, Batracien Anoure vivipare. Archives d'Anatomie, Microscopie et Morphologie expérimentale 53: 179-224.

Laubmann, W. 1926. Die Entwicklung der Hypophyse bei *Hypogeophis rostratus*. Zeitschrift für Anatomie und Entwicklung 79-103.

Lawson, R. 1959. The anatomy of *Hypogeophis rostratus* Cuvier. Amphibia: Apoda or Gymnophiona. PhD Dissertation, University of Durham, King's College, U.K.

Leclercq, B. 1995. Contribution à l'étude du complexe pinéal des Amphibiens actuels. Diplome E.P.H.E., Lille, France.

Leclercq, B., Martin-Bouyer, L. and Exbrayat, J.-M. 1995. Embryonic development of pineal organ in *Typhlonectes compressicaudus* (Dumeril and Bibron, 1841), a viviparous Gymnophonan Amphibia. Pp. 107-111. In G. A. Llorente, A. Montosi, X. Santos and M. A. Carretero (eds), *Scientia Herpetologica*. Proceedings of the 7th General Meeting of the Societas Europaea Herpetologica, Barcelona, Spain.

Lofts, B. 1974. Reproduction. Pp. 108-218. In B. Loft, (ed), *Physiology of the Amphibia*, vol. II. Academic Press, New York, London.

Lupo, C., Zerani, M., Carnevali, O., Gobetti, A. and Polzonetti-Magni, A.M. 1988. Testosterone binding protein in the encephalon and plasma sex hormones during the annual cycle in *Rana esculenta* complex (Amphibia, Ranidae). Monitore Zoologico Italiano 22: 133-144.

Lynch, L. C. and Blackburn, D. G. 1995. Effects of testosterone administration and gonadectomy on nuptila pad morphology in overwintering male leopard frog, *Rana pipiens*. Amphibia Reptilia 16: 113-121.

Malonza, P.K. and Measey, G.J. 2005. Life history of an African caecilian: *Boulengerula taitanus* Loveridge 1935 (Caeciilidae: Amphibia: Gymnophiona). Tropical Zoology 18: 49-66.

Marcus, H. 1939. Beitrag zur kenntnis der Gymnophionen. Ueber keimbahn, keimdruusen, Fettkörper und Urogenitalverbindung bei *Hypogeophis*. Biomorphosis 1: 360-384.

Masood Parveez, U. 1987. Some aspects of reproduction in the female Apodan Amphibian *Ichthyophis*. PhD., Karnatak University, Dharwad, India.

Masood-Parveez, U. and Nadkarni, B. 1991. Morphological, histological, histochemical and annual cycle of the oviduct in *Ichthyophis beddomei* (Amphibia: Gymnophiona). Journal of Herpetology 25: 234-237.

Masood-Parveez, U. and Nadkarni, B. 1993a. The ovarian cycle in an oviparous gymnophione amphibian, *Ichthyophis beddomei* (Peters). Journal of Herpetology 27: 59-63.

Masood-Parveez, U. and Nadkarni, B. 1993b. Morphological, histological and histochemical studies of the ovary of an oviparous Caecilian, *Ichthyophis beddomei* (Peters). Journal of Herpetology 27: 63-69.

Masood-Parveez, U., Bhatta, G.K. and Nadkarni, V.B. 1992. Interrenal of a female gymnophione amphibian, *Ichthyophis beddomei*, during the annual reproductive cycle. Journal of Morphology 211: 201-206.

Maurer, F. 1887. Schildrüse, Thymus und Kiemenreste der Amphibien. Morphologisches Jahrbuch 13: 296-382.

Mazzi, V., Vellano, C. and Toscano, C. 1967. Antigonadal effects of prolactin in adult male newt (*Triturus cristatus carnifex* Laur.). General and Comparative Endocrinology 8: 320-324.

Moodie, G. E. E. 1978. Observations on the life history of the caecilian *Typhlonectes compressicaudus* (Dumeril and Bibron) in the Amazon Basin. Canadian Journal of Zoology 56: 1005-1008.

Oyama, J. 1952. Microscopical study of the visceral organs of Gymnophiona, *Hypogeophis rostratus*. Kumamoto Journal of Science 1B: 117-125.

Ozon, R. and Xavier, F. 1968. Biosynthèse in vitro des stéroïdes par l'ovaire de l'Anoure vivipare *Nectophrynoides occidentalis* au cours du cycle sexuel. Comptes rendus des Séances de l'Académie des Sciences de Paris, série. D 266: 1173-1175.

Parker, H. W. 1934. Reptiles and Amphibians from Southern Ecuador. Annals and Magazine of Natural History 10: 264-273.

Parker, H. W. 1936. The Caecilians of the Mamfe Division, Cameroons. Proceedings of Zoological Society of London 1936: 135-163.

Pasteels, J. L. and Herlant, M. 1962. Les différentes catégories de cellules chromophiles dans l'hypophyse des Amphibiens. Anatomischer Anzeiger 109: 764-767.

Pillay, K. V. 1957. The hypothalamo-hypophyseal neurosecretory system of *Gegenophis carnosus* Beddome. Zeitschrift für Zellforschung und mikroskopische Anatomie 46: 577-582.

Rastogi, R. K. and Chieffi, G. 1970. Cytological changes in the pars distalis of pituitary of the green frog *Rana esculenta* L. during the reproductive cycle. Zeitschrift für Zellforschung und Mikroskopische Anatomie 111: 505-518.
Rathke, H. 1852. Bemerkungen uber mehrere Korpert heile der *Coecilia annulata*. Mullers Arkiv 1852: 334-350.
Redshaw, M. R. 1972. The hormonal control of the amphibian ovary. American Zoologist 12: 289-306.
Sarasin, P. and Sarasin, F. 1887-1890. Ergebnisse Naturwissenschaftlicher Forschungen auf Ceylon. Zur Entwicklungsgeschichte und Anatomie der Ceylonischen Blindwuhle *Ichthyophis glutinosus*. C. W. Kreidel's Verlag, Wiesbaden.
Schubert, C. and Welsch, U. 1979. Elektronmikroskopische Beobachtungen ander Adenohypophyse von Gymnophionen (*Chthonerpeton indistinctum*). Zoologisches Jahrbuch für Anatomie 101: 105-112.
Schubert, C., Welsch, U. and Goos, H. 1977. Histological, immuno and enzyme-histochemical investigations on the adenohypophysis of the Urodeles, *Mertensiella caucasica* and *Triturus cristatus* and the Caecilian, *Chthonerpeton indistinctum*. Cell and Tissue Research 185: 339-349.
Semon, R. 1892. Studien über den Bauplan des Urogenitalsystem der Wirbeltiere. Dargelegt an der Entwicklung dieses organysystems bei *Ichthyophis glutinosus*. Jena Zeitschrift. Naturwissenschaften 26: 89-203.
Seshachar, B. R. 1936. The spermatogenesis of *Ichthyophis glutinosus* (Linn.) I. The spermatogonia and their division. Zeitschrift für Zellforschung und Mikroskopische Anatomie 24: 662-706.
Seshachar, B. R. 1937. The spermatogenesis of *Ichthyophis glutinosus* (Linn.). II. The meiotic divisions. Zeitschrift für Zellforschung und Mikroskopische Anatomie 27: 133-158.
Seshachar, B. R. 1943. The spermatogenesis of *Ichthyophis glutinosus* (Linn.), III. Spermateleosis. Proceedings of National Institute of Sciences of India 9: 271-285.
Spengel, J. W. 1876. Das Urogenitalsystem der Amphibien. I. Theil. Der Anatomische Bau des Urogenitalsystem. Arbeitenaus dem Zoologzootom Institute Wurzburg 3: 51-114.
Stendell L. W. 1914. Die Hypophysis Cerebri. In Oppel's Lehrbuch der Vergleichender Mikropischen Anatomie, Fisher (ed.), Jena, 8: 1-168.
Taylor, E. H. 1968. *The Caecilians of the World. A Taxonomic Review*. University of Kansas Press, Lawrence, 848 pp.
Taylor, E. H. 1969. A new family of African Gymnophiona. University of Kansas Scientific Bulletin 48: 297-305.
Tonutti, E. 1931. Beitrag zur Kenntnis der Gymnophionen. XV. Das Genital-system. Morphologisches Jahrbuch 68: 151-292.
Tonutti, E. 1932. Vergleichende morphologische Studen uber Eddarm und Kopulations-organe. Morphologisches Jahrbuch 70: 101-130.
Tonutti, E. 1933. Beitrag zur Kenntnis der Gymnophionen. XIX. Kopulations-organe bei Weiteren Gymnophionenarten. Morphologisches Jahrbuch 72: 155-211.
Van Oordt, P. G. W. 1974. Cytology of the adenohypophysis. Pp 53-106. In B. Loft, (ed), *Physiology of the Amphibia*, vol. II, Academic Press, New York, London.
Van Oordt, P. G. W., Van Dongen, W.J. and Loft, B. 1968. Seasonal changes in endocrine organs of the male commn frog *Rana temporaria*. I. The pars distalis of the adenohypophysis. Zeitschrift für Zellforschung und Mikroskopische Anatomie 88: 549-559.

Vellano, C., Peyrot, A. and Mazzi, V. 1967. Effects of prolactin on the pituito-thyroid axis, integument and behavior of the adult male crested newt. Monitore Zoologico Italiano 1: 207-227.

Vilter, A. and Vilter, V. 1962. Rôle de l'oviducte dans l'uniparité utérine de la Salamandre noire des Alpes orientales (*Salamandra atra*). Compte Rendus de la Société de Biologie 156: 49-51.

Vilter, V. 1966. Histochimie de l'oviducte chez *Salamandra atra* mature, Urodèle totalement vivipare de haute montagne. Comptes Rendus de la Société de Biologie 161: 260-264.

Vilter, V. 1968. Histochimie quantitative des sécrétions mucipares acides de l'oviducte mature de *Salamandra atra* des Alpes vaudoises. Comptes Rendus de la Société de Biologie 162: 76-80.

Vilter, V. 1986. La reproduction de la salamandre noire. Pp 487-495. In P.P. Grassé et M. Delsol (eds), *Traité de Zoologie*, tome XIV, fasc.I B, Masson, Paris.

Vilter, V. and Vilter, A. 1960. Sur la gestation de la salamandre noire des Alpes, *Salamandra atra* Laur. Comptes Rendus de la Société de Biologie 154 : 290-294.

Wake, M. H. 1968a. The comparative morphology and evolutionary relationships of the urogenital system of Caecilians. PhD. Dissertation, University of California.

Wake, M. H. 1968b. Evolutionary morphology of the Caecilian urogenital system. Part I: the gonads and fat bodies. Journal of Morphology 126: 291-332.

Wake, M. H. 1970. Evolutionary morphology of the caecilian urogenital system. Part II: the kidneys and urogenital ducts. Acta Anatomica 75: 321-358.

Wake, M. H. 1972. Evolutionary morphology of the caecilian urogenital system. Part IV: the cloaca. Journal of Morphology 136: 353-366.

Wake, M. H. 1977a. Fetal maintenance and its evolutionary significance in the Amphibia: Gymnophiona. Journal of Herpetology 11: 379-386.

Wake, M. H. 1977b. The reproductive biology of Caecilians. An evolutionary perspective. Pp. 73-100. In D. H. Taylor and S. I. Guttman (eds), *The Reproductive Biology of Amphibians*, Miami University, Oxford, Ohio.

Wake, M. H. 1980. Reproduction, growth and population structure of the Central American Caecilian *Dermophis mexicanus*. Herpetologica 36: 244-256.

Wake, M. H. 1981. Structure and function of the male Mullerian gland in Caecilians (Amphibia: Gymnophiona), with comments on its evolutionary significance. Journal of Herpetology 15: 17-22.

Wake, M.H. 1995. The spermatogenic cycle of *Dermophis mexicanus* (Amphibia: Gymnophiona). Journal of Herpetology 29: 119-122.

Wake, M. H. and Dickie, R. 1998. Oviduct structure and function and reproductive modes in Amphibians. The Journal of Experimental Zoology 282: 477-506.

Welsch, U. and Schubert, C. 1975. Observations on the fine structure, enzyme histochemistry and innervation of parathyroid gland and ultimobranchial body of *Chthonerpeton indistinctum* (Gymnophiona, Amphibia). Cell and Tissue Research 164: 105-119.

Welsch, U., Schubert, C. and Storch, V. 1974. Investigation on the thyroid gland of embryonic, larval and adult *Ichthyophis glutinosus* and *Ichthyophis kohtaoensis* (Gymnophiona, Amphibia). Histology, fine structure and studies with radioactive iodide (I 131). Cell and Tissue Research 155: 245-268.

Welsch, U., Schubert, C. and Siak Hauw Tan 1976. Histological and histochemical observations on the neurosecretory cells in the diencephalon of *Chthonerpeton indistinctum* and *Ichthyophis paucisulcus* (Gymnophiona, Amphibia). Cell and Tissue Research 175: 137-145.

Wiedersheim, R. 1879. *Die Anatomie der Gymnophionen*. Jena, Gustav. Fisher, 101 pp.

Xavier, F. 1986. La reproduction des *Nectophrynoïdes*. Pp. 497-513. In P.- P. Grassé et M. Delsol (eds), *Traité de Zoologie*, Tome XIV, fasc. I-B, Masson, Paris, France.

Xavier, F. et Ozon, R. 1971. Recherches sur l'activité endocrine de l'ovaire de *Nectophrynoïdes occidentalis* Angel (Amphibien Anoure vivipare). 2—Synthèse in vitro des stéroïdes. General and Comparative Endocrinology 16: 30-40.

Xavier, F., Zuber-Vogeli, M. and Le Quang Trong, Y. 1970. Recherches sur l'activité endocrine de l'ovaire de *Nectophrynoïdes occidentalis* Angel (Amphibien Anoure vivipare). 1—Etude histochimique. General and Comparative Endocrinology 15: 425-431.

Zuber-Vogeli, M. 1968. Les variations cytologiques de l'hypophyse distale des femelles de *Nectophrynoides occidentalis*. General and Comparative Endocrinology 11: 495-514.

Zuber-Vogeli, M. and Doerr-Schott, J. 1976. L'ultrastructure de quatre catégories cellulaires de la pars distalis de *Nectophrynoides occidentalis* Angel (Amphibien Anoure vivipare). General and Comparative Endocrinology 28: 299-312.

Zuber-Vogeli, M. and Doerr-Schott, J. 1981. Description morphologique et cytologique de l'hypophyse de *Typhlonectes compressicaudus* (Duméril et Bibron) (Amphibien Gymnophione de Guyane française). Comptes rendus des Séances de l'Académie des Sciences de Paris, série D 292: 503-506.

CHAPTER 6

Caecilian Spermatogenesis

Mathew Smita [1], George M. Jancy [2], Mohammad A. Akbarsha[2], Oommen V. Oommen[1] and Jean-Marie Exbrayat[3]

6.1 INTRODUCTION

Caecilians constitute an interesting group of anamniote vertebrates with several aspects of biology that are unique. Such uniqueness has made scientists interested in caecilians divide themselves into two groups, one considering caecilians to be degenerate and the other as a specialized entity. We have the fortunate experience of dealing directly with the biology of caecilians of the Western Ghats of India in general and the biology of male reproduction in particular. There are reasons to believe that caecilians are more advanced than other amphibians especially in the context of reproduction which involves internal fertilization and partial or complete viviparity accompanied by, in several cases, omission of an aquatic larval phase. It is the internal fertilization which made us inquisitive about *understanding* the biology of the male reproduction in caecilians, thanks to the rediscovery of their abundance in the Western Ghats of Kerala, India (Oommen *et al.* 2000). We provide an outline presentation of the findings, against the background of the already available literature on spermatogenesis of caecilians. In this attempt, we have tried our best not to repeat the material already covered in recent reviews.

6.2 HISTORICAL PERSPECTIVES

Different aspects of the reproductive biology and life cycles of Gymnophiona have only been poorly investigated. The earliest review on male genital organs and spermatogenesis of Gymnophiona is that of Spengel (1876). Work

---

[1] Department of Zoology, University of Kerala, Kariavattom 695 581, Thiruvananthapuram, Kerala, India
[2] Department of Animal Sciences, Bharathidasan University, Tiruchirappalli 620 024, Tamil Nadu, India
[3] Laboratoire de Biologie générale, Université catholique de Lyon et Laboratoire de reproduction et Développement des Vertébrés, Ecole Pratique des Hautes Etudes, 25 rue du Plat, 69288 Lyon Cedex 02, France.

on the anatomy and development of *Ichthyophis glutinosus* by Sarasin and Sarasin (1887-1890) paved the way for numerous later investigations. A review by Ballowitz (1913) denied any knowledge of the sperm of Apoda. Since Tonutti's (1931) account of the structure of the testis in *Hypogeophis* based on the testis material not properly fixed, Seshachar (1936) began research on the history and development of germ cells in *Ichthyophis glutinosus*. The pioneering work of what little has been done on caecilian male reproductive biology was that of Seshachar, who during 1930s and 1940s published several articles pertaining to male reproduction, anatomy, spermatogenesis (1936, 1937, 1939, 1940, 1942a), spermateleosis, sperm morphology (1943 a, b, 1945) and eggs and embryos of Gymnophiona of South India (1942b). He gave a description of the mature sperm of *Ichthyophis glutinosus*, *Uraeotyphlus narayani* and *Siphonops annulatus*. Subsequently he described the spermatogenesis in *Siphonops annulatus*, *Dermophis gregorii* (1942a) *Uraeotyphlus narayani* and *Gegeneophis carnosus* (1945). Seshachar also briefly described intra-locular oocytes in males (1942d). He necessarily restricted his study to light microscopy. Exbrayat and Sentis (1982) and Exbrayat (1986a, b) examined the annual testicular cycle of *Typhlonectes compressicauda*. De Sa and Berois (1986) described the spermatogenesis of *Chthonerpeton indistinctum*. Scheltinga *et al.* (2003) have given an account of spermatozoal ultrastructure in the families Ichthyophiidae (*Ichthyophis beddomei* and *I. tricolor*), Uraeotyphliidae (three forms, probably distinct species, of *Uraeotyphlus*), Caeciliidae (*Gegeneophis ramaswamii*) and *Typhlonectes natans* (Typhlonectidae) (chapter 7 of this volume). Wake (1970 a, b, 1972, 1977, 1981, 1994, 1995) published a series of papers dealing with reproduction in several species of caecilians. She has also attempted to trace the evolutionary morphology of caecilian urogenital system, gonads and fat bodies (Wake 1968).

6.3 SEASONALITY OF REPRODUCTION

Sarasin and Sarasin (1887-1890), who studied the reproductive biology of *Ichthyophis* in natural surroundings, concluded that *Ichthyophis* breeds from July to September. However, in Seshachar's (1936) opinion, *Ichthyophis* breeds from December to February or March, which should be reflections of the breeding activity of the females. The embryonic period is very long and the gilled stage of development is passed inside the eggshell. He found tiny larvae in May and June in a small stream. According to Seshachar (1936), the testis of *Ichthyophis glutinosus* shows signs of inactivity in December, indicating the winter regression after contributing sperm for the breeding activity for the year. The winter testis lobes are more flattened and the locules are smaller in size due to the development of a large quantity of connective tissue between the locules, which take on a spongy appearance because of the presence of a number of lymph spaces. This condition lasts till March of the following year when the spermatogonia spring into activity again and divide as a result of which the locule enlarges and assumes the conditions that have been described towards contributing sperm for the breeding activity for the following year.

Wake (1968) made a classical study of spermatogenesis of *Gymnopis multiplicata* over a period from July to December and identified six phases in the spermatogenic cycle. In contrast, she found that *Dermophis mexicanus* undergoes continuous spermatogenesis. Exbrayat (1986a, b) and Exbrayat and Sentis (1982) reported that the reproductive cycle of *Typhlonectes compressicauda* was not continuous, with spermatogenesis occurring at the beginning of the rainy season in January-February, followed by regression during the next rainy season (April-May). Our observations on the activity of the testis based on histological procedures in *Ichthyophis tricolor* reveal three phases of spermatogenetic activity *viz.* active spermatogenesis (July-November), early regression (December-March) and spermatogenetic quiescence (April-June). In *Ichthyophis malabarensis* there is inactivity in July while *Ichthyophis tricolor* is active in spermatogenesis during the same period (unpublished data).

6.4 ANATOMY AND MORPHOLOGY OF THE TESTIS

Though the different taxa show variation in the pattern of gross morphology of the testis, the general disposition is that lobes are strung along a connective tissue strand and occupy both sides of the alimentary canal, through almost two-thirds the length of the body. The number of testicular lobes in each side varies among various species of caecilians (see also chapter 3 of this volume) and only a very brief description is attempted in this review in this regard as little contribution has been made since Wake's (1968) review. In the entire study on specimens of *Dermophis gregorii* that Seshachar (1942a) dissected, the lowest number of testes lobes was found to be three pairs. In our work on species belonging to three different families, we found *Ichthyophis tricolor* of the family Ichthyophiidae and *Uraeotyphlus* cf. *narayani* of the family Uraeotyphliidae to have lobes between 9 and 14 on each side (Smita *et al.* 2003, 2004a) whereas in *Gegeneophis ramaswamii* of the family Caeciliidae it was 8 in most cases. African caecilians *Geotrypetes seraphinii*, *Gymnopis multiplicata*, *Schistometopum gregorii*, *S. thomensis* and *Herpele squalostoma* have variable numbers and sizes of testis lobes. Wake (1968) considered *Scolecomorphus kirkii*, *S. uluguruensis* and *S. vittatus* to reflect a condition intermediate for caecilians. The diminutive *Idiocranium russeli* has reduced and highly fused testis. In the New World caecilians, *Rhinatrema* sp. has the highest number of nearly uniform testis lobes. *Siphonops braziliensis*, *Chthonerpeton viviparum* and *Typhlonectes compressicauda* have enlarged anterior lobes. Members of *Gymnopis* and *Dermophis* have only a few lobes. *Dermophis mexicanus* has the greatest degree of testis fusion (Wake 1968). Thus, the above description makes it clear that there is variation in the number of testis lobes in various caecilian species reflecting an evolutionary trend towards reduction of lobes through fusion in the derived species.

6.5 HISTOLOGY OF THE TESTIS

Spengel (1876) in his description of spermatogenesis in *Caecilia gracilis, Ichthyophis glutinosus* and *Siphonops annulatus,* described the testis lobes as containing a number of chambers which he called locules. Seshachar adopted the same terminology in describing the testis of *Ichthyophis glutinosus* (1936), *Uraeotyphlus narayani* (1939), *Siphonops annulatus* and *Dermophis gregorii* (1942a). Wake (1968, 1977) considered locules described by Spengel (1876) as lobules and Seshachar (1936, 1939, 1942a) treated the cell nests inside the lobules as locules. The description of Wake (1968, 1977) appears more appropriate and, thus, is adopted in the present review.

Our study reveals that the lobules are somewhat circular to hexagonal in outline and that each lobule is encased by the basement membrane and a framework of fibers and cells which give support to germ cells and Sertoli cells (Smita *et al.* 2004a) (Fig. 6.1A). It was pointed out by Seshachar (1936) that interstitial tissue occupies the triangular areas between the roughly hexagonal locules (lobules, in our usage) and also similar areas in the periphery of the testis. He made the interesting observation that there are more interstitial cells in the resting testis than in the active testis in *Ichthyophis glutinosus*, indicating that the abundance of interstitial cells is inversely related to spermatogenic activity. Our study (Smita *et al.* 2005) corroborates the views of Seshachar (1941) (Fig. 6.1B). A major portion of the interstitial tissue consists of Leydig cells, but little is known regarding the role of Leydig cells in the secretion of male hormones and their role in the regulation of spermatogenesis. Therefore, there is a pertinent need to investigate the structure and function of Leydig cells in caecilians. Another component of lesser importance present in the interstitial areas is fibroblasts. Likewise, sperm collecting ducts constitute yet another component in the interstitial area, which assume importance in the context of the generation of spermatogonia from the epithelium (Seshachar 1936). Evidence that we obtained for *Ichthyophis tricolor* and *Uraeotyphlus* cf. *narayani*, confirms the origin of primary spermatogonium from the epithelium of collecting ductule (Fig. 6.1C) (unpublished data).

According to Seshachar (1936) the "fatty matrix" in the testis lobule of *Ichthyophis glutinosus* is essentially fat, held in a general matrix of mucus. He considered the product to have nutritional value and, hence, suggested that the origin of fat globules is to be traced to the interstitial cells. Alternatively, according to him, the fatty matrix could be in the nature of a degeneration product. Exbrayat has made notable comments on 'matrix' in a comprehensive contribution to reproductive anatomy of caecilians. His light microscopic observations suggest that each lobe is filled with a matrix in which a group of germ cells is floating. Electron microscopic studies give a clear picture of the concept of matrix as 'matrix filaments' and suggests a true cytoplasm in which classical organelles, voluminous vacuoles and osmiophilic inclusions (fat droplets) are observed (Exbrayat and Dansard 1994). Our observation supports the views of Exbrayat (1986a, b, 2000) that portions of Sertoli cells rich in lipid droplets, the so-called filamentous portions are only apical portions of Sertoli cells irrespective of the species (Smita *et al.* 2003) (Fig. 6.1D).

6.6 CELLULAR ORGANIZATION OF THE TESTIS LOBULE

One feature common to caecilian and other amphibian spermatogenesis is that all the cells in a single cyst/locule represent the same stage of differentiation and maturation (Lofts 1974, 1984). In Anura and Urodela, all the locules or the cell nests in the testis at a given point of time represent the same stage of differentiation and maturation unlike in caecilian testis, where locules/cysts in a lobule represent different stages of maturation (Fig. 6.1A). However, Wake (1968), in her review, has made a statement to the effect that groups of spermatozoa present in a single locule represent different stages of differentiation and maturation. In our observation in TBO-stained semi-thin sections of the testis of *Ichthyophis tricolor* and *Uraeotyphlus* cf. *narayani*, all the germ cells in an association (a cyst) are at the same stage of maturation; some loose irregular shaped cells with a dense heterochromatic nucleus appear in the lumen of the cysts containing primary spermatocytes (Stage IIIb) (Figs. 6.1E, F). Their association continues in all stages and even among sperm that are spermiated. Ultrastructural analysis revealed these irregular cells to possess blunt and lobose pseudopodial protuberances with a heterochromatic nucleus and a cytoplasm which is very different from the organization of Sertoli cells. We call these ameboid cells (Fig. 6.2A) (Smita *et al.* 2005).

Sertoli cells, the somatic cells of the testis, are a consistent feature in vertebrates. In the case of Anura and Urodela, Sertoli cells appear typical of the pattern in higher vertebrates, i.e. adherent to the wall of the cyst, with sperm heads embedded in them (Lofts 1984; Saidapur 1989). According to Seshachar (1942c), in *Ichthyophis glutinosus* and *Uraeotyphlus narayani* Sertoli cells occur at the periphery of the lobules, just beneath the septum. He could not find an association between Sertoli cells and the differentiating spermatids. While raising the question as to how the sperm become attached and are sustained, he indicated that the necessity of an association between the differentiating spermatids and the Sertoli cells is obviated as the spermatids are embedded in the "fatty matrix". The peripheral giant cells and the 'degenerative germ cells' which Seshachar (1936) had described are true Sertoli cells according to the interpretations given by Exbrayat (chapter 3 of this volume). His observation on the evolution of Sertoli cells suggests their origin from follicular cells that surround primary spermatogonia and this corroborates Seshachar's view (1942a). He also suggests that by completion of development, Sertoli cells degenerate, their volume reduces and they fall into the evacuative ductule, among spermatozoa, as round isolated cells. He critically observed the changes in Sertoli cells during the breeding cycle in *Typhlonectes compressicauda*. However, in a more recent paper van der Horst and van der Merwe (1991)

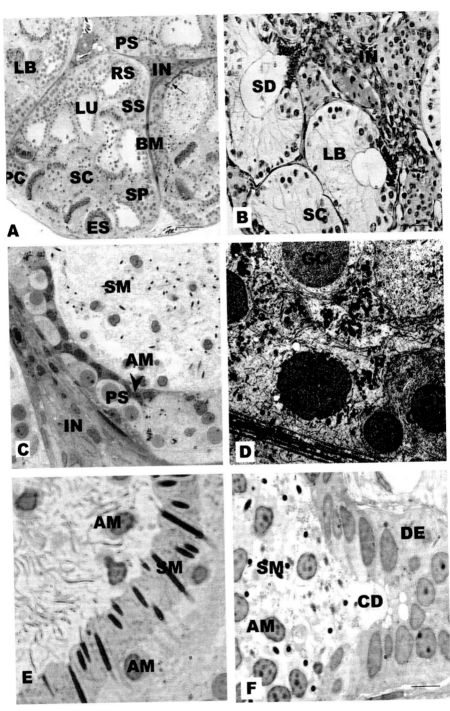

Fig. 6.1 contd

on *Typhlonectes natans*, claim that late spermatids are closely associated with the Sertoli cells with the cell membrane of the spermatids and Sertoli cells remaining closely adherent. The spermatid midpieces are situated in slight indentations around the Sertoli cells. In our observation on the testis of *Ichthyophis tricolor* and *Uraeotyphlus* cf. *narayani* from TBO-stained semi-thin sections and transmission electron microscopy, we find that Sertoli cells have a basally located pleiomorphic nucleus with a thin patch of dense heterochromatin just underneath the nuclear envelope and dispersed patches of heterochromatin more internally. The cytoplasm extends deep into the core of the lobule (Fig. 6.1D). The study throws light on their intimate association with germ cells at different stages of differentiation and suggests the permanency of Sertoli cells (Smita *et al.* 2003).

6.7 SPERMATOGENESIS

6.7.1 Stages of Spermatogenesis

Germ cells in the lobules of the caecilian testis, active in spermatogenesis, consist of cells and cell associations at various stages of differentiation and maturation. Spengel (1876), in his description of spermatogenesis in *Caecilia gracilis*, *Ichthyophis glutinosus* and *Siphonops annulatus*, identified seven stages of differentiation of the germ cells. Seshachar (1936) described six stages in the male germ cell development namely primary spermatogonia, secondary spermatogonia, primary spermatocytes, secondary spermatocytes, spermatids and sperm.

Fig. 6.1 contd

Fig. 6.1. A. Testis of *Uraeotyphlus* cf. *narayani* in active spermatogenesis showing spherical lobules containing the germ cell cyst (GC) and the Sertoli cells (SC). Interstitial areas (IA), basement membrane (BM) and cell cysts are also shown. Cell cysts with different primary spermatogonia (PS), secondary spermatogonia (SS), primary spermatocyte (PC), secondary spermatocyte (SP), round spermatids (RS), elongated spermatids (ES) are also shown. LM Toluidine blue-stained semi thin section. Scale bar = 100 µm. **B.** Regressed testis of *Ichthyophis tricolor* showing regressed lobule with wider interstitial area. Sertoli cells are found inside the lobules (LB) at the boundary Abbreviations: IN, Interstitial area, LB, lobule, SC, Sertoli cell; SD, spermiating duct. LM, Haematoxylin and eosin (stained section). Scale bar = 100 µm. **C.** Testis of *Uraeotyphlus* cf. *narayani* showing the origin of ameboid cells and primary spermatogonia from the epithelium of the collecting ductule. Progenitor ameboid cell (AM) and progenitor primary spermatogonia (PS) passing into the lobule (arrowheads). Also note the loose ameboid cells in the lumen, which match in organization to the progenitor ameboid cell. AM, ameboid cell; IN, interstitial area; PS, progenitor primary spermatogonia; SM, sperm; TBO stained semi-thin section. Scale bar = 25 µm. **D.** Testis of *Ichthyophis tricolor* showing Sertoli cells adherent to the basal lamina of the lobule supporting germ cells. The cytoplasm extends deep into the core of the lobule. BM, basement membrane; GC, germ cell cysts; SC, Sertoli cell (CY) cytoplasm. TEM. Scale bar = 20 µm. **E.** *Uraeotyphlus* cf. *narayani* testis section showing the association of ameboid cells with sperm cyst. AM, ameboid cell; SM, sperm. TBO stained semi-thin section. Scale bar = 20 µm. **F.** *Ichthyophis tricolor* testis section showing the association of ameboid cell with the sperm even in the collecting duct. AM, ameboid cell; CD, collecting duct; DE, duct epithelium; SM, sperm; TBO stained semi-thin section. Scale bar = 20 µm. Original.

6.7.2 Primary Spermatogonia

The primary spermatogonium was described by Spengel (1876) and De Sa and Berois (1986) as being found in the region where the sperm collecting duct widens into the lobule of the testis. In the observations of Seshachar (1936), the nucleus of the primary spermatogonium in the resting condition consists of a number of chromatin blocks connected together to form a network. Our observation identifies primary spermatogonia possessing basophilic nucleus with granular chromatin lying singly in the periphery of the lobule, bounded on all sides by Sertoli cells (Smita et al. 2004a).

Primary spermatogonia pass through four rounds of mitotic divisions to produce cell nests containing 2-16 secondary spermatogonia. The topography of the cytoplasmic inclusions of the primary spermatogonium was described briefly in the articles of Seshachar. According to Seshachar (1936) the primary spermatogonium undergoes six to eight divisions to result in the formation of spermatids. He counted up to 256 cells in the spermatid cysts of the caecilians he worked with. There are references in Seshachar's (1942a) study on *Siphonops annulatus* and *Dermophis gregorii* in which he referred to centrioles that are clearly defined and occupy the center of the archoplasmic area, which bears the same relationships with the nucleus and the cell, in general, as in *Ichthyophis* and *Uraeotyphlus*. The cell division brings about change in the topography of the nucleus. Seshachar (1936) described the chromosome changes and related the conversion of the polymorphic nucleus to a regular spherical condition, which marks the onset of division, in detail.

6.7.3 Secondary Spermatogonia

Division of primary spermatogonia produces secondary spermatogonia. The latter differ from the former in that they are smaller. A polymorphic condition was not observed in the spermatogonia of *Chthonerpeton indistinctum* by De Sa and Berois (1986). They hypothesized that it could be due either to the fact that spermatogonia in this species do not exhibit change in the nuclear morphology or that because of the high metabolic state of the primary spermatogonia it is very short and hence not observed.

6.7.4 Primary Spermatocytes

Division of secondary spermatogonia produces primary spermatocytes. The primary spermatocytes are arranged along the periphery of the locule in two rows and form compact masses. After a brief period of rest, the primary spermatocytes embark on the meiotic prophase (Seshachar 1937). This is followed by leptotene and pachytene after which the nucleus is marked by a "diffuse condition" when chromosome bivalents lose their identity temporarily. The nucleus presents the appearance of a resting state. When the bivalents emerge from this network, their chiasmata are clear and in the larger bivalents they are numerous but are probably reduced later as in *Ichthyophis* and *Uraeotyphlus*. The first, reductional division of meiosis gives two secondary spermatocytes. De Sa and Berois (1986) could not trace out the cellular limits of the spermatocytes but they noted that spermatocytes are smaller than spermatogonia.

6.7.5 Spermatids

After a brief period of interkinesis, the second, non-reductional, meiotic division occurs, giving rise to the spermatids.

6.7.6 Cell Nests

We have made an interesting observation in the distribution of the cell nests in the lobules (locules) in the testis of *Ichthyophis tricolor* and *Uraeotyphlus* cf. *narayani*. This is mainly that cysts of primary spermatogonia and secondary spermatogonia, until the 16/32 cell stage, are adherent to the basement membrane/septum and are associated in some way with the Sertoli cells of the lobule boundary. We have also noticed cell nests associating with the Sertoli cells until spermiation (Smita *et al.* 2003). The observation of van der Horst and van der Merwe (1991) that the differentiating spermatids remain associated with the Sertoli cells is comprehensible in the light of the topography of the Sertoli cells and spermatid cell nests.

6.8 STAGES IN SPERMATOGENESIS

6.8.1 Cell Associations

There have been successful attempts to identify stages in spermatogenesis in vertebrates in general where the term stage means a specific 'cell association', and is very different from the usage of Spengel (1876) and Seshachar (1936). In the case of non-cystic spermatogenesis, it is seen that cellular differentiation progresses along the seminiferous tubule in the form of a wave. Leblond and Clermont (1952) identified fourteen cell associations in the cycle of the seminiferous epithelium of rat. This study has been extended to several other higher vertebrates including reptiles, birds and mammals. These various species differ in the number of cell associations in the wave (de Kretser and Kerr 1994). In cystic spermatogenesis, the first successful attempt at such a classification was that of van Oordt (1956) in *Rana temporaria*. He identified six stages, each representing an exclusive cell association. The six stages are Stage 0 (primary spermatogonia in resting phase), Stage I (less than 10 secondary spermatogonia), Stage II (more than 10 secondary spermatogonia), Stage III a and b are two phases of primary spermatocytes, Stage IV (secondary spermatocytes) and Stage V a and b (spermatids in different phases of differentiation). Although this classification has been adopted for Anura and Urodela in general without any modification, a slight modification of this scheme combining Stage I and II as Stage I, making Stage III as II, Stage IV as III, Stage V a and b as IV and introducing a new Stage V referring to the sperm bundles attached to the Sertoli cells has also been introduced to identify the stages (Saidapur 1989).

6.8.2 Stages in Caecilian Spermatogenesis

There has been no previous attempt to identify corresponding stages in spermatogenesis of caecilians. In our study on the spermatogenesis of *Ichthyophis tricolor* and *Uraeotyphlus* cf. *narayani*, we identified the same number of stages, each being a cell association (Fig. 6.1A) matching the classification of van Oordt (1956). Cells at stage O are the primary spermatogonia characterized by spherical or oval nucleus and cytoplasm rich in lipid inclusions and lysosomes. They are found along the basal aspect of the Sertoli cells along the lobule wall. Stage I secondary spermatogonia have dense heterochromatic nuclei and cytoplasm rich in mitochondria, Golgi apparatus, lysosomes and lipid inclusions. They are sandwiched between Sertoli cells. Stage II represents spherical to oblong secondary spermatogonia with a densely vacuolated cytoplasm associated with Sertoli cells. Stage III a and b are cell nests of primary spermatocytes. Stage IIIa are found closer to the boundary of the lobule whereas stage III b cell nests sink deeper into the stroma. The primary spermatocytes arrive at the apices of the Sertoli cells and are enwrapped by Sertoli cell cytoplasm. The cytoplasm contains vacuoles and spherical mitochondria. Stage IIIb cell nests mark the formation of a lumen. Rod-shaped and filamentous mitochondria are present in the cytoplasm. Stage IV secondary spermatocytes are arranged in tiers around an expanded lumen and have nuclei with dense chromatin. Stage V spermatids are small round cells with nuclei orientated along the pole of the cell and a cytoplasm with vacuoles, mitochondria, SER and Golgi apparatus. Stage V b spermatids are transforming towards spermateleosis. The details are published elsewhere (Smita *et al.* 2004a). The developmental sequences of germ cells towards spermatogenesis appear to differ little between typical amphibians and the caecilians.

6.9 SPERMATELEOSIS

Spermateleosis is the transformation of round spermatids into mature sperm, a term synonymous to spermiogenesis in the higher vertebrates. Spermateleosis in caecilians was described using light microscopy by Seshachar (1943a, 1945), Exbrayat and Sentis (1982) and Exbrayat (1986a, b). Ours is the first ultrastructural description of spermateleosis in caecilians, and reveals new details about the formation of spermatozoa from round spermatids (Smita *et al.* 2004b). The stages are portrayed as early mid and late phases concurrently depicting the changes in acrosome, centrioles, nucleus, and mitochondria. The acrosomal vesicle forms as a diffuse material as in anurans and suggests an increase in the size of the acrosomal vesicle through fusion of small Golgi-derived vesicles. Accumulation of a fibrous material between the inner and outer sheets of the acrosomal seat of the nucleus is noticed in our study. The greater part of the width of the nuclear tip forms a straight-sided bowl-like indentation closely fitting the posterior end of the acrosome. The perforatorium fits into the endonuclear canal. We also noticed the first appearance of the centriole much before the round spermatid elongates. The axoneme, which connects to the distal centriole likely, extends towards the proximal centriole. Jamieson (1999), while describing the sperm of caecilians, has shown that the proximal centriole lies ahead of the distal centriole in the posterior nuclear fossa (see also chapter 7 of this volume). We also noticed that during late phase, the condensation of the nuclear chromatin begins at the core, concomitant with the establishment of the endonuclear canal. The midpiece has mitochondria arranged in the form of a spiral aggregate around the axial filament and is entirely posterior to the centrioles. According to Jamieson (1999), the centrioles and anterior part of the sperm tail are surrounded by 35-40 spherical mitochondria (chapter 7 of this volume). We observed that the arrangement of mitochondria is spiral and cristae become concentrically lamellate in *Ichthyophis tricolor* or array-like in *Uraeotyphlus* cf. *narayani*. Seshachar (1943a, 1945) reported a 90% reduction in volume of the nucleus during transformation of the nucleus of the spermatid to that of ripe sperm. We have observed the nuclear size of the mature sperm is greatly reduced to about 4% that of the nucleus of the round spermatids (Smita *et al.* 2004b).

In our study, we found that in the sperm of *Ichthyophis tricolor* the tip of the acrosome is pointed whereas in *Uraeotyphlus* cf. *narayani* it is blunt and bulbous (Figs. 6.2B, C). A comparison of the sperms of the two species is also reported by Smita *et al.* 2004b, which can be a tool in phylogenetic analysis (see also chapter 7 of this volume).

6.10 SPERM DUCTS

6.10.1 Testicular Ducts

According to Seshachar (1936), the main duct follows a tortuous route around the loculi, giving off ducts to the loculi, rather than a straight duct with straight branches as figured by Spengel (1876). Wake (1968) in her study in *Gymnopis* presents convincing evidence of transverse ducts that clearly drain the longitudinal duct above and below the smaller posterior testis lobes. But the *Rhinatrema* species has a distinct duct with branches to locules in the active testis, and Wake agrees with Spengel's (1876) views that the duct pattern is typical for all caecilians.

6.10.2 Egress of Mature Spermatozoa

The course followed by a mature sperm from the locule to the intromittent organ is a rigorous one and Wake's (1968) observations corroborate those of Spengel (1876), Semon (1892), Wiedersheim (1879) and Tonutti (1931). Transverse ducts connect the longitudinal duct in the lobule to the kidney

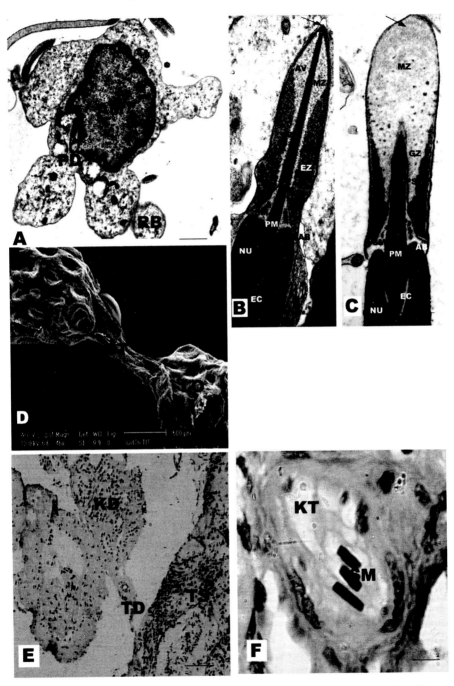

Fig. 6.2. A. *Ichthyophis tricolor* testis section showing ameboid cell possessing pseudopodia and establishing continuity with a residual body. AM, ameboid cell; PD, pseudopodia; RB, residual body. TEM. Scale bar = 20 µm. **B** and **C.** Longitudinal sections of anterior portion of sperm of the two species

Fig. 6.2 contd

then to the mesonephric duct and to the cloaca. In our observation on the three species of caecilians we have studied, the connection between the chain of testicular lobule is only a connective tissue strand, and not a longitudinal duct (unpublished data) (Fig. 6.2D). The sperm ducts, on leaving the lobules, run transversely and enter the kidney (Fig. 6.2E). Wake's (1970a) report on the urogenital system of caecilians describes that the sperm-bearing kidney tubules apparently remain as fully functional osmoregulatory units. This is in contrast to the situation found in many frogs and salamanders where the anterior part of the kidney loses its urinary role and functions exclusively for the sperm transport.

6.10.3 Role of Connective Tissue Strands

A unique and interesting observation in our study was the presence of collecting ducts, containing sperm, in the connective tissue strands between the lobes of the testis (Fig. 6.2D). The role of the longitudinal connective tissue strand, earlier mistaken as a longitudinal duct, in sperm transport remains enigmatic. This is remarkable because the connective tissue bridge between the testis lobes is a new report and the occurrence of the collecting ducts in the strand suggests an attribute unique to caecilians. The collecting ducts are comparable to the ducts present in the connective tissue between the lobes. They may have a role different from such collecting ducts directly connecting the testis lobes and the kidney. We propose that perhaps these ducts serve a purpose in storage of sperm before ejaculation. This is highly relevant in the context of the seasonality of reproduction in caecilians. A meticulous search of several sections of the kidney enabled us to locate spermatozoa within the urinary system. Such parts as seen under light microscope (Fig. 6.2F) reflect a histological picture of the collecting duct. Our observations on spermatogenesis in three species of caecilians, while putting to rest some of the issues, have opened up newer issues.

We have already started addressing these issues in caecilian spermatogenesis.

Fig. 6.2 contd

compared. TEM. **B**. *Ichthyophis tricolor*. **C**. *Uraeotyphlus* cf. *narayani* sperm. Acrosome (arrow) contains acrosomal vesicle (AV) with three zones, electron-dense homogeneous zone (EZ), moderately homogeneous zone (MZ), and granular zone (GZ). Periperforatorial subacrosomal material (PS) is shown. Perforatorium (PM) is lodged in the endonuclear canal (EC) within the nucleus (NU). Subacrosomal material (SM) and acrosomal base plate (AB) are also shown. Scale bar = 2 µm. **D**. *Ichthyophis tricolor*. Testis lobes (LO) are connected by a connective tissue strand (CT) SEM. Scale bar = 500 µm. **E**. *Ichthyophis tricolor*. Transverse ducts (TD) or vasa efferentia connect testis (TS) to kidney (KD). LM. Haematoxylin and eosin stained paraffin section. Scale bar = 100 µm. **F**. *Ichthyophis tricolor*. Collecting ducts with sperms (SM) found inside the kidney (KT). LM. Haematoxylin and eosin stained paraffin section. Scale bar = 100 µm. Original.

6.11 ACKNOWLEDGMENTS

We acknowledge financial assistance from the University Grants Commission, New Delhi to Oommen V. Oommen and M. A. Akbarsha through major research projects (grants no. F.3-33/2002 (SR-II)), DST, New Delhi for FIST programme, and academic support from Dr. Mark Wilkinson and Dr. David Gower, Natural History Museum, London. We thank the University of Kerala and Bharathidasan University for all laboratory facilities.

6.12 LITERATURE CITED

Ballowitz, E. 1913. Sperma, spermien, spermatozoen, spermatogenese. Handworterbuch der Naturwissenschaften 9: 251.

De Krester, D. M. and Kerr, J. B. 1994. The cytology of testis. Pp. 1177-1290. In E. Knobil and J. D. Neill (eds), *The Physiology of Reproduction,* Raven Press Ltd., New York.

De Sa, R. and Berois, N. 1986. Spermatogenesis and histology of the testes of the caecilian, *Chthonerpeton indistinctum*. Journal of Herpetology 20: 510-514.

Exbrayat, J.-M. 1986a. Le testicule de *Typhlonectes compressicaudus*: Structure, ultrastructure, croissance et cycle de reproduction. Mémoires de la Société Zoologique de France 43: 122-132.

Exbrayat, J.-M. 1986b. Quelques aspects de la biologie de la reproduction chez *Typhlonectes compressicaudus* (Duméril et Bibron, 1841). Résumé de thèse. Bulletin de la Société Herpétologique de France 38: 19-22.

Exbrayat, J.-M. 2000. Les Gymnophiones, ces curieux Amphibiens, Editions Boubée, Paris. 443 pp.

Exbrayat, J.-M. and Dansard, C. 1994. Apports de techniques complémentaires à la connaissance de l'histologie du testicule d'un Amphibien Gymnophione. Revue Française d'Histotechnologie 7: 19-26.

Exbrayat, J.-M. et Sentis, P. 1982. Homogénéité du testicule et cycle annuel chez *Typhlonectes compressicaudus* (Dumeril et Bibron, 1841), amphibien apode vivipare. Comptes Rendus des Séances de l'Académie des Science de Paris 24: 757-762.

Jamieson, B. G. M.1999. Spermatozoal phylogeny of the Vertebrata. Pp. 304-331. In Claude Gagnon (ed.). *The Male Gamete. From Basic Science to Clinical Applications.* Cache River Press, USA

Leblond, C. P. and Clermont, Y. 1952. Definition of the stages of the cycle of the seminiferous epithelium in the rat. Annals of New York Academy of Sciences 55: 548-573.

Lofts, B. 1974. Reproduction. Pp. 107-218. In: B. Lofts (ed.) *Physiology of the Amphibia,* Vol. 2. Academic Press, New York.

Lofts, B. 1984. Reproductive cycles of vertebrates-amphibians. Pp. 127-205. In G. E. Lamming (ed.) *Marshall's Physiology of Reproduction,* Vol. I, Churchill Livingstone, London.

Oommen, O. V., Measey, J. G., Gower, D. J and Wilkinson, M. 2000. Distribution and abundance of the caecilian *Gegeneophis ramaswamii*. Current Science 79: 1386-1389.

Saidapur, S. K. 1989. Reproductive cycles of amphibians. Pp. 116-224. In S. K. Saidapur (ed.) *Reproductive Cycles of Indian Vertebrates.* Allied Publishers Ltd., Ahmedabad, India.

Sarasin, R. and Sarasin, P. 1887-1890. Ergebnisse Naturwissenschaftlicher Forschungen auf Ceylon. Zur Entwicklungsgeschichte und Anatomie der

Ceylonischen Blindwuhle Ichthyophis glutinosus. C. W. Kreidel's Verlag, Wiesbaden.
Scheltinga, D. M., Wilkinson, M., Jamieson, B. G. M. and Oommen, O. V. (2003). Ultrastructure of the mature spermatozoa of caecilians (Amphibia: Gymnophiona). Journal of Morphology 258: 179-192.
Semon, R. 1892. Studien über den Bauplan des Urogenital System der Wirbeltiere. Dargelegt an der Entwicklung dies Organ Systems. Jena Zeitschrift Naturwissenschaften 25: 80-203.
Seshachar, B. R. 1936. The spermatogenesis of *Ichthyophis glutinosus* L. The spermatogonia and their division. Zeitschrift für Zellforschung und mikroskopische Anatomie 24: 662-706.
Seshachar, B. R. 1937. Spermatogenesis of *Ichthyophis glutinosus*. II. The meiotic divisions. Zeitschrift für Zellforschung und mikroskopische Anatomie 27: 133-158.
Seshachar, B. R. 1939. The spermatogenesis of *Uraeotyphlus narayani* Seshachar. La Cellule 48: 63-76.
Seshachar, B. R. 1940. The apodan sperm. Current Science 11: 439-441.
Seshachar, B. R. 1941. The interstitial cells in the testis of *Ichthyophis glutinosus* Linn. Proceedings of the Indian Academy of Sciences 13B: 244-254.
Seshachar, B. R. 1942a. Stages in the spermatogenesis of *Siphonops annulatus* and *Dermophis gregorii* (Boul). Proceedings of the Indian Academy of Sciences 15B: 263-277.
Seshachar, B. R. 1942b. Eggs and embryos of *Gegenophis carnosus*. Current Science 11: 439-441.
Seshachar, B. R. 1942c. The Sertoli cells in apoda. Half Yearly Journal, Mysore University 3: 65-71.
Seshachar, B. R. 1942d. Origin of intralocular oocytes in male apoda. Proceedings of the Indian Academy of Sciences 15B: 278-279.
Seshachar, B. R. 1943a. Spermatogenesis of *Ichthyophis glutinosus* Lin. III. Spermateleosis. Proceedings of the National Institute of Sciences of India 9: 271-286.
Seshachar, B. R. 1943b. Volume relations of the nucleus in apodan spermateleosis. Proceedings of the Indian Academy of Sciences 17: 138.
Seshachar, B. R. 1945. Spermateleosis in *Uraeotyphlus narayani* Sesh. and *Gegenophis carnosus* Bedd. Proceedings of the National Institute of Sciences of India 11: 36-40.
Smita, M., Oommen, O. V., Jancy, M. G. and Akbarsha, M. A. 2003. Sertoli cells in the testis of Caecilians, *Ichthyophis tricolor* and *Uraeotyphlus* cf. *narayani* (Amphibia: Gymnophiona): Light and electron microscopic perspectives. Journal of Morphology 258: 317-326.
Smita, M., Oommen, O. V., Jancy, M. G. and Akbarsha, M. A. 2004a. Stages in spermatogenesis of two species of caecilians *Ichthyophis tricolor* and *Uraeotyphlus* cf. *narayani* (Amphibia: Gymnophiona): Light and electron microscopic study. Journal of Morphology 261: 92-104.
Smita, M., Jancy, M. G., Girija, R., Akbarsha, M. A. and Oommen, O. V. 2004b. Spermiogenesis in caecilians *Ichthyophis tricolor* and *Uraeotyphlus* cf. *narayani* (Amphibia: Gymnophiona) Analysis by light and transmission electron microscopy. Journal of Morphology 262: 484-499.
Smita, M., Jancy, M.G., Akbarsha, M.A., Oommen, O.V. 2005. Ameboid cells in Spermatogenic Cysts of Caecilian Testis. Journal of Morphology 263: 340-355.
Spengel, J. W. 1876. Das Urogenital system der Amphibien. 1. Theil. Der anatomische Bau des Urogenitalsystem. Arbeiten aus dem Zoologzootom Institute Würzburg 3: 51-114.

Tonutti, E. 1931. Beitrag zur Kenntnis der Gymnophionen. XV. Das Genitalsystem. Morphologisches Jahrbuch 68: 151-292.
Van der Horst, G. and van der Merwe, L. 1991. Late spermatid-sperm/Sertoli cell association in the caecilian, *Typhlonectes natans* (Amphibia: Gymnophiona). Electron Microscopy Society of Southern Africa 21: 247-248.
Van Oordt, P. G. W. J. 1956. Regulation of spematogenetic cycle in the common frog (*Rana temporaria*) Thesis, Utrecht University, G. W. Van-der Weil and Company, 116 pp.
Wake, M. H. 1968. Evolutionary morphology of the caecilian urogenital system. Part I. The gonads and fat bodies. Journal of Morphology 126: 291-332.
Wake, M. H. 1970a. Evolutionary morphology of the caecilian urogenitial system. Part II. The kidneys and urinogenital ducts. Acta Anatomica 75: 321-358.
Wake, M. H. 1970b. Evolutionary morphology of the caecilian urogenital system. Part III. The bladder. Herpetologica 28: 120-128.
Wake, M. H. 1972. Evolutionary morphology of the caecilian uroogenital system. Part IV. The cloaca. Journal of Morphology 136: 353-366.
Wake, M. H. 1977. The reproductive biology of caecilians: An evolutionary perspective. Pp. 73-101. In D. H. Taylor and S. I. Guttman (eds), *The Reproductive Biology of Amphibians*, Plenum Press, New York.
Wake, M. H. 1981. Structure and function of the male Mullerian gland in caecilians (Amphibia: Gymnophiona), with comments on its evolutionary significance. Journal of Herpetology 15: 17-22.
Wake, M. H. 1994. Comparative morphology of caecilian sperm (Amphibia: Gymnophiona). Journal of Morphology 221: 261-276.
Wake, M. H. 1995. The spermatogenic cycle of *Dermophis mexicanus* (Amphibia: Gymnophiona). Journal of Herpetology 29: 119-122.
Wiedersheim, R. 1879. Die Anatomie der Gymnophionen. Verlag von Gustav Fischer. Jena.101 pp.

CHAPTER 7

Ultrastructure and Phylogeny of Caecilian Spermatozoa

David M. Scheltinga and Barrie G. M. Jamieson

7.1 INTRODUCTION

The interrelationships of the three groups of extant Amphibia (frogs, salamanders and caecilians) have, as yet, not been definitively resolved by molecular and/or morphological data and relationships among caecilians are far from fully understood (see chapter 2). Interpretation of comparative sperm data has provided useful evidence of relationships of amphibians at a variety of levels (see Scheltinga 2002 and references therein).

Light microscope observations of caecilian sperm were reported for *Ichthyophis glutinosus* (Ichthyophiidae), *Uraeotyphlus narayani* (Uraeotyphlidae), *Siphonops annulatus* (Caeciliidae), and *Gegeneophis carnosus* (Caeciliidae) in pioneering work by Seshachar (1939, 1940, 1943, 1945), and in *Chthonerpeton indistinctum* (Typhlonectidae) by de Sa and Berois (1986). Owing to changes in caecilian taxonomy and an absence of known voucher specimens the specific, though not the generic, identity of the material referred to by Seshachar as *I. glutinosus*, *G. carnosus* and perhaps also *U. narayani* must be considered uncertain. Wake (1994), in a major contribution, greatly augmented the number of caecilians (22 genera, 29 species, representing all families) of which spermatozoa have been examined by light microscopy (Table 7.1). Because of the age and fixation of some samples, major features of spermatozoal morphology, such as the presence or absence of an undulating membrane, were often uncertain. Furthermore, features such as the structure of the acrosome or the occurrence of axial and/or juxta-axonemal fibers were not determinable by the techniques employed.

The first ultrastructural descriptions of caecilian spermatozoa were for *Typhlonectes natans* (Typhlonectidae) (van der Horst and van der Merwe 1991; van der Horst *et al.* 1991). Scheltinga *et al.* (2003) have augmented our knowledge through ultrastructural descriptions of mature spermatozoa of representatives of the caecilian families Ichthyophiidae (*Ichthyophis beddomei*

The School of Integrative Biology, University of Queensland, Brisbane, Qld, 4072, Australia and *I. tricolor*), Uraeotyphlidae (three forms, probably distinct species, of *Uraeotyphlus*), and Caeciliidae (*Gegeneophis ramaswamii*). They also reexamined the spermatozoal structure of *Typhlonectes natans* (Typhlonectidae).

Table 7.1. Caecilian spermatozoa examined

Species	Method used	Stage examined	Author
Caeciliidae			
Afrocaecilia taitana	LM	M	Wake 1994
Boulengerula boulengeri	LM; TEM	M	Present study
Caecilia occidentalis	LM	M	Wake 1994
Caecilia orientalis	LM; TEM	M	Present study
Dermophis mexicanus	LM	M	Wake 1994
Gegeneophis carnosus	LM	M	Seshachar 1945
Gegeneophis ramaswamii	LM; TEM	M	Wake 1994; Scheltinga 2002; Scheltinga et al. 2003
Geotrypetes seraphini	LM	M	Wake 1994
Grandisonia alternans	LM	M	Wake 1994
Gymnopis multiplicata	LM	M	Wake 1994
Hypogeophis rostratus	LM	M	Wake 1994
Idiocranium russelli	LM	M	Wake 1994
Microcaecilia albiceps	LM	M	Wake 1994
Mimosiphonops vermiculatus	LM	M	Wake 1994
Minascaecilia sartoria	LM	M	Wake 1994
Siphonops annulatus	LM	M	Seshachar 1940; Wake 1994
Siphonops paulensis	LM	M	Wake 1994
Ichthyophiidae			
Caudacaecilia asplenia	LM	M	Wake 1994
Ichthyophis beddomei	LM; TEM	M	Scheltinga 2002; Scheltinga et al. 2003
Ichthyophis glutinosus	LM	Sd; M	Seshachar 1943; Wake 1994
Ichthyophis kohtaoensis	LM	M	Wake 1994
Ichthyophis orthoplicatus	LM	M	Wake 1994
Ichthyophis tricolor	LM; TEM	M	Scheltinga 2002; Scheltinga et al. 2003
Rhinatrematidae			
Epicrionops bicolor	LM	M	Wake 1994
Epicrionops petersi	LM	M	Wake 1994
Scolecomorphidae			
Scolecomorphus uluguruensis	LM	M	Wake 1994
Typhlonectidae			
Chthonerpeton indistinctum	LM; TEM	Sd; M	de Sa and Berois 1986; Wake 1994; Present study
Typhlonectes natans	LM; TEM	M	van der Horst et al. 1991; Wake 1994; Scheltinga 2002; Scheltinga et al. 2003
Uraeotyphlidae			
Uraeotyphlus sp. A	LM; TEM	M	Scheltinga 2002; Scheltinga et al. 2003
Uraeotyphlus sp. B	LM; TEM	M	Scheltinga 2002; Scheltinga et al. 2003
Uraeotyphlus narayani	LM	Sd; M	Seshachar 1945; Wake 1994

Method used: LM—Light microscopy; TEM—Transmission electron microscopy. Stage examined: Sd—Spermatid; M—Mature spermatozoon.

Spermatozoal ultrastructure is therefore known for representatives of four of the six currently recognized caecilian families and has allowed a preliminary evaluation of caecilian interrelationships based on spermatozoal characters (Scheltinga *et al.* 2003). Caecilian species which have been investigated for spermatozoal ultrastructure are listed in Table 7.1.

Low-level caecilian taxonomy is difficult and unstable because of a paucity of external characters and a lack of understanding of their variation (Nussbaum and Wilkinson 1989). Comparative sperm ultrastructure provides characters of potential use for elucidating evolutionary relationships within the Gymnophiona. Variation in the proportions of caecilian sperm may be of considerable assistance in distinguishing caecilian taxa at low taxonomic levels, more especially in the context of testing hypotheses that particular populations represent distinct species, than in the context of practical identification (Scheltinga *et al.* 2003).

7.2 SPERMATOZOAL ULTRASTRUCTURE

7.2.1 General

Examination of the spermatozoa of the three putative species of *Uraeotyphlus*, and the two species of *Ichthyophis*, revealed no difference in ultrastructure within a genus, however, differences in sperm dimensions (Table 7.2) were detected between species (Scheltinga *et al.* 2003). Within the limits of the number of species examined, spermatozoal characters distinguish caecilian genera from each other.

Caecilian spermatozoa are filiform, being composed of a distinct head (acrosome and nucleus), midpiece and tail (with undulating membrane). Under light microscopy the acrosome and nucleus appear as distinct structures in many species (*Ichthyophis* (Fig. 7.1E), *Boulengerula* (Fig. 7.1A) and *Uraeotyphlus* (Fig. 7.1F)) but indistinct in others (*Chthonerpeton* (Fig. 7.1B), *Gegeneophis* (Fig. 7.1D) and *Typhlonectes* (Fig. 7.1C)). In *Ichthyophis* and *Uraeotyphlus* species the tip of the acrosome is rounded (Fig. 7.1E, F), whereas in *Boulengerula*, *Chthonerpeton*, *Gegeneophis* and *Typhlonectes* the acrosome is pointed (Fig. 7.1A-D). At the light microscope level, distinction can be made between the midpieces of the examined *Ichthyophis* species with those of other examined caecilians. The midpiece of *Ichthyophis* appears homogeneous while those of *Boulengerula*, *Chthonerpeton*, *Uraeotyphlus*, *Gegeneophis* and *Typhlonectes* have spherical mitochondria which resemble a cluster of grapes (racemose arrangement) (Figs. 7.1, 7.5C). Dimensions of the sperm are provided in Table 7.2. The spermatozoa of *Gegeneophis*, *Ichthyophis* and *Uraeotyphlus* are represented semi-diagrammatically in Figures 7.2, 7.3 and 7.4, respectively, and should be referred to throughout.

7.2.2 Acrosome Complex

In all caecilians the acrosome complex is composed of an acrosome vesicle surrounding an electron-dense acrosome rod, termed the perforatorium

Table 7.2. Dimensions of gymnophionan spermatozoa taken from light and transmission electron microscopy

Species	Total length	Head length	Acrosome vesicle length	Nucleus length	Nucleus width	Midpiece length	Tail length	Endonuclear canal length	Perforatorium length	Author
Caeciliidae										
Afrocaecilia taitana	97	23	10	13	2	10	64	-	-	Wake 1994[++]
Boulengerula boulengeri	73	10	3.6	6.7	1.6	5.2	58	1.26	-	Present study
Caecilia occidentalis	162	13-15	3-4	10-11	2	7	60-140	-	-	Wake 1994[++]
Caecilia orientalis	-	9	3	6	1.5 anterior 1.3 base	7	-	1.38	-	Present study
Dermophis mexicanus	186	16-25	3-6	13-19	3-5	5-8	105-153	-	-	Wake 1994[++]
Gegeneophis carnosus	100	28	7.5	20	1.2	7	66	-	-	Seshachar 1945
Gegeneophis ramaswamii	102	27	7	20	2.5	12	63	-	-	Wake 1994[++]
Gegeneophis ramaswamii	99	27	-	-	0.69 anterior 0.97 base	5.7	69	0.25	>3.5	Scheltinga et al. 2003
Geotrypetes seraphini	116	10-11	2	8-9	2-3	4-6	92-99	-	-	Wake 1994[++]
Grandisonia alternans	148	14-18	2-5	12-13	2-3	10-13	109-117	-	-	Wake 1994[++]
Gymnopis multiplicata	129	16-21	2-4	14-17	3-5	7-8	86-100	-	-	Wake 1994[++]
Hypogeophis rostratus	123	19-20	4	15-16	2-3	8-9	56-94	-	-	Wake 1994[++]
Idiocranium russelli	121	18	5	13	2.5	5-8	55-95	-	-	Wake 1994[++]
Microcaecilia albiceps	110	39	4	35	1.3	11	60	-	-	Wake 1994[++]
Mimosiphonops vermiculatus	112	21	6	15	2.5	9	82	-	-	Wake 1994[++]
Minascaecilia sartoria	132	17	4	13	2.5	8	107	-	-	Wake 1994[++]
Siphonops annulatus	-	17	4	13	-	6	-	-	-	Seshachar 1940 (Fig. 6.3)
Siphonops annulatus	97	21-26	4-8	17-18	2.5-3	10	61	-	-	Wake 1994[++]
Siphonops paulensis	162	14-27	5-8	9-19	3.5	10-13	81-122	-	-	Wake 1994[++]
Ichthyophiidae										
Caudacaecilia asplenia	85	21-27	3	18-24	3-4	7-9.6	37-48	-	-	Wake 1994[++]
Ichthyophis beddomei	99	13	4.8	9	1.4 anterior 1.3 base	6.1	81	1.2	5.1	Scheltinga et al. 2003
Ichthyophis glutinosus	110	13	5	8	2	5.5	92	-	-	Seshachar 1943

Ultrastructure and Phylogeny of Caecilian Spermatozoa 251

Species									Reference
Ichthyophis kohtaoensis	151	21-24	4-6	17-18	3-5	6-7	100-120	-	Wake 1994++
Ichthyophis orthoplicatus	142	21-25	4-6.5	17-18	3-5	-	64-117	-	Wake 1994++
Ichthyophis tricolor	254	47	4	43	2	7	200	-	Wake 1994++
	85	14	4.7	9.4	1.5 anterior 1.1 base	4.1	68	1.5	Scheltinga et al. 2003
Rhinatrematidae									
Epicrionops bicolor	156	38-40	5	-	1.3-2	10	105-106	-	Wake 1994++
Epicrionops petersi	112	35-38	3.7-5	-	1-2.3	8.5-10	52-64	-	Wake 1994++
Scolecomorphidae									
Scolecomorphus uluguruensis	129	11	3	8	3-4	7-9	105-109	-	Wake 1994++
Typhlonectidae									
Chthonerpeton indistinctum	≈70	17	-	-	-	-	≈52	-	de Sa and Berois 1986
	157	38-50	7-15	31-35	1.5-2.6	7	86-100	-	Wake 1994++
	112	32	11	19	0.99 anterior 1.36 base	10	70	1.75	Present study
Typhlonectes natans	-	≈18	6†	12†	1.2†	5.8†	-	-	van der Horst et al. 1991 (from Fig. 6.2)
	184	35-37	10-11	25-28	2	6-10	120-137	≈10	Wake 1994++
	105	26	8.5	18	1.1 anterior 1.6 base	7.5	71	1.4	Scheltinga et al. 2003
Uraeotyphlidae									
Uraeotyphlus A	95	19	4.9	14	1.5 anterior 1.4 base	4.6	71	1.8	Scheltinga et al. 2003
Uraeotyphlus B	98	20	5.1	15	1.2 anterior 1.1 base	5.6	73	2	Scheltinga et al. 2003
Uraeotyphlus narayani	120	17	5.5	11	1.4	5	98	4.8	Seshachar 1945
	120	21-26	3	18-23	2.5-2.6	6-12	48-82	-	Wake 1994++

Values are in μm. ++Wake (1994) recorded the length of the nucleus as "head length" for all species she examined; it has here been shown as nucleus length. Therefore, the value given here for head length is the sum of Wake's "acrosome length" and "head" (=nucleus) length. The measurement of total length given here for species examined by Wake is the sum of the maximum values recorded for "acrosome", "head" (=nucleus), "midpiece" and "tail" length by Wake (1994).

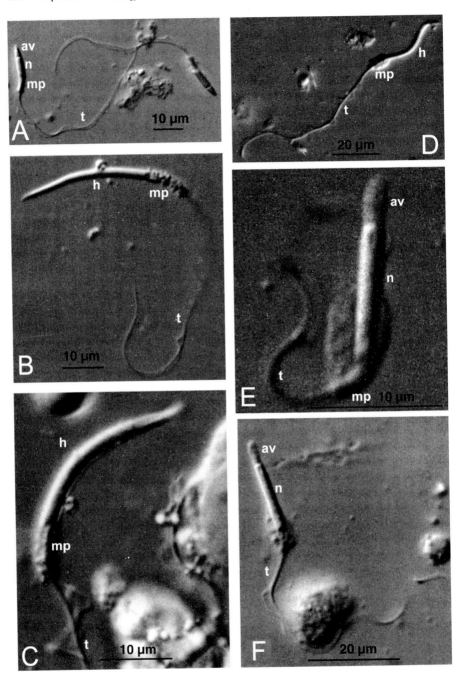

Fig. 7.1. Light micrographs of the spermatozoa of **A.** *Boulengerula boulengeri*, **B.** *Chthonerpeton indistinctum*, **C.** *Typhlonectes natans*, **D.** *Gegeneophis ramaswamii*, **E.** *Ichthyophis tricolor*, **F.** *Uraeotyphlus* sp. A. Scale as indicated. av, acrosome vesicle; h, head (acrosome and nucleus); mp, midpiece; n, nucleus; t, tail. Original.

(Figs. 7.2, 7.3, 7.4, 7.5A, 7.7A, 7.8A, 7.9A, F, G). In *Ichthyophis* and *Uraeotyphlus* species the acrosome vesicle is cylindrical and consists of three distinct regions: apically, a moderately electron-dense homogeneous zone, a granular zone, and basally an electron-dense homogenous zone (Fig. 7.9F). The basal homogenous zone of *Ichthyophis* is larger than that of *Uraeotyphlus*. The opposite is true for the granular zone, which is larger in *Uraeotyphlus*, while the apical homogenous zone is of a similar size in both genera. The acrosome is circular in transverse section throughout most of its length, becoming slightly irregular at its apex.

In *Gegeneophis ramaswamii* the acrosome vesicle is sharply attenuated apically and is initially circular in transverse section (Fig. 7.9A, B). Posteriorly, the acrosome becomes laterally flattened, then spatulate before again appearing circular. It is composed of moderately electron-dense material throughout its length, lacking the distinct zones of *Chthonerpeton, Ichthyophis, Typhlonectes* and *Uraeotyphlus*. However in *Gegeneophis*, for an indeterminate length a second rod shaped fiber, the acrosome vesicle fiber, lies longitudinally within the acrosome parallel to the perforatorium (Fig. 7.9A, B). A thin veneer of electron-lucent material occurs just within the outer limits of the acrosome vesicle. When observed in transverse section this thin veneer does not form a complete circle (Fig. 7.9A). Both an acrosome vesicle fiber and the electron-lucent veneer are absent from all other examined caecilians.

The acrosome vesicle of *Boulengerula boulengeri* is attenuated apically and is circular in transverse section throughout its length (Fig. 7.5A, E-G). It is composed of moderately electron-dense material throughout its length.

In *Typhlonectes natans* and *Chthonerpeton indistinctum* the acrosome vesicle is composed basally of granular material of differing electron densities and anteriorly of less electron dense homogenous material (Fig. 7.7A, G-I). In *T. natans* and *C. indistinctum* the contents of the acrosome vesicle are divided into two zones.

Because of inadequate formalin fixation it was not possible to determine the structure of the entire acrosome complex of *Caecilia orientalis*. However, the basal portion of the acrosome vesicle is composed of granular material (Fig. 7.8A, D).

In all caecilians examined by TEM, the base of the acrosome conforms in shape to the flattened anterior tip of the nucleus but is separated from it by a thin disc of granular material, the acrosomal base-plate (Figs. 7.5B, 7.7C, D, J, 7.8A, B, D, 7.9F).

7.2.3 Perforatorium

In all caecilians the perforatorium extends from the anterior of the acrosome to within the endonuclear canal (Figs. 7.5A, B, 7.6G, 7.7C, D, J, 7.8B, E, 7.9G). In *Ichthyophis* and *Uraeotyphlus* it does not extend to the apical tip of the acrosome vesicle but in *Gegeneophis* and *Typhlonectes* it does. The endonuclear canal is wider in *Ichthyophis* and *Uraeotyphlus* than in

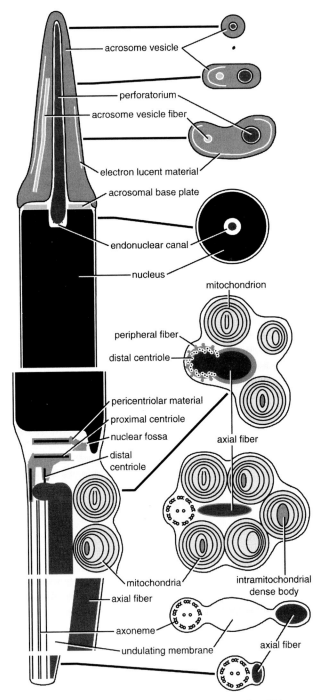

Fig. 7.2. Highly diagrammatic representation of the sperm of the Caeciliidae as exemplified by *Gegeneophis ramaswamii*. From Scheltinga D.M., Wilkinson, M., Jamieson, B.G.M. and Oommen, O.V. 2003. Journal of Morphology 258: 179-192, Fig. 5, © 2003 Wiley-Liss, Inc.

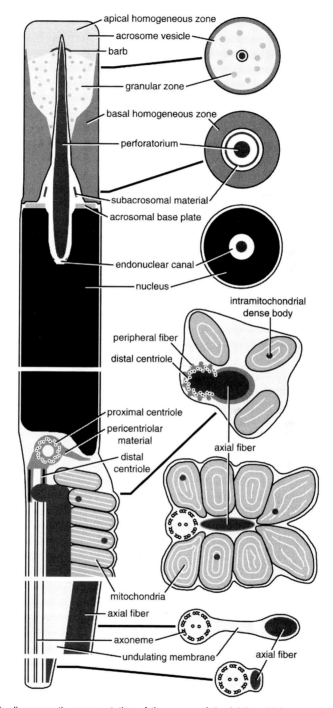

Fig. 7.3. Highly diagrammatic representation of the sperm of the Ichthyophiidae as exemplified by *Ichthyophis beddomei* and *I. tricolor*. From Scheltinga *et al.* (2003). Journal of Morphology 258: 179-192, Fig. 2, © 2003 Wiley-Liss, Inc.

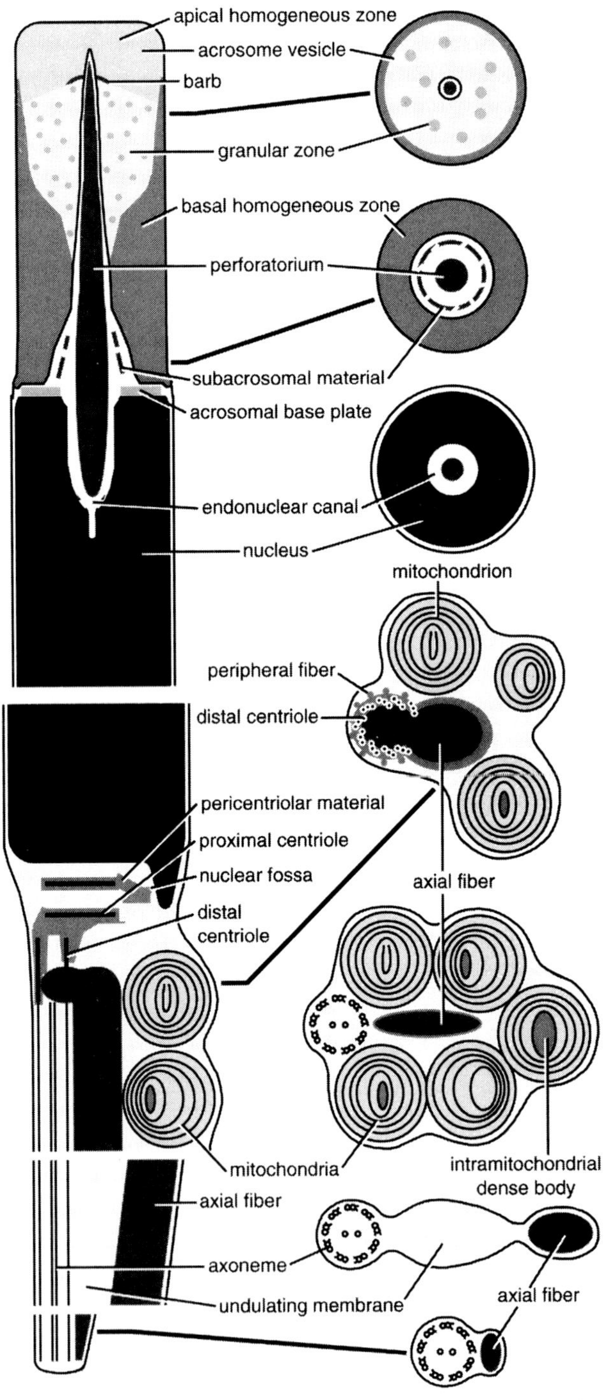

Fig. 7.4. Highly diagrammatic representation of the sperm of the Uraeotyphlidae as exemplified by *Uraeotyphlus* species. Original.

Boulengerula, Caecilia, Chthonerpeton and *Typhlonectes*. Anteriorly, in *Ichthyophis* and *Uraeotyphlus* the perforatorium becomes closely associated with the 'inner' membrane of the acrosome vesicle, from which distinct lateral projections, which appear barb-like in longitudinal section, are observed (Fig. 7.9F). This barb like extension occurs between the apical homogeneous and granular zones of the acrosome vesicle. An internal barb has not been observed in other caecilians.

7.2.4 Subacrosomal Granular Material

Basally, in *Boulengerula, Ichthyophis* and *Uraeotyphlus*, within the subacrosomal space, between the perforatorium and the acrosome vesicle there is a ring of granular material of a similar texture to that of the acrosomal base-plate, but separate from it (Figs. 7.5A, B, G, 7.9F, G). In *Boulengerula* and *Ichthyophis* the granular subacrosomal material forms a continuous ring around the perforatorium, whereas in *Uraeotyphlus* it forms a discontinuous ring of distinct parts. Subacrosomal material, either as a continuous or discontinuous ring, appears to be absent from *Caecilia, Gegeneophis, Chthonerpeton* and *Typhlonectes* sperm.

7.2.5 Nucleus

The nucleus is in the form of a relatively short cylinder of constant diameter in *Boulengerula, Caecilia, Ichthyophis* and *Uraeotyphlus* or an elongate cylinder which is slightly narrower anteriorly in *Chthonerpeton, Gegeneophis* and *Typhlonectes*. The nucleus is composed of strongly condensed chromatin in all species (Figs. 7.5C, 7.6C, 7.7D, 7.8C, F, 7.9E, G) and is circular in cross section, with the exception of *Caecilia orientalis* in which it is distinctly egg shaped (Fig. 7.8F). The anterior tip of the nucleus is flat and indented medially for a short distance as an anterior nuclear fossa, the endonuclear canal (Figs. 7.5A, B, 7.6G, 7.7D, J, 7.8B, E, 7.9G). The endonuclear canal is well developed, though short, and contains the base of the perforatorium for most of its length in all species. Posteriorly, the base of the canal is rounded in all caecilians; *Uraeotyphlus* differs in possessing a distinct narrow posterior extension of the canal (Fig. 7.9G). Lengths of the endonuclear canal are given in Table 7.2. Basally, the nucleus ends with an asymmetrical nuclear fossa (Figs. 7.6F, 7.8C, 7.9E).

7.2.6 Midpiece

The midpiece consists of the centrioles, anterior part of the tail, and the mitochondria. The proximal centriole is close to the base of the nucleus and is surrounded by pericentriolar material which connects it to the nuclear fossa and the distal centriole (Figs. 7.6F, 7.9E). It lies at a right angle to the long axis of the nucleus. The distal centriole lies in the long axis of the spermatozoon and forms the basal body of the axoneme. Basally, the distal centriole is penetrated by the anterior portion of the axial fiber (Figs. 7.6B, F, 7.7E, 7.9H). In *Ichthyophis* and *Uraeotyphlus* short peripheral fibers are

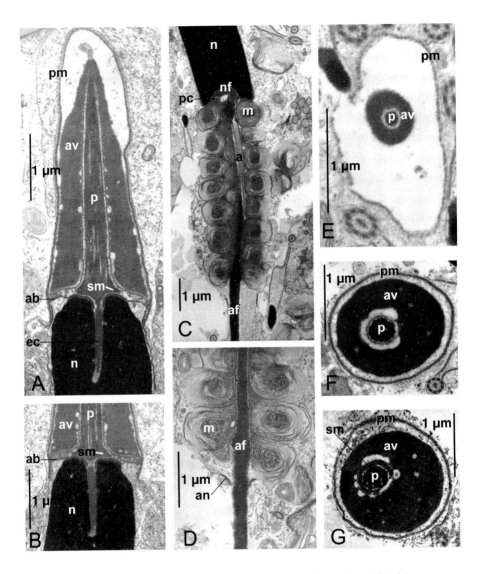

Fig. 7.5. TEM of *Boulengerula boulengeri* spermatozoa. **A.** Longitudinal section (L.S.) of the acrosome complex and anterior nucleus. **B.** L.S. through the anterior nucleus showing the endonuclear canal and acrosomal base plate. **C.** L.S. through the entire length of the midpiece. **D.** L.S. through the base of the midpiece showing the annulus and delicate concentric cristae of the mitochondria. **E.** Transverse section (T.S.) through the anterior acrosome vesicle. **F.** T.S. through the mid-region of the acrosome vesicle. **G.** T.S. through the base of the acrosome vesicle showing a portion of the complete granular ring of subacrosomal material. Scale as indicated. a, axoneme; ab, acrosomal base-plate; af, axial fiber; an, annulus; av, acrosome vesicle; ec, endonuclear canal; m, mitochondrion; n, nucleus; nf, nuclear fossa; p, perforatorium; pc, proximal centriole; pm, plasma membrane; sm, subacrosomal material. Original.

associated with the nine triplets of the distal centriole (Fig. 7.9H). Eight, or occasionally only seven, peripheral fibers can be seen in transverse section through the distal centriole, however, it is not possible to determine if a 'ninth' fiber exists due to the presence of the axial fiber, or whether the axial fiber is actually a greatly enlarged 'ninth' fiber. The peripheral fibers of *Uraeotyphlus* are well developed compared to those of *Ichthyophis*. In both genera the peripheral fibers do not continue beyond the length of the centriole. The peripheral fibers, of undetermined number, appear as little more than indistinct swellings of the pericentriolar material in *Gegeneophis* and *Boulengerula*. The presence or absence of peripheral fibers in *Chthonerpeton*, *Caecilia* and *Typhlonectes natans* has not been established.

The centrioles, axial fiber, and the anterior part of the axoneme are surrounded by mitochondria. The mitochondria of *Boulengerula*, *Caecilia*, *Chthonerpeton*, *Uraeotyphlus*, *Gegeneophis* and *Typhlonectes* are spherical, have an extensive array of delicate concentric cristae, number approximately 35 per spermatozoon (undetermined total number in *Caecilia* and *Chthonerpeton*), and occur in a racemose arrangement; in transverse section the number of mitochondria seen alternates between four and five per layer (Figs. 7.6E, F, 7.7K, 7.9C). In contrast, the mitochondria of *Ichthyophis*, when viewed in longitudinal section, appear flattened and to spiral around the tail (see Scheltinga *et al.* 2003), and although the cristae are concentric they do not form the delicate array seen in other caecilians (e.g. see delicate array in *Boulengerula*, Fig. 7.6E, F). In transverse section a maximum of eight mitochondria is seen in *Ichthyophis*. The mitochondria of all genera contain dense bodies (Figs. 7.6C, 7.7K, 7.9C), never completely surround the axoneme (Figs. 7.6F, 7.7K, 7.9C, H), and clearly define the length of the midpiece (Figs. 7.5C).

7.2.7 Annulus

A thin, but distinct, annulus is present at the base of the midpiece and defines the beginning of the tail in *Boulengerula*, *Chthonerpeton*, *Ichthyophis* and *Uraeotyphlus* (Figs. 7. 5C, 7.6D, 7.7L, 7.9E). Although the presence of an annulus was not determined for the mature spermatozoa of *Caecilia*, *Typhlonectes* and *Gegeneophis*, it is clearly present in late spermatids of *Gegeneophis*.

7.2.8 Tail

The tail consists of the 9+2 axoneme and axial fiber, enclosed by a plasma membrane, an arrangement which is typical of amphibian sperm (Figs. 7.6I, J, 7.8H, 7.9D). Anteriorly, within the midpiece, the axoneme and elongate axial fiber run closely adjacent to each other (Figs. 7.6E, 7.7K, 7.8G, 7.9C). More posteriorly, within the tail, the axoneme is separated from the round/oval shaped axial fiber by the undulating membrane (Figs. 7.6I, J, 7.8H, 7.9D). For much of the length of the tail the two faces of the plasma membrane of the undulating membrane are not closely apposed but are

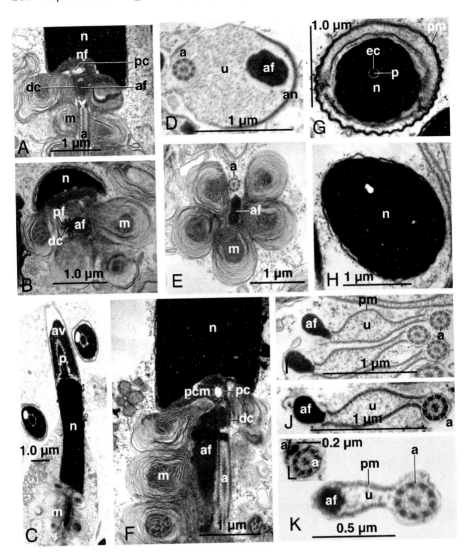

Fig. 7.6. TEM of *Boulengerula boulengeri* spermatozoa. **A.** Longitudinal section (L.S.) through the centriolar region showing the axial fiber within the distal centriole. **B.** L.S. through the distal centriole showing the axial fiber within the distal centriole and peripheral fibers associated with the triplets. **C.** L.S. through the sperm head and anterior midpiece. **D.** Transverse section (T.S.) through the annulus. **E.** T.S. through the midpiece showing the delicate concentric cristae of the mitochondria. **F.** L.S. through the centriolar region. **G.** T.S. through the anterior nucleus showing the endonuclear canal containing the perforatorium. **H.** T.S. through the mid-region of the nucleus. **I** and **J.** T.S. through the principal piece of the tail. **K.** T.S. through the posterior region of tail. **L.** T.S. through the end of the tail, note the absence of the undulating membrane. Scale as indicated. a, axoneme. af, axial fiber; an, annulus; av, acrosome vesicle; dc, distal centriole; ec, endonuclear canal; m, mitochondrion; n, nucleus; p, perforatorium; pc, proximal centriole; pcm, pericentriolar material; pf, peripheral fiber; pm, plasma membrane; u, undulating membrane. Original.

widely separated by cytoplasm. Immediately beneath the plasma membrane there is a dense layer, which is little more in appearance than a thickening of the membrane (Figs. 7.6J, K, 7.9D). The width of the axial fiber varies from between one to two times that of the axoneme for most of the length of the tail. Towards the posterior end of the tail the axial fiber decreases in size and again becomes closely associated with the axoneme (Figs. 7.6K, L, 7.8I). No juxta-axonemal fibers are present in the mature spermatozoon of caecilians (Figs. 7.6I, J, 7.8H, 7.9D), however, in *Ichthyophis tricolor* what appears to be a small juxta-axonemal fiber associated with doublet 3 is occasionally seen in the tail of spermatids (Scheltinga *et al.* 2003).

7.3 PHYLOGENETIC CONSIDERATIONS

7.3.1 Amphibian Apomorphies of Caecilian Sperm

The unilateral location of the mitochondria relative to the axoneme, the unilateral undulating membrane, and axial fiber occur in all three amphibian orders, and as suggested by Jamieson *et al.* (1993) and Jamieson (1999), appear to be amphibian autapomorphies. There are no apomorphic characters seen in caecilian sperm that suggest a closer relationship to either the Urodela or Anura (but see below).

7.3.2 Position of Gymnophiona within the Lissamphibia

Sperm characters were not found to provide much that was phylogenetically informative for resolving relationships between the Gymnophiona, Urodela and Anura (Scheltinga *et al.* 2003). However, the extensive delicate array of concentric cristae and dense bodies of the mitochondria are absent from the sperm of Urodela and Anura whereas they occur in some amniotes. If homologous, their retention in caecilians and loss in other amphibians would support, albeit weakly, a basal position for caecilians in the Lissamphibia. Concentric cristae are present in the mitochondria of Urodela sperm but not in the same form of a delicate array.

7.3.3 Caecilian Sperm Autapomorphies

Autapomorphies recognized by Scheltinga *et al.* (2003) for the Gymnophiona are: 1) penetration of the distal centriole by the axial fiber; 2) the presence of an acrosomal base-plate; 3) "acrosome seat" i.e. junction between acrosome complex and the flattened anterior end of the nucleus, although the perforatorium and endonuclear canal are themselves symplesiomorphies; and 4) absence of juxta-axonemal fibers throughout the length of the axoneme of mature spermatozoa.

Penetration of the distal centriole by the axial fiber and the presence of an acrosomal base-plate are not observed in the sperm of any other vertebrate. The acrosome seat is considered apomorphic (Scheltinga *et al.* 2003) because the acrosome complex of those few fish possessing one (Jamieson 1991), Urodela (Picheral 1979; Selmi *et al.* 1997; Scheltinga 2002;

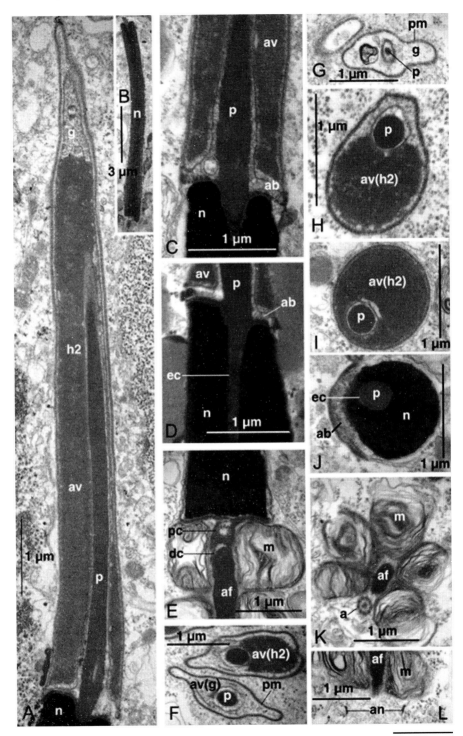

Fig. 7.7 contd

Scheltinga and Jamieson 2003a), Anura (Pugin-Rios 1980; Kwon and Lee 1995; Scheltinga 2002; Scheltinga and Jamieson 2003b) and amniotes (Jamieson 1995) caps the pointed nucleus. Juxta-axonemal fibers are absent from the flagellum of all caecilians so far examined. In contrast, a juxta-axonemal fiber is associated with doublet 8 in Urodela (Picheral 1979; Lee and Jamieson 1993; Selmi *et al.* 1997; Scheltinga 2002; Scheltinga and Jamieson 2003a) and doublet 3 in Anura (Pugin-Rios 1980; Lee and Jamieson 1992, 1993; Kwon and Lee 1995; Scheltinga 2002; Scheltinga and Jamieson 2003b) with the exception of *Leiopelma* (Scheltinga *et al.* 2001), and occasional *Bufo marinus* (Swan *et al.* 1980) spermatozoa, which have juxta-axonemal fibers at both 3 and 8. Juxta-axonemal fibers are also prefigured in Dipnoi (Jamieson 1999).

Another potential apomorphy of the Gymnophiona, tentatively suggested by Jamieson (1999) and endorsed by Scheltinga *et al.* (2003), is the wide separation of the plasma membrane of the two faces of the undulating membrane by a considerable amount of cytoplasm. In the Urodela, Dipnoi and most Anura the plasma membrane is closely apposed for most of the length of the flagellum. In those Anura where the two faces of the undulating membrane are widely separated, they are separated by a paraxonemal rod (Jamieson *et al.* 1993; Meyer *et al.* 1997; Jamieson 1999) and not by cytoplasm as in caecilian sperm. Thus the separation of the two faces of the undulating membrane in caecilians and some frogs does not show detailed similarity and the anuran and caecilian conditions are considered most likely to be independent apomorphies.

The presence of an endonuclear canal containing a perforatorium is a plesiomorphic feature of caecilian sperm that is shared with Urodela (Fawcett 1970; Picheral 1979; Selmi *et al.* 1997; Scheltinga and Jamieson 2003a), and some basal Anura (Sandoz 1970; Furieri 1975; Jamieson *et al.* 1993; Scheltinga and Jamieson 2003b). It is also seen in some amniotes (turtles, crocodiles and tuatara, Healy and Jamieson 1992; Jamieson 1995), and in lampreys, and sarcopterygian fish (see Jamieson 1991). However, the

---

Fig. 7.7 contd

Fig. 7.7. TEM of *Chthonerpeton indistinctum* spermatozoa. **A.** Longitudinal section (L.S.) of the acrosome complex, note the acrosome vesicle is composed of an anterior granular zone (g) and a basal homogenous zone (h2). **B.** L.S. through the entire nucleus. **C.** L.S. through the base of the acrosome showing the acrosome base plate. **D.** L.S. through the anterior nucleus showing the endonuclear canal. **E.** L.S. through the centriolar region. **F.** Transverse section (T.S.) through the anterior region the acrosome complex. **G.** T.S. through the apical tip of the acrosome. **H.** T.S. through the mid-region of the acrosome. **I.** T.S. through the base of the acrosome. **J.** T.S. through the anterior nucleus showing the endonuclear canal. **K.** T.S. through the midpiece showing the mitochondria with delicate concentric cristae. **L.** L.S. through the base of the midpiece showing the annulus. Scale as indicated. a, axoneme; ab, acrosomal base-plate; af, axial fiber; an, annulus; av, acrosome vesicle; dc, distal centriole; ec, endonuclear canal; g, acrosome vesicle—granular zone; h2, acrosome vesicle—electron-dense basal homogenous zone; m, mitochondrion; n, nucleus; p, perforatorium; pc, proximal centriole; pm, plasma membrane. Original.

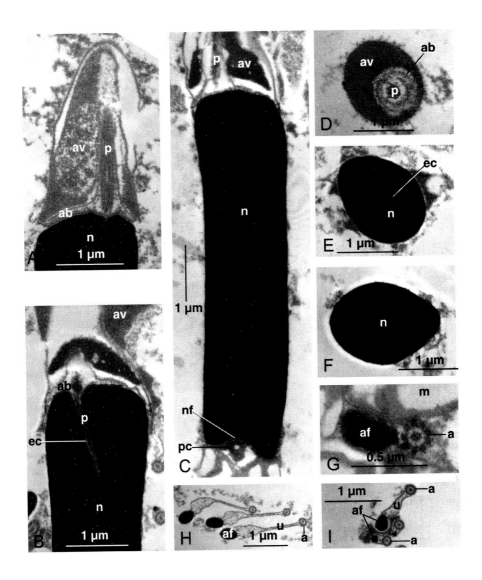

Fig. 7.8. TEM of *Caecilia orientalis* spermatozoa. **A.** Longitudinal section (L.S.) of the acrosome complex. **B.** L.S. through the anterior nucleus showing the endonuclear canal and acrosomal base plate. **C.** L.S. through entire nucleus. **D.** Transverse section (T.S.) through the anterior region the acrosome complex showing a complete ring of granular subacrosomal material. **E.** T.S. through the anterior nucleus showing the endonuclear canal. **F.** T.S. through the mid-region of the nucleus, note the 'egg' shape T.S. **G.** T.S. through the midpiece, note the axial fiber has an electron-dense inner medulla surrounded by a less electron dense (electron-moderate) cortex. **H.** T.S. through the principal piece of the tail. **I.** T.S. through various regions of the tail, note the short and absent undulating membrane at the base of the tail. Scale as indicated. a, axoneme; ab, acrosomal base-plate; af, axial fiber; av, acrosome vesicle; ec, endonuclear canal; m, mitochondrion; n, nucleus; nf, nuclear fossa; p, perforatorium; pc, proximal centriole; sm, subacrosomal material; u, undulating membrane. Original.

endonuclear canal of amphibians differs from those observed in the sarcopterygian fish and amniotes in being singular and containing only one perforatorium. With the exception of some plethodontid salamanders, the endonuclear canal in caecilians is also distinctive in penetrating only the apical region of the nucleus. This canal was first described in caecilians, for *Typhlonectes natans*, by van der Horst *et al.* (1991: pg. 445) as 'an indentation at the anterior wall of the nucleus' that is 'identical to the cup-shaped depression' described for other caecilians by Seshachar (1945). Thus the precise form of the endonuclear canal in caecilians, rather than its presence, appears apomorphic with a degree of convergence with some plethodontids (Scheltinga 2002; Scheltinga *et al.* 2003).

7.3.4 Putative Plesiomorphy of Concentric Cristae

Mitochondria with concentric cristae occur in the spermatozoa of the Gymnophiona and Urodela (van der Horst *et al.* 1991; Selmi *et al.* 1997; Scheltinga 2002) although the mitochondria of Urodela differ in being smaller and containing fewer cristae which are not in the form of an extensive array. Concentric cristae are interpreted by Scheltinga *et al.* (2003) as plesiomorphic in salamanders and caecilians because they also occur in turtles, crocodiles, tuatara (Jamieson and Healy 1992), and some marsupials, opossums (Fawcett 1970; Phillips 1970; Temple-Smith and Bedford 1980) and the macropod *Lagorchestes hirsutus* (Jamieson 1999), and therefore appear to be an autapomorphy of the Tetrapoda, though multiple homoplastic origin of the concentric condition cannot categorically be dismissed. Similarly, intramitochondrial dense bodies are considered symplesiomorphic because they occur in caecilians, salamanders (Selmi *et al.* 1997), turtles, crocodiles and the tuatara (Jamieson and Healy 1992). The short peripheral dense fibers associated with the distal centriole observed in *Ichthyophis, Uraeotyphlus* and *Gegeneophis ramaswamii* sperm are not seen in any other amphibian but similar structures occur along the axoneme in some fish (see Jamieson 1991), molluscs (see Healy 1996) and amniote sperm (see Jamieson 1995). That they are strictly homologous is debatable but the combination of mitochondria with concentric cristae, intramitochondrial dense bodies, annulus, and peripheral fibers is not inconsistent with homology of these characters with those of lower amniotes. This suggests (Scheltinga *et al.* 2003) that, of the extant Amphibia, caecilian sperm are the most similar (albeit greatly modified) in these and perhaps other respects to those of the common ancestor of Amphibia and Amniota. However, it has previously been suggested that the presence of bilateral fibers in the dipnoan sperm axoneme indicates that bilateral fibers are plesiomorphic for amphibian sperm (Jamieson 1999). That nine fibers are seen in the sperm axonemes of lampreys and cephalopods, reveals that this pattern has been acquired repeatedly in the axonemes of animal spermatozoa. Thus caution is required in interpreting the peripheral fibers in caecilians as homologous with those of amniotes.

266 Reproductive Biology and Phylogeny of Gymnophiona

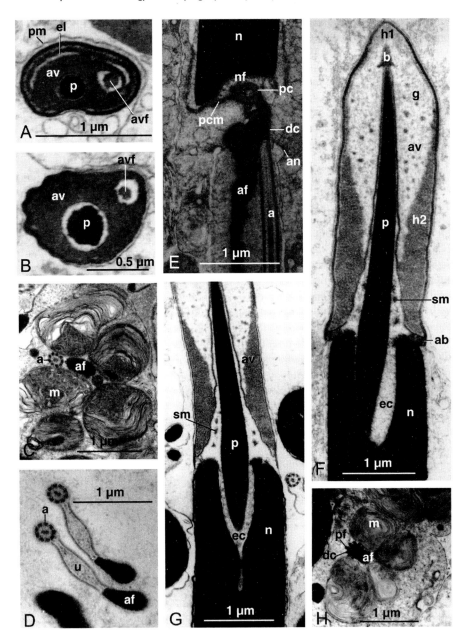

Fig. 7.9. TEM of caecilian spermatozoa. **A-C.** *Gegeneophis ramaswamii*. **A.** Transverse section (T.S.) through the mid-region of the acrosome complex showing a thin veneer of electron lucent material and a second fiber within the acrosome vesicle (avf). **B.** T.S. through the base of the acrosome. **C.** T.S. through the midpiece showing the delicate concentric mitochondrial cristae. **D.** *Typhlonectes natans*. T.S. through the principal piece of the tail, note the absence of any juxta-axonemal fibers. **E.** *Ichthyophis*

Fig. 7.9 contd

7.3.5 Homology of the Annulus

In many metazoan sperm there is a post-mitochondrial dense ring which is termed an annulus, though homology in disparate taxa is doubtful. The presence of an annulus may be a tetrapod apomorphy (or an apomorphic re-acquisition) as it is not seen in fish but is seen in caecilians, salamanders and amniotes. However, a 'ring body' (Jespersen 1971) or 'retronuclear body' (Jamieson 1999) is present in the lungfish *Neoceratodus* where it has the form of a post mitochondrial ring and is considered to be the precursor to the tetrapod annulus (Scheltinga *et al.* 2003). An annulus (or annulus-like structure) is not known in other Dipnoi. Against this homology, the retronuclear body observed in the sperm of the lungfish *Protopterus* appears to be the homologue of the 'neck' or 'connecting piece' of Urodela sperm (Jamieson 1999). An annulus is absent from anuran sperm and highly modified in some salamander sperm (Picheral 1979), though a simple ring-like annulus is present in basal Urodela (Scheltinga and Jamieson 2003a). The annulus seen in caecilians appears more similar to those occurring in basal amniotes than it does to the elongate annulus (*sensu* Picheral) of higher Urodela. Scheltinga *et al.* (2003) tentatively suggest that the absence of the annulus in anurans and its modification in salamanders are independent apomorphies while acknowledging that evolutionary history and homologies of the annulus are unclear.

7.3.6 Sperm Synapomorphies of Uraeotyphlidae and Ichthyophiidae

The following unique shared character states are recognizable (Scheltinga *et al.* 2003) for the spermatozoa of Uraeotyphlidae and Ichthyophiidae: 1) barbed (lateral extensions) acrosome membrane associated with the perforatorium tip; 2) acrosomal vesicle divided into three distinct zones; 3) wide, blunt-ended endonuclear canal: round ended in *Ichthyophis*; round ended but with a narrow, axial extension in *Uraeotyphlus*; and 4) cylindrical and apically blunt-ended acrosome vesicle. That Ichthyophiidae are more

---

Fig. 7.9 contd

tricolor. Longitudinal section (L.S.) through the centriolar region of a late spermatid showing the axial fiber penetrating the distal centriole. **F** and **G.** *Uraeotyphlus* sp. B. **F.** L.S. through the acrosome complex showing the 'internal' barb, three different zones of the acrosome vesicle (h1, g and h2), the granular subacrosomal material (which forms an incomplete ring) and the acrosomal base plate, **G.** L.S. through the anterior nucleus showing the endonuclear canal. **H.** *Uraeotyphlus* sp. A. T.S. through the distal centriole showing the axial fiber penetrating the distal centriole and peripheral fibers associated with the triplets. Scale as indicated. a, axoneme; ab, acrosomal base-plate; af, axial fiber; an, annulus; av, acrosome vesicle; avf, acrosome vesicle fiber; b, 'internal' barb of acrosome vesicle; dc, distal centriole; ec, endonuclear canal; el, veneer of electron-lucent material; g, acrosome vesicle—granular zone; h1, acrosome vesicle—moderately electron-dense apical homogeneous zone; h2, acrosome vesicle—electron-dense basal homogenous zone; m, mitochondrion; n, nucleus; nf, nuclear fossa; p, perforatorium; pc, proximal centriole; pcm, pericentriolar material; pf, peripheral fiber; pm, plasma membrane; sm, subacrosomal material; u, undulating membrane. Original.

closely related to the Uraeotyphlidae than either is to the Typhlonectidae is strongly supported by molecular and other morphological data (Wilkinson and Nussbaum 1996; Wilkinson 1997; Wilkinson *et al.*, 2002). Scheltinga *et al.* (2003) also reported that the presence of subacrosomal granular material: a continuous ring in *Ichthyophis*; a 'discontinuous' ring in *Uraeotyphlus* was unique to Uraeotyphlidae and Ichthyophiidae sperm. However, it has here been shown that a continuous ring of subacrosomal granular material is present in the caeciliid *Boulengerula boulengeri*.

The polarity of characters one and four only, can be readily determined. These characters are unique to the Uraeotyphlidae and Ichthyophiidae within the Gymnophiona, and as they are not found in any outgroup they are here considered to be synapomorphies. The polarities of characters two and three are uncertain as a shortening of the endonuclear canal and subdivision of the acrosome vesicle into different zones occurs, though with notable differences, in all caecilians. For example, it has not been possible to determine whether division of the acrosome vesicle into two (as in Typhlonectidae) or three (as in Uraeotyphlidae and Ichthyophiidae) zones is the plesiomorphic condition. The acrosome of several salamanders have been described as being barbed (Picheral 1979; Wortham *et al.* 1982; Selmi *et al.* 1997), but these differ from the condition observed here in that the barb of salamanders is an 'external' outgrowth along one side of the acrosome vesicle whereas that observed in *Ichthyophis* and *Uraeotyphlus* is a lateral extension of the 'internal' acrosome vesicle membrane associated with perforatorium tip. van der Horst *et al.* (1991) considered the curved tip of the acrosome in *Typhlonectes natans* to be comparable with the barbed condition in higher Urodela. However, a distinct 'external' barb structure was not observed by Scheltinga *et al.* (2003) who consider homology between the caecilian and salamander conditions doubtful.

Although the mitochondria of all caecilian sperm examined to date possess concentric cristae, those of Ichthyophiidae differ from those of the Uraeotyphlidae, Caeciliidae, and Typhlonectidae in several respects. The mitochondria of Ichthyophiidae appear flattened and to spiral around the tail, do not possess a delicate array of cristae, and are not in a racemose arrangement. The flattened shape appears to be apomorphic because spherical mitochondria are widespread in many distant and more proximate outgroups. The absence of the delicate array of concentric cristae is more difficult to interpret given that this is present in some amniotes (turtles and tuataras) but absent in salamanders and frogs.

7.3.7 Sperm Synapomorphies of Caeciliidae and Typhlonectidae

Scheltinga *et al.* (2003) reported that the spermatozoa of the Caeciliidae and Typhlonectidae display a single synapomorphic spermatozoal character, the spatulate/flattened acrosome vesicle. However, this does not now appear to be the case as the acrosome vesicle of the caeciliids *Boulengerula* and

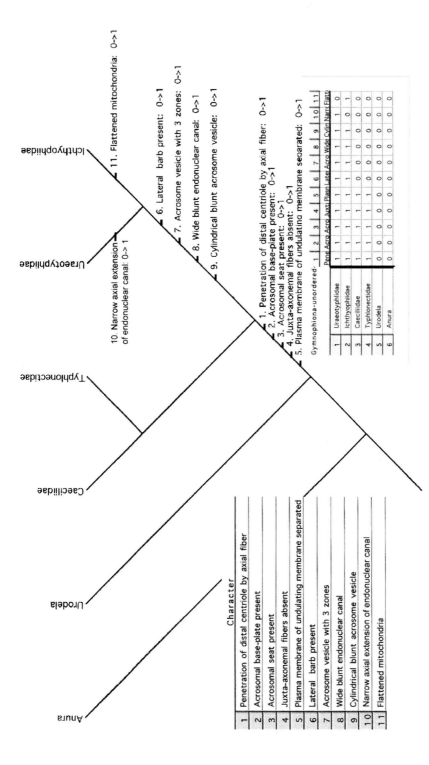

Fig. 7.10. Dendrogram indicating tentative synapomorphies within the Gymnophiona. The relationships between the Anura and Urodela and of these with the Gymnophiona are to be considered unresolved.

Caecilia, and the typhlonectid *Chthonerpeton* are round in transverse section throughout. The acrosomes of Dipnoi, the few other fish possessing them, Urodela, Anura, and basal amniotes are round in transverse section (Picheral 1979; Pugin-Rios 1980; Jamieson 1991, 1999; Healy and Jamieson 1992). The acrosome of *Ichthyophis glutinosus*, *Uraeotyphlus narayani*, and *Siphonops annulatus* have also been described as spatulate by Seshachar (1940), from light microscopy. However, this is highly doubtful for *I. glutinosus* and *U. narayani* as it is clearly shown as circular here for other species of *Ichthyophis* and *Uraeotyphlus*.

The presence of the acrosome vesicle fiber and a veneer of electron-lucent material within the acrosome vesicle are unique to *Gegeneophis ramaswamii* sperm and are presumably derived characters within caecilians. Their presence in other Caeciliidae requires examination though they are clearly absent from at least *Boulengerula* and *Caecilia* sperm.

The spermatozoa of Uraeotyphlidae, Caeciliidae and Typhlonectidae share only the plesiomorphic racemose arrangement of spherical mitochondria with an extensive array of delicate concentric cristae.

A cladistic, parsimony analysis of relationships within the Gymnophiona and between these and other Amphibia would be premature in view of our still limited knowledge of the ultrastructure of gymnophionan spermatozoa. Nevertheless, the tentative synapomorphies reported here, within the Gymnophiona are indicated on a dendrogram in Fig. 7.10. The relationships between the Anura and Urodela and of these with the Gymnophiona are to be considered unresolved.

7.3.8 Size and Shape Correlates

Wake (1994), from light microscopic examination of caecilian sperm morphology, considered that caecilians could be divided into two groups on the shape and size of the head. One group contained long heads with a pointed acrosome, the other having wider short heads with a blunt-ended acrosome. Using the criteria of Wake, sperm of all species, from four families, examined by Scheltinga *et al*. (2003) had mid-length acrosomes (5-8 µm), short (<8 µm) midpieces, and short (<100 µm) tails. As observed by Scheltinga *et al*. (2003) and noted by Wake (1994), tail and midpiece length is variable within genera and does not correlate with higher systematic relationships. The acrosome length and shape as well as the head size and shape is more consistent within genera and appears to be more phylogenetically informative.

7.3.9 Intrageneric Variation

Examination of the spermatozoa of the three putative species of *Uraeotyphlus* and of the two species of *Ichthyophis* by Scheltinga *et al*. (2003), revealed no discernible difference in ultrastructure within a genus, thus indicating that the characters derived from sperm ultrastructure are relatively constant at generic levels within the Gymnophiona. In contrast, considerable differences in sperm dimensions occur between the species of *Ichthyophis* and in *Uraeotyphlus* suggested distinct species. Similarly, substantial differences exist between the sperm measurements reported by Seshachar and Scheltinga *et al.* from light microscopic investigations for *Ichthyophis glutinosus, I. beddomei, I. tricolor, Uraeotyphlus narayani* and the three putative *Uraeotyphlus* species examined (Seshachar 1943, 1945; Scheltinga *et al.* 2003).

7.4 SUMMARY

The sperm of Gymnophiona show the following autapomorphies: 1) penetration of the distal centriole by the axial fiber; 2) presence of an acrosomal base-plate; 3) presence of an acrosome seat (flattened apical end of nucleus); and 4) absence of juxta-axonemal fibers. The wide separation of the plasma membrane bounding the undulating membrane is here also considered to be apomorphic. Two plesiomorphic sperm characters are recognized which are not seen in other Amphibia but occur in basal amniotes: 1) presence of mitochondria with a delicate array of concentric cristae (concentric cristae of salamander sperm differ in lacking the delicate array); and 2) presence of peripheral dense fibers associated with the triplets of the distal centriole. The presence of a plesiomorphic simple annulus is also observed in basal Urodela (a highly modified, elongate annulus is present in higher salamander sperm). The presence of an endonuclear canal containing a perforatorium is a plesiomorphic feature of caecilian sperm that is shared with Urodela, some basal Anura, lampreys, sarcopterygian fish, and some amniotes. Spermatozoal synapomorphies are identified for the Uraeotyphlidae and Ichthyophiidae, suggesting that the members of these families are more closely related to each other than to other caecilians. Although caecilian sperm exhibits clear amphibian synapomorphies, they have no apomorphic characters that suggest a closer relationship to either the Urodela or Anura.

7.5 ACKNOWLEDGMENTS

The research on which much of this chapter is based were made possible by a Department of Zoology and Entomology Research Grant to DMS, an Australian Research Council grant to BGMJ and was supported in part by a Natural Environment Research Council GST/02/832 grant to Mark Wilkinson. The assistance, contributions and enthusiasm of Mark Wilkinson and Oommen V. Oommen in the collection and initial fixation of *Boulengerula, Chthonerpeton, Gegeneophis, Ichthyophis, Typhlonectes* and *Uraeotyphlus* testes and sperm, and their input into the original 2003 publication upon which this chapter is based are greatly appreciated. We thank J. Sheps, D. Gower, J. Measey, S. Vish, and the many local people who helped locate, collect or process specimens. G. R. Zug and the staff at the National Museum of Natural History, Smithsonian Institution (Washington D.C., USA) are thanked for their assistance in allowing access to specimens of *Caecilia orientalis* in their collection.

7.6 LITERATURE CITED

de Sa, R. and Berois, N. 1986. Spermatogenesis and histology of the testes of the caecilian, *Chthonerpeton indistinctum*. Journal of Herpetology 20: 510-514.

Fawcett, D. W. 1970. A comparative view of sperm ultrastructure. Biology of Reproduction (Supplement) 2: 90-127.

Furieri, P. 1975. The peculiar morphology of the spermatozoon of *Bombina variegata* (L.). Monitore Zoologico Italiano 9: 185-201.

Healy, J. M. 1996. Molluscan sperm ultrastructure: correlation with taxonomic units within the Gastropoda, Cephalopoda and Bivalvia. Pp. 99-113. In J. D. Taylor (ed), *Origin and Evolutionary Radiation of the Mollusca*. Oxford University Press, Oxford.

Healy, J. M. and Jamieson, B. G. M. 1992. Ultrastructure of the spermatozoon of the tuatara (*Sphenodon punctatus*) and its relevance to the relationships of the Sphenodontida. Philosophical Transactions of the Royal Society of London B 335: 193-205.

Jamieson, B. G. M. 1991. *Fish Evolution and Systematics: Evidence from Spermatozoa*. Cambridge University Press, Cambridge, 319 pp.

Jamieson, B. G. M. 1995. Evolution of tetrapod spermatozoa with particular reference to amniotes. In B. G. M. Jamieson J. Ausio and J. -L. Justine (eds), *Advances in Spermatozoal Phylogeny and Taxonomy*. Mémoires du Muséum national d'Histoire naturelle, Paris 166: 343-358.

Jamieson, B. G. M. 1999. Spermatozoal Phylogeny of the Vertebrata. Pp. 303-331. In C. Gagnon (ed), *The Male Gamete: From Basic Science to Clinical Applications*. Cache River Press, Vienna (USA).

Jamieson, B. G. M. and Healy, J. M. 1992. The phylogenetic position of the tuatara, *Sphenodon* (Sphenodontida, Amniota), as indicated by cladistic analysis of the ultrastructure of spermatozoa. Philosophical Transactions of the Royal Society of London B 335: 207-219.

Jamieson, B. G. M., Lee, M. S. Y. and Long K. 1993. Ultrastructure of the spermatozoon of the internally fertilizing frog *Ascaphus truei* (Ascaphidae: Anura: Amphibia) with phylogenetic considerations. Herpetologica 49: 52-65.

Jespersen, Å. 1971. Fine structure of the spermatozoon of the Australian lungfish *Neoceratodus forsteri* (Krefft). Journal of Ultrastructure Research 37: 178-185.

Kwon, A. S. and Lee, Y.H. 1995. Comparative spermatology of anurans with special references to phylogeny. In: B. G. M. Jamieson, J. Ausio and J. -L. Justine (eds), *Advances in Spermatozoal Phylogeny and Taxonomy*. Mémoires du Muséum national d'Histoire naturelle, Paris 166: 321-332.

Lee, M. S. Y. and Jamieson, B. G. M. 1992. The ultrastructure of the spermatozoa of three species of myobatrachid frogs (Anura, Amphibia) with phylogenetic considerations. Acta Zoologica (Stockholm) 73: 213-222.

Lee, M. S. Y. and Jamieson, B. G. M. 1993. The ultrastructure of the spermatozoa of bufonid and hylid frogs (Anura, Amphibia): Implications for phylogeny and fertilization biology. Zoologica Scripta 22: 309-323.

Meyer, E., Jamieson, B. G. M. and Scheltinga, D. M. 1997. Sperm ultrastructure of six Australian hylid frogs from two genera (*Litoria* and *Cyclorana*): phylogenetic implications. Journal of Submicroscopic Cytology and Pathology 29: 443-451.

Nussbaum, R. A. and Wilkinson, M. 1989. On the classification and phylogeny of caecilians (Amphibia: Gymnophiona), a critical review. Herpetological Monographs 3: 1-42.

Phillips, D. M. 1970. Ultrastructure of spermatozoa of the woolly opossum *Caluromys philander*. Journal of Ultrastructure Research 33: 381-397.

Picheral, B. 1979. Structural, comparative, and functional aspects of spermatozoa in urodeles. Pp. 267-287. In D. W. Fawcett and J. M. Bedford (eds), *The Spermatozoon*. Baltimore: Urban and Schwarzenberg.
Pugin-Rios, E. 1980. Étude Comparative sur la Structure du Spermatozoïde des Amphibiens Anoures. Comportement des Gamètes lors de la Fécondation. Ph.D. Thèse, Université de Rennes, France.
Sandoz, D. 1970. Étude ultrastructurale et cytochimique de la formation de l'acrosome du discoglosse (Amphibien Anoure). Pp. 93-113. In B. Baccetti (ed), *Comparative Spermatology*. Rome: Accademia Nazionale dei Lincei.
Scheltinga, D. M. 2002. Ultrastructure of Spermatozoa of the Amphibia: Phylogenetic and Taxonomic Implications. Ph.D. Thesis, University of Queensland, Australia.
Scheltinga, D. M. and Jamieson B. G. M. 2003a. The mature spermatozoon. Pp. 203-274. In D. M. Sever (ed) *Reproductive Biology and Phylogeny of Urodela*. Science Publishers USA.
Scheltinga, D. M. and Jamieson B. G. M. 2003b. Spermatogenesis and the mature spermatozoon: form, function and phylogenetic implications. Pp. 119-251. In B. G. M. Jamieson (ed), *Reproductive Biology and Phylogeny of Anura*. Science Publishers, USA.
Scheltinga, D. M., Jamieson, B. G. M., Eggers, K. E. and Green, D. M. 2001. Ultrastructure of the spermatozoon of *Leiopelma hochstetteri* (Leiopelmatidae, Anura, Amphibia). Zoosystema 23: 157-171.
Scheltinga, D. M., Wilkinson, M., Jamieson, B. G. M. and Oommen, O. V. 2003. Ultrastructure of the mature spermatozoa of caecilians (Amphibia: Gymnophiona). Journal of Morphology 258: 179-192.
Selmi, M. G., Brizzi, R. and Bigliardi, E. 1997. Sperm morphology of salamandrids (Amphibia, Urodela): implications for phylogeny and fertilization biology. Tissue and Cell 29: 651-664.
Seshachar, B. R. 1939. The spermatogenesis of *Uraeotyphlus narayani* Seshachar. La Cellule 48: 63-76.
Seshachar, B. R. 1940. The apodan sperm. Current Science, Bangalore 10: 464-465.
Seshachar, B. R. 1943. The spermatogenesis of *Ichthyophis glutinosus* Linn. Part 3. Spermateleosis. Proceedings of the National Institute of Sciences, India 9: 271-285.
Seshachar, B. R. 1945. Spermateleosis in *Uraeotyphlus narayani* Seshachar and *Gegeneophis carnosus* Beddome (Apoda). Proceedings of the National Institute of Sciences, India 11: 336-340.
Swan, M. A., Linck, R. W., Ito, S. and Fawcett, D. W. 1980. Structure and function of the undulating membrane in spermatozoan propulsion in the toad *Bufo marinus*. Journal of Cell Biology 85: 866-880.
Temple-Smith, P. and Bedford, J. M. 1980. Sperm maturation and the formation of sperm pairs in the epididymis of the opossum, *Didelphis virginiana*. Journal of Experimental Biology 214: 161-171.
van der Horst, G. and van der Merwe, L. 1991. Late spermatid-sperm/sertoli cell association in the caecilian, *Typhlonectes natans* (Amphibia: Gymnophiona). Electron Microscopic Society of Southern Africa 21: 247-248.
van der Horst, G., Visser, J. and van der Merwe, L. 1991. The ultrastructure of the spermatozoon of *Typhlonectes natans* (Gymnophiona: Typhlonectidae). Journal of Herpetology 25: 441-447.
Wake, M. H. 1994. Comparative morphology of caecilian sperm (Amphibia: Gymnophiona). Journal of Morphology 221: 261-276.

Wilkinson M. 1997. Characters, congruence and quality: a study of neuroanatomical and traditional data in caecilian phylogeny. Biological Review 72: 423-470.

Wilkinson M. and Nussbaum R. A. 1996. On the phylogenetic position of the Uraeotyphlidae (Amphibia: Gymnophiona). Copeia 1996: 550-562.

Wilkinson M., Richardson M. K., Gower D. J. and Oommen O. V. 2002. Extended embryo retention, caecilian oviparity and amniote origins. Journal of Natural History 36: 2185-2198.

Wortham J. W. E., Jr, Murphy J. A., Martan J. and Brandon R. A. 1982. Scanning electron microscopy of some salamander spermatozoa. Copeia 1982: 52-60.

CHAPTER 8

Oogenesis and Folliculogenesis

Jean-Marie Exbrayat

8.1 THE OVARIES OF GYMNOPHIONA

The female genital tract of Gymnophiona has been studied in several species (Müller 1832, Rathke 1852, Spengel 1876, Sarasin and Sarasin 1887-1890, Chatterjee 1936, Tonutti 1931, Garg and Prasad 1962).

The comparison of the ovaries and urogenital ducts of numerous species of Gymnophiona has been the object of a series of publications (Wake 1968, 1970a, b, 1972, 1977). The female genital tract of *Dermophis mexicanus* has been more particularly studied (Wake 1980). The ovaries have been described for *Chthonerpeton indistinctum* (Berois and De Sa 1988), *Typhlonectes compressicauda* (Exbrayat 1983, 1986, Exbrayat and Collenot 1983) and *Ichthyophis beddomei* (Masood-Parveez 1987, Masood-Parveez and Nadkarni 1993a, b).

In caecilians, the ovaries are paired elongated sac-like organs (see also chapter 3 of this volume). They are parallel to the gut and the kidneys, to which they are bound by a sheet of mesovarium. They are also bound to the fat bodies by another sheet of connective tissue. Some transverse blood vessels are observed between each ovary and the corresponding kidney and fat body. Each ovary contains several follicles at different stages of growth according to the period of the sexual cycle. After laying of the oocytes, some corpora lutea can be observed in oviparous as well as viviparous species, at the same time as the eggs and embryos lie into oviduct.

Very few works have been devoted to a precise description of oogenesis and folliculogenesis but by examination of published data, it is possible to give a general idea of oogenesis and follicle evolution in caecilians.

8.2 DEVELOPMENT OF OVARIES FROM BIRTH TO MATURATION

The increase of germ cells and follicles from birth to maturation has been studied in *Typhlonectes compressicauda* (Exbrayat 1986, Exbrayat and Anjubault 2003, 2005, Anjubault and Exbrayat 2000, 2004).

Laboratoire de Biologie générale, Université Catholique de Lyon and Laboratoire de Reproduction et Développement des Vertébrés, Ecole Pratique des Hautes Etudes, 25, rue du Plat, F-69288 Lyon Cedex 02.

In young animals, the ovaries are limited by a double connective wall containing germ cells that are not surrounded by follicle cells. At birth, the ovaries can vary from one individual to another. In certain ovaries, there are only some oogonia and young primary oocytes. The oogonia are 20 µm in diameter, with a 10 µm nucleus and their cytoplasm are not stained by the usual dyes. The oocytes are 30 µm in diameter with a central nucleus. Most of them are naked; a single layer of follicle cells surrounds the others. It is also possible to observe several somatic cells migrating towards naked oocytes.

Other ovaries contain some dividing oogonia. These ovaries also contain primary oocytes that may be surrounded by follicle cells (Fig. 8.1A). More voluminous oocytes (150 to 300 µm in diameter) with cytoplasm filled by granules stained with fuchsine or aniline blue (proteic) have central or slightly eccentric nuclei, 75 × 60 µm in diameter, with several nucleoli contained often in a perinuclear space without granules. These cells are always surrounded by a granulosa composed of a single layer of follicle cells.

Some oocytes contain more voluminous peripheral vacuole-like granules. These oocytes are separated from the granulosa by a membranous structure. The granulosa is composed of a single layer of columnar cells with a voluminous nucleus that are pressed one against the other.

In all the animals studied, a proliferation of oogonia is observed throughout the ovary. The largest follicles eject the oogonia and young oocytes. Proliferative oocytes form several "germinal nests" or proliferation areas, disposed in a segmented fashion throughout the ovary.

In one-year-old animals (July-August), some oogonia and primary oocytes are found in the ovaries. The primary oocytes are or are not surrounded by a single layer of follicle cells. Some primary oocytes present premeiotic characteristics. They are disposed in the proliferation areas. Numerous oocytes are more developed but the follicle diameter never exceeds 400 µm.

In 18-month-old animals (January), certain ovaries contain follicles that may reach 600 µm in diameter (Fig. 8.1B). The oocyte contains a nucleus measuring 50 to 100 µm in diameter, and several nucleoli. Granules stained by fuchsine or aniline blue fill their cytoplasm. A perinuclear space without granules is unstained. Some peripheral vacuoles are sometimes observed. A membrane and a single layer of prismatic follicle cells surround the oocyte. Other ovaries have the appearance previously described.

Between February and May (19 to 22-month-old animals), some larger oocytes measuring 600 to 750 µm in diameter are observed. These oocytes contain a eccentric nucleus (100 µm in diameter) with several nucleoli. In the cytoplasm, the granules are increasingly voluminous and numerous. They become PAS positive indicating their carbohydrate nature. The region of cytoplasm surrounding the nucleus does not contain granules. The oocyte is separated from follicle cells by a very thin zona pellucida, PAS positive and stained in grey-blue by azan staining that reveal its glucidic

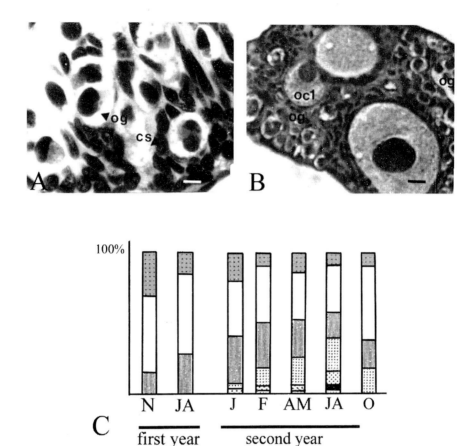

Fig. 8.1. *Typhlonectes compressicauda.* **A.** Germinal nest of ovary in new-born. Scale bar = 10 μm. **B.** Germinal zone of ovary in one-year-old animal with oogonia and young oocytes. Scale bar = 30 μm. **C.** Relative quantity of follicle types during the growth of the ovary. og, oogonium; cs, somatic cell; oc1, oocyte 1; ▦ 20 to 150 μm follicles; ☐ 150 to 300 μm follicles; ▥ 300 to 400 μm follicles; ▦ 400 to 600 μm follicles; ▦ 600 to 750 μm follicles; ■ 750 to 1200 follicle; ▦ atretic follicles. From Exbrayat 1986. Unpublished D. Sci. Thesis, Université Paris VI, France, Pl. IX, Figs. 1, 2; Fig. 57.

nature. The follicle cells are 10 × 20 μm in diameter. The surrounding connective tissue begins to be organised as a theca. Several atretic follicles are observed.

From January until May, some proliferative areas are particularly abundant and they contain several growing germ cells.

In July-August (24 to 28-month-old animals), certain follicles reach 750 μm in diameter: they are previtellogenic oocytes (Exbrayat 1986). The follicle cells are often separated from the oocyte by an increasing zona pellucida. Some atretic follicles are also observed. The proliferative areas are less abundant but growing oocytes are always observed in germinal nests.

These observations show that the oogonia multiply and develop into primary oocytes after birth (and perhaps just before). During the first year, they continue to grow and yolk reserves begin to fill the cytoplasm. During the second year, oocytes and follicles continue to increase in number. Nevertheless, some oocytes can degenerate, their follicles becoming atretic. All two-year-old females (July-August for *Typhlonectes*) have previtellogenic oocytes with a diameter measuring more than 750 μm. The adult state is reached in January-February of the third year (Fig. 8.1C, and see also chapter 5 of this volume).

8.3 OOCYTES AND FOLLICLES IN ADULTS

No precise study of the stages of oogenesis is available for the Gymnophiona comparable with what exists for spermatogenesis (see chapters 3 and 5 of this volume). A general description of oocytes and follicles has been given by Wake (1968). Oocytes and follicles have also been described for *Chthonerpeton indistinctum* (Berois and De Sa 1988), *Tyhlonectes compressicauda* (Exbrayat, 1983, 1986, Exbrayat and Collenot 1983) and *Ichthyophis beddomei* (Masood-Parveez 1987, Masood-Parveez and Nadkarni 1993a, b). Each author has given a different denomination for the stages of development of follicles. We therefore give here the main descriptions published and, for the first time, a synthesis of them to generalize the stages of development of follicles and oocytes for all Gymnophiona.

8.3.1 Oocytes and Follicles in *Chthonerpeton indistinctum*

The germinal cells are scattered among the epithelial cells of the medulla. The smallest cells are the oogonia with a round nucleus containing only one nucleolus. The largest cells are the oocytes. Their nuclei contain one or several nucleoli. Follicle cells surround the oocytes. Three stages of ovarian follicles have been described (Berois and De Sa 1988) (Fig. 8.2A).

The primary follicles develop into the external layer of ovarian epithelium (Fig. 8.2B). The oocyte contains a central nucleus with one to several nucleoli, one of them being bigger than the others. At this stage, the cytoplasm is weakly stained with some PAS positive granulations (e.g. containing carbohydrates). A discontinuous single layer of some flattened epithelial cells that are applied directly against the oocyte surrounds these cells.

The secondary follicles are located deeper in the ovary (Fig. 8.2C). The oocyte is more or less large according to the stage of development. Its cytoplasm contains some small yolk granules that are slightly stained by PAS and solochrom blue. The oval nucleus becomes peripheral. It is weakly stained and contains a large number of nucleoli. The chromatin is granular. A single layer of flattened follicle cells that is not always continuous surrounds the oocyte. A zona pellucida begins to appear between the oocyte and the follicle cells as a PAS positive membrane.

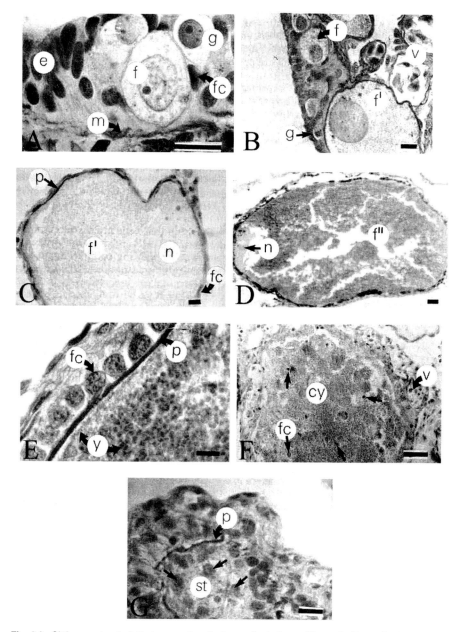

Fig. 8.2. *Chthonerpeton indistinctum* ovaries. **A**. Ovary. Scale bar = 20 μm. **B**. Ovary. Scale bar = 20 μm. **C**. Secondary follicle. Scale bar = 50 μm. **D**. Tertiary follicle. Scale bar = 50 μm. **E**. Detail of a tertiary follicle. Scale bar = 10 μm. **F**. Atretic follicle. Scale bar = 50 μm. **G**. Corpus luteum. Scale bar = 20 μm. cy, cytoplasm with clustered yolk; e, epithelial cells; m, basal membrane; f, secondary follicle; f", tertiary follicle; fc, follicular cells; g, oogonium; n, nucleus; p, zona pellucida; st, stroma; v, blood vessel; y, yolk. In F, arrows show lymphocytes; in G, arrows show follicular cells in the stroma. From Berois, N. and De Sa, R. 1988. Journal of Herpetology 22: 146-151, Figs.1b, c, d, e, f, g, h.

The tertiary follicles are the largest and they are the deepest in the ovary (Fig. 8.2D). The cytoplasm of the oocyte is filled with two types of granules. Some fine granules, already described, are peripheral, adjacent to the membrane of the oocyte. The largest inclusions are well stained by PAS. The nucleus of the oocyte is flat and compressed against the oocyte wall. The chromatin is homogenous and nuclei are numerous. A PAS positive zona pellucida and a layer of columnar follicle cells with a central oval nucleus surround the oocyte. The cytoplasm of follicle cells is granular and basophilic. The follicle is surrounded by an external theca of connective tissue in which numerous blood vessels can be observed (Fig. 8.2E).

Besides these follicles, some "atypical" follicles have been described (Berois and de Sa 1988). They have the same size as the tertiary follicles but neither a follicle layer nor a zona pellucida is observed. Some large PAS positive yolk granules are aggregated into clusters, and the connective theca is highly vascularised and disorganised. The oocyte belonging to this follicle contains numerous lymphocytes (Fig. 8.2F). This description is reminiscent of the atretic follicles also observed in *Typhlonectes compressicauda* (see below, in this chapter, and chapter 5 of this volume).

Some additional structures without yolk have been observed in the ovaries of *Chthonerpeton indistinctum*. These structures are masses constituted by a stroma without yolk granules, mixed with cells originating from peripheral follicle cells, surrounded by a PAS positive membrane and a layer of follicle cells and a vascularised connective theca. These structures are observed at the surface of the ovarian epithelium to which they are linked by a narrow pedicel or a large contact zone. They can be compared with the corpora lutea observed in other species (Fig. 8.2G) (Wake 1968, Exbrayat 1983, 1986, Exbrayat and Collenot 1983).

8.3.2 Oocytes and Follicles in *Typhlonectes compressicauda*

Oocytes and the follicles of this viviparous species have been described (Exbrayat 1983, 1986, Exbrayat and Collenot 1983). The germ cells develop in several areas that constitute germinal nests, as in the young females, as we have previously described. The germinal nests are disposed in a segmental manner throughout the length of the ovaries. In each ovary, the oocytes and follicles are more or less developed. Some atretic follicles have been observed in all the ovaries observed. Corpora lutea at different stages of development have been also found in pregnant females.

The oogonia are small cells measuring 20 μm in diameter, as in the young females (Fig. 8.1A). They are still not surrounded by any follicle cells. The follicles measuring 150 to 600 μm and 600 μm to 750 μm are also similar to those of young animals (Fig. 8.1B). Two others categories of follicles based on their size and aspect have been found: 750 to 1200 μm in diameter (that are the equivalent to the secondary follicles described in *Chthonerpeton indistinctum* and also, we will see, *Ichthyophis beddomei*), and more than 1200 μm (equivalent to the tertiary follicles of *Chthonerpeton*).

In the secondary follicles of *Typhlonectes* (750 to 1200 µm), the cytoplasm of the oocyte contains PAS positive granules that are also deeply stained by molybdic orange G, showing their glycoproteic nature. The vesicular nucleus, well stained by haematoxylin, migrates to the periphery of oocyte. A zona pellucida stained by aniline blue, fast green and PAS positive, and a layer of follicle cells surround the oocyte. The peripheral connective tissue is organised as a theca. This stage corresponds to the previtellogenic oocytes.

In the oocyte belonging to a tertiary follicle, the nucleus is applied to the wall; the cytoplasm is filled by yolk platelets measuring 5 to 10 µm in diameter. These platelets are PAS positive, stained by orange G and eosin, indicating their glycoproteic nature. The cortical granules can be observed in a peripheral disposition. The oocyte is surrounded by a thick zona pellucida and a layer of follicle cells (see also chapter 5 of this volume). The follicles are surrounded by a connective theca. This stage is that of vitellogenic oocytes.

The structure of the ovaries has been studied throughout the sexual cycle (Fig. 8.3A, B). In January, the beginning of the breeding period, the germinal nests are voluminous. They contain numerous oogonia and primary oocytes. Few dividing oogonia are observed. The chromosomes are not often individualised and a membrane surrounds the nuclei. Most of the primary oocytes do not present any premeiotic figures. The oocytes begin to be surrounded by follicle cells. More voluminous oocytes with follicle cells are also included in the germinal areas. In February, the period of breeding, the proliferation areas are always developed. The oogonia are dividing and the chromosomes are individualised. In the young oocytes, premeiotic figures can be observed. Some germ cells are growing. In pregnant and non-pregnant females, the germinal nests are still well developed in April. The oogonia multiply, the premeiotic oocytes are numerous. From May until October (end of breeding and sexual inactivity), the germinal nests are reduced, but one can still observe some dividing oogonia and premeiotic oocytes. These cells are sometimes lesser numerous in pregnant females than in others. From October until January, period of preparation for breeding, the germinal nests develop and they may be very extensive in December. The dividing oogonia and premeiotic oocytes are particularly numerous. Some growing oocytes and follicles are always observed. Several kinds of atretic follicles have been observed in the ovaries of *Typhlonectes compressicauda*. They have been classified into four categories, based on their size and appearance: (1) 1200 to 2000 µm atretic follicles, (2) 750 to 1200 µm atretic follicles, (3) 600 µm to 750 µm and (4) less than 600 µm atretic follicles.

The first category of atretic follicles is found especially at the beginning of the breeding cycle, just after ovulation. These follicles originate from degeneration of the most voluminous oocytes filled by yolk platelets (belonging to the tertiary follicles), that have not been laid. Some vacuoles are observed in the cytoplasm of the oocyte. The granulosa develops to produce cells (80 µm in diameter) that migrate into the oocyte after have destroyed its plasma membrane. The yolk platelets are phagocytozed (Fig. 5.8E, chapter 5 of this volume). The contents of follicle cells become stained by Soudan black B, showing their lipid nature (Fig. 5.8G, chapter 5 of this volume). Some PAS positive granules are also observed in these cells.

The second category of atretic follicles can be divided into two subcategories. The first corresponds to previous atretic follicles that have decreased in size. The cells are now 60 µm in diameter and they resemble adipocytes. The second subcategory consists of atretic follicles originating from the degenerating previtellogenic follicles (Fig. 5.8F, chapter 5 of this volume). The granules of the cytoplasm of the oocyte are phagocytozed by the cells originating from the granulosa.

In the third category, cells measuring 60 µm in diameter resemble an adipose tissue. In the fourth category, the atretic follicles contain degenerative cells, some of them being still filled by fat substance. These cells finally disappear.

Some corpora lutea are also observed in the ovaries of pregnant females. At the beginning of pregnancy, they possess a large central cavity filled by fluid (Fig. 5.8A, chapter 5 of this volume), then they become increasingly compact, the cavity becoming invaded by the proliferation of cells providing from the granulosa and the external theca, with numerous blood vessels (Fig. 5.8B, chapter 5 of this volume). At the end of pregnancy, these corpora lutea degenerate, giving bodies that resemble to the most degenerated atretic follicles (Fig. 5.8C, D). A more detailed description is given in chapter 5 of this volume.

From a study of the percentage of the follicles in *in toto* preparations, it is possible to define three categories of ovaries: (1) vitellogenic ovaries, mainly found in December and January, (2) pregnant ovaries (February until August-September) and (3) non-pregnant ovaries (February until October-November). In December and January, 40% of follicles are 1200 to 2000 µm in diameter. 10% of these follicles are atretic (Fig. 8.3A, B).

After egg laying, there are no longer any 750 to 1200 µm follicles. Numerous atretic follicles corresponding to the degeneration of these latter are observed. After ovulation, the corpora lutea comprise 15 to 20% of the total follicles. After parturition, they make up only 6 to 10%. During this period, 10 to 15% of the follicles are degenerative (Fig. 8.3B).

In pregnant females, the percentage of the smallest follicles (150 to 600 µm) increases in February (35%) and April (60%), then it is constant until parturition. During pregnancy, the percentage of the 600 to 750 µm follicles is constant.

In non-pregnant females, the percentage of 150 to 600 µm follicles increases in April (25 to 30%) and May (60%). The percentages of 600 to 750 µm follicles and 750 to 1200 µm follicles show little variation. The vitellogenic follicles degenerate to bring on atretic structures that progressively decrease in volume. There are always 20% of atretic follicles in the ovaries.

From October until December, the percentage of vitellogenic follicles increases, whereas the other categories decrease in percentage. Nevertheless, follicles measuring 600 to 750 µm and 750 to 1200 µm are constant in percentage.

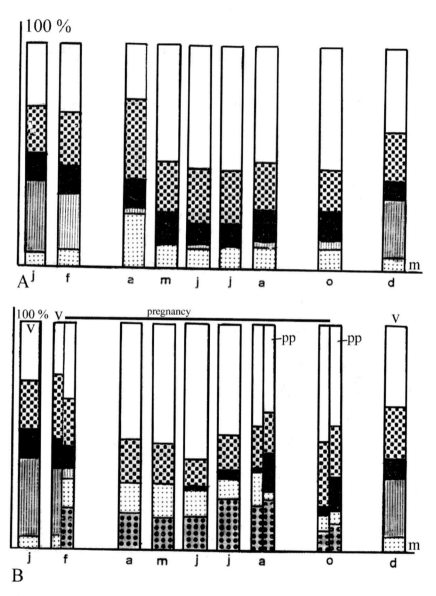

Fig. 8.3. Variation of relative quantities of follicles during the reproductive cycle in female *Typhlonectes compressicauda*. **A**. Year with sexual quiescence; **B**. Year with pregnancy. Abbreviations. ☐ 150 to 600 µm follicles; ▨ 600 to 750 µm follicles; ■ 750 to 1200 µm follicles; ▥ more than 1200 µm follicles; ▦ corpora lutea; ▦ atretic follicles; pp, post parturition; V, vitellogenesis. From Exbrayat, J.-M. 1986. Unpublished D. Sci. Thesis, Université Paris VI, Fig. 66.

In this species, whatever the period of the year, some dividing oogonia are always found in both non-pregnant and pregnant females. There is a continuous oogenesis. The germinal nests are the most developed from February until April, correlated with the increase in the number of the 150 to 600 µm follicles. Throughout the year, the stock of oocytes is constantly renewed with a permanent increase in the numbers of follicles. These oocytes and follicles degenerate just before previtellogenesis when the females become pregnant, and just before vitellogenesis in resting females.

In non-pregnant females, the previtellogenic follicles continue to develop from June and especially October to become vitellogenic. From August until October, after parturition, the small, 150 to 600 µm, follicles become previtellogenic then vitellogenic. In December and January, all the ovaries are at the same stage.

Ovulation is observed from February until April in 50 to 60% of adult females. Some corpora lutea are then obtained from follicles having an ovulated oocyte. The vitellogenic follicles that have not ovulated degenerate and result in atretic follicles.

To conclude, oogenesis is continuous in *Typhlonectes compressicauda*. The female sexual cycle is biennial. During the first year, some vitellogenic oocytes are laid from February until April. Then during pregnancy, some corpora lutea are observed. After parturition, the small follicles and oocytes increase in number. The first vitellogenic oocytes are observed in October. By the following February, ovulation no longer occurs, and vitellogenic oocytes degenerate in the ovaries. Their follicles become atretic. By next October, or sometimes before, a new increase of the follicles is observed, several vitellogenic oocytes appear and are ready to be laid by the following February and a new sexual cycle starts.

8.3.3 Oocytes and Follicles in *Ichthyophis beddomei*

The histological structure of ovaries and the ovarian cycles of this species have been described in three works (Masood-Parveez 1987, Masood-Parveez and Nadkarni 1993a, b).

In *Ichthyophis beddomei*, germinal areas are observed as islets of germ cells that are regularly distributed in the ovaries (Fig. 8.4A). This disposition is very similar to the germinal nests found in female *Typhlonectes compressicauda*. Between germinal nests, some ovarian follicles at different stages of their development are found. The authors signal that the developing oogonia and secondary oogonia are observed only during the breeding period (January and February). In later months, they remain in a resting phase. Four categories of follicles have been described according to their size and cytological appearance.

The previtellogenic follicles (or stages I) are the smallest (Fig. 8.4B). Their diameter ranges from 40 µm to 100 µm. In these follicles, the oocyte contains a large vesicular nucleus that becomes irregularly shaped. A single layer of follicle cells and a very thin connective theca surrounds the oocyte.

The vitellogenic follicles are 200 to 7500 µm in diameter. These follicles have been divided into five categories based upon their size: 200 to 1500 µm, 1600 to 3000 µm, 3100 to 3500 µm, 3600 to 6500 µm and 6600 to 7500 µm. The oocyte contains some yolk platelets that are increasingly numerous during the growth of the oocyte. The follicle layers always have the same appearance (Fig. 8.4C).

In postovulatory follicles, the granulosa has hypertrophied and some cells are found in the central space left after loss of the oocyte. The follicle cells have a vacuolated cytoplasm and an irregularly shaped nucleus. The theca has also hypertrophied (Fig. 8.4D).

Atretic follicles have also been observed. Two types of atresia have been described. In the first type, there is a gap between the follicle layers and the shrinking oocyte (Fig. 8.4E). In the second type, follicle cells become larger; they proliferate and invade the oocyte. These cells become phagocytic and they destroy the oocyte. Masood-Parveez and Nadkarni (1993a, b) divided this second type of atresia into five stages: (1) hypertrophy of follicle cells; (2) multiplication of these cells that invade the oocyte (Fig. 8.4F); (3) phagocytic cells observed throughout the oocyte (Fig. 8.4G), (4) degenerating oocyte that is marked by a brownish-red pigmentation of phagocytic cells and hypertrophy of the theca; (5) degeneration of granulosa and theca.

The structure of the ovary has been studied throughout the annual sexual cycle (Masood-Parveez and Nadkarni 1993a). The smallest follicles are found in April. The follicles increase from October to reach a maximal size in February. In March, the follicle size decreases. The oocytes and follicles are the greatest in size by December, January and February.

Several previtellogenic oocytes are found throughout the year. Their percentage is minimal by February and increases by August. From September until April, this percentage does not vary. The percentage of vitellogenic oocytes (stages II to V) is minimal in May. It increases from August to reach a maximal value in November. The preovulatory follicles (stage VI) are observed from September until February; when they are at their maximal value. As in other species, several atretic follicles are observed throughout the year. Their percentage increases from February until April, at which they reach their maximal value, and decreases by May. The largest atretic follicles are observed from November until January. The number of postovulatory follicles correlates with the number of eggs in the clutches.

A yearly sexual cycle divided in three phases has been described in *Ichthyophis beddomei* (Masood-Parveez and Nadkarni 1993a). The first phase is the pre-breeding phase, in August until November. The first is the recruitment of 70 to 80% of increasing vitellogenic oocytes and a decrease in the percentage of atretic follicles. The second phase is that of breeding. It occurs from December until February and is characterised by the presence of a great number of large vitellogenic oocytes and a minimal number of previtellogenic oocytes, with many of dividing oogonia. The

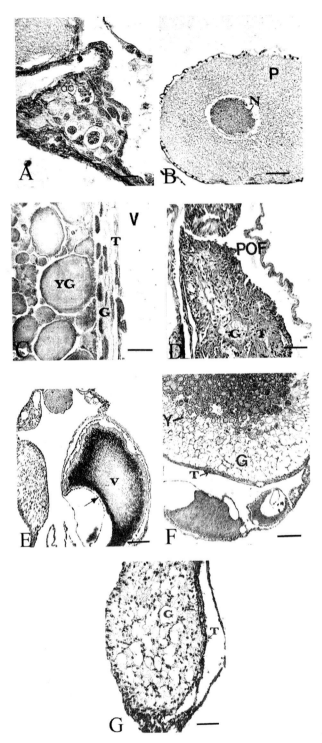

Fig. 8.4 contd

third phase, in the post-breeding period, occurs from March until July. It is characterised by the decrease in follicle diameter, number of vitellogenic follicles, increase in the percentage of atretic follicles and the presence of postovulatory follicles (corpora lutea).

8.4 CONCLUSIONS

The study of oocytes and follicles in several species of Gymnophiona has provided data that can be synthesized. In Gymnophiona, several stages of development of germ cells and follicles can be described. I propose here a general classification of follicles and oocytes of caecilians ovaries.

The first stage (stage A) is that of oogonia that are grouped into germinal nests and has been observed in all the species studied (Wake 1968, Exbrayat 1983, 1986, Exbrayat and Collenot 1983, Masood Parveez 1987, Berois and De Sa 1988, Masood Parveez and Nadkarni 1993 a, b).

The second stage (stage B) is that of the young primary oocytes that are also observed in germinal nests. These oocytes can be naked or surrounded by a small layer of flattened follicle cells.

The third stage (stage C) is that of the previtellogenic oocytes and follicles. The size of follicle varies according to the degree of development of follicle, but they are always small. The oocyte is round, with a central rounded nucleus containing one to several nucleoli. The chromatin is granular. The cytoplasm is stained by basic dyes, they contain tiny granules and cortical granules can be observed. Numerous flattened cells that are separated from the oocyte by a more or less conspicuous zona pellucida compose the granulosa.

The fourth stage (stage D) is that of vitellogenic oocytes and follicles. The cytoplasm of the oocyte is filled with large yolk platelets that contribute to adpressing the nucleus against the oocyte wall. Some cortical granules are always seen. The oocyte is surrounded by a PAS positive zona pellucida and a layer of prismatic follicle cells that greatly increase when the period of ovulation is approaching (see chapter 5 of this volume).

The fifth stage (stage E) is that of atretic follicle with several stages according to the type of follicle.

The sixth stage (stage F) is that of corpora lutea. It is observed in oviparous as well as viviparous species, and it is related to the presence of eggs, or of young or old embryos in the oviduct of these species that use internal fertilisation (see chapter 11 of this volume).

Fig. 8.4 contd

Fig. 8.4. *Ichthyophis beddomei* ovaries. **A**. Germinal epithelium with oogonia. Scale bar = 5 μm. **B**. Previtellogenic oocyte. Scale bar = 5 μm. **C**. Detail of a vitellogenic follicles. Scale bar = 5 μm. **D**. Postovulatory follicle. Scale bar = 5 μm. **E**. Vitellogenic oocyte showing a type I atresia. Scale bar = 20 μm. **F**. Stage 2 of type II atresia. Scale bar = 20 μm. **G**. Stage 3 of type II atresia. Scale bar = 20 μm. G, granulosa; N, nucleus; OC, secondary oogonia; P, previtellogenic oocyte; POF, postovulatry follicle; T, theca; V, vitellogenic oocyte; Y, yolk; YG, yolk granule. From Masood-Parveez, U. and Nadkarni, B. 1993a. Journal of Herpetology 27: 59-63, Figs. 1c, d, e, f; 2a, c, d.

In all species, the germinal nests are disposed in a segmented manner, that has been also observed during the development of gonads (Anjubault and Exbrayat 2000, chapter 9 of this volume). This disposition is reminiscent to the structure of testis (chapters 3 and 6 of this volume).

The germinal nests are always observed during the sexual cycle, with several variations according to the species. In *Chthonerpeton indistinctum*, the germinal nests have been observed together with corpora lutea (Berois and De Sa 1988) but the authors lacked data to complete their study of individuals captured throughout the sexual cycle. In *Typhlonectes compressicauda*, the germinal nests are particularly well developed during the biennial sexual cycle, in pregnant females as well as in non-pregnant ones (Exbrayat 1983, 1986). Nevertheless, they are more or less developed according to the month. These germinal nests are particularly developed during the breeding period (February until April). Conversely, in *Ichthyophis beddomei*, dividing oogonia are observed only during the breeding period.

These observations indicate that in *Typhlonectes compressicauda*, oogenesis is continuous throughout the sexual cycle with an increase at the beginning of reproduction. In *Ichthyophis beddomei*, oogonia divide and give oocytes only during the breeding period. Nevertheless, whatever the period of division, oogenesis is continuous for oogonia divide at each sexual cycle. In all the species studied, several oocytes degenerate before laying, especially during the non-breeding periods. The atretic follicles have several characteristics according to the state of degeneration of the oocytes.

Some corpora lutea have also been found in all these species. In addition, in pregnant *Typhlonectes compressicauda*, these structures have an endocrine function (Exbrayat 1986, Exbrayat and Collenot 1983, and chapter 5 of this volume).

We can conclude that, in Gymnophiona, oogenesis is continuous throughout the year. When oocytes attain a critical stage, they can be laid if it is the time. If not, they do not reach this stage and degenerate. In *Typhlonectes compressicauda*, we have shown that the previtellogenic oocytes degenerate when it is not the time of ovulation. The data given by various authors indicate that the follicles have the same pattern of evolution.

From birth to the first year, the ovaries increase. Then, during the second year, they are prepared for the first ovulation that occurs by the third year.

8.5 ACKNOWLEDGEMENTS

The author thanks Catholic University of Lyon (U.C.L.) and Ecole Pratique des Hautes Etudes (E.P.H.E.) that freed him to organize his research fields and supported the works on Gymnophiona. He also thanks Singer-Polignac Foundation that supported the missions necessary to collect the material of study. The author also thanks more especially, Michel Delsol, Marie-Thérèse Laurent, who made thousand and thousand of sections. He thanks Barrie Jamieson, who invited him to contribute to this book series and also for his patience in reviewing the manuscript. He thanks Jean-François Exbrayat for his help in preparing the illustrations.

8.6 LITERATURE CITED

Anjubault, E. and Exbrayat J.-M. 2000. Development of gonads in *Typhlonectes compressicauda* (Amphibia, Gymnophiona). XVIIIth International Congress of Zoology, Athens, August-September 2000: 51.

Anjubault, E. and Exbrayat, J.-M. 2004. Contribution à la connaissance de l'appareil génital de *Typhlonectes compressicauda* (Duméril et Bibron, 1841), Amphibien Gymnophione. I. Gonadogenèse. Bulletin Mensuel de la Société Linnéenne de Lyon. 73: 379-392.

Berois, N. and De Sa, R. 1988. Histology of the ovaries and fat bodies of *Chthonerpeton indistinctum*. Journal of Herpetology 22: 146-151.

Chatterjee, B. K. 1936. The anatomy of *Uraeotyphlus menoni* Annandale. Part I: digestive, circulatory, respiratory and urogenital systems. Anatomischer Anzeiger 81: 393-414.

Exbrayat, J.-M. 1983. Premières observations sur le cycle annuel de l'ovaire de *Typhlonectes compressicaudus* (Duméril et Bibron, 1841), Batracien Apode vivipare. Comptes Rendus des Séances de l'Académie des Sciences de Paris 296: 493-498.

Exbrayat, J.-M. 1986. Quelques aspects de la biologie de la reproduction chez *Typhlonectes compressicaudus* (Duméril et Bibron, 1841), Amphibien Apode. Doctorat ès Sciences Naturelles, Université Paris VI, France.

Exbrayat, J.-M. and Anjubault, E. 2003. Development, differentiation and growth of gonads in *Typhlonectes compressicauda* (Amphibia, Gymnophiona). 12th Meeting of Societas Europaea Herpetologica, Saint-Petersburg, Russia, August 2003.

Exbrayat, J.-M. and Anjubault, E. 2005. The development, differentiation and growth of gonads in *Typhlonectes compressicauda* (Amphibia, Iymnophiona). Pp 136-139. In N. Ananjeva and O. Tsinenko (eds), *Herpetologica Petropolitana*. Proceedings of the 12th Ordinary General Meeting of the Societas Europaea. Herpetologica, August 12-16, 2003, St. Petersburg, Russia. Russian Journal of Herpetology, 12 (Supplement).

Exbrayat, J.-M. and Collenot, G. 1983. Quelques aspects de l'évolution de l'ovaire de *Typhlonectes compressicaudus* (Duméril et Bibron, 1841), Batracien Apode vivipare. Etude quantitative et histochimique des corps jaunes. Reproduction, Nutrition, Développement 23: 889-898.

Garg, B. L. and Prasad, J. 1962. Observations of the female urogenital organs of limbless amphibian *Uraeotyphlus oxyurus*. Journal of Animal Morphology and Physiology 9: 154-156.

Masood-Parveez, U. 1987. Some aspects of reproduction in the female Apodan Amphibian *Ichthyophis*. PhD., Karnatak Univ., Dharwad, India.

Masood-Parveez, U. and Nadkarni, B. 1993a. The ovarian cycle in an oviparous gymnophione amphibian, *Ichthyophis beddomei* (Peters). Journal of Herpetology 27: 59-63;

Masood-Parveez, U. and Nadkarni, B. 1993b. Morphological, histological and histochemical studies of the ovary of an oviparous Caecilian, *Ichthyophis beddomei* (Peters). Journal of Herpetology 27: 63-69.

Müller, J. 1832. Beitrag zur Anatomie und Naturgeschichte der Amphibien. Zeitschrift Jahrbur für Physiologie 4: 195-275.

Rathke, H. 1852. Bemerkungen uber mehrere Korpertheile der *Coecilia annulata*. Mullers Arkiv 1852: 334-350.

Sarasin, P. and Sarasin, F. 1887-1890. Ergebnisse Naturwissenschaftlicher Forschungen auf Ceylon. Zur Entwicklungsgeschichte und Anatomie der Ceylonischen Blindwuhle *Ichthyophis glutinosus*. C. W. Kreidel's Verlag, Wiesbaden.

Spengel, J. W. 1876. Das Urogenitalsystem der Amphibien. I. Theil. Der Anatomische Bau des Urogenitalsystem. Arbeitenaus demm Zoologzootom Institute Wurzburg 3: 51-114.

Tonutti, E. 1931. Beitrag zur Kenntnis der Gymnophionen. XV. Das Genital-system. Morphologisches Jahrbuch 68: 151-292.

Wake, M. H. 1968. Evolutionary morphology of the Caecilian urogenital system. Part I: the gonads and fat bodies. Journal of Morphology 126: 291-332.

Wake, M. H. 1970a. Evolutionary morphology of the caecilian urogenital system. Part II: the kidneys and urogenital ducts. Acta Anatomica 75: 321-358.

Wake, M. H. 1970b. Evolutionary morphology of the caecilian urogenital system. Part III: the bladder. Herpetologica 26: 120-128.

Wake, M. H. 1972. Evolutionary morphology of the caecilian urogenital system. Part IV: the cloaca. Journal of Morphology 136: 353-366.

Wake, M. H. 1977. The reproductive biology of Caecilians. An evolutionary perspective. Pp. 73-100. In D. H. Taylor and S. I. Guttman (eds), *The Reproductive Biology of Amphibians*, Miami Univ., Oxford, Ohio, S. A.

Wake, M. H. 1980. Reproduction, growth and population structure of the Central American Caecilian *Dermophis mexicanus*. Herpetologica 36: 244-256.

CHAPTER 9

Development of Gonads

Elisabeth Anjubault and Jean-Marie Exbrayat

9.1 INTRODUCTION

The development of gonads is very little known in Gymnophiona. Development has been described in some species (Spengel 1876, Sarasin and Sarasin 1887-1890, Brauer 1902, Tonutti 1931, Marcus 1939). Morphogenesis of the testis has been particularly studied in *Ichthyophis* (Seshachar 1936). More recently, general data have also been published (Wake 1968, Exbrayat and Anjubault 2003, Anjubault and Exbrayat 2004a).

The gonads develop from thickenings of the ventral part of the pronephritic mesoderm. The buds arising from these thickenings give bulb-shaped structures that are linked to the kidneys by a delicate layer of connective tissue. These structures were called "primary gonads" by Tonutti (1931) and have been described as hermaphrodite (Marcus 1939). When the hermaphroditism disappears, two secondary gonads develop and the primary gonads become the fat bodies.

In this chapter, we give observations on the development of the gonads in *Typhlonectes compressicauda*, in which the urogenital tracts and reproductive cycles are well known (Wake 1981, Exbrayat and Sentis 1982, Exbrayat 1983, 1984a, b, 1985, 1986a, b, 1988a, b, 1989, 1992, 1996, Exbrayat and Collenot 1983, Exbrayat and Delsol 1985, Exbrayat and Flatin 1985, Exbrayat and Dansard 1994, Exbrayat *et al*. 1998, Anjubault and Exbrayat 1999a, b, 2000, 2004a, b).

9.2 DEVELOPMENT OF *TYPHLONECTES COMPRESSICAUDA*

Typhlonectes compressicauda is a viviparous species with a six to seven months long pregnancy. Development can be divided in several stages. Stage I groups the first stages of development (segmentation, gastrulation and neurulation). In stage II, the embryos with a more or less abundant yolk mass are surrounded by a mucous envelope. This stage is divided into 9 (II_8 to II_{16} after

---

Laboratoire de Biologie Générale, Université Catholique and Laboratoire de Reproduction et Développement des Vertébrés, Ecole Pratique des Hautes Etudes, 25 rue du Plat, F-69288 Lyon Cedex 02, France.

Delsol *et al.* 1981) or 12 subdivisions (14 to 25 after Sammouri *et al.* 1990). In stage III, the yolk mass has been resorbed, the fetuses are no longer surrounded by a mucous envelope and are free into the uterus where they can move and grasp the wall. This stage is divided into 5 (III$_1$ to III$_5$ after Delsol *et al.* 1981) or 6 subdivisions (26 to 31 after Sammouri *et al.* 1990). In stage IV, the intra uterine larvae resemble small adults. This stage is divided into 3 subdivisions (IV$_1$ to IV$_3$ after Delsol *et al.* 1981 or 32 to 34 after Sammouri *et al.* 1990).

Hatching occurs at stages 25 or 26, metamorphosis approximately from stages 30 to 32, and birth at stage 34.

9.3 DEVELOPMENT OF GONADS IN *TYPHLONECTES COMPRESSICAUDA*

In *Typhlonectes compressicauda*, development of gonads can be divided in four phases: 1) the segregation and migration of the primordial germ cells (PGC or gonocytes) between stages 23 and 25 of Sammouri *et al.* (1990), 2) establishment of the median crest, between stages 26 to 28 or 29; 3) the stage of the bipotential undifferentiated gonadic glands, between stages 29 to 30 and 4) the sexual differentiation and growth of the gonads. Because the development of embryos is not perfectly synchronous from one individual to another, gonads can be more or less developed at a given stage.

9.3.1 Segregation and Migration

The observations begin at stage 23, before hatching. This stage is that of the caudal bud. The mesoblast is divided into two parts: the external somatopleure that is applied to the ectoderm and the internal splanchnopleure that is applied to the endoderm. At the mid-length of the ventral side of the intestine, an abundant yolk mass with numerous platelets is observed.

The PGC are observed in the dorsal part of the endoderm. They present some characteristics that distinguish them from the endodermal cells. They are larger than that of the other cells, are oval in shape, have numerous dense cytoplasmic granules, and the chromatin is more or less scattered in the nucleus. Mitoses are observed, particularly during migration. The PGC never contain yolk platelets.

Stage 23. Beginning of segregation. In the second anterior half of the primitive intestine, the PGC are observed among the endodermal cells constituting the dorsal part of the gut. The PGC start to move and they group into an endodermal dorsal crest. Some of them leave the endoderm by successive waves and enter the dorsal mesentery, between the collagen fibers that seems to guide their migration. At this time, the yolk is progressively resorbed in the mid-intestine. Moreover, it appears that the migration of the PGC from the endodermal dorsal crest to the genital ridges is characterised by an increase in the number of germ cells. Many PGC exhibit some morphological differentiation. Such gonocytes are rounded, differing from the surrounding endodermal somatic cells that are polygonal. The spaces observed between the PGC and the endodermal cells are larger than between the endodermal cells. During this phase, prophases of mitosis can be observed in several PGC.

Stages 24 and 25. Formation of temporary median crest (Fig.9.1A, B). The gonocytes have a large granular cytoplasm. The gonocytes are observed in a more dorsal part of the endoderm. This primordial median ridge is formed from the dorsal mesenterium that is united with the median part of the endodermal roof. The gonocytes included in the roof of the intestine are pinched off from the endoderm, consecutively to the penetration of the lateral mesoderm. Some mesenchymal filaments arising from the mesodermal tissue, that will give the smooth muscular layer of the gut, possibly form the sustaining tissue of the median crest. The median crest is located under the dorsal aorta. It is characteristic of a phase at in which the endoderm and the mesoderm exhibit close interactions favoring migration of PGC. Then, in stage 25, the PGC continue to migrate, leaving the median crest to reach the two future genital glands, that form two small ridges projecting into the coelomic cavity, on each side of the dorsal aorta. At this stage, the nuclei of the germ cells increase in volume. The median ridge progressively collapses. During this period of migration, while some PGC migrate, others continue to segregate from the endodermal roof.

9.3.2 Formation of Genital Gland Ridges

Stages 26 and 27. At these stages, the PGC continue to reach the genital glands. This period is characterised by a continuous flow of PGC. Only some regressive tiny connections are observed between the lateral and the median genital ridge. At this phase of development, the anterior intestine begins to differentiate (Hraoui-Bloquet and Exbrayat 1992), tubular elements of the kidneys appear (Sammouri et al. 1990) lungs extend to the posterior end of the body.

Stage 28 or 29 (Fig.9.2A, B). The genital glands enlarge. As the PGC reach the ridges, gliding on the coelomic walls on both sides of the aorta, the ridges become increasingly evident. They are pea-like structures narrowly connected to such anatomical structures as the kidneys and their adjacent Wolffian ducts, the lungs and the aorta. The development of these ridges is linked to the continuous arrival of the PGC. At this stage, the ridges are no longer homogeneous. The gonocytes are received in their median region and become gonia. The lateral sides are subjected to the formative movement of the coelomic walls and the genital ridges become a groove. The folds that form the sides of this new structure will develop into fat bodies.

9.3.3 Bipotential Undifferentiated Gonadic Glands

Stage 29 or 30. Anatomical structures are observed in several regions. The genital glands are situated near the mesonephros and the Wolffian ducts. The fat bodies enlarge. At this stage, the gonocytes migrate into the genital

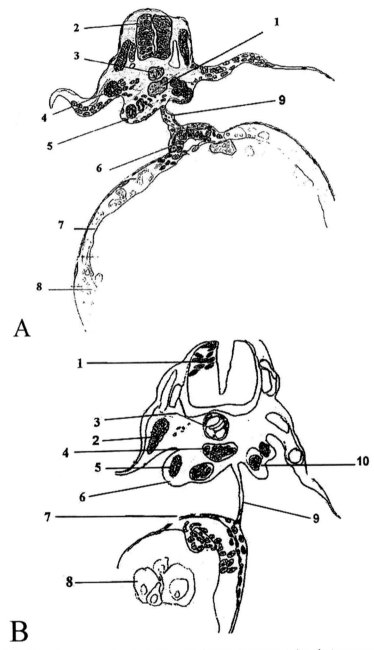

Fig. 9.1. *Typhlonectes compressicauda.* **A.** Stage 25. Schematic representation of a transverse section (TS). PGC are situated among the endodermal cells on the top of intestine (6); they migrate using the median ridge (9) to reach a position under the aorta (1). 1, aorta; 2, neural tube; 3, notochord; 4, nephrotome; 5, coelome; 6, endodermal epithelium; 7 and 8, yolk; 9, median ridge. **B.** Stage 25. Schematic representation of a TS. PGC have reached the two genital glands (10) that begin to develop on each side of the aorta (1). 1, neural tube; 2, Somite 3:,notochord; 4, aorta; 5, nephrotome; 6, coelome; 7, endodermal epithelium; 8, yolk; 9, median ridge; 10, genital gland. Original.

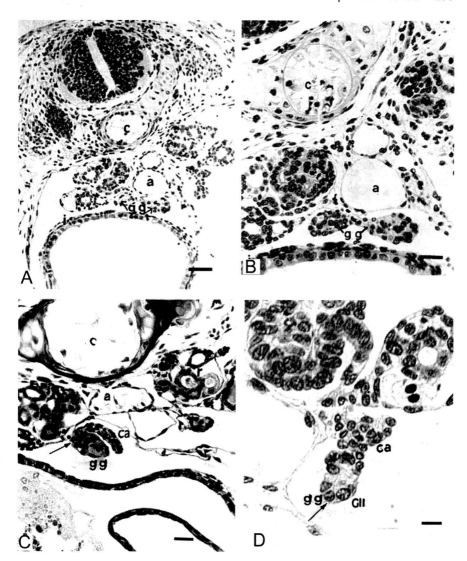

Fig. 9.2. *Typhlonectes compressicauda*. **A**. Stage 29. TS at level of mid-intestine. Genital glands are constituted under the aorta with migrating PGC. The median ridge is a small structure situated between the two genital glands. Scale bar = 20 μm. **B**. Stage 29. TS at level of mid-intestine. The median ridge has collapsed and it is no longer observed on the sections. Scale bar = 20 μm. **C**. Advanced stage 30. TS at level of mid-intestine. stage 30. Scale bar = 20 μm. **D**. Young stage 30. TS at level of mid-intestine. Scale bar = 40 μm. a, aorta; c, notochord; ca, fat body; gg, genital gland; I, intestine. ! gonocyte; GII, developed gonocyte; r, kidney. From Anjubault, E. and Exbrayat, J.-M. 2004a. Bulletin mensuel de la Société Linnéenne de Lyon, 73: 379-392. Pl. I, Figs 1-4.

gland. Two areas are observed in the genital gland: a peripheral area with large cells containing a large granular nucleus and a ventral fibrous part with some smaller germ cells. These two areas herald the future bipotential gonads with a medulla and a cortex but the border between the two regions is poorly defined.

Stage 30 (Fig. 9.2C, D). The medulla and cortex are increasingly defined. The gonocytes migrate into the gonad in a centripetal direction, some being larger than others. Some voluminous follicular cells are situated in the periphery of the gonads. The central fibrous part delimits a stroma. The gonia are not particularly situated in the medulla or the stroma (cortex), i.e. there is no gonadic differentiation in testis or ovary. The fat bodies continue to develop.

9.3.4 Sexual Differentiation

Stage 31 (Fig. 9.3A, B). The cortex and medulla begin to be distinctly differentiated. The center of the gonad contains some ovoid cells with a granular nucleus. At the periphery of the medulla, the cells are larger. Several small germ cells and follicular cells are found in the cortical region At this stage, certain follicular cells begin to surround smaller gonia.

Stage 32 (Fig. 9.3C, D). The gonads are surrounded by the mesonephros with Wolffian ducts, the aorta, the intestine and the fat bodies and contain some blood vessels. The gonia show various mitotic stages (Fig. 9.3D). The follicular cells persist near the peripheral elongated cells. Groups of gonia are observed in some part of the gonad, but do not occupy a distinct peripheral or central location.

Stage 33. Several gonia are in anaphase (Fig. 9.3E, F). But it is not still possible to recognize sexual differentiation of the gonads according to the position of the germ cells in the cortex or the medulla (Fig. 9.4A). The gonads are more or less developed and are reminiscent of the hermaphrodite structures of von Marcus (1939).

Stage 34. At birth, the gonads are differentiated into testes or ovaries.

9.4 VISUALISATION OF TESTOSTERONE DURING GONADOGENESIS

From stage 31 (beginning of metamorphosis) to birth, we could obtain a signal by use of a serum directed against testosterone, by immunocytochemistry. The signal was observed in the connective tissue and more particularly near the gonia that are surrounded by follicular cells that may be Sertoli cells if the animal is a male (Fig. 9.4B). In the young and adult males, the signal has been observed in the Leydig and Sertoli cells (see chapter 3 of this volume). In adult females, we observed the presence of δ5 3β hydroxysteroid dehydrogenase in the germinal nests, with a distribution resembling that of testosterone in the embryonic gonads.

9.5 CONCLUSION

In *Typhlonectes compressicauda*, development of the gonads has been described. After migration of the PGC from the endodermal dorsal ridge to the unpaired

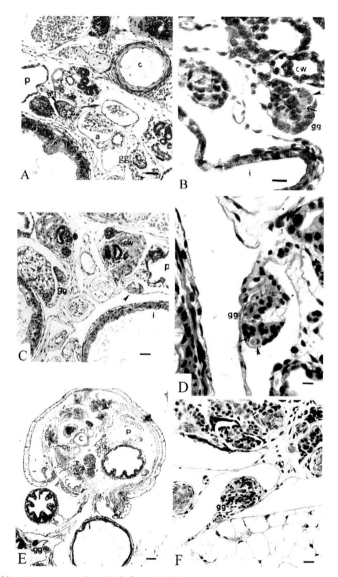

Fig. 9.3. *Typhlonectes compressicauda.* **A**. Stage 31. Transverse section (TS). Developed fat bodies are observed under the genital gland and the lung, on the left part of the picture. Scale bar = 20 μm. **B**. Stage 31. Differentiation of genital glands. Germ cells may be disposed into the medulla or the cortex. Scale bar = 40 μm. **C**. Stage 32. TS. Cortical and medullar zones. Fat bodies are particularly developed. Scale bar = 20 μm. **D**. Stage 33. At level of mid-intestine. Scale bar = 10 μm. **E**. Stage 33. TS. Undifferentiated gonads. Fat bodies that can be seen under the genital gland have now a definitive structure. Scale bar = 20 μm. **F**. Stage 33. TS. Medulla and cortex. Gonia are surrounded by somatic cells (follicle cells). Scale bar = 40 μm. C. Stage 31: TS. Visualisation of testosterone by immunocytochemistry. Scale bar = 40 μm. a, aorta; c, notochord; cw, Wolffian duct; gg, genital gland; I, intestine; p, lung. r, kidney; small arrow, cortex; black arrow, gonia; white arrow, medulla. **A-D**, **F**, From Anjubault, E. and Exbrayat, J.-M. 2004a. Bulletin Mensuel de la Société Linnéenne de Lyon. 73: 379-392, Pl. II, Fig. 5-8. Pl. III, Fig. 10. **E**, original.

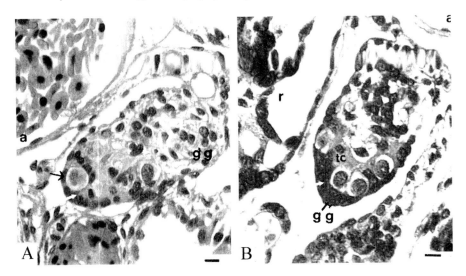

Fig. 9.4. *Typhlonectes compressicauda*. **A**. Stage 33. Germinal nest. Scale bar = 40 μm. **B**. Stage 34: TS. Visualisation of testosterone by immunocytochemistry. Scale bar = 40 μm. a: aorta; gg: gonad; r: kidney; tc: connective tissue; black arrow: follicle cells (future Sertoli cells); white arrow: cortex. **A**, original; **B**, From Anjubault, E. and Exbrayat, J.-M. 2004b. Bulletin Mensuel de la Société Linnéenne de Lyon 73: 393-405. Fig. 5.

gonadic crest and the paired primordial genital glands, two secondary bipotential gonads are progressively formed. These genital glands are each divided into a medulla and a cortex. By comparison with the Anura and Urodela, sexual differentiation is delayed. It is only at birth, or a little before or after, that it is possible to distinguish if gonads are testes or ovaries.

In the Urodela, the PGC are of mesodermal origin (Humphrey 1928, Nieuwkoop 1947, Blackler 1958, Houillon 1967, Capuron 1972, Satasurja and Nieuwkoop 1974, Wakahara 1996). In the urodele *Pleurodeles waltl*, some experimental studies have shown the inductive action of the endoderm on the multiplication of PGC (Maufroid and Capuron 1977). In Anura, the origin of PGC is endodermal (Bounoure *et al*. 1954, Smith 1965; 1966, Blackler 1970, Gipouloux 1970a, b, Lamotte and Xavier 1972). In Gymnophiona, the PGC have been located at a young stage in the endoderm (Marcus 1939). In *Typhlonectes compressicauda*, our own works confirm this observation. In this species, the genital ridges are constituted from coelomic epithelium that is at the origin of the cortex and medulla. In the Anura *Rhacophora arboreus*, similar observations have been made (Tanimura and Iwasawa 1989).

Sexual differentiation of Anura and Urodela is observed before or just after the beginning of the metamorphosis (Lofts 1974). In the Urodela *Hynobius retardatus*, the maturation of gonads occurs independently of metamorphosis (Wakahara 1996). Our own study reveals that the sexual differentiation in *Typhlonectes compressicauda* is delayed by comparison with the Anura and Urodela. It is observed at birth, two months after the metamorphosis (Exbrayat 1986a)

It is possible that this delayed sexual differentiation is related to the presence of Mullerian glands in adult males (Wake 1981, Exbrayat 1985). In Anura, the Mullerian ducts generally persist until the sexual differentiation of the male embryo. After this period, an anti-Mullerian hormone is secreted by the Sertoli cells. This hormone stops the development of the Mullerian ducts and these degenerate, though rudiments persist in some adult male anurans. In females, these ducts persist and become the oviducts (Duellman and Trueb 1986). In *Typhlonectes compressicauda*, the gonads have a delayed development and the Sertoli cells are present well after metamorphosis (see chapters 11 and 12 of this volume). It is possible that an anti-Mullerian hormone exists but at a period in which the Mullerian ducts may not react to it perhaps because receptors are not available.

Our observations confirm that of Marcus (1939). To generalise these conclusions, it would be useful to study the development of the gonads in other Gymnophiona and especially in some oviparous species.

9.6   ACKNOWLEDGEMENTS

The authors thank Catholic University of Lyon (U.C.L.) and Ecole Pratique des Hautes Etudes (E.P.H.E.) that freed them to organize their research fields and supported the works on Gymnophiona. They also thank Singer-Polignac Foundation that supported the missions necessary to collect the material of study. They also thank Barrie Jamieson, who invited them to contribute to this collection devoted to reproduction and phylogeny of Amphibia and also for his patience in reviewing the manuscript. They thank Jean-François Exbrayat for his help in preparing the illustrations.

9.7   LITERATURE CITED

Anjubault, E. and Exbrayat, J.-M. 1999a. Yearly cycle of Leydig-like cells in testes of *Typhlonectes compressicaudus* (Amphibia, Gymnophiona). Pp 53-58. In C. Miaud and R. Guyetant (eds), *Current Studies in Herpetology*, Proceedings of the 9th General Meeting of the Societas Europaea Herpetologica, Le Bourget du Lac, France.

Anjubault, E. and Exbrayat, J.-M. 1999b. Cycle annuel du tissu interstitiel des testicules chez *Typhlonectes compressicaudus* (Amphibia, Gymnphiona) Bulletin de la Société Zoologique de France 125: 133.

Anjubault, E. and Exbrayat, J.-M. 2000. Development of gonads in *Typhlonectes compressicauda* (Amphibia, Gymnophiona). Abstract. XVIIIth International Congress of Zoology, Athens, Greece, August-September 2000: 51.

Anjubault, E. and Exbrayat, J.-M. 2004a. Contribution à la connaissance de l'appareil génital de *Typhlonectes compressicauda* (Duméril et Bibron, 1841), Amphibien Gymnophione. I. Gonadogenèse. Bulletin Mensuel de la Société Linnéenne de Lyon 73: 379-392.

Anjubault, E. and Exbrayat J.-M. 2004b. Contribution à la connaissance de l'appareil génital de *Typhlonectes compressicauda* (Duméril et Bibron, 1841), Amphibien Gymnophione. II. Croissance des gonades et maturité sexuelle des mâles. Bulletin Mensuel de la Société Linnéenne de Lyon 73: 393-405.

Blackler, A. W. 1958. Contribution to the study of germ cells in the Anura. Journal of Embryology and Experimental Morphology 6: 491-503.

Blackler, A. W. 1970. The integrity of the reproductive cell line in the amphibia. Current Topics in Developmental Biology 5: 71-88.
Bounoure, L., Aubry, R. and Huck, M. L. 1954. Nouvelles recherches expérimentales sur les origines de la lignée reproductrice chez la grenouille rousse. Journal of Embryology and Experimental Morphology 2: 245-263.
Brauer, A. 1902. Beitrage zur Kenntniss der Entwicklung und Anatomie der Gymnophionen. III. Die Entwicklung der Excretionsorgane. Zoologisches Jahrbuch für Anatomie 16: 1-176.
Capuron, A. 1972. Mise en évidence de gonocytes dans divers territoires isolés au stade du bourgeon caudal et cultivés in vitro chez *Pleurodeles waltlii* Michah (Amphibien Urodèle). Comptes rendus des Séances de l'Académie des Sciences de Paris 274: 277-279.
Delsol, M., Flatin, J., Exbrayat, J.-M. and Bons, J. 1981. Développement de *Typhlonectes compressicaudus* Amphibien Apode vivipare. Hypothèse sur sa nutrition embryonnaire et larvaire par un ectotrophoblaste. Comptes Rendus des Séances de l'Académie des .Sciences de Paris, série III 293: 281-285.
Duellman, W. E. and Trueb, L. 1986. *Biology of Amphibians*. McGraw-Hill Inc., U.S.A. 670 pp.
Exbrayat, J.-M. 1983. Premières observations sur le cycle annuel de l'ovaire de *Typhlonectes compressicaudus* (Duméril et Bibron, 1841), Batracien Apode vivipare. Comptes Rendus des Séances de l'Académie des Sciences de Paris 296: 493-498.
Exbrayat, J.-M. 1984a. Cycle sexuel et reproduction chez un Amphibien Apode: *Typhlonectes compressicaudus* (Duméril et Bibron, 1841). Bulletin de la Société Herpétologique de France 32: 31-35.
Exbrayat, J.-M. 1984b. Quelques observations sur l'évolution des voies génitales femelles de *Typhlonectes compressicaudus* (Duméril et Bibron, 1841), Amphibien Apode vivipare, au cours du cycle de reproduction. Comptes Rendus des Séances de l'Académie des Sciences de Paris 298: 13-18.
Exbrayat, J.-M. 1985. Cycle des canaux de Müller chez le mâle adulte de *Typhlonectes compressicaudus* (Duméril et Bibron, 1841), Amphibien Apode. Comptes Rendus des Séances de l'Académie des Sciences de Paris 301: 507-512.
Exbrayat, J.-M. 1986a. Quelques aspects de la biologie de la reproduction chez *Typhlonectes compressicaudus* (Duméril et Bibron, 1841), Amphibien Apode. Doctorat ès Sciences Naturelles, Université Paris VI.
Exbrayat, J.-M. 1986b. Le testicule de *Typhlonectes compressicaudus*; structure, ultrastructure, croissance et cycle de reproduction. Mémoires de la Société Zoologique de France 43: 121-132.
Exbrayat, J.-M. 1988a. Croissance et cycle des voies génitales femelles chez *Typhlonectes compressicaudus* (Duméril et Bibron, 1841), Amphibien Apode vivipare. Amphibia Reptilia 9: 117-137.
Exbrayat, J.-M. 1988b. Variations pondérales des organes de réserve (corps adipeux et foie) chez *Typhlonectes compressicaudus*, Amphibien Apode vivipare au cours des alternances saisonnières et des cycles de reproduction. Annales des Sciences Naturelle, Zoologie, Paris, 13éme série, 9: 45-53.
Exbrayat, J.-M. 1989. Quelques observations sur les appareils génitaux de trois Gymnophiones; hypothéses sur le mode de reproduction de *Microcaecilia unicolor*. Bulletin de la Société Herpétologique de France 52: 34-44.
Exbrayat, J.-M. 1992. Appareils génitaux et reproduction chez les Amphibiens Gymnophiones 117: 291-296.
Exbrayat, J.-M. 1996. Croissance et cycle du cloaque chez *Typhlonectes compressicaudus* (Duméril et Bibron, 1841), Amphibien Gymnophione. Bulletin de la Société Zoologique de France 121: 99-104.

Exbrayat J.-M. and Anjubault, E. 2003. Development, differentiation and growth of gonads in *Typhlonectes compressicauda* (Amphibia, Gymnophiona). 12th Meeting of Societas Europaea Herpetologica, Saint-Petersburg, August 2003.

Exbrayat, J.-M. and Collenot, G. 1983. Quelques aspects de l'évolution de l'ovaire de *Typhlonectes compressicaudus* (Duméril et Bibron, 1841), Batracien Apode vivipare. Etude quantitative et histochimique des corps jaunes. Reproduction, Nutrition, Développement 23: 889-898.

Exbrayat, J.-M. and Dansard, C. 1994. Apports de techniques complémentaires à la connaissance de l'histologie du testicule d'un Amphibien Gymnophione. Bulletin de l'Association Française d'Histotechnologie 7: 19-26.

Exbrayat, J.-M. and Delsol, M. 1985. Reproduction and growth of *Typhlonectes compressicaudus*, a viviparous Gymnophione. Copeia 1985: 950-955.

Exbrayat, J.-M. and Flatin, J. 1985. Les cycles de reproduction chez les Amphibiens Apodes. Influence des variations saisonnières. Bulletin de la Société Zoologique de France 110: 301-305.

Exbrayat, J.-M. and Sentis, P. 1982. Homogénéité du testicule et cycle annuel chez *Typhlonectes compressicaudus* (Duméril et Bibron, 1841), Amphibien Apode vivipare. Comptes Rendus des Séances de l'Académie des Sciences de Paris 294: 757-762.

Exbrayat, J.-M., Pujol, P. and Leclercq, B. 1998. Quelques aspects des cycles sexuels et nycthéméraux chez les Amphibiens. Bulletin de la Société Zoologique de France 123: 113-124.

Gipouloux, J.-D. 1970a. Recherches expérimentales sur l'origine, la migration des cellules germinales et l'édification de crêtes génitales chez les Amphibiens Anoures. Bulletin de Biologie de la France et de la Belgique 104: 21-93.

Gipouloux, J.-D. 1970b. Précisions sur le stade auquel débute la migration des cellules germinales chez le Crapaud commun (*Bufo bufo*). Comptes Rendus des Séances de l'Académie des Sciences de Paris 270: 533-535.

Hraoui-Bloquet, S. and Exbrayat, J.-M. 1992. Développement embryonnaire du tube digestif chez *Typhlonectes compressicaudus* (Duméril et Bibron, 1841), Amphibien Gymnophione vivipare. Annales des Sciences Naturelles, Zoologie, Paris, 13éme série 13: 11-23.

Houillon, C. 1967. *Embryologie*, coll. Méthodes, Hermann, Paris, 183 pp.

Humphrey, R. R. 1928. Sex differentiation in gonads developed from transplants of intermediate mesoderm of *Ambystoma*. Biological Bulletin of Marine Biology of Woods Hole 55: 367-399.

Lamotte, M. and Xavier, F. 1972. Recherches sur le développement embryonnaire de *Nectophrynoides occidentalis* Angel, Amphibien Anoure vivipare. l. Les principaux traits morphologiques et biométriques du développement. Annales d'Embryologie et de Morphologie 5: 315-340.

Lofts, B. 1974. Reproduction. Pp. 53-106. In B. Lofts, (ed), *Physiology of the Amphibia*. II, Academic Press, New York, London.

Marcus, H. 1939. Beitrag zur kenntnis der Gymnophionen. Ueber keimbahn, keimdruusen, Fettkörper und Urogenitalverbindung bei *Hypogeophis*. Biomorphosis 1: 360-384.

Maufroid, J. P. and Capuron, A. 1977. Induction du mésoderme et de cellules germinales primordiales par l'endoderme chez *Pleurodeles waltlii* (Amphibien Urodèle): évolution au cours de la gastrulation. Comptes rendus des Séances l'Académie des Sciences de Paris 284: 1713-1716.

Nieuwkoop, P. D. 1947. Experimental investigations on the origin and determination of the germ cells and on the development of the lateral plates and germ ridges in Urodeles. Archives Néerlandaises de Zoologie 8: 1-205.

Sammouri, R., Renous, S., Exbrayat, J.-M. and Lescure, J. 1990. Développement embryonnaire de *Typhlonectes compressicaudus* (Amphibia, Gymnophiona). Annales des Sciences Naturelles, Zoologie, 13ème série 11: 135-163.

Sarasin, P. and Sarasin, F. 1887-1890. Ergebnisse Naturwissenschaftlicher Forschungen auf Ceylon. Zur Entwicklungsgeschichte und Anatomie der Ceylonischen Blindwuhle *Ichthyophis glutinosus*. C. W. Kreidel's Verlag, Wiesbaden.

Satasurja, L. A. and Nieuwkoop, D. 1974. The induction of the primordial germ cells in the Urodeles. Wilhelm Roux's Archives 175: 199-220.

Seshachar, B. R. 1936. The spermatogenesis of *Ichthyophis glutinosus* (Linn.) I. The spermatogonia and their division. Zeitschrift für Zellforschung und mikroskopische Anatomie 24: 662-706.

Seshachar, B. R. and Srinath, K. V. 1946. Studies on the nucleolus. I. The nucleolus of the Apodan Sertoli cell. Proceedings of Indian Scientific Congress 3: 96.

Smith, J. W. 1965. Transplantation of the nuclei of primordial germ cells into enucleated eggs of *Rana pipiens*. Proceedings of National Academy of Sciences, U.S.A. 54: 101-107.

Smith, J. W. 1966. The role of a "germinal plasm" in the formation of primordial germ cells in *Rana pipiens*. Developmental Biology 14: 330-347.

Spengel, J. W. 1876. Das Urogenitalsystem der Amphibien. I. Theil. Der Anatomische Bau des Urogenitalsystem. Arbeitenaus demm Zoologzootom Institute Wurzburg 3: 51-114.

Tanimura, A. and Iwasawa, H. 1989. Origin of somatic cells and histogenesis in the primordial gonad of the Japanese tree frog *Racophorus arboreus*. Anatomy Embryology 180: 165-173.

Tonutti, E. 1931. Beitrag zur Kenntnis der Gymnophionen. XV. Das Genital-system. Morphologische Jahrbuch 68: 151-292.

Wakahara, M. 1996. Heterochrony and neotenic salamanders: possible clues for understanding the annual development and evolution. Zoological Science 13: 765-766.

Wake, M. H. 1968. Evolutionary morphology of the Caecilian urogenital system. Part I: the gonads and fat bodies. Journal of Morphology 126: 291-332.

Wake, M. H. 1981. Structure and function of the male mullerian gland in Caecilians (Amphibia: Gymnophiona), with comments on its evolutionary significance. Journal of Herpetology 15: 17-22.

CHAPTER 10

Modes of Parity and Oviposition

Jean-Marie Exbrayat

10.1 EGGS AND EGG-LAYERS IN GYMNOPHIONA

In caecilians, a variable number of oocytes are laid according to the species. In the oviparous species, the eggs are laid externally after having traversed the oviduct in which they are surrounded by several envelopes. The eggs are united to each other to form a string (Figs. 12.3A and 12.4A, chapter **12** of this volume). In viviparous species, the eggs are surrounded by a delicate envelope secreted by the anterior part of the oviduct (*Typhlonectes compressicauda*, Exbrayat 1986, 1988a); then they develop in the uterus, and 20 to 50 oocytes are usually laid (Exbrayat and Delsol 1988). In oviparous species the number of deposited oocytes is greater than in viviparous species (Wake 1977b). Some species produce a low number of eggs: *Idiocranium russelii* (Wake 1977b), *Siphonops annulatus* (Goeldi 1899), *S. paulensis* (Gans 1961) and the direct developing *Boulengerula taitanus* (Malonza and Measey 2005) lay only 5 to 6 eggs. In contrast, *Ichthyophis malabarensis* lays about 100 eggs at a time (Balakrishna *et al*, 1983) (Table 10.1).

In viviparous species, 4 to 30 mature oocytes are observed in a single female of *Geotrypetes seraphinii* (Wake 1968b, 1977b), in *Gymnopis multiplicata*, *Dermophis mexicanus* (Wake 1977b) and *Typhlonectes compressicauda* (Exbrayat 1986), 20 to 30 mature oocytes are counted in a female. Generally, the diameter of oocyte in oviparous species is between 4 and 10 mm. In viviparous species, this diameter does not exceed 2 to 3 mm (Wake 1977b, Exbrayat and Delsol 1988) (Table 10.1).

In the oviparous species, the eggs are attached to each other by a gelatinous substance (*Ichthyophis glutinosus*, Sarasin and Sarasin 1887-1890, *Gegenophis carnosus*, Seshachar 1942, *I. malabarensis*, Seshachar *et al*, 1982,

---

Laboratoire de Biologie générale, Université Catholique de Lyon and Laboratoire de Reproduction et Développement des Vertébrés, Ecole Pratique des Hautes Etudes, 25, rue du Plat, F-69288 Lyon Cedex 02.

Table 10.1. Mode of reproduction with number of eggs and oocytes and sizes in several gymnophionan species (after data given in Exbrayat and Delsol 1988)

Species	Reproductive mode	Egg or oocyte number	Egg or oocyte size	Author
Uraeotyphlus menoni	oviparous	4 to 15	7 × 4 mm	Wake 1968a, b
Siphonops annulatus	oviparous		10 × 8 ?5 mm	Goeldi 1899
Siphonops paulensis	oviparous		4.3 × 4.5 mm	Gans 1961
Ichthyophis glutinosus	oviparous	54	9 × 3 mm	Sarasin and Sarasin 1887-1890
		12-20	9 × 6 mm	Gundappa et al. 1961, Wake 1977
		41		Breckenridge and Jayasinghe 1979
		25-40		
Ichthyophis beddomei	oviparous		7.5 mm	Masood-Parveez 1987
Idiocraniu russeli	oviparous	6		Wake 1977a, b
Ichthyophis malabarensis	oviparous		8.5 mm	Seshachar et al. 1982
Gegenophis carnosus	oviparous	15		Seshachar 1942
Hypogeophis rostratus	oviparous	5-20	10 mm	Tonutti 1931
		15 - 20	9 × 6 mm	Wake 1977a, b
Geotrypetes seraphini	viviparous	4 – 15	2 × 5 mm	Wake 1968a, b
		18-30	3.5 × 2 to 3.7 mm	Wake 1977a, b
Caecilia ochrocephala	viviparous	4-15	2 to 5 mm	Wake 1968a, b
Gymnopis multiplicata	viviparous	18-30	2 mm	Wake 1968a, b
				Wake 1977a, b
Schistometopum thomensis	viviparous		3 × 2 mm	Wake 1977a, b
Typhlonectes compressicauda	viviparous	20	3 × 25 mm	Wake 1977a, b
Afrocaecilia taitana	?		2 mm	Exbrayat 1986
(Boulengerula taitanus)		4 to 15	2 to 5 mm	Wake 1968a, b

Balakrishna *et al.* 1983, *Boulengerula taitanus*, Malonza and Measey 2005). After egg laying, the eggs are separated after having broken the elastic string in *Siphonops annulatus* (Goeldi 1899) and *I. glutinosus* (Breckenridge and Jayasinghe 1979). The eggs of *I. glutinosus* are united to form a cluster. Each egg bears a mucous projection on one side and at a point diametrically opposed; the string uniting the eggs is severed on each side of the egg (Breckenridge and Jayasinghe 1979). Each egg is contained in a rough and elastic envelope. The external part of the envelope consists of numerous fibers that are disposed in several concentric layers that react positively to stains specific for elastic tissue. An extension of this part constitutes the string linking each egg to the others. The internal zone is weakly stained and lacks fibers. This envelope surrounds a cavity filled with a fluid in which the embryo develops (Figs. 12.3, 12.4, chapter 12 of this volume).

In *Typhlonectes compressicauda*, a viviparous species, oocytes are surrounded by a delicate mucous envelope that is PAS positive and is stained by alcian blue (staining of acidic carbohydrates), secreted by the anterior part of oviduct (Exbrayat 1986). This envelope persists during development and will be resorbed at intra-uterine hatching.

10.2 VIVIPARITY IN GYMNOPHIONA

50% of Gymnophionan species are estimated to be viviparous (Wake 1977b, 1993) but this number has been debated (Wilkinson and Nussbaum 1998). It is known that Scolecomorphidae and Typhlonectidae are obligatory viviparous families (Wake 1993). Whatever the exact number of viviparous species, very few of them have been deeply studied. The features of pregnancy have been particularly studied in *Dermophis mexicanus* (Wake 1980b) and *Typhlonectes compressicauda* (Exbrayat *et al*, 1981. Exbrayat and Delsol 1985, Exbrayat 1986, Exbrayat and Hraoui-Bloquet 1992a, b, Hraoui-Bloquet and Exbrayat 1992, 1994, 1996, Hraoui-Bloquet 1995). The fetal teeth of the embryos of viviparous species have been well studied in several papers (Parker 1956, Parker and Dunn 1964, Wake 1970a, 1976, 1980a, Hraoui-Bloquet 1995, Hraoui-Bloquet and Exbrayat 1996). The gills of *Gymnopis* embryos have been studied (Wake 1967, 1969). In *T. compressicauda*, the gills are transformed into vesicular structures. These were first reported in the 19th century (Peters 1874a, b, 1875), and later studied by several authors (Delsol *et al.* 1986, Exbrayat 1986, Wake 1969, Sammouri *el al.* 1990, Exbrayat and Hraoui-Bloquet 1991, Hraoui-Bloquet and Exbrayat 1994, Hraoui-Bloquet 1995). Some hypotheses have been advanced suggesting that they function as organs involved in exchanges between embryo and mother.

10.2.1 Number of Eggs and Embryos

In *Typhlonectes compressicauda*, all the live-bearing females possess both fertilized oocytes and eggs in the oviducts, throughout pregnancy. Certain oocytes, found in the anterior part of oviduct, are not surrounded by any envelope. Other oocytes or eggs, in a posterior position, are surrounded by a mucous envelope secreted by mucous glands of the anterior part of the oviduct. In the median and posterior part of the oviduct, differentiated as a uterus, several developing embryos are observed. Sometimes, some damaged oocytes and undeveloped embryos are also found among healthy ones.

In one and the same female, all the embryos of the two uteri are at first at very similar stages of development but by the end of pregnancy the embryos may be at different stages (Delsol et al. 1981). In this case, the less developed embryos are dead and more or less dehydrated. At the beginning of development, each embryo is surrounded by a mucous envelope. They are situated in the anterior part of uterus. In the same uterus they are separated by a distance of 10 mm or more. Then embryos lose the mucous envelope and become fetuses. They are scattered within the uterus to occupy the greatest amount of space. These embryos, now freed from their envelopes, are free in the uterus where they can move. At this period, the yolk mass is resorbed and the embryos feed from the secretions of the uterine wall, probably seeking the areas in which secretions are the most abundant. When they become intra-uterine larvae (*sensu* Exbrayat 1986), they occupy the entire uterine lumen. Their very voluminous gills are spread out on each side and applied to the wall of the uterus until the uterine wall is entirely covered by larval gills (Fig. 11.6C, D, chapter 11 of this volume). These larvae are very elongated, and they are folded (U or S shaped) into the space limited by the gills. The gills resemble a cocoon: one gill is disposed as a coat, and the other as a hood. At the end of pregnancy, a uterus measuring 10 mm in length can contain three or four embryos measuring 150 to 200 mm each.

In *Typhlonectes compressicauda*, the number of oocytes and embryos has been counted in the oviducts (Exbrayat 1986). To know the number of oocytes produced by each ovary, the number of corpora lutea has been counted. 11 to 28 corpora lutea have been found in each ovary of each pregnant female. The oocytes or eggs lacking a mucous envelope are particularly numerous during the period of ovulation. Some of them do not develop and degenerate while some oocytes persist in the genital tract throughout pregnancy. At the beginning of pregnancy, several eggs, with an envelope are found in the oviducts.

The average number of embryos in each uterus decreases as pregnancy advances. For all the pregnant females examined, we counted 5 embryos at the beginning of pregnancy, then 4 and finally 1 at the end of the gestation. At this time, certain females contained 4 larvae in each uterus, other contained only 1 embryo per uterus. The youngest embryos are more often lost. We found, on average, 5 embryos in April and in May. When the presence of larvae is observed, the total number of the embryos further decreases. These data indicate that all the fetuses do no reach the larval stage. Finally, a large loss of ovular material is observed, first in the anterior part of the oviduct, then in the uterus. At the beginning of pregnancy, 6 eggs or embryos are observed in the oviducts for 11 to 28 oocytes originating from each ovary. Just before birth, only one to three embryos are found in each uterus. Some dead fetal stages have also been often found in the oviducts with larval stages.

In *Dermophis mexicanus*, each ovary may carry 13 to 36 oocytes that are mature when breeding occurs. The number of mature oocytes is always higher than the number of embryos that will be found during pregnancy (4 to 12 embryos according to the period of observation) (Wake 1980b).

In *Chthonerpeton indistinctum*, each female can give birth to 6 to 10 young animals (Barrio 1969), 11 embryos have been observed (Prigioni and Langone 1983 a, b) and three young in *Schistometopum thomense* (see chapter 2 of this volume, Fig. 2.5).

In conclusion, very few data are available about number of eggs and embryos in viviparous Gymnophiona. It seems that females of viviparous species can give birth to 8 to 11 young and that the number of eggs is much higher than number of young animals. We will see that the degenerative oocytes and eggs, and also the dead embryos, can be used as nourishment for the viable developing animals.

10.2.2 Intra-uterine Feeding and Materno-Fetal Exchanges

The development of all species of Gymnophiona occurs from a telolecithal egg. In the oviparous species, the eggs are 4 to 10 mm in diameter. The development of the embryos proceeds from the yolk mass to a stage at which the larvae, that are free in the field, are able to feed by themselves (*Ichthyophis glutinosus*, Breckenridge *et al*. 1987).

In viviparous species, for animals with a comparable size, the diameter of the egg does not exceed 3 mm at ovulation (Exbrayat and Delsol 1988). Yolk of viviparous species is less abundant than in oviparous species and it is utilised at the beginning of development. Then, as already noted, the embryo is released from its mucous envelope and is free in the uterine lumen. It can move in the uterus and feeds on uterine secretions. During pregnancy, small eggs (2 to 3 mm in diameter) will develop to give large new-borns (150 mm in *Typhlonectes compressicauda* and *Dermophis mexicanus*, Exbrayat *et al*. 1981, Wake 1980b). *T. compressicauda* has been studied in some details (Delsol *et al*. 1981, 1983, 1986, Exbrayat *et al*. 1981, Exbrayat 1983, 1986, Exbrayat and Hraoui-Bloquet 1991, 1992b, Hraoui-Bloquet and Exbrayat 1992, 1994, 1996, Hraoui-Bloquet 1995). Some data also exist for other species.

Very few works have been devoted to materno-fetal relationships in the viviparous species. Two aspects have been the chief objects of study: the fetal teeth (Parker 1956, Parker and Dunn 1964, Largen *et al*. 1972, Wake 1976, 1977a, 1978, 1980b) and the gills (Parker 1956, Wake 1967, 1969). We will examine successively the importance of the yolk, feeding by oophagy and adelphophagy, the fetal teeth and the gills. In a separate chapter, a detailed study of viviparity in *Typhlonectes compressicauda* is presented (see chapter 11 of this volume).

10.2.2.1 Yolk mass

In viviparous species, the development of embryos begins from the yolk previously elaborated in oocytes. The eggs of these species are small and the yolk is quickly exhausted, more rapidly than in the oviparous species. In several species, the yolk is resorbed by the first phase of embryogenesis (Parker and Dunn 1964, Wake 1967, 1977a, 1980a). During this first phase of development, the embryo is protected by a mucous envelope secreted from the wall of the anterior part of the oviduct. When the yolk is exhausted, the embryo must find other sources of nutrition to continue its development. The embryo also possesses a pair of triradiate gills that are presumably involved in gas exchange with its mother across the intra-uterine fluid.

10.2.2.2 Oophagy and adelphophagy

Very few data have been published about these subjects. In numerous species, a considerable proportion of the oocytes found in the oviducts are not fertilized. For instance, in *Dermophis mexicanus*, twenty oocytes have been observed in the lumen of the oviduct but only three to ten develop (Wake 1980b). Generally, the number of unfertilized oocytes is greater than the number of healthy embryos. Some dead embryos and fetuses are sometimes found in the oviducts with embryos with a normal development. It is possible that oophagy and/or adelphophagy exist in the viviparous Gymnophiona. We will see that these phenomena have been shown in *Typhlonectes compressicauda* in which small embryos have been observed in the mouth or the gut of healthy fetuses and larvae (Exbrayat 1984, Exbrayat and Hraoui-Bloquet 1992b, Hraoui-Bloquet 1995) (Figs. 11.4G, 11.6F, chapter 11 of this volume).

10.2.2.3 Fetal teeth

A characteristic fetal dentition has been demonstrated in the embryo and fetuses with intra-uterine development. The first genera observed were *Geotrypetes*, *Schistometopum* and *Chthonerpeton* (Parker and Dunn 1964). More recently, the dentition has been described for young oviparous Gymnophiona (*Siphonops annulatus*, *S. paulensis*, and some *Caecilia*) (Wilkinson and Nussbaum 1998). Nevertheless, fetal teeth have been mainly observed in viviparous species.

In *Geotrypetes seraphinii*, some small embryos (17 mm) with a voluminous yolk mass and still protected by a mucous envelope do not possess teeth. At 26 mm, the yolk mass is much reduced; certain embryos are still contained in their envelope, other are free. At this size, several spoon-shaped fetal teeth are disposed in two parallel rows, on the lower jaw. At birth (animal measuring 73 to 77 cm in length), spoon-shaped teeth are numerous. They are disposed in quincunx, in parallel rows on the inner aspect of the lower jaw. In the upper jaw, only adult teeth are observed. Four rows of fetal teeth disposed in quincunx were observed on the lower jaw of an embryo of *Geotrypetes seraphinii*, measuring 38 mm in length, with a resorbed yolk mass, and free in the uterus. The vomer and palatine parts lacked teeth. Several characteristic adult teeth were observed on maxillary and premaxillary bones.

Several *Schistometopum thomense* embryos were observed. At 32 mm and 36 mm, the yolk mass was resorbed and the embryos were not surrounded by any envelope. Three rows of spoon-shaped teeth were observed on the lower jaw. At 88 mm, the embryos possessed only a single row of fetal teeth. At birth (117 mm), the teeth had the adult form (Parker and Dunn 1964). In *Scolecomorphus uluguruensis*, the embryos less than 30 mm long lack teeth. In the biggest embryos, fetal teeth were always observed. In new-born (85 mm in length), the dentition was that of an adult. In several subspecies of *Dermophis mexicanus* and *Gymnopis multiplicata*, some large embryos (100 to 115 mm), sometimes with a yolk mass, possess several rows of spoon-shaped fetal teeth. Just before birth and in certain newborn, the fetal teeth may persist besides the definitive teeth.

The evolution of the fetal dentition has been well studied in *Gymnopis multiplicata* (Wake 1976). In the embryos measuring 10 mm, there is a series of tooth buds on the lower jaw (Fig. 10.1A). On each side of the jaw, a bud already mineralised, situated at the antero-lateral angle of the head was particularly developed. The cells of the jaw were differentiated as procartilaginous cells. In 12 mm embryos, the teeth were resorbed. The other buds continued to develop. For the first time, the teeth begin to be formed on the upper jaw. In some embryos, the dental buds were numerous in several rows (Fig. 10.1B, C). Some teeth begin to develop, more particularly on the upper jaw. In 45 mm embryos, fetal teeth were observed on both upper and lower jaws (Fig. 10.1D). These teeth were elongated with several denticles. From 54 mm embryos, ossification of the jaw occurs. New fetal teeth were now unmineralised but were constant in appearance.

The fetal teeth of several species of Typhlonectidae have been studied (Parker and Dunn 1964, Wake 1978), and particularly in *Typhlonectes compressicauda* (Wake 1976, Hraoui-Bloquet 1995, Hraoui-Bloquet and Exbrayat 1996). Six stages of development have been described. Each tooth develops from a pulp cavity that is constituted by the odontoblasts in continuity with the mesenchyme. At stage (a), the enamel organ resembles a cup (Fig. 10.2A, 10.3A). At stage (b), it is differentiated into an external layer of flattened cells and an internal layer of columnar cells (Fig. 10.2B). The odontoblasts penetrate into the cup of the enamel organ. At stage (c), the odontoblasts separate from dental pulp and the space is progressively filled with a substance corresponding to predentine (Fig. 10.2C). At stage (d), the top of the cup flattens (Fig. 10.2D). Anteriorly, in the part directed to the pulp, the enamel cells lengthen and accumulate a substance resembling enamel. At this stage, the crown is observed. At stage (e), teeth are elongated (Fig. 10.2E). At stage (f), the top of the tooth is near the surface of the jaw (Fig. 10.2F). Under the crown, the odontoblasts secrete a substance (predentine) forming the collar. Posteriorly the pedicle develops and attaches to the dental bone. The tooth pierces the epidermis (Hraoui-Bloquet and Exbrayat 1996).

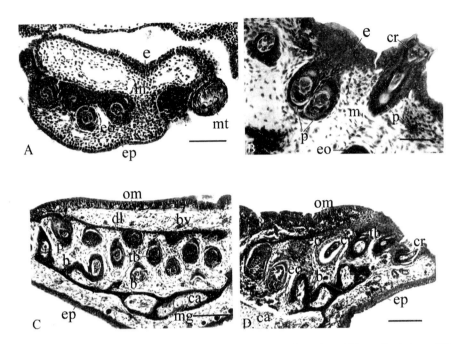

Fig. 10.1. Fetal teeth in *Gymnopis multiplicata*. **A**. Tooth buds of lower jaw of 10 mm. Scale bar = 120 µm. **B**. Medial tooths of lower jaw of 30 mm embryo. Scale bar = 370 µm. **C**. Tooth buds on the lower jaw of a 54 mm fetus. Scale bar = 150 µm. **D**. Tooth buds on the lower jaw of an 84 mm fetus. Scale bar = 150 µm. b, bone; bv, blood vessel; ca, cartilage; cr, crown; dl, dental lamina; e, epithelium; eo, enamel organ; ep, epidermis; m, mesenchyme; mg, mucous gland; mt, mineralized tooth; om, oral mucosa; p, pedicel; tb, tooth bud. From Wake, M. H. 1980a. Journal of Morphology 166: 203-216, Plate 1, Fig. 11; Plate 2, Fig. 13B; Plate 3; Figs. 14, 15.

In stages 23 and 24, just before the hatching, the first buds of the fetal teeth develop (Fig. 11.1A, chapter 11 of this volume). At this time, two new rows develop. At the end of the fetal phase (stages 25 to 31), the teeth are disposed in quincunx, in seven or eight rows (Fig. 10.3A-F). In each row, teeth at different stages of development are observed. The more developed teeth are situated in the more ventral rows, i.e. the more posterior in the mouth (Fig. 10.4A-F). On the upper jaw, only three rows develop, but no tooth emerges. Developing buds and worn teeth are observed together. At birth, the pedicels of fetal teeth are resorbed, though some persist for several days. In new-born, dentition characteristic of the adult is observed.

The 18 mm embryos of *Typhlonectes compressicauda* always possess two encapsulated mineralised teeth on the two sides of the lower jaw (Wake 1976). The study of the fetal dentition in *Typhlonectes compressicauda* has shown that several replacement waves occur: From hatching to birth, some tooth buds are observed in all stages of development.

In *Chthonerpeton viviparum* and *C. petersi*, the embryos lacking a yolk mass possess a fetal dentition. Teeth are spoon-shaped and disposed in quincunx in three parallel rows. Fusion of the tooth buds may occur, and

Modes of Parity and Oviposition 311

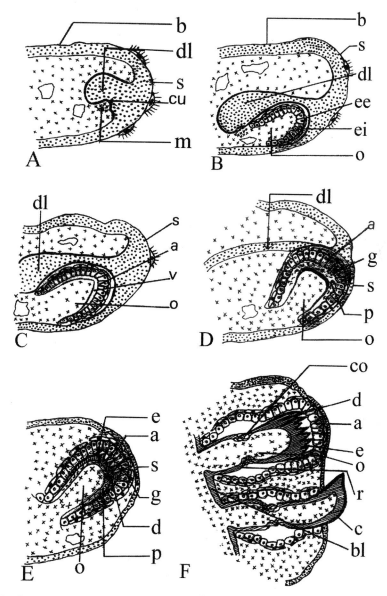

Fig. 10.2. Schematic representation of fetal tooth evolution in *Typhlonectes compressicauda*. **A.** Stage (a) corresponding to the invagination of dental lamina. **B.** stage (b), differentiation of enamel organ. **C.** stage (c), odontoblastic cells are separated from adamental ones by a gap. **D.** stage (d), secretion and accumulation of predentine. **E.** stage (e), beginning of mineralization of predentine into dentine and of enamel secretion. **F.** stage (f), end of dental ontogeny. The tooth is ready to merge. a, enamel organ;. b, floor of mouth; bl, bone plate; co, collecting duct; cu, dental cup; d, dentine; dl, dental lamina; e,. enamel; ee, external part of enamel organ; ei, internal part of enamel organ; g, granulations in the cells of enamel organ; m, mesenchymal condensation; o, odontoblast; p, predentine; r, root; s, surface of the mandible; v, gap between enamel organ and dental pulp. From Hraoui-Bloquet, S. and Exbrayat, J.-M. 1996. Annales des Sciences Naturelles, Zoologie, 13éme série 17: 11-23, Pl. II.

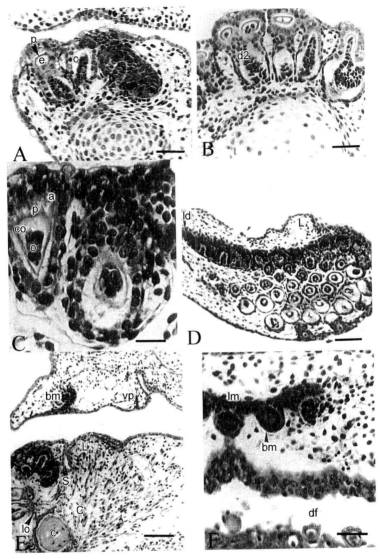

Fig. 10.3. Development of fetal teeth in *Tyhlonectes compressicauda*. **A.** Lower jaw in a stage 25-26 embryo. Scale bar = 85 μm. **B.** Lower jaw in a stage 25-26 embryo with a double tooth. Scale bar = 140 μm. **C.** dental buds in stage 26 embryo. Scale bar = 40 μm. **D.** Lower jaw in stage 28 embryo; the teeth are disposed in quincunx on several rows. Scale bar = 130 μm. **E.** Lower and upper jaws in a stage 29 embryo. Scale bar = 140 μm. **F.** Upper jaw in a stage 28 embryo. Scale bar = 80 μm. a, enamel cell; bm, tooth bud on the upper jaw; c, fetal tooth at stage (c); co, crown; C, Cartilage; df, fetal tooth on mandible; d2, double tooth; e, fetal tooth at stage (e); lo, bone plate; L, anterior tip of the tongue; ld, dental lamina; lm, upper dental lamina; o, odontoblast; p, point of a fetal tooth; s, gap between the tongue and lower jaw; vp, vomer-palatine dental lamina. **A**, from Hraoui-Bloquet, S. and Exbrayat, J.-M. 1996. Annales des Sciences Naturelles, Zoologie, 13éme série 17: 11-23, Pl. I, Fig. 4. **B-F**, from Hraoui-Bloquet, S. 1995. Unpublished Ph.D. thesis, Ecole Pratique des Hautes Etudes, Paris, France, Pl. IV, Fig. 5; Pl. V, Figs. 3, 5, 6; Pl. VI, Fig. 2.

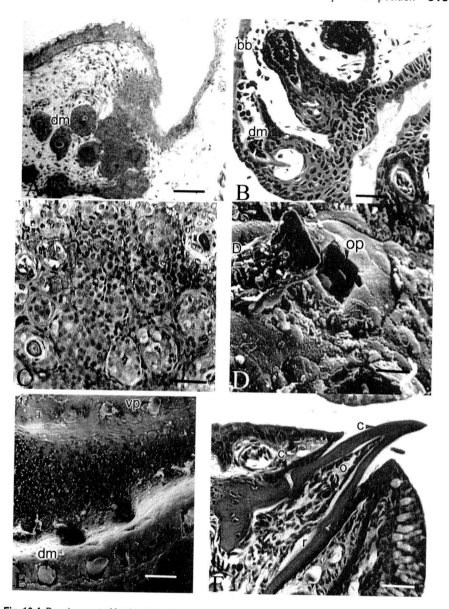

Fig. 10.4. Development of fetal teeth in *Typhlonectes compressicauda*. **A.** Upper jaw in a stage 31 fetuse. Scale bar = 135 µm. **B.** Details of teeth on upper jaw in a stage 32 larva. Scale bar = 55 µm. **C.** Fetal teeth scars on the mandible of a new-born. Scale bar = 85 µm. **D.** MEB view of fetal teeth scars on the mandible of a stage 32 larva. Scale bar = 20 µm. **E.** MEB view of upper jaw of new-born. Scale bar = 130 µm. **F.** Dental tooth in a new-born. Scale bar = 85 µm. Abbreviations: bb, bony blade; c, scar; dm, maxillary tooth; o, odontoblast; op, opening indicating the place of a tooth; D, degradation of tooth; r, root; vp, vomer-palatine tooth. **A, C, D,** From Hraoui-Bloquet, S. 1995. Unpublished Ph.D. thesis, Ecole Pratique des Hautes Etudes, Paris, France, Pl. VI, Figs. 3, 5; Pl. VII, Fig. 4. **B, E, F,** From Hraoui-Bloquet, S., and Exbrayat, J.-M. 1996. Annales des Sciences Naturelles, Zoologie, 13éme série 17: 11-23, Pl. I, Figs. 2, 3, 6.

the appearance is that of a bony plate. In *C. petersi*, the teeth of the upper jaw are very pointed and disposed in quincunx in about twelve rows. At birth, dentition of *Chthonerpeton* is that of an adult (Parker and Dunn 1964).

In *Dermophis mexicanus*, the fetal dentition has been studied in detail (Wake 1980a). The 22 mm embryos possess minute mineralised teeth. At 35 mm, the dental buds are disposed in four rows. Teeth are small and they emerge very little above the epidermis of the mouth. Two series of unmineralised multidenticulate teeth are also observed in the buccal roof. In 52 mm embryos they are mineralised on the lower jaw. The disposition of the rows with dental buds suggests successive waves. The dentition continues to develop on the lower as well as the upper jaw (in contrast to *Typhlonectes compressicauda*). At 110 mm, several transitional teeth intermediate between fetal and adult teeth appear. In a 123 mm embryo, the adult teeth are in place. At birth, the animals possess their adult teeth; but certain fetal teeth are still observed before they disappear.

The presence of fetal teeth is constant in all the viviparous species that have been examined. From Parker and Dunn (1964), all the amphibia could have a genetic factor causing the development of these teeth in several rows. At a moment in development, this factor is suppressed and the implantation of the teeth is no longer caused. According to these authors, these teeth have no feeding function for the embryos feed on a fluid secreted by the uterine wall. However Wake (1980a) considers that this specific dentition is used by the embryos to scrape secretions off the uterine wall, a view accepted by the present author. The fetal teeth are mainly specific to viviparous species though such a dentition has been reported in the larvae of some oviparous species (Wilkinson and Nussbaum 1998). They appear just after resorption of the yolk mass, when the animal must find a new source of food. At this same period, the uterus secretes abundant glycoproteic substances that are grasped by fetuses. The epithelial cells found in the digestive tract of the embryos are evidence of this action. During development of *Typhlonectes compressicauda*, we could observe wear of the teeth, after their action. We also observed modification of the uterine wall during this period (Exbrayat 1986, 1984, Hraoui-Bloquet, 1995, Hraoui-Bloquet and Exbrayat 1996, Hraoui-Bloquet *et al.* 1994).

10.2.2.4 The gills

The embryos of the viviparous Gymnophiona possess a pair of external triradiated gills resembling those of the larvae of oviparous species (Parker and Dunn 1964 Wake 1967, 1969, 1977a) (Fig. 10.5A, B). Yet, in Typhlonectidae, the gills have a very particular appearance. They develop as a pair of vesicular structures. A major study on the gills of viviparous species has been published by Parker and Dunn (1964).

In *Geotrypetes seraphinii*, gills are found in embryos measuring 17 to 25 mm, but they are absent in the new-born (75 to 77 mm in length). In *G. angelii*, the authors did not observe any gill in an intra-uterine embryo measuring 58 mm.

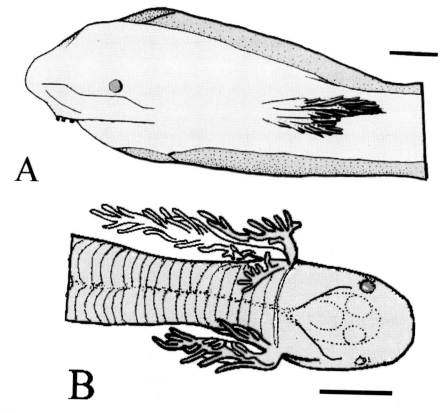

Fig. 10.5. **A**. Head of a 52 mm *Gymnopis multiplicata* embryo, with fetal teeth and gills. Scale bar = 1 mm. **B**. Head of a 40 mm *Dermophis mexicanus* fetus, with gills. Scale bar = 5 mm. **A,** modified after Wake, M. H. 1967, Bulletin of South California Academy of Sciences 66: 109-116, Fig. 1. **B**, modified after Wake, M. H. 1977a. Journal of Herpetology 11: 379-386, Fig. 5, modified.

In *Schistometopum thomense*, embryos measuring 33 to 36 mm in length have triradiate gills that are in resorbtion. In the same species, the 88 to 90 mm fetuses and new-born do not possess gills. In *Scolecomorphus uluguruensis*, the gills are observed in the small embryos (24 to 26 mm) but there are no longer gills in the new-born. Resorption of the gills has been studied in *Gymnopis multiplicata proxima* (Wake 1967, 1969) (Fig. 10.5A). In the young embryos (10.5 mm) still covered by a mucous envelope, two gills are observed. One of them is out of the envelope, in contact with the lumen of uterus. The other gill is still covered by the envelope. Wake proposes that the need for oxygen increases when development starts. The gills could be used for oxygenation. In the same species, some free embryos measuring between 45 and 55 mm have been observed in one of the two uteri of a female. The less developed animals (45 and 52 mm) have the most developed gills. For each gill, the three branches are united at a bud situated on each side of the head. The central branch is twice the length of the others. It bears four filaments on each side, two on the other, and a tuft of three other filaments at the top. The other branches of the gill also bear filaments in a similar arrangement. In the same female, a 54 mm embryo, posterior to the two others, possesses normal gills on the left side of the head, but on the right side, gills are resorbing and supported only by a long strip of connective tissue. This gill contains a great number of blood cells that indicate its degeneration. The gills are not supported by any bony or cartilaginous tissue. The branches are covered by a single or stratified epithelium, in continuity with that of the head. In the central part of each branch, the gaps are filled by red blood cells. Certain filaments also contain blood cells. The gills are connected to three arterial arches that drive blood through the head. In all cases, in this species, the most posterior embryo, measuring 54.5 mm does not possess gills. Wake suggests that the stage at which the degeneration of gills is found, corresponds to the larval stage of the oviparous genus *Hypogeophis* in which such a degeneration of gills has also been observed (Marcus 1908).

Typhlonectes compressicauda and other Typhlonectidae possess a pair of large vesicular gills. The development of gills has been deeply studied (Delsol *et al.* 1986, Exbrayat 1986, Exbrayat and Hraoui-Bloquet 1991, 1992a, b, Hraoui-Bloquet and Exbrayat 1994, Hraoui-Bloquet 1995) (Fig. 11.6 B, C, chapter 11 of this volume). During their development, there is a differentiation between the embryonic face (in front of the body of the embryo) and the uterine face that will be applied to the uterine wall at the end of pregnancy. Finally, at the end of development, the larvae are protected by their giant gills that envelope them, giving a cocoon-like appearance. The uterine face of the gill is closely applied to the uterine wall. Both the maternal and fetal tissues have a placenta-like structure. This point is, developed in chapter 11 of this volume, devoted to viviparity in *Typhlonectes compressicauda*.

In addition to these structures, the body of *Typhlonectes compressicauda* is covered by absorptive buttons that are increasingly numerous during the phase when the yolk mass is present (Sammouri *et al.* 1990). At the same stages, an area situated on the ventral skin, the ectotrophoblast (Delsol *et al.* 1981, 1983) seems to be more particularly used in exchanges between the embryo and the uterine fluid. These structures are described with details in chapter 11 of this volume.

In *Typhlonectes compressicauda*, some works have also been devoted to the diffusive exchanges between the mother and the embryo (Exbrayat and Hraoui-Bloquet 1992b, Hraoui-Bloquet 1995), the respiratory properties of the embryos (Toews and MacIntyre 1977), the maternal organs involved in the materno-fetal exchanges (Exbrayat 1986, 1988b), and the calcification of the embryo (Wake *et al.* 1985, Paillot 1995).

10.2.2.5 Materno-fetal immunological relationships
The sole study concerning materno-fetal immunological relationships is that for *Typhlonectes compressicauda* (Exbrayat *et al.* 1995, Hraoui-Bloquet 1995, chapter 3 of this volume).

10.3 CONCLUSIONS: HOMOGENEITY OF VIVIPARITY IN GYMNOPHIONA

In the most studied viviparous gymnophonian species (*Dermophis mexicanus* and *Typhlonectes compressicauda*), each female with a weight about 100 g and measuring 300 to 500 mm in length, gives birth to 4 to 6 (in *Typhlonectes*) or 6 to 7 (in *Dermophis*) new-born. Each new-born is 15 to 20 mm and 5 to 6 g. The yolk reserves of a small egg (several mg and 1.5 to 2.5 mm in diameter) are not sufficient to provide all the materials and energy necessary for such a growth. Therefore, supply by an intra-uterine feeding is necessary for the development of these animals.

After resorption of the yolk mass, the embryo must find its food in the uterus. Several sources of nourishment are available. One of them is constituted by some secretions of the anterior part of the oviduct, and yolk originating from abortive eggs; these substances are the components of the "uterine milk". The dissolution of the dead embryos could also be one of the elements of the feeding medium in which the fetuses develop. The uterine wall also possesses a direct feeding function that can be related to the presence of the fetal teeth, even if the presence of such a dentition has also been reported in a few oviparous species (Wilkinson and Nussbaum 1998). Adelphophagy seems also to be an important element of intra-uterine feeding. Some traces of ingested embryos have been observed in the gut of several large larvae, at least in *Typhlonectes compressicauda*. Furthermore, osmotic and diffusive mechanisms certainly also exist between the mother and its embryos. These exchanges could be performed by means of the gills that are observed in all the oviparous species at a relatively young stage. In *Typhlonectes*, the absorptive buttons are observed at a young stage of the development. These buttons are scattered all over the body and become increasingly dense (Sammouri *et al.* 1990). In same species, a ventral ectotrophoblast is also observed. Such structures have not been reported in the other viviparous species but that is perhaps mainly due to the lack of data and not to negative observations. In *Typhlonectes compressicauda* and other Typhlonectidae, the gills have a very peculiar shape and they are certainly important organs implicated in the materno-fetal exchanges. More particularly, they participate, with the uterine wall, in building a true placenta by the end of pregnancy.

It is interesting to compare the pregnancy of Gymnophiona to that of the other viviparous Amphibia. The characteristics of pregnancy in Gymnophiona are relatively closed to that of *Salamandra atra* (Vilter 1986). In this urodele, as in *Dermophis* and in *Typhlonectes*, for each female, the number of mature oocytes is initially high (Vilter and Vilter 1960). After laying of 40 to 60 eggs in uterus, only two embryos (rarely 3 or 4) develop. Other eggs and oocytes degenerate in the oviduct and become food for the embryos. The duration of the gestation is very long in *Salamandra atra*: 3 to 4 years. After one year, the embryo reaches 20 mm in length; it is 35 mm by 2 years, 45 mm by 3 years and 51 mm by 4 years. During the development, the embryo eats the uterine secretions, the epithelial cells of the uterine wall (that regenerate in contrary of *Typhlonectes compressicauda*) and the abortive eggs.

It is also interesting to compare the pregnancy of Gymnophiona to that of *Salamandra salamandra*. In this species, several degrees of pregnancy have been observed according to the altitude. In mountain populations, the most viviparous form, the number of abortive eggs is high (30%) and 35% of embryos die during development. These abortive eggs and dead embryos are used as food by the live intra-uterine larvae that have no other intra-uterine supply. One larva can eat one to four eggs or young embryos during its development. The larval gills are implicated in osmotic and diffusive exchanges between the larvae and the intra-uterine medium (Greven 1980). This mechanism is close to some aspects of materno-fetal exchanges in Gymnophiona.

There are some comparative points with the modes of viviparity in Anura, between Gymnophiona and the anuran genus *Nectophrynoides* genera (Wake 1980c, 1993, Xavier 1986, Wake and Dickie 1998). In *Nectophrynoides occidentalis*, the eggs are small (600 to 700 µm in diameter) and poor in yolk. Each of them will give birth to a young animal, the size of which will be one third that of the mother. The increase of weight during the intra-uterine growth is 300 times the egg mass (Lamotte and Tuchman-Duplessis 1948, Lamotte and Xavier 1972). The embryos are free in the uterus, no placentation being observed. In *N. occidentalis*, the digestive tract differentiates early and is filled by feeding masses (Vilter and Lamotte 1956, Vilter and Lugand 1959). The embryos are maintained in a fluid secreted by the uterine wall. This fluid has a different appearance between and after the formation of the gut, showing that this liquid constitutes by itself the food of embryos. Therefore, in the case of *N. occidentalis*, the feeding of embryos is also provided by the uterine wall.

It is also interesting to compare the case of *Typhlonectes compressicauda* to the anuran species that sequester their embryos in dorsal alveolae (*Pipa pipa*) or in a dorsal pouch (*Gastrotheca*). In these few anuran species studied, the embryonic gills develop as a veil or a bell and are applied against the highly vascularised wall of the maternal pouch or alveola. This phenomenon is close to that is observed in *Typhlonectes*.

Comparison between the different viviparous amphibian species shows that this mode of reproduction appeared several times in Amphibia but only the Gymnophiona are homogenous in this respect. The modes of pregnancy seem to be very similar in the different species, even though duration of development is variable, and viviparity appears to have occurred independently several times (Wake 1977a, b, Wilkinson and Nussbaum 1998). Nevertheless, it seems possible to observe an evolution in viviparity in Gymnophiona in which the family Typhlonectidae presents the highest degree.

10.4 PARENTAL CARE

Parental care in Caecilians has not been the subject of publications, it is known that in the oviparous forms the female surrounds the eggs during development (*Ichthyophis glutinosus*, Breckenridge and Jayasinghe 1979; Sarasin and Sarasin 1887-1890 (Fig.12.3B); *Boulengerula taitanus*, Malonza and Measey 2005; *I. malabarensis*, Bhatta and Nadkarni, personal communication). Furthermore, young viviparous *Geotrypetes seraphinii* feed on skin secretions of their mother (O' Reilly et al. 1998).

10.5 ACKNOWLEDGEMENTS

The author thanks Catholic University of Lyon (U.C.L.) and Ecole Pratique des Hautes Etudes (E.P.H.E.) that freed him to organize his research fields and supported the works on Gymnophiona. He also thanks Singer-Polignac Foundation that supported the missions necessary to collect the study material. The author thanks Michel Delsol, Honorary Professor at he U.C.L. and at E.P.H.E. who is at the origin of these works in Gymnophiona, Marie-Thérèse Laurent, who made thousand and thousand of sections. He also thanks Barrie Jamieson, who invited him to contribute to this collection devoted to reproduction and phylogeny of Amphibia and also for his patience in reviewing the manuscript. He thanks his son, Jean-François Exbrayat for his help in preparing the illustrations.

10.6 LITERATURE CITED

Balakrishna, T. A., Gundappa, K. R. and Katre Shakuntala. 1983. Observations on the eggs and embryos of *Ichthyophis malabarensis* (Taylor) (Apoda: Amphibia). Current Science 52: 990-991.

Barrio, A. 1969. Observaciones sobre *Chthonerpeton indistinctum* (Gymnophiona, Caecilidae) y su reproduccion. Physis 28: 499-503.

Breckenridge, W. R. and Jayasinghe, S. 1979. Observations on the eggs and larvae of *Ichthyophis glutinosus*. Ceylon Journal of Sciences (Biological Sciences) 13: 187-202.

Breckenridge, W. R., Shirani, N. and Pereira, L. 1987. Some aspects of the biology and development of *Ichthyophis glutinosus* (Amphibia, Gymnophiona). Journal of Zoology 211: 437-449.

Delsol, M., Flatin, J., Exbrayat, J.-M. and Bons, J. 1981. Développement de *Typhlonectes compressicaudus* Amphibien Apode vivipare. Hypothèse sur sa nutrition embryonnaire et larvaire par un ectotrophoblaste. Comptes Rendus des Séances de l'Académie des Sciences de Paris, série III 293: 281-285.

Delsol, M., Flatin, J., Exbrayat, J.-M. and Bons, J. 1983. Développement embryonnaire de *Typhlonectes compressicaudus* (Duméril et Bibron, 1841): constitution d'un ectotrophoblaste. Bulletin de la Société Zoologique de France 108: 680-681.

Delsol, M., Exbrayat, J.-M., Flatin, J. and Gueydan-Baconnier, M. 1986. Nutrition embryonnaire chez *Typhlonectes compressicaudus* (Duméril et Bibron, 1841) Amphibien Apode vivipare. Mémoires de la Société Zoologique de France 43: 39-54.

Exbrayat, J.-M. 1983. Premières observations sur le cycle annuel de l'ovaire de *Typhlonectes compressicaudus* (Duméril et Bibron, 1841), Batracien Apode vivipare. Comptes Rendus des Séances de l'Académie des Sciences de Paris 296: 493-498.

Exbrayat, J.-M. 1984. Quelques observations sur l'évolution des voies génitales femelles de *Typhlonectes compressicaudus* (Duméril et Bibron, 1841), Amphibien Apode vivipare, au cours du cycle de reproduction. Comptes Rendus des Séances de l'Académie des Sciences de Paris 298: 13-18.

Exbrayat, J.-M. 1986. Quelques aspects de la biologie de la reproduction chez *Typhlonectes compressicaudus* (Duméril et Bibron, 1841), Amphibien Apode. Doctorat ès Sciences Naturelles, Université Paris VI, France.

Exbrayat, J.-M. 1988a. Croissance et cycle des voies génitales femelles chez *Typhlonectes compressicaudus* (Duméril et Bibron, 1841), Amphibien Apode vivipare. Amphibia-Reptilia 9: 117-137.

Exbrayat, J.-M. 1988b. Variations pondérales des organes de réserve (corps adipeux et foie) chez *Typhlonectes compressicaudus*, Amphibien Apode vivipare au cours des alternances saisonnières et des cycles de reproduction. Annales des Sciences Naturelles, Zoologie, 13éme série 9: 45-53.

Exbrayat, J.-M. and Delsol, M. 1985. Reproduction and growth of *Typhlonectes compressicaudus*, a viviparous Gymnophione. Copeia 1985: 950-955.

Exbrayat, J.-M. and Delsol, M. 1988. Oviparité et développement intra-utérin chez les Gymnophiones. Bulletin de la Société Herpétologique de France 45: 27-36.

Exbrayat, J.-M. and Hraoui-Bloquet, S. 1991. Morphologie de l'épithélium branchial des embryons de *Typhlonectes compressicaudus* (Amphibien Gymnophione) étudié en microscopie électronique à balayage. Bulletin de la Société Herpétologique de France 57: 45-52.

Exbrayat, J.-M. and Hraoui-Bloquet, S. 1992a. Evolution de la surface branchiale des embryons de *Typhlonectes compressicaudus*, Amphibien Gymnophione vivipare, au cours du développement. Bulletin de la Société Zoologique de France 117: 340.

Exbrayat, J.-M. and Hraoui-Bloquet, S. 1992b. La nutrition embryonnaire et les relations foeto-maternelles chez *Typhlonectes compressicaudus*, Amphibien Gymnophione vivipare. Bulletin de la Société Herpétologique de France 61: 53-61.

Exbrayat, J.-M., Delsol, M. and Flatin, J. 1981. Premières remarques sur la gestation chez *Typhlonectes compressicaudus* (Duméril et Bibron, 1841) Amphibien Apode vivipare. Comptes rendus des Séances de l'Académie des Sciences de Paris 292: 417-420.

Exbrayat, J.-M., Pujol, P. and Hraoui-Bloquet, S. 1995. First observations on the immunological materno-fetal relationships in *Typhlonectes compressicaudus*, a viviparous Gymnophionan Amphibia. Pp. 271-273. In G. A. Llorente, A. Montori, X. Santos and M. A. Carretero (eds), *Scientia Herpetologica*, Proceedings of the 7th General Meeting of the Societas Europaea Herpetologica, Barcelona, Spain.

Gans, C. 1961. The first record of egg laying in the Caecilian *Siphonops paulensis* Boettger. Copeia 1961: 490-491.

Goeldi, E. 1899. Ueber die Entwicklung von *Siphonops annulatus*. Zoologische Jahrbucher Systematische 12: 170-173.

Greven, H. 1980. Ultrahistochemical and autoradiographic evidence for epithelial transport in the uterus of the ovoviviparous salamander *Salamandra salamandra* (L.) (Amphibia, Urodela). Cell and Tissue Research 212: 147-162.

Gundappa, K. R., Balakrishna, T. A. and Katre Shakuntala. 1981. Ecology of *Ichthyophis glutinosus* (Linn.) (Apoda, Amphibia). Current Science 50 : 480-483.

Hraoui-Bloquet, S. 1995. Nutrition embryonnaire et relations materno-fetales chez *Typhlonectes compressicaudus* (Duméril et Bibron, 1841), Amphibien Gymnophione vivipare. Thèse de Doctorat E.P.H.E., Lyon, France.

Hraoui-Bloquet, S. and Exbrayat, J.-M. 1992. Développement embryonnaire du tube digestif chez *Typhlonectes compressicaudus* (Duméril et Bibron, 1841),

Amphibien Gymnophione vivipare. Annales des Sciences Naturelles, Zoologie, 13éme série 13: 11-23.

Hraoui-Bloquet, S. and Exbrayat, J.-M. 1994. Développement des branchies chez les embryons de *Typhlonectes compressicaudus*, Amphibien Gymnophione vivipare. Annales des Sciences naturelles, Zoologie, 13éme série 15: 33-46.

Hraoui-Bloquet, S. and Exbrayat, J.-M. 1996. Les dents de *Typhlonectes compressicaudus* (Amphibia, Gymnophiona) au cours du développement. Annales des Sciences Naturelles, Zoologie, 13éme série 17: 11-23.

Hraoui-Bloquet, S., Escudié, G. and Exbrayat, J.-M. 1994. Aspects ultrastructuraux de l'évolution de la muqueuse utérine au cours de la phase de nutrition orale des embryons chez *Typhlonectes compressicaudus*, Amphibien Gymnophione vivipare. Bulletin de la. Société Zoologique de France 119: 237-242.

Lamotte, M. and Tuchman-Duplessis, H. 1948. Structure et transformations gravidiques du tractus génital femelle chez un Anoure vivipare (*Nectophryoides occidentalis* Angel). Comptes Rendus des Séances de l'Académie des Sciences de Paris 226: 597-599.

Lamotte, M. and Xavier, F. 1972. Recherches sur le développement embryonnaire de *Nectophrynoides occidentalis* Angel, Amphibien Anoure vivipare. I. Les principaux traits morphologiques et biométriques du développement. Annales d'Embryologie, Morphologie 5: 315-340.

Largen, M. J., Morris, P. A. and Yalden, D. W. 1972. Observations on the Caecilian *Geotrypetes grandisonae* Taylor (Amphibia Gymnophiona) from Ethiopia. Monitore Zoologico Italiano 8: 185-205.

Malonza, P.K. and Measey, G.J. 2005. Life history of an African caecilian: *Boulengerula taitanus* Loveridge 1935 (Caeciilidae: Amphibia: Gymnophiona). Tropical Zoology 18: 49-66.

Marcus, H. 1908. Beitrage zur Kenntniss der Gymnophionen. I. Über das Schlundspaltengebiet. Archiv für Mikroskopie und Anatomie 71: 695-774.

Masood Parveez 1987. Some aspects of reproduction in the female Apodan Amphibian *Ichthyophis*. PhD., Karnatak University., Dharwad, India, 205 pp.

O'Reilly, J.C., Fenolio, D., Rania, L.C. and Wilkinson, M. 1998. Altriciality and extended parental care in the West African caecilian *Geotrypetes seraphini* (Gymnophiona: Caeciliidae). American Zoologist 38: 187A.

Paillot, R. 1995. Variations de la calcification osseuse chez *Typhlonectes compressicaudus* (Duméril et Bibron, 1841) au cours de la gestation. Place du tissu osseux des gymnophiones parmi les Vertébrés inférieurs. Mémoire de fin d'études, Spécialisation en histologie, Université catholique de Lyon, France.

Parker, H. W. 1956. Viviparous caecilians and amphibian phylogeny. Nature 178: 250-252.

Parker, H. W. and Dunn, E. R. 1964. Dentitional metamorphosis in the Amphibia. Copeia 1964: 75-86.

Penhos, J. C. 1953. Rôle des corps adipeux du *Bufo arenarum*. Boletin de la Sociedad argentina de Biologia 28: 1095-1096.

Peters, W. 1874a. Observations sur le développement du *Caecilia compressicauda*. Annales des Sciences Naturelles, Zoologie, série 5: article 13.

Peters, W. 1874b. Derselbe las ferner über die Entwicklung der Caecilien und besonders der *Caecilia compressicauda*. Monatsberichte der Deutschen Akademie der wissenschaften zu Berlin 1874: 45-49.

Peters, W. 1875. Uber die Entwicklung der Caecilien. Monatsberichte der Deutschen Akademie der wissenschaften zu Berlin 1875: 483-486.

Prigioni, M. Y and Langone, J. A. 1983a. Notas sobre *Chthonerpeton indistinctum* (Amphibia, Typhlonectidae). IV Notas complementarias. Jornadas de Ciencias naturales, Montevideo 19-24 set. 1983, 3: 81-83

Prigioni, M. Y and Langone, J. A. 1983b. Notas sobre *Chthonerpeton indistinctum* (Amphibia, Typhlonectidae). V Notas complementarias. Jornadas de Ciencias naturales, Montevideo 19-24 set. 1983, 3: 97-99.

Sammouri, R., Renous, S., Exbrayat, J.-M. and Lescure, J. 1990. Développement embryonnaire de *Typhlonectes compressicaudus* (Amphibia, Gymnophiona). Annales des Sciences Naturelles, Zoologie, 13ème série 11: 135-163.

Sarasin, P. and Sarasin, F. 1887-1890. Ergebnisse Naturwissenschaftlicher Forschungen auf Ceylon. Zur Entwicklungsgeschichte und Anatomie der Ceylonischen Blindwuhle *Ichthyophis glutinosus*. C. W. Kreidel's Verlag, Wiesbaden.

Seshachar, B. R. 1942. The eggs and embryos of *Gegenophis carnosus* (Bedd.). Proceedings of Indian Academy of Sciences B15 278: 439-441.

Seshachar, B. R., Balakrishna, T. A., Katre Shakuntala and Gundappa, K. R. 1982. Some unique feature of egg laying and reproduction in *Ichthyophis malabarensis* (Taylor) (Apoda: Amphibia). Current Science 51: 32-34.

Toews, D. and Macintyre, D. 1977. Blood respiratory properties of a viviparous Amphibian. Nature 266: 464-465.

Tonutti, E. 1931. Beitrag zur Kenntnis der Gymnophionen. XV. Das Genital-system. Morphologisches Jahrbuch 68: 151-292.

Vilter, V. 1986. La reproduction de la salamandre noire. Pp. 487-495. In P.P. Grassé et M. Delsol (eds), *Traité de Zoologie*, tome XIV, fasc.I B, Masson, Paris.

Vilter, V. and Lamotte, M. 1956. Evolution post-gravidique de l'utérus chez *Nectophrynoides occidentalis* Ang., Crapaud totalement vivipare de la Haute-Guinée. Comptes Rendus de la Société de Biologie 150: 2109-2113.

Vilter, V. and Lugand, A. 1959. Trophisme intra-utérin et croissance embryonnaire chez *Nectophrynoides occidentalis* Ang., Crapaud totalement vivipare du mont Nimba (Haute-Guinée). Comptes Rendus de la Société de Biologie 153: 29-32.

Vilter, V. and Vilter, A. 1960. Sur la gestation de la salamandre noire des Alpes, *Salamandra atra* Laur. Comptes rendus de la Société de Biologie 154: 290-294.

Wake, M. H. 1967. Gill structure in the Caecilian genus *Gymnopis*. Bulletin of South California Academy of Sciences 66: 109-116.

Wake, M. H. 1968a. The comparative morphology and evolutionary relationships of the urogenital system of Caecilians. Ph.D. Dissertation, University of California.

Wake, M. H. 1968b. Evolutionary morphology of the Caecilian urogenital system. Part I: the gonads and fat bodies. Journal of Morphology 126: 291-332.

Wake, M. H. 1969. Gill ontogeny in embryos of *Gymnopis* (Amphibia : Gymnophiona). Copeia, 1969: 183-184.

Wake, M. H. 1970a. Evolutionary morphology of the caecilian urogenital system. Part II: the kidneys and urogenital ducts. Acta Anatomica 75: 321-358.

Wake, M. H. 1976. The development and replacement of teeth in viviparous Caecilians. Journal of Morphology 148: 33-63.

Wake, M. H. 1977a. Fetal maintenance and its evolutionary significance in the Amphibia : Gymnophiona. Journal of Herpetology 11: 379-386.

Wake, M. H. 1977b. The reproductive biology of Caecilians. An evolutionary perspective. Pp. 73-100. In D. H. Taylor and S. I. Guttman (eds), *The Reproductive Biology of Amphibians*, Miami Univ., Oxford, Ohio.

Wake, M. H. 1978. Comments on the ontogeny of *Typhlonectes obesus* particulary its dentition and feeding. Papeis avulsos de Zoologia 32: 1-13.

Wake, M. H. 1980a. Fetal tooth development and adult replacement in *Dermophis mexicanus* (Amphibia : Gymnophiona): Fields versus clones. Journal of Morphology 166: 203-216.
Wake, M. H. 1980b. Reproduction, growth and population structure of the Central American Caecilian *Dermophis mexicanus*. Herpetologica 36: 244-256.
Wake, M. H. 1980c. The reproductive biology of *Nectophrynoides occidentalis malcolmi* (Amphibia: Bufonidae) with comments on the evolution of reproductive modes in the genus *Nectophrynoides*. Copeia 1980: 193-209.
Wake, M. H. 1981. Structure and function of the male mullerian gland in Caecilians (Amphibia: Gymnophiona), with comments on its evolutionary significance. Journal of Herpetology 15: 17-22.
Wake, M. H. 1993. Evolution of oviductal gestation in Amphibians. The Journal of Experimental Zoology 266: 394-413.
Wake, M. H. and Dickie, R. 1998. Oviduct structure and function and reproductive modes in Amphibians. The Journal of Experimental Zoology 282:477-506.
Wake, M. H., Exbrayat, J.-M. and Delsol, M. 1985. The development of the chondrocranium of *Typhlonectes compressicaudus* (Gymnophiona), with comparison to other species. Journal of Herpetology 19: 568-577.
Wilkinson, M. and Nussbaum, R. 1998. Caecilian viviparity and amniote origins. Journal of Natural History 32: 1403-1409.
Xavier, F. 1986. La reproduction des *Nectophrynoïdes*. Pp. 497-513. In P.-P. Grassé et M. Delsol (eds), *Traité de Zoologie, Amphibiens*, Tome XIV, fasc. I-B, Masson, Paris.

CHAPTER 11

Viviparity in *Typhlonectes compressicauda*

Jean-Marie Exbrayat[1] and Souad Hraoui-Bloquet[2]

11.1 SOME ASPECTS OF DEVELOPMENT

Viviparity of *Typhlonectes compressicauda* has been well studied (Delsol *et al.* 1981, 1983; Exbrayat *et al.* 1981, 1994, 1995; Exbrayat and Delsol 1985; Exbrayat 1986, 1984, 1988a, b, 1990, 1996; Exbrayat and Hraoui-Bloquet 1991, 1992a, b; Hraoui-Bloquet and Exbrayat 1992, 1994, 1996; Hraoui-Bloquet *et al.* 1994; Hraoui-Bloquet 1995). In this species several feeding modes have been observed during intra-uterine life.

Viviparous Gymnophiona have been the subject of several works of descriptive embryology. In an earlier work, five main stages were described (Exbrayat *et al.* 1981). Stage 0 corresponds to eggs surrounded by a mucous envelope. Stage I groups the first stages of development (segmentation, gastrulation and neurulation). Stage II is that of embryos during morphogenesis still surrounded by a mucous envelope, with a relatively developed yolk mass. Stage III is that of fetuses, without yolk mass and envelope, found free in the uterus where they are able to move. Stage IV is that of intra-uterine larvae that resemble young new-born but are enveloped by a pair of vesicular gills.

In the first table of development that has been given (Delsol *et al.* 1981, and chapter **12** of this volume), stage II is divided into 9 subdivisions (II_8 to II_{16}); hatching occurs at stage II_{16}. Stage III is divided into 5 subdivisions (stages III_1 to III_5) and stage IV into 3 subdivisions (stages IV_1 to IV_3). Stage IV_1 corresponds to the end of metamorphosis and IV_3 to birth.

Description of development has been definitively given by a closer study (Sammouri *et al.* 1990, and chapter **12** of this volume) in which numeration of stages is given by Arabic number exclusively. The development of *Typhlonectes compressicauda* is now divided into 21 stages (stages 24 to 34).

---

[1] Laboratoire de Biologie Générale, Université Catholique and Laboratoire de Reproduction et Développement des Vertébrés, Ecole Pratique des Hautes Etudes, 25 rue du Plat, F-69288 Lyon Cedex 02, France. [2] Université Libanaise, Faculté des Sciences II, BP 26 11 02 17. Fanar El Maten, Lebanon.

Certain of the stages given by the oldest table have been further divided. Stages 14 to 25 correspond to the category II, stages 26 to 31 to category III and stages 32 to 34 to category IV.

Intra-uterine hatching is observed at stages 25-26, metamorphosis occurs approximately from stage 30 to stage 32 and birth occurs at stage 34.

11.2 YOLK AND ECTOTROPHOBLAST

For a viviparous species, eggs of *Typhlonectes compressicauda* are relatively rich in yolk. Fertilisation occurs in the anterior region of the oviduct (Hraoui-Bloquet and Exbrayat 1993, and chapter 12 of this volume). The egg is surrounded by a mucous envelope secreted by the wall of oviduct. First phases of development (to stages 25-26) occur within this mucous envelope that isolates the egg from the intra-uterine medium. This first part of development seems to utilize yolk. The weight increases very little at this period but the embryo lengthens greatly. It is possible (and even probable) that gas and fluid exchanges occur between embryo and uterine fluid. During this period, the surface of embryo is covered by absorption buds that are increasingly numerous (Fig. 11.1A). These buds are found all over the body of the embryo. In addition, an area situated on the skin of the venter, called ectotrophoblast by Delsol *et al.* (1981, 1983), seems to be more particularly involved in these exchanges.

At the beginning of development, the intestine (still possessing yolk), is floating in a pouch that is limited ventrally by the ectoderm (Fig. 11.1B). This pouch is the coelomic cavity limited by splanchnopleure and somatopleure; the ectoderm is thin and translucent. The ventral part of embryo constitutes the ectotrophoblast that is clearly differentiated at stage 23 (Fig. 11.1B, D). At stage 26, when the embryo loses its mucous envelope, the appearance of the ectotrophoblast changes. It becomes thicker. Its cells possess cilia, that reveal a brush border by scanning electron microscopy, and an apical bulge contains the nucleus (Fig. 11.1C, E, F).

The ultrastructure of the ectotrophoblast has been studied (Hraoui-Bloquet 1995). At stages 23 and 25, the wall of ectotrophoblast consists of a layer of ectodermal cells (embryonic epidermis) and a layer of mesodermal cells, the somatopleure. The epidermis can be stratified (Fig. 11.1G, H). Ventrally, the ectotrophoblast is thinner (10 to 15 µm) than laterally (30 µm). This variation of thickness is due to collagen fibers that are situated between epidermis and somatopleure. These fibers are synthesized by the somatopleure. Cells of the ectotrophoblast are very flattened, and there are several ciliated cells at the periphery of the organ. Some cells are inflated. Mitoses are numerous. Some epithelial cells contain degenerative yolk platelets. Epidermis that is situated on the side of ectotrophoblast is thicker (45 µm). It is separated from somatopleure by a narrow collagen layer (Fig. 11.1H).

At stage 25 or 26, cells of the peripheral layer are united by desmosomes wich soon disappear (Fig. 11.2A). A large vacuole may be

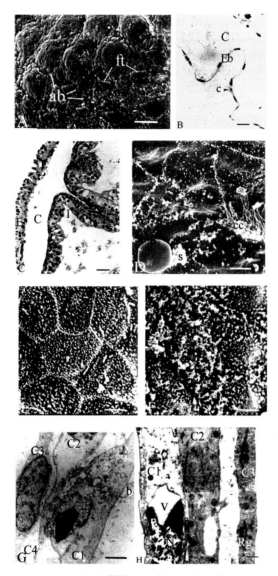

Fig. 11.1. *Typhlonectes compressicauda.* **A.** SEM view of the surface of integument during the embryonic development. Stage 22, absorption buds. Scale bar = 50 μm. **B.** LM view of ectotrophoblast. Stage 25. Scale bar = 50 μm. **C.** LM view of ectotrophoblast. Stage 26. Scale bar = 50 μm. **D.** SEM view of ectotrophoblast. Stage 23. Scale bar = 10 μm. **E.** SEM view of ectotrophoblast. Stage 28. Scale bar = 10 μm. **F.** SEM view of ectotrophoblast. Stage 30. Scale bar = 10 μm. **G.** TEM view of the ectotrophoblast during the embryonic development of stage 25. Scale bar = 1.5 μm. **H.** TEM Details showing the three cell layers. Scale bar = 0.5 μm. ab, absorptive buds; b, balloon-like microvilli; C, coelome; c, cell of ectotrophoblast; cc, ciliated cell; C1, epidermal cell layer; C2, epidermal basal layer; C3, somatopleure; C4, collagen; E, ectoderm; Eb, ectotrophoblast; ft, fetal tooth; I, intestine; N, nucleus; P, yolk platelet; Rg, rough endoplasmic reticulum; S, secretion; V, vacuole. From Hraoui-Bloquet, S. 1995. Unpublished Ph.D. thesis, Ecole Pratique des Hautes Etudes, Paris, France, Pl. VII, Fig. 3 ; Pl. XV, Figs. 1, 2; Pl. XVI, Figs. 1, 4-6; Pl. XVII, Fig. 1.

present. Cells of the discontinuous basal layer and of the external layer are linked by some desmosomes. A wide gap separates the two layers. Connective tissue is reduced with few collagen fibers, lacking fibroblasts, blood vessels or melanocytes. The external layer consists of pavement cells with microvilli that become sparse during this stage. Nuclei of cells have masses of condensed chromatin disposed against the nuclear membrane. Endoplasmic reticulum is formed by dilated cisternae. Cytoplasm is clear. Mitochondria are sparse in the supranuclear zone, but abundant in lateral parts of the cell. Thin fibrils and numerous pinocytotic vesicles are found. The apical regions of microvilli often inflate as vesicles. In certain zones, the plasma membrane loses its condensed appearance. The periphery of some cells is dark and electron-dense granules are accumulated against the apical membrane. The nucleus then becomes vacuolated and the cells degenerate, probably by apoptosis.

Basal cells are vacuolated and contain pinocytotic vesicles. Mitochondria are scarce and localised in the lateral cytoplasm. Chromatin is less condensed than in apical cells.

Somatopleure cells synthesise collagen of the chorion and have extensions into the coelomic cavity. Mitochondria are abundant. Endoplasmic reticulum, endocytotic and exocytotic vesicles are present.

At stages 26 and 27, in the anterior part of the ectotrophoblast, the skin is thick. The epidermis is constituted by two deep cells layers. External cells peripheral to the ectotrophoblast become club-shaped. The ectotrophoblast then quickly thickens (at stage 28) and reaches 40 µm in thickness. It consists of club-shaped cells with a voluminous apical nucleus. These cells seem to be linked to each other only by their slender bases. Some of them possess oval nuclei. The chorion begins to differentiate into dermis with mesodermal cells and collagen fibers.

At stage 28, the skin of the ectotrophoblast is constituted by a bistratified epithelium with flattened cells that are no longer ciliated. The chorion consists of a thick middle region with numerous fibroblasts, collagen fibres and blood-vessels. The latero-ventral skin is constituted by a thick epithelium (80 µm) with club-shaped cells, collagen fibers, abundant fibroblasts and numerous blood-vessels. In the ectotrophoblast, light and dark cells are observed (Fig. 11.2B, C). Dark cells are the most abundant with numerous mitochondria, and a dense Golgi apparatus. Electron-dense granules are found against the apical membrane. Endocytotic and exocytotic vesicles are less abundant than at younger stages. Microvilli are observed on the apical surface and more particularly near cell junctions (Fig. 11.2D). Each nucleus contains two to three nucleoli and a small amount of dark chromatin against the membrane. By scanning electron microscopy, budded extensions are seen to develop and become increasingly abundant. The light cells are not interdigitated. They are smaller than the dark cells but could be a younger stage in their differentiation and possibly originate from the basal layer. Apical secretory granules are present in light cells.

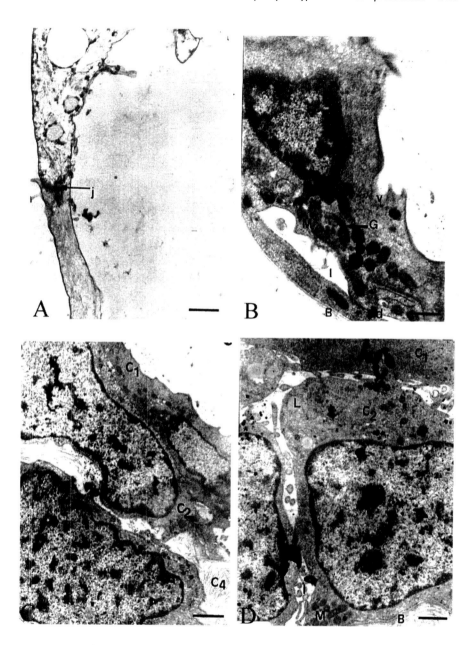

Fig. 11.2. A. Stage 26. MET view of junction between two epithelial cells of the ectotrophoblast in *T. compressicauda*. Scale bar = 0.5 µm. **B.** Stage 28. Scale bar = 0.5 µm. **C.** Stage 28, dark cells of the ectotrophoblast. Scale bar = 1 µm. **D.** Stage 28: latero-dorsal view of the skin. Scale bar = 28 µm. B, basal lamina; C1, epidermal cell layer; C2, epidermal basal layer; C4, collagen; G, Golgi apparatus; I, intercellular space; j, junction between two cells; L, lipids; M, mitochondria; v, vesicle. From Hraoui-Bloquet, S. 1995. Unpublished Ph.D. thesis, Ecole Pratique des Hautes Etudes, Paris, France, Pl. XVIII, Figs. 1- 4.

The cytoplasm of somatopleural cells is darker than previously. Nuclear membranes are indented. In the latero-dorsal region of the ectotrophoblast, interdigitated cells contain fat droplets, tonofilaments and numerous microvilli.

At stages 29 and 30, the epidermis of the median region of the embryo is discontinuous in thickness between the region of the heart and that of the posterior intestine, at the ectotrophoblast level. The chorion is very thick. Collagen appears as very thick and dense masses that are strongly stained by toluidine blue.

Subsequently, a very narrow medio-ventral groove appears. It consists of thickened epidermis with three layers of cells. The external cells possess numerous long extensions. The connective tissue is very dense. On each side of this zone, dermis (100 µm in thickness) is differentiated into loose and dense layers. Capillaries and melanophores are observed. At stage 32, skin glands differentiate.

The ectotrophoblast is probably implicated in absorption but the different tissue and cellular structures could have various functions. It has been suggested that the thin ectotrophoblast found in youngest stages could be a filter which may be modified to allow a direct and active absorption. Elements originating from uterus termed uterine fluid or uterine milk may enter the coelomic cavity to reach the organs. Conversely, wastes could be eliminated in a similar manner (Delsol *et al.* 1981).

In all the cases, one can find in the ectotrophoblast, bud structures that may also be implicated in filtration. Histologically, that of *Typhlonectes compressicauda* resembles the bistratified epidermis of embryos of other Gymnophiona (Welsch and Storch 1973, Fox 1986). In young stages, cells of this region are greatly elongated and possess numerous villosities. These observations suggest that this transient structure is implicated in exchange of soluble substances between the embryo and its mother. It is interesting to compare these observations with those for *Chthonerpeton indistinctum*, a viviparous species, and *Ichthyophis glutinosus*, an oviparous one (Fox 1986). In the viviparous species, cutaneous glands of embryos have more villosities than in oviparous larvae. This difference could be related to parity. There would be an increase of absorptive surfaces in a viviparous species that lives in a regulated and enclosed medium, in comparison with free larvae living in the field.

11.3 OOPHAGY

This mode of feeding is observed at the stage at which the digestive tract begins to function, i.e. in fetuses and larvae. As already noted, the number of oocytes that enter each oviduct is higher than the number of embryos produced. We have also reported that oocytes and degenerating eggs were found in the anterior part of oviducts. Free yolk platelets are also found throughout the oviducts. They are probably used as food by embryos when they are free into the uterus. On the other hand, cell masses provided by the anterior oviduct can include these platelets to give "uterine magma" that is also probably used as food (Fig. 5.6D, chapter 5 of this volume). These observations are reminiscent of oophagy in *Salamandra atra* (Vilter and Vilter 1960, Vilter 1986).

At this period, the uterine wall emits abundant secretions (Exbrayat 1988a). In parallel, the lower jaw is covered by fetal teeth used by embryos to grasp these secretions and also to remove epithelial cells from the uterine wall. The gut is filled by a substance stained by aniline blue in which cells in a more or less normal state are observed (Hraoui-Bloquet and Exbrayat 1994). This can be considered as further evidence of the role of uterus in embryonic feeding.

11.4 ADELPHOPHAGY

At the beginning of development, the average number of embryos in the uterus is always higher than the average number of new-born. There is a deletion of some embryos and fetuses during pregnancy. In females bearing embryos or fetuses, desiccated corpses are often observed in the uterus. When embryos reach the larval stage, these corpses are rarely observed. It is deduced that these latter are eaten by larvae and it has been possible to observe dead fetuses partly swallowed in old larvae. In histological studies, we have observed large degenerated cell masses with an organised aspect of a fetus in the stomach or intestine of healthy larvae (Fig. 11.5A).

11.5 DEVELOPMENT OF THE GUT

Development of the gut in *Typhlonectes compressicauda* reveals phases that are linked to intra-uterine development. Because studies are so few, before describing specific development of the digestive tract of this species, we will give some indications concerning early development of this organ in *Hypogeophis rostratus*, an oviparous species (Brauer 1897a, b). More recent works are devoted only to *T. compressicauda* (Hraoui-Bloquet and Exbrayat 1992, Hraoui-Bloquet 1995).

11.5.1 First Stages

In *Hypogeophis rostratus* gastrulae, the endoblast has the appearance of a gutter that is opened beneath the dorsal notochord. The two free sides of this lamina will approach each other beneath the cord, to give the gut (Fig. 11.3A, B, C). Posterior stages will be described here for *Typhlonectes compressicauda*.

11.5.2 Development of Mouth, Pharynx and Fetal Teeth

By stage 23, a stomodeum has developed. The opening of the pharynx is initially closed by a thin membrane that is pierced by small holes. This membrane entirely disappears during stage 25. During this period, buds of fetal teeth appear and the embryo is able to grasp the uterine wall.

Development of fetal teeth has been previously described (chapter 10 of this volume).

11.5.3 Development of Esophagus and Stomach

In the earliest stages that have been studied (stage 23), the anterior part of the digestive tract (esophagus and stomach) opens posteriorly into the intestine. The floor of the intestine is constituted by the yolk mass surrounded by endoblast cells.

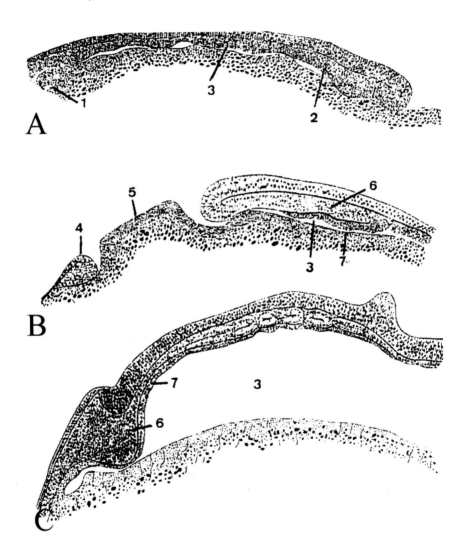

Fig. 11.3. A, B, C. First stages of development in *Hypogeophis alternans*. 1, vegetal cells; 2, animal cells; 3, primary gut; 4, blastoporal dorsal lip; 5, vitelline bud; 6, mesoderm; 7, endoderm. Modified after Brauer, A. 1897a. Zoologisches Jahrbuch für Anatomie 10: 277-279, Fig. 1.

11.5.3.1 Esophagus

In stages 23 and 24, the lumen of the esophagus progressively narrows. Its wall is bordered by a pseudostratified epithelium with columnar cells containing a basal nucleus and chorion with loose connective tissue which is differentiated from stage 25, blood vessels are still rare (Fig. 11.4A). The muscle layer is very thin with a single layer of muscle fibers. The wall thickens in the posterior part of esophagus but never in the stomach.

At stage 26, chorion and muscularis mucosae are also observed around the stomach and anterior intestine. These tissues are particularly developed in stages 31 to 33 and they are also highy vascularised. The muscularis mucosae progressively thickens but the connective tissue is thin, sometimes being absent around the mid and posterior intestine. From stage 26, the esophageal wall becomes a little folded; folds are supported by connective tissue villosities (Fig. 11.4B). Connective tissue is denser than previously and blood vessels are more numerous. The musculature is particularly thick in certain zones with two layers of muscle fibers. The thickness of the epithelium decreases. All the cells become more or less rounded. In the neighboring pharynx, two layers of cells of different appearance are observed. Cells bordering the lumen are ciliated and have a rounded central nucleus with scattered chromatin.

Cells of the basal layer of the epithelium have a small flattened nucleus with condensed chromatin. The esophagus wall has a single layer of cells. Ciliated cells are less numerous and disappear near the stomach where they are replaced by mucous cells with PAS positive secretions that are stained by eosin, aniline blue and alcian blue, showing that they are acidic carbohydrates. These mucous secretions will become more and more abundant as the embryos develop.

In stages 28 to 33, the connective tissue layer is more and more dense and the musculature progressively thickens. From stage 31, a third layer of muscule fibers appears (Fig. 11.4C).

Goblet cells have been observed in the anterior region of the esophagus from stage 31. In embryos at stages 32 and 33, the epithelium is pseudostratified or stratified and secretions are particularly abundant. The posterior region of the esophagus is composed only of cells with a closed mucous pole, resembling gastric cells (Fig. 11.4D).

In stage 25, the mid-region of the esophagus of all the embryos examined contains cells forming an organised structure in the connective tissue (Fig. 11.4A). This structure is dorsally in contact with the ventral epithelium of the esophagus and, ventrally, with musculature. In the youngest individuals, this structure appears in a discontinuous manner, as two or three islets of cells scattered around a central lumen. In stage 25, this structure is relatively important by comparison with the diameter of the esophagus. Subsequently increase in size of this structure is very small in comparison with other organs. In stages 31 to 33, the posterior part of this duct is connected with the esophageal lumen. We did not observe it in the

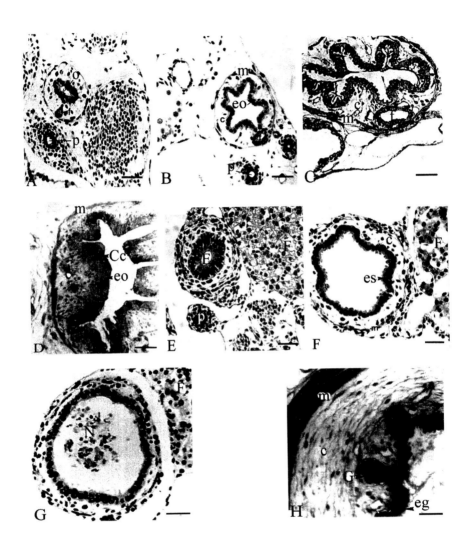

Fig. 11.4. *Typhlonectes compressicauda* during embryonic development. **A.** LM views of the esophagus, stage 25. Scale bar = 90 μm. **B.** LM views of the esophagus, stage 29. Scale bar = 90 μm. **C.** LM views of the esophagus, stage 31. Scale bar = 90 μm. **D.** LM views of the esophagus, stage 32. Scale bar = 130 μm. **E.** LM views of the stomach, stage 25. Scale bar = 90 μm. **F.** LM views of the stomach, stages 26-27. Scale bar = 90 μm. **G.** LM views of the stomach, stage 28. Scale bar = 90 μm. **H.** LM views of the stomach, stage 33: pylorus part. Scale bar = 95 μm. c, chorion; Cc, goblet cells; E, stomach; eg, gastric epithelium; eo, epithelium of the esophagus; es, single gastric epithelium; f, epithelial formation; F, liver; G, gastric gland; m, muscularis; N, stomach contents; o, esophagus; p, lung. From Hraoui-Bloquet, S. 1995. Unpublished Ph.D. thesis Ecole Pratique des Hautes Etudes, Paris, France, Pl. XI, Fig. 1-6. Pl. XXI, Fig. 1. Pl. XII, Fig. 1.

adult but it would be interesting to investigate the role of this structure or if it is a relict organ.

11.5.3.2 Stomach

In stages 23 and 24, the stomach is a tube with a narrow circular lumen, bordered by a pseudostratified epithelium. In stage 25, the epithelium is entirely constituted by mucous cells with an apical pole covered by small amount of PAS positive substances, stained by eosin, aniline blue and alcian blue (Fig. 11.4E). In stage 26, the lumen enlarges. The wall is covered by numerous folds without glands (Fig. 11.4F, G). The epithelium becomes monostratified by stage 30. In stage 26, chorion and musculature are differentiated.

The very thin muscularis mucosae contains one or two layers of muscle fibers. Blood vessels are not very abundant. At stage 29, the muscularis becomes thick, mucous is dense and blood vessels are numerous. Gastric glands appear at stage 30. They will become more and more numerous and developed by the end of development (Figs. 11.4H, 11.5A).

11.5.4 Development of the Intestine: Stages with Yolk Mass

Anterior region. In stages 23 to 25, the intestinal lumen contains yolk platelets. The dorsal epithelium becomes stratified with columnar cells containing a basal nucleus. Certain cells contain yolk. The ventral wall is thick, deriving from the yolk mass. It consists of cells that are different from the dorsal ones (Fig. 11.5B). The ventral part is surrounded by numerous blood vessels.

Mid intestine. This part of intestine consists dorsally of a stratified epithelium and ventrally of an unsegmented yolk mass, that is already becoming depleted. It is surrounded by large endoblast cells with rounded or irregular nuclei; some nuclei are vacuolated. The cytoplasm of these cells contains degenerated yolk platelets. The lateral and ventral aspects of the yolk mass are surrounded by splanchnopleure with numerous blood vessels. The yolk mass suspended in a large coelomic cavity that is limited by ectotrophoblast (Delsol et al. 1981). During development, the yolk is exhausted, digested by endoblast cells. Then, the lateral and ventral walls close dorsally, giving the definitive intestine (Fig. 11.5D).

Posterior intestine. In the posterior part of embryos at stages 23 to 25, there is a short portion of gut. In one examined embryo, the lumen was very narrow (Fig. 11.5C, E), in other embryos, it was larger. The wall is thick, epithelial cells disposed in several layers are large and their cytoplasm is vacuolated. Nuclei are often small and disposed against the plasma membrane, displaced by the vacuoles. In a rostral direction, the ventral part of this duct thickens progressively by addition of others endoblast cells often containing degenerated yolk platelets originating from the wall of the yolk mass.

Just in front of cloaca, the gut lumen is reduced (12 to 25 µm in diameter).

11.5.5 Development of the Intestine: Stages without Yolk

Yolk is entirely resorbed at stage 26 though platelets are still observed in some cells of the intestinal epithelium. The diameter of the lumen increases during development. Mucosa and muscularis are observed. In the anterior part, the chorion is well developed and highly vascularised. Its thickness decreases posteriorly and it is much reduced around the posterior part of the intestine.

Enterocytes are observed in the dorsal epithelium of the intestine at stage 25. In embryos at stages 26 to 29, the epithelium contains cells with striated border and goblet cells. Weakly developed folds are observed in the anterior and mid parts but it is smooth and thin posteriorly. In stages 28, 29 and 30, numerous epithelial cells, isolated or grouped, are released from the wall and remain free in the lumen (Fig. 11.5F).

In stages 31 to 33, the wall of the intestine is covered by villosities that are very well developed in the anterior intestine, but less so in the mid intestine. A new pseudostratified or stratified epithelium with columnar cells containing a basal nucleus is observed in anterior and mid intestine. Enterocytes are numerous. Goblet cells are increasingly abundant in a caudal direction. They elaborate a mucus that is PAS positive, stained by alcian blue, aniline blue and eosin. In the posterior region, the epithelium is thin and smooth and the cells appear to be flattened. The lumen is wide and opens into the cloaca. The cloaca is bordered by a stratified epithelium with 2 to 3 layers of cuboidal cells surrounded by a thick connective tissue and musculature.

Activity of alkaline phosphatase studied by histoenzymology has given negative results in enterocytes at stages 26 to 29. The reaction is slightly positive at stage 30 and is more intense at stages 32 and 33. Numerous dividing cells have been observed throughout the gut, more particularly in stages 23 to 31. They are rare at stages 31 and 32.

Opening of the cloaca has been observed in all examined embryos.

11.5.6 Development of the Liver

At stage 23, the liver is still poorly developed. Hepatocytes are united in rows separated by numerous blood vessels and some are organised in strings with a duct (Fig. 11.5G). The gall-bladder is observed on the posterior region of the liver.

At stage 30, the liver is surrounded by lymphoid cells that will constitute an envelope formed by two or three layers at stages 32 and 33 (Fig. 11.5H). At stages 26 to 29, hepatocytes are PAS-negative, i.e. lack glycogen. At stage 31, small quantities of carbohydrates are observed in the hepatocytes. At stage 32, glycogen is abundant.

11.5.7 Conclusion

Study of embryonic development of the gut in *Typhlonectes compressicauda* permits an understanding of one of the aspects of embryonic feeding and

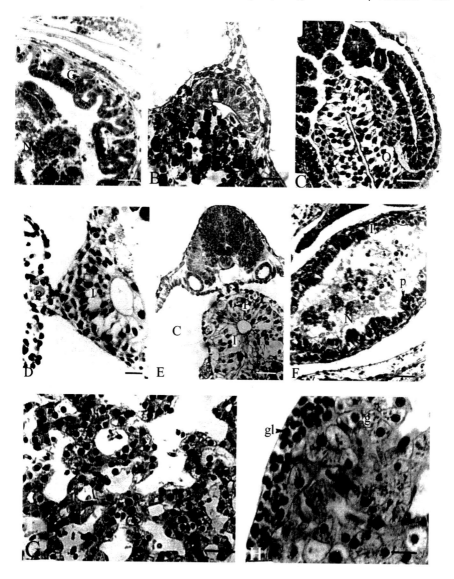

Fig. 11.5. *Typhlonectes compressicauda* during embryonic development. **A.** LM views of the stomach stage 33. Scale bar = 80 μm. **B.** LM views of anterior part of the mid-intestine during the embryonic development. Stage 23. Scale bar = 60 μm. **C.** Stage 24, LM views of the cloaca. Scale bar = 70 μm. **D.** LM views of posterior part of the mid-intestine during the embryonic development. Stage 25. Scale bar = 50 μm. **E.** LM views of posterior part of the mid-intestine during the embryonic development. Stage 23. Scale bar = 80 μm. **F.** Views of posterior part of the intestine during the embryonic development. Stage 29. Scale bar = 80 μm. **G.** LM view of liver at stages 26-27. Scale bar = 7 μm. **H.** LM view of liver at stage 33. Scale bar = 40 μm. C, coelomic cavity; g, glycogen; gl, haematopoietic layer; G, gland; I, intestine; N, nutritive elements; O, cloacal opening; P, yolk platelets in a cell; v, yolk. From Hraoui-Bloquet 1995, S. Unpublished Ph.D. Thesis, Ecole Pratique des Hautes Etudes, Paris, France, Pl. XII, Figs. 2, 4-6. Pl. XII, Figs. 3, 4. Pl. XIV, Fig. 1.

materno-fetal exchanges in a gymnophionan. In several viviparous vertebrates, the gut of the embryo becomes functional during development. In Gymnophiona, very few works have been published, but studies on oviparous species suggest that gut becomes functional at the same stage in development. In *Ichthyophis glutinosus*, the animal possesses a reduced yolk mass at hatching that is stage 26 in *T. compressicauda* (Breckenridge and Jayasinghe 1979, Breckenridge *et al.* 1987). From this period, *Ichthyophis* larvae feed via the mouth, indicating that the digestive tract is functional to allow intestinal absorption. Development of the gut in *T. compressicauda* probably fellows the standard pattern of development seen in all Gymnophiona, even if this development occurs into uterus. We have seen that the gut is functional because secretions, cells and dead embryos have been found in it.

11.6 GILLS

Typhlonectes compressicauda embryos have a pair of large floating gills that are observed at stages 18 to 20, i.e. in very young animals; their growth is correlated with that of the body (Exbrayat 1986). Growth of gill is faster than that of the animal, before the final phase of development. The histology of gills has been studied throughout development (Delsol *et al.* 1986). First, on each side of head, two gill buds are observed. At stage 21, the two first buds have entirely merged (Fig. 11.6A). They increase to give, on each side of the head, a palette-like structure that continues to grow (Fig. Fig. 11.6B). Finally, the two gills are at least as long as the body of the embryo (Fig. 11.6C). They envelop the embryo and their external surface is applied to the uterine wall (Fig. 11.6D). Cytoplasmic bridges are observed between the two. The internal structure of the gills is a very discontinuous connective tissue in which there are numerous blood vessels. The epithelium type is different according to the level and the embryonic stage. Structural and ultrastructural studies of gills have been described (Exbrayat and Hraoui-Bloquet 1991, 1992a. Hraoui-Bloquet and Exbrayat 1994. Hraoui-Bloquet 1995).

11.6.1 At Hatching

At stages 25 to 27, the gills are covered by two layers of epithelial cells on both the embryonic and uterine sides (Fig. 11.7A, B). The apical layer is composed of ciliated cells and cells with microvillosities and numerous mitochondria. A vacuole may occupy a large part of the cytoplasm in both cell types (Fig. 11.7D, E). Chromatin is broken up. These characteristics are those of apoptotic cells. In the basal layer, some cells have a small amount of cytoplasm, few mitochondria and a nucleus with very scattered chromatin and a large vacuole is present in the cytoplasm. Other cells belonging to this layer possess fewer mitochondria and a nucleus with masses of chromatin. Apical cells are linked by numerous desmosomes. The basal lamina is discontinuous and collagen very sparse.

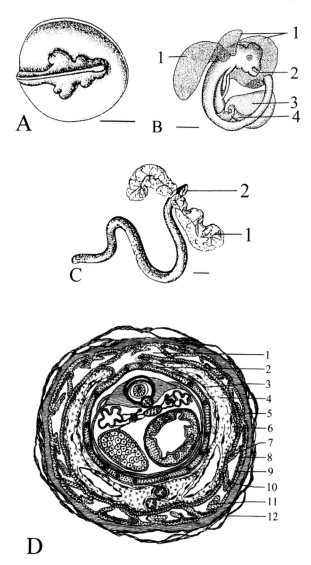

Fig. 11.6. Development of gills in *Typhlonectes compressicauda*. **A.** Stage 20, two pairs of gill buds can be observed on the posterior part of the head. Scale bar = 1 mm. **B.** Stage 26, gills begins to develop as a pair of vesiculous structure. Scale bar = 1 mm. **C.** Stage 34, presence of a pair of vesiculous gills on each side of the head. Scale bar = 5 mm. Abbreviations: 1. gill. 2. mouth. 3. ectotrophoblast. 4. intestine. **D.** Schematic representation of a section of *T. compressicauda* embryo in a uterus. gills are narrowly applied against the uterine wall. 1, lumen of uterus; 2, larval epidermis; 3, gill epithelium on the larval side; 4, gill connective tissue; 5, uterus muscle; 6, gill epithelium on the uterine side; 7, blood vessels; 8, uterine crypt; 9, uterine epithelium at the bottom of a crypt; 10, uterine connective tissue; 11, flattened tip of a uterine crest; 12, serosa. **A, B,** From Delsol, M., Flatin, J. Exbrayat, J.-M. and Bons, J. 1981. Comptes Rendus des Séances de l'Académie des .Sciences de Paris, série III 293: 281-285, Figs 1, 4. **C,** modified after Toews and MacIntyre 1977. Nature 266: 464-465, Fig. 1. **D,** from Hraoui-Bloquet, S. 1995. Unpublished Ph.D. Thesis, Ecole Pratique des Hautes Etudes, Paris, France, Pl. ILIII.

11.6.2 At Stages 28-29

At stages 28-29, uterine and embryonic faces are differentiated. On the uterine side, epithelial cells possess lengthened and particularly dense microvillosities (Figs. 11.7C, F, 11.8C). These cells have a very developed Golgi apparatus with numerous vesicles and electron-dense granules. Mitochondria are abundant. Some microfilaments are found. Fat droplets are sometimes observed. Large vacuoles are also observed in some cells.

Between these cells, another cell type, which is lighter, with many mitochondria, also possesses abundant microvillosities (Fig. 11.8E). A third cell type with a triangular shape, possesses very dense cytoplasm and a well developed Golgi apparatus.

Basal cells are disposed as a discontinuous layer (Fig. 11.8F). They contain very few mitochondria. Sometimes, vacuoles with yolk platelets are observed. The space between peripheral layers of cells and the basal layer is greatly distended. Interdigitations are observed between basal cells and light cells.

On the fetal (or embryonic) side (Figs. 11.7G, 11.8A, B), apical cells with extensions contain few mitochondria but numerous tonofilaments. These cells are interdigitated with lateral and basal cells that are also rich in tonofilaments (Fig. 11.8C). Other flattened cells containing peripheral electron-dense granules, are very poor in organelles (Fig. 11.8D). The basal lamina is thicker than on the uterine side. Capillary walls are thin, often discontinuous. Endothelial cells are covered by pinocytotic vesicles.

11.6.3 In Larval Stages

At stages 30 to 33, gill differentiation continues. On the anterior part of uterine side, club shaped cells are equipped with numerous microvillosities that are supported by microfilaments (Figs. 11.8G, H, 11.9A, B, G, H, 11.10A- H, 11.11G, H). They possess numerous pinocytotic vesicles. The Golgi apparatus is developed with many vesicles. Rough and smooth endoplasmic reticulum are abundant and sometimes contain electron-dense substances. Numerous large cavities communicate with saccules. Multilamellar bodies, originating by stacking of reticulum vesicles, develop. Degradation vesicles containing mitochondria and reticulum saccules are also observed (Fig. 11.10C, H). Some of them separate from the base of the cell to penetrate into the adjacent cell with a pale cytoplasm (Fig. 10I). Numerous fat droplets are observed in certain cells. Electron-dense granules are also observed in certain cells and in the extracellular space. These cells are in contact with the uterine lumen and sometimes in direct contact with the basal lamina of the uterine wall.

Two basal cell types can be distinguished. The first type has a reduced cytoplasm with a very electron-dense nucleus, mitochondria and little rough endoplasmic reticulum. Cells of the second type have a rounded nucleus, several mitochondria, a reduced Golgi apparatus and very little endoplasmic

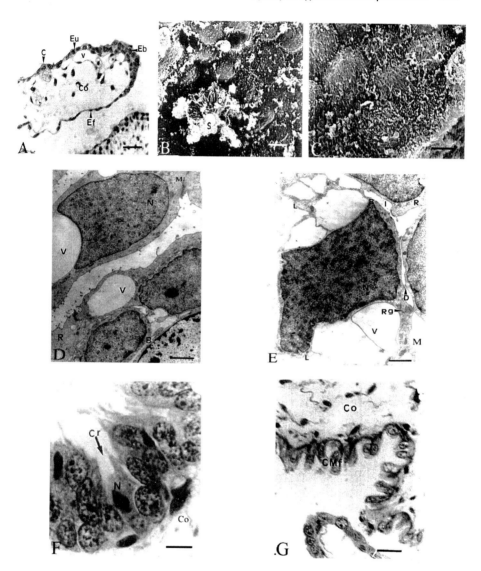

Fig. 11.7. *Typhlonectes compressicauda* during embryonic development. **A.** LM overview of a gill. Stage 24. Bar = 80 μm. **B.** SEM view of a gill: epithelial cells of the uterine side. Stages 26-27. Scale bar = 7.5 μm. **C.** Stage 28, surface of anterior epithelium of gill. Scale bar = 5 μm. **D.** Stage 25. Vacuoles in epithelial cells of gill. Scale bar = 1.8 μm. **E.** Stage 25. Junctions between the cells of epidermis peripheral layer and the cells of basal layer. Scale bar = 1 μm. **F.** stage 28. LM view of crypts in the anterior epithelium of gill, uterine side. Scale bar = 25 μm. **G.** stage 29. Cells of gill epithelium, fetal side. Scale bar = 60 μm. B, basal cell; C, capillary; CMf, club-shaped cells on the fetal side; Co, connective tissue; Cr, crypt; D. desmosome; Eb, epithelium of the tip of gill; Ef, epithelium on the fetal side; Eu, epithelium of uterine side; I, intercellular space; M, cell with microvilli; N, nucleus; R, covering epithelial cells; Rg, rough endoplasmic reticulum; S, secretions; v, vacuole. From Hraoui-Bloquet, S. 1995. Unpublished Ph.D. Thesis, Ecole Pratique des Hautes Etudes, Paris, France, Pl. XXVII, Figs. 1, 2, 5. Pl. XXIX, Figs. 2, 3. Pl. XXXI, Figs. 3, 4.

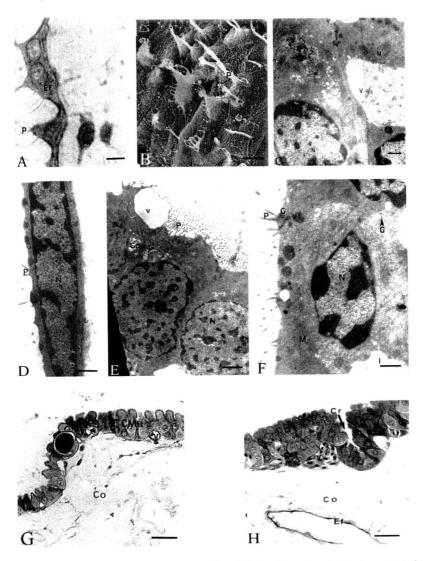

Fig. 11.8. Views of gills in *Typhlonectes compressicauda* during the embryonic development. **A.** Stage 29, LM view of anterior lining of gill epithelium, fetal side. Scale bar = 2.5 µm. **B.** Stage 29, SEM view of cells in the anterior lining of gill epithelium, fetal side. Scale bar = 14 µm. **C.** Stage 29, TEM view of epithelial cells on the uterine side. Scale bar = 0.5 µm. **D.** Stage 29 TEM view of a squamous cell of gill epithelium, fetal side. Scale bar = 0.7 µm. **E.** Stage 29, uterine side; microvillosities are numerous. Scale bar = 1.3 µm. **F.** Stage 29, squamous cell of the mid-region, uterine side. Scale bar = 0.5 µm. **G.** Stage 31, club-shaped cells on the anterior epithelium, uterine side. Scale bar = 80 µm. **H.** Stage 32, anterior part, on the uterine side, the epithelium is thick with crypts, on the fetal side, the epithelium is squamous. Scale bar = 80 µm. Cmu, club-shaped cells; Co, connective tissue; Cr, crypt; Ef, epithelium of fetal side; F, fibroblast; G, electron-dense granules; Go, Golgi apparatus; M, mitochondria; N, nucleus; P, cytoplasmic extensions; v, vacuole. From Hraoui-Bloquet, S. 1995. Unpublished Ph.D. Thesis, Ecole Pratique des Hautes Etudes, Paris, France, Pl. XXVII, Figs. 3, 4. Pl. XXVIII, Fig. 3. Pl. XXIX, Fig. 6. Pl. XXXII, Figs. 1-4.

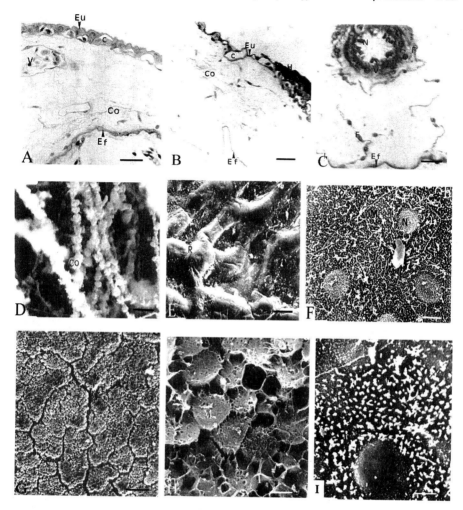

Fig. 11.9. A. Stage 30, mid-region. Scale bar = 80 μm. **B.** Stage 30, posterior part. Scale bar = 80 μm. **C.** Stage 32, connective tissue and epithelium, larval side. Scale bar = 40 μm. **D.** Stage 32, SEM view of connective tissue in gill. Scale bar = 1.4 μm. **E.** Stage 30, SEM view of distal part of gill surface, fetal side. Scale bar = 38 μm. **F.** Stage 30, gill surface, fetal side. Scale bar = 9 μm. **G.** Stage 32, anterior part of gill epithelium, uterine side. Scale bar = 90 μm. **H.** Stage 32, surface of epithelial cells. Scale bar = 12 μm. **I.** Stage 32. gill epithelium, larval side. Scale bar = 6.6 μm. Co, connective tissue; Ef, epithelium on the fetal side; Eu, epithelium on the uterine side; F, fibroblast; M, cell with microvillosities; N, nucleus; R, peripheral blood vessels; V, blood vessel. From Hraoui-Boquet, S. 1995. Unpublished Ph.D., Thesis, Ecole Pratique des Hautes Etudes, Paris, France, Pl. XXVIII, Figs. 1, 2, 4, 6. Pl. XXIX, Figs. 4, 5. Pl. XXX, Figs. 2-4.

reticulum (Fig. 11.10E, F). Electron-dense granules can be observed in the cytoplasm. The cells are linked to the basal lamina by hemidesmosomes.

In the mid-region, external cells (Fig. 11.10I) have very few organelles but the Golgi apparatus is well developed. These cells interdigitate with the sole

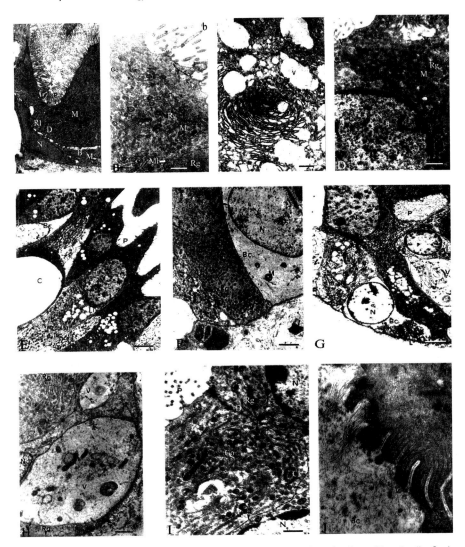

Fig. 11.10. TEM views of gills. **A.** Stage 32, uterine side with club-shaped cells and basal cells. Scale bar = 2.8 μm. **B.** Stage 32, apical part of cell, uterine side. Scale bar = 0.2 μm. **C.** Stage 33, degradation of an epithelial cell, uterine side. Scale bar = 1.3 μm. **D.** Stage 31, epithelial cell, uterine side. Scale bar = 0.4 μm. **E.** Stage 32, epithelium of gill, anterior part, uterine side. Scale bar = 2.7 μm. **F.** Stage 31, basal cytoplasm of an epithelial cell and clear basal cell, anterior part of gill, uterine side. Scale bar = 2 μm. **G.** Stage 32, anterior epithelium of gill, uterine side. Scale bar = 2 μm. **H.** Stage 32, catalytic vacuoles in the clear basal cells, anterior part of gill, uterine side. Scale bar = 0.8 μm. **I.** Stage 31, apical cytoplasm of an epithelial cell in the mid-region of the gill, uterine side. Scale bar = 0.5 μm. **J.** Stage 30, contact between an epithelial cell and a clear basal cell. Scale bar = 0.2 μm. b, basal cell; Bc, clear basal cell; c, capillary; Cl, lamellar body; D, desmosome; Go, Golgi apparatus; L, basal lamina; Li, lipids; M, mitochondria; Ml, lateral membrane; N, nucleus; P, cytoplasme extensions; R, club-shaped cells; Rl, smooth endoplasmic reticulum; Rg, rough endoplasmic reticulum; V, vacuole. From Hraoui-Bloquet, S. 1995. Unpublished Ph.D. Thesis, Ecole Pratique des Hautes Etudes, Paris, France, Pl. XXXIV, Figs. 1-4. Pl. XXXV, Figs. 1-4. Pl. XXXVI, Figs. 1, 2.

type of basal cell. Apical cells are sometimes phagocytosed by basal cells and degraded by lysosomes (Fig. 11.11A, B).

In the posterior region, the epithelium becomes thicker. Apical cells are flattened, often electron-dense, making observation of organelles difficult. Numerous pinocytotsic vesicles can be observed (Fig. 11.11C, D). Basal cells, always of a pale type, are also flattened, with a fibrillar cytoplasm and are linked to the basal lamina by hemidesmosomes (Fig. 11.11D).

On the fetal side, the epithelium is also bi-stratified. Anteriorly, cells are flattened or club-shaped with very long ramified extensions with microvillosities (Fig. 11.9F, I) covered by large mass of secretion. These cells contain numerous mitochondria and tonofilaments (Fig. 11.11I). Basal and lateral surfaces interdigitate with other cells.

In mid and posterior parts, cells have few organelles. The apical membrane is covered by short microvillosities (Fig. 11.11J). Basal and lateral membranes are interdigitated, with a large intercellular space. Cells of the basal layer are flattened and linked to the basal lamina by hemidesmosomes.

The basal lamina is thicker on the fetal side of the gill than on the uterine side. Fibroblasts are present near this lamina. Blood vessels are not observed on the fetal side, but are abundant on the uterine side. In the posterior half of the gills, blood vessels are separated from the uterine lumen by two thin layers of epithelial cells and a thin layer of collagen. Pale basal cells sometimes surround these capillaries of which endothelium is very thin with numerous microvillosities. The wall of some blood vessels is discontinuous. These capillaries are often surrounded by pericytes. Large blood vessels are surrounded by a thick layer of collagen (Fig. 11.9C). Characteristic fibroblasts with numerous bud-shaped extensions (seen by SEM and TEM and with various fixatives) are observed near these vessels. Numerous gaps limited by fibroblasts are also observed (Figs. 11.9D, E, 11.11D-F).

11.6.4 Conclusions

The very peculiar structure of gills in *Typhlonectes compressicauda* indicates that these organs have a respiratory function and they are certainly involved in other exchange types between mother and fetus. At the first stage of development, gill buds are poorly developed. From stage 25, a differentiation begins to be observed between embryonic and uterine faces. Then the epithelium differentiates several times, especially on the uterine side. At stage 31, that is the beginning of metamorphosis, two cell types replace the first type. Cells with numerous vesicles containing unknown substances are observed. Data suggest that these substances are caught by the gill surface, then introduced by phagocytosis, into the epithelial cells, and even into the connective tissue (after having moved across epithelial cells), particularly in the oldest stages. These substances were also observed on the epithelial surface. They have the characteristic stainings of oviductal secretions, yolk platelets or even hemocytes. Abundant superficial blood

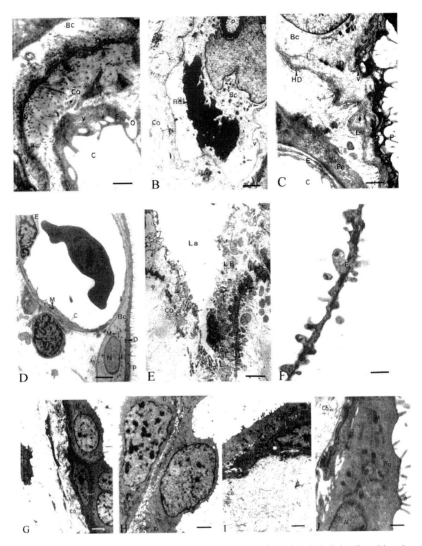

Fig. 11.11. A. Stage 31, pinocytotic vesicles in the clear basal cells and endothelial cells, mid-region of gill. Scale bar = 0.4 μm. B. Stage 31, degradation of an epithelial cell into the cytoplasm of a clear basal cell, mid-region of gill; uterine wall. Scale bar = 1.4 μm. C. Stage 31, gill epithelium and subepithelial capillary, posterior part of gill, uterine side. Scale bar = 0.5 μm. D. Stage 32, gill epithelium and subepithelial capillary, posterior part of gill, uterine side. Scale bar = 2 μm. E. Stage 31, gap into the connective tissue. Scale bar = 1.8 μm. F. Stage 31, fibroblasts. Scale bar = 1 μm. G. Stage 31, fetal side, anterior part. Scale bar = 2 μm. H. Stage 32, anterior part of gill, larval side. Scale bar = 0.6 μm. I. Stage 31: mid-region of gill, fetal side. Scale bar = 0.6 μm. J. Stage 31, mid-part of gill, fetal side. Scale bar = 0.4 μm. Bc, clear basal cell; C, capillary; Co, collagen; D, desmosome; E, endothelium; F, fibroblast; Go, Golgi apparatus; HD, hemi-desmosome; I, intercellular space; L, basal lamina; La, gap in the connective tissue; LB, border of the gap; M, mitochondria; N, nucleus; O, opening of a capillary; P, cytoplasmic extensions; Rg, rough endoplasmic reticulum; v, vesicle. From Hraoui-Bloquet, S. 1995. Unpublished Ph.D. Thesis, Ecole Pratique des Hautes Etudes, Paris, France, Pl. XXXVI, Figs. 3, 4. Pl. XXXVII, Figs. 1-4. Pl. XXXIX, Figs. 1-4.

vessels in the gills show that exchanges occur between the fetuses and the mother.

Presence of ciliated cells on gills of *Typhlonectes compressicauda* embryos observed at stage 29, is reminiscent of the observation for the gills of the oviparous species *Ichthyophis kohtaoensis* (Fox 1986) and *I. paucisulcus* (Welsch 1981) or even the gills of the viviparous urodele *Salamandra* (Greven 1980a, b). In these species, it has been shown that these structural types of cell are involved in respiration. Club-shaped cells resemble chloride cells of certain fishes. On the other hand, these cells contain lamellar bodies as in the pulmonary epithelium of *Chthonerpeton indistinctum* (Welsch 1981).

The gills of *Typhlonectes compressicauda* have a respiratory function and they are also organs involved in the absorption of exogenous substances for intra-uterine feeding. Their association with the uterine wall, at a certain stage of development, acts as a true placenta.

During development, the parts of the gill that are nearest the insertion on the embryo's head, are always composed of pavement cells resembling ectotrophoblast.

In fetal and larval stages, the wall of gills presents three epithelial types. Type I, situated near insertion of the gill, consists of one or two layers of cells situated on a connective layer, sometimes separated by a basal lamina. This epithelium is ciliated with a brush aspect in SEM. Connective tissue resembles the connective tissue of the umbilical cord of mammals. The type I epithelium is restricted to the gill part anteriorly in the embryo. Type II is also situated near the insertion but never in contact with the embryo (it is on the uterine side of the gill). It is a unilayered ciliated epithelium with club-shaped cells. Nuclei are basal and the apical part of the cells is prominent, with cilia or villosities. At certain stages, this epithelium presents relations with uterine cells. At this level, blood cells are particularly abundant in the connective tissue of the gill. Cells of type III form a stratified or pseudo-stratified epithelium. At its surface, folds delimit three crypt types: open crypts containing granules (type I), closed crypts also with granules (type II) and closed crypts lacking granules (type III). Blood vessels are abundant in these areas. This last epithelial type is situated near the point of gill insertion, first in islets, then in a continuous manner to the end of the gill.

Types II and III are situated on the external face of the gill, in contact with the uterine wall. Distribution of these epithelia varies according to the stage of development. In the youngest stages, only types I and II are observed. Type I covers the largest surface area of each gill (70 to 90%), type II appears at stage 28 and, at the end, it is 20 to 30% of gill surface (Delsol *et al.* 1986).

Tight junctions and cytoplasmic bridges are established between the gill epithelium and connective tissue of the uterus at the end of pregnancy. They may favor gas and food exchanges. The gill epithelium is composed of cells with numerous extensions, as is characteristic of cells that are specialised for exchange. The absence of uterine epithelium reduces the distance between blood vessels of the uterus and those of the gill. The morphological relationships that are observed between these two organs are reminiscent of the mesochorial placenta of certain mammals.

11.7 DIFFUSIVE EXCHANGES

After injection of tritiated thymidin into pregnant females, in a histo-auto-radiographic study, presence of labeled nuclei in intra-uterine larvae suggests that a diffusion of thymidine occurs from mother to embryo. This transport occurs via the gills or, more probably, through the skin.

11.8 RESPIRATORY PHYSIOLOGY IN EMBRYOS

Very few works have been devoted to embryonic respiration. Certain authors reveal that exchanges occur between intra-uterine larvae and the mother in *Typhlonectes compressicauda* (Toews and McIntyre 1977).

11.9 MATERNAL ORGANS INVOLVED IN MATERNO-FETAL EXCHANGES

In some species, viviparity requires exchanges between embryo and mother. In viviparous amphibia, few studies have been devoted to maternal reserves and their transfert to embryos during pregnancy. In Gymnophiona, only *Typhlonectes compressicauda* has been the subject of such studies (Exbrayat 1986, 1988a). Weights of female body, fat bodies and liver have been studied throughout the reproductive cycle, and particularly during pregnancy. By comparison with variations of weight in males as controls, it is possible to show the importance of reserves in the female breeding cycle.

In this chapter we do not describe the variations of oviducts according to the pregnancy, a description has been given in previous chapters of this volume.

11.9.1 Total Weight of Animal

The size of individuals being very variable (300 mm to more than 500 mm), it has been difficult to estimate the exact weight variations of individuals during the reproductive cycle and pregnancy. Therefore, weights have been replaced by an arbitrary "density" that was calculated in the following way:

$$d \text{ (density)} = W \text{ (animal weight in g)} / [L \text{ (animal length in mm)}]^3$$

Comparison of these values throughout the reproductive cycle indicated that they are at their highest level during the rainy season and more particularly in February (beginning of pregnancy). At this period, non-pregnant females also show particularly high densities. From August to October, at the end of pregnancy, density abruptly decreases to a minimal value. Then it rapidly increases again at the beginning of the rainy season. Thus, density progressively decreases during pregnancy (Fig. 11.12A).

In males used as controls, density progressively increases from October until December. The higher value is reached in August (dry season) then it decreases before next October when the cycle starts again (Fig. 11.12B). In general, the male variation curves are different from those of females and the amplitudes of variations are narrower than in females (0.9 to 2.0 for males and 0.9 to 3.0 for females). It is confirmed that variations of weight in females are mainly linked to pregnancy.

11.9.2 Fat Bodies

As stated in chapter 3, all amphibians possess a pair of fat bodies associated with the gonads. Several authors have observed fat bodies in Gymnophiona but they did not make a detailed study (Sarasin and Sarasin 1887-1890. Fuhrmann 1914. Tonutti 1931. Wake 1968b). Gymnophiona have a pair of segmented fat bodies that extend from the posterior end of the liver to the cloaca. As in the total weight study, we calculated the "density" of fat bodies:

d_{FB} (density) = W_{FB} (weight of 2 fat bodies in g)/[L (animal length in mm)]^3

During sexual quiescence, this value is relatively constant (8 to 10 × 10^{-3}). An increase is observed in August and fat bodies have reached a large size (110 × 200 mm in sections). In the middle and at the end of pregnancy, these organs strongly decrease. In females with embryonic stages, fat bodies are still developed (d = 14 × 10^{-3}). In females with larval stages, density decreases to 9 × 10^{-3} with a minimal value of 3 × 10^{-3} (Fig. 11.12C). Examination of relative weight (fat body weight/animal weight) show similar results with particularly significant differences between pregnant females and the others.

In males, some variations of fat bodies have also been shown but they are less pronounced (Fig. 11.12D). At the end of the dry season values are low. They become higher in the rainy season (February to June), during breeding.

In *Typhlonectes compressicauda* female, fat bodies are involved in embryonic development, the greatest variations being observed during pregnancy and they are correlated with variations in the size of the embryo.

11.9.3 Liver

Density of the liver has been also calculated according to the same method:

d_L (density) = W_L (weight of liver in g)/[L (animal length in mm)]^3

It is low during the dry season (1 × 10^{-3}) but increases from December to February. In pregnant females, liver density significantly increases at the beginning of the breeding period, reaching a maximal value in April-May. Then, it decreases and becomes minimal in July. In general, there is a decrease of liver weight when embryos transform from embryonic to larval stages. In non-pregnant females, after a maximal value in April, liver density decreases to reach a minimal value that will stay constant until December

(Fig. 11.12E). Variations of liver relative weight (liver weight/animal weight) are similar.

In males, some variations of liver density have also been shown, but in a different manner. Values are minimal from October to May-June but increase to reach a maximal value in August (always lower than that of the female). Then it decreases (Fig. 11.12F). The relative weight of the male liver varies very little, suggesting that it follows the general variation of organismal weight.

11.9.4 Conclusions

In *Typhlonectes compressicauda*, significant variations of reserves have been observed in both males and females, but with different modalities. In males, minimal values are observed during the dry season only, explicable by the

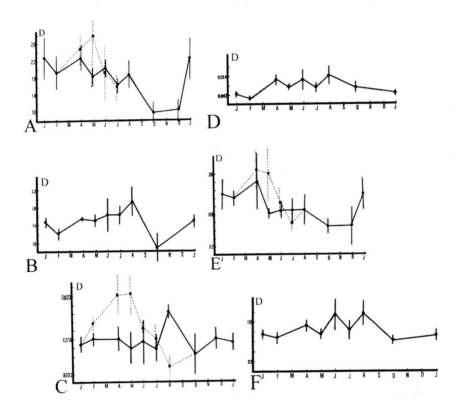

Fig. 11.12. Variations of reserve organs during breeding cycle in *Typhlonectes compressicauda*. **A.** Variations of total density in females. Full line: non-pregnant females; interrupted line: pregnant females. **B.** Variations of total density in males. **C.** Variations of fat bodies density in females. Full line: non-pregnant females; interrupted line: pregnant females. **D.** Variations of fat bodies density in males. **E.** Variations of liver density in females. Full line: non-pregnant females; interrupted line: pregnant females. **F.** Variations of liver density in males. From Exbrayat, J.-M. 1988. Annales des Sciences Naturelles, Zoologie, 13éme série, 9: 45-53, Figs. 1, 2a, 3a, 4, 5a, 6a.

paucity of food in the field at this period. According to this explanation, metabolic reserves would be progressively exhausted, thus entailing a loss of weight of the animal. During the rainy season, animals have an increase of food but they also need a supplementary source of energy, linked to sexual activity. Some of the reserves could be used and other part accumulated in fat bodies. After breeding, the liver could also accumulate reserves for the dry season.

In females, variations in weight and densities seem to be linked to breeding and pregnancy. Increase of all the organs that were studied occurs at the beginning of the rainy season. In non-pregnant females, fat bodies do not vary notably over the year. Only the increase of weight in August could be related to an accumulation of reserves before the dry season. The increase of fat bodies is important at the beginning of pregnancy. These organs abruptly decrease by the birth of young animals. The variations can be very well explained by development of the embryos' need for food. At the beginning of their development, embryos use uterine secretions. It is from this period that decrease of reserve organs begins to be observed. At the end, intrauterine larvae being at maximal growth, they need to feed actively. At this period, the most important decrease of reserve organs is observed.

The increase of liver weight at the beginning of the rainy season can be related to preparation for pregnancy. At the beginning of pregnancy, accumulation is maximal. Decrease in liver weight can be related to increase in size of the embryo, as is the case for fat bodies.

In addition to these variations related to pregnancy, it is possible that variations of weight could be related to variations in hydration of the animal.

The observations on fat body variations in *Typhlonectes compressicauda*, suggest several ideas as to the role and regulation of fat bodies in Amphibia. In general, fat bodies of all Amphibia seem to be reserves that vary with seasonal alternation and breeding cycles. The role of these organs is certainly different according to species. In the anuran *Bufo arenarum*, fat bodies are implicated in the increase of ovarian mass (Penhos 1953). In *Rana esculenta*, variations of weight and of chemical composition of fat bodies are linked to hibernation and sexual activity (Roca et al. 1970, Mauget et al. 1971). Certain studies have shown that hibernation was the main factor affecting variation (Schlaghecke and Blum 1978a, b). In *R. cyanophlyctis*, the weight of fat bodies decreases when that of ovaries increases (Pancharatna and Saidapur 1985). In *R. esculenta*, the ablation of fat bodies is accompanied by decrease of the ovarian mass and atrophy of vitellogenic oocytes (Pierantoni et al. 1983). In *R. tigrina*, the ablation of fat bodies results in a delayed vitellogenesis (Sumittrapa Pramoda and Saidapur 1984). In oviparous Anura with much yolk, fat bodies are involved in accumulation of ovarian reserves.

In males, fat bodies have another function. All observations in Anura as in Gymnophiona, show that they are polyvalent reserve organs that can be used in several circumstances.

In Anura, it has been suggested that reserve movements in fat bodies were under hypophyseal control (*Bufo bufo*, Larsen 1963. *Rana hexadactyla*, Kasinathan *et al.* 1978. *R. esculenta*, Chieffi *et al.* 1995). In *Typhlonectes compressicauda*, variations of fat bodies seem to be closely correlated to variations of size of gonadotropic and lactotropic cells in the pituitary (chapter 5 of this volume). Control by ovarian hormones is also certainly implied in viviparous species as well as in oviparous ones, because fat bodies are reduced in female with only very small oocytes (Wake 1968a, b, Masood Parveez 1987).

11.10 CALCIFICATION OF EMBRYOS

Gymnophionan bony tissue is not well known. Our own works have shown the presence of lamellar bony tissue in the skull and vertebrae, a fibrous bony tissue near integument and sinew insertion and between the plates of the skull. A fibrous bony tissue with a woven structure has been observed, in the deep parts of the calvaria (the roof of the skull), lower jaw and vertebrae. This bone contains all three types of cells: rare osteoclasts, osteocytes that are numerous in lamellar zones and scattered in fibrous or woven bone, and a layer of cells observed around the rachidian duct and certain areas of the calvaria. In the female, bony tissue is subject to mineral variations according to the breeding cycle.

At the beginning of pregnancy, bony structures of the embryo are particularly hard. At the end of pregnancy, and just after birth, they are very soft. In pieces preserved in ethanol, no variations of phosphocalcic ratio or calcium and phosphorus concentrations have been observed. In pieces preserved in Bouin's fuid that is considered as a demineralizing substance, clear variations of phosphocalcic ratio and in concentrations of calcium and phosphorus have been observed throughout pregnancy. These variations heve been shown using histology with a special staining of bone (Fig. 11.13A, B), histochemistry, biochemistry and X-ray microanalysis (Fig. 11.13C) (Paillot 1995). All the results do not suggest an abrupt demineralization of female bones during gestation, but a certain lability of mineral elements that would be more and more mobile during pregnancy. At the end of pregnancy, this lability is maximal, at the time when the chondrocranium becomes particularly calcified (Wake *et al.* 1985). These results indicated that the female is involved in mineralization of the embryo.

11.11 MATERNO-FETAL IMMUNOLOGICAL RELATIONSHIPS

We previously described (chapter 3 of this volume) that *Typhlonectes compressicauda* embryos are protected against immunological rejection, by the pregnant female (Exbrayat *et al.* 1994, 1995).

11.12 CONCLUSIONS

In *Typhlonectes compressicauda*, several features show pronounced adaptation to viviparous development. Development of the species can be described in

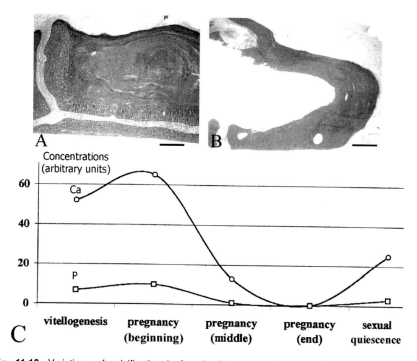

Fig. 11.13. Variations of calcification in female during breeding cycle. in female *Typhlonectes compressicauda*. **A.** Section of calvarias tined with cyanin solochrome in non-pregnant females; bone is showing presence of calcium. Scale bar = 30 µm. **B.** Section of calvarias tined with cyanin solochrome in pregnant females; bone is showing a total decalcification. Scale bar = 30 µm. **C.** X-ray microanalysis of bone in females during breeding. The highest line shows variations of Ca; the lowest line shows variations of P. Original.

three main phases. At each phase, morphological and physiological structures aid the growth of embryos in the uterus. First the animal relies on its yolk mass. When this yolk mass is exhausted, the animal is freed into the uterus where it is able to grasp the secretions and cells originating from the uterine wall. For this, it is equipped with teeth and the gut is now functional. Concomitantly, female genital ducts become secretory. During the last phase, a pair of giant gills surrounds each embryo giving a cocoon-like appearance. These gills are closely applied to the uterine wall, constituting a placenta-like structure. In the mother, reserve organs are closely linked to embryonic development. On the other hand, an immunological mechanism protects embryos against rejection by the mother. Considering all these features, it is clear that this species is the most adapted to viviparity of the investigated Gymnophiona.

11.13 ACKNOWLEDGMENTS

The authors thank Catholic University of Lyon (U.C.L.) and Ecole Pratique des Hautes Etudes (E.P.H.E.) that allowed them freedom in organizing their research fields and supported the works on Gymnophiona. They also thank Singer-Polignac Foundation that supported the missions necessary to collect the material of study. The authors also thank Michel Delsol, Honorary Professor at the U.C.L. and at E.P.H.E. who is at the origin of these works in Gymnophiona, Marie-Thérèse Laurent, who made thousands of sections. They also thank Barrie Jamieson, who invited them to contribute to this collection devoted to reproduction and phylogeny of Amphibia and also for his patience in reviewing the manuscript. They thank Jean-François Exbrayat for his help in preparing the illustrations.

11.14 LITERATURE CITED

Brauer, A. 1897a. Beitrage zur kenntniss der Entwicklungsgeschichte und der Anatomie der Gymnophionen. Zoologisches Jahrbuch für Anatomie 10: 277-279.

Brauer, A. 1897b. Beitrage zur kenntniss der Entwicklungsgeschichte und der Anatomie der Gymnophionen. Zoologisches Jahrbuch für Anatomie 10: 389-472.

Breckenridge, W. R. and Jayasinghe, S. 1979. Observations on the eggs and larvae of *Ichthyophis glutinosus*. Ceylon Journal of Sciences (Biological Sciences) 13: 187-202.

Breckenridge, W. R., Shirani, N. and Pereira, L. 1987. Some aspects of the biology and development *of Ichthyophis glutinosu*s (Amphibia, Gymnophiona). Journal of Zoology 211: 437-449.

Chieffi, G., Rastogi, R. K., Iela, L. and Milone, M. 1975. The function of fat bodies in relation to the hypothalamo-hypophyseal-gonadal axis in the frog *Rana esculenta*. Cell and Tissue Research 161: 157-165.

Delsol, M., Flatin, J., Exbrayat, J.-M. and Bons, J. 1981. Développement de *Typhlonectes compressicaudus* Amphibien Apode vivipare. Hypothèse sur sa nutrition embryonnaire et larvaire par un ectotrophoblast. Comptes Rendus des Séances de l'Académie des .Sciences de Paris, série III 293: 281-285.

Delsol, M., Flatin, J., Exbrayat, J.-M. and Bons, J. 1983. Développement embryonnaire de *Typhlonectes compressicaudus* (Duméril et Bibron, 1841): constitution d'un ectotrophoblaste. Bulletin de la Société Zoologique de France 108: 680-681.

Delsol, M., Exbrayat, J.-M., Flatin, J. and Gueydan-Baconnier, M. 1986. Nutrition embryonnaire chez *Typhlonectes compressicaudus* (Duméril et Bibron, 1841) Amphibien Apode vivipare. Mémoires de la Société Zoologique de France 43: 39-54.

Exbrayat, J.-M. 1984. Quelques observations sur l'évolution des voies génitales femelles de *Typhlonectes compressicaudus* (Duméril et Bibron, 1841), Amphibien Apode vivipare, au cours du cycle de reproduction. Comptes Rendus des Séances de l'Académie des Sciences de Paris 298: 13-18.

Exbrayat, J.-M. 1986. Quelques aspects de la biologie de la reproduction chez *Typhlonectes compressicaudus* (Duméril et Bibron, 1841), Amphibien Apode. Doctorat ès Sciences Naturelles, Université Paris VI, France.

Exbrayat, J.-M. 1988a. Croissance et cycle des voies génitales femelles chez *Typhlonectes compressicaudus* (Duméril et Bibron, 1841), Amphibien Apode vivipare. Amphibia-Reptilia 9: 117-137.

Exbrayat, J.-M. 1988b. Variations pondérales des organes de réserve (corps adipeux et foie) chez *Typhlonectes compressicaudus*, Amphibien Apode vivipare au cours des alternances saisonnières et des cycles de reproduction. Annales des Sciences Naturelles, Zoologie, 13éme série, 9: 45-53.

Exbrayat, J.-M. 1990. Quelques observations sur l'ostium et la paroi coelomique chez deux Amphibiens Gymnophiones femelles au cours des différentes périodes du cycle de reproduction. Bulletin de la Société Zoologique de France 115: 199-200.

Exbrayat, J.-M. 1996. Croissance et cycle du cloaque chez *Typhlonectes compressicaudus* (Duméril et Bibron, 1841), Amphibien Gymnophione. Bulletin de la Société Zoologique de France 121: 99-104.

Exbrayat, J.-M. and Delsol, M. 1985. Reproduction and growth of *Typhlonectes compressicaudus*, a viviparous Gymnophione. Copeia 1985: 950-955.

Exbrayat, J.-M. and Hraoui-Bloquet, S. 1991. Morphologie de l'épithélium branchial des embryons de *Typhlonectes compressicaudus* (Amphibien Gymnophione) étudié en microscopie électronique à balayage. Bulletin de la Société Herpétologique de France 57: 45-52.

Exbrayat, J.-M. and Hraoui-Bloquet, S. 1992a. Evolution de la surface branchiale des embryons de *Typhlonectes compressicaudus*, Amphibien Gymnophione vivipare, au cours du développement. Bulletin de la Société Zoologique de France 117: 340.

Exbrayat, J.-M. and Hraoui-Bloquet, S. 1992b. La nutrition embryonnaire et les relations foeto-maternelles chez *Typhlonectes compressicaudus*, Amphibien Gymnophione vivipare. Bulletin de la Société Herpétologique de France 61: 53-61.

Exbrayat, J.-M., Delsol, M. and Flatin, J. 1981. Premières remarques sur la gestation chez *Typhlonectes compressicaudus* (Duméril et Bibron, 1841) Amphibien Apode vivipare. Comptes rendus de des Séances de l'Académie des Sciences de Paris 292: 417-420.

Exbrayat, J.-M., Pujol, P. and Hraoui-Bloquet, S. 1994. Premières observations sur les relations immunitaires materno-foetales chez *Typhlonectes compressicaudus*, Amphibien Gymnophione vivipare. Bulletin de la Société Zoologique de France 119: 386.

Exbrayat, J.-M., Pujol, P. and Hraoui-Bloquet, S. 1995. First observations on the immunological materno-foetal relationships in *Typhlonectes compressicaudus*, a viviparous Gymnophionan. Amphibia. Pp. 271-273. In G. A. Llorente, A. Montori, X. Santos and M. A. Carretero (eds), *Scientia Herpetologica*, Proceedings of the 7th General Meeting of the Societas Europaea Herpetologica, Barcelona, Spain.

Fox, H. 1986. Early development of caecilian skin with special reference to the epidermis. Journal of Herpetology 20: 154-157.

Fuhrmann, O. 1914. Le genre *Typhlonectes*. Mémoires de la Société Neuchâteloise de Sciences Naturelles 5: 112-138.

Greven, H. 1980a. Ultrastructural investigations of the epidermis and the gill epithelium in the intrauterine larvae of *Salamandra salamandra* (L.) (Amphibia, Urodela). Zeitschrift für Mikroskopie und Anatomie Forschungen 94: 196-208.

Greven, H. 1980b. Ultrahistochemical and autoradiographic evidence for epithelial transport in the uterus of the ovoviviparous salamander *Salamandra salamandra* (L.) (Amphibia, Urodela). Cell and Tissue Research 212: 147-162.

Hraoui-Bloquet, S. 1995. Nutrition embryonnaire et relations materno-foetales chez *Typhlonectes compressicaudus* (Duméril et Bibron, 1841), Amphibien Gymnophione vivipare. Thèse de Doctorat E.P.H.E., Lyon, France.

Hraoui-Bloquet, S. and Exbrayat, J.-M. 1992. Développement embryonnaire du tube digestif chez *Typhlonectes compressicaudus* (Duméril et Bibron, 1841), Amphibien Gymnophione vivipare. Annales des Sciences Naturelles, Zoologie, 13éme série 13: 11-23.

Hraoui-Bloquet, S. and Exbrayat, J.-M. 1993. La fécondation chez *Typhlonectes compressicaudus* (Duméril et Bibron, 1841), Amphibien Gymnophione. Bulletin de la Société Zoologique de France 118: 356-357.

Hraoui-Bloquet, S. and Exbrayat, J.-M. 1994. Développement des branchies chez les embryons de *Typhlonectes compressicaudus*, Amphibien Gymnophione vivipare. Annales des Sciences naturelles, Zoologie, 13éme série 15: 33-46.

Hraoui-Bloquet, S. and Exbrayat, J.-M. 1996. Les dents de *Typhlonectes compressicaudus* (Amphibia, Gymnophiona) au cours du développement. Annales des Sciences Naturelles, Zoologie, 13éme série 17: 11-23.

Hraoui-Bloquet, S., Escudié, G. and Exbrayat, J.-M. 1994. Aspects ultrastructuraux de l'évolution de la muqueuse utérine au cours de la phase de nutrition orale des embryons chez *Typhlonectes compressicaudus*, Amphibien Gymnophione vivipare. Bulletin de la Société Zoologique de France 119: 237-242.

Kasinathan, S., Guna Singh, A. and Basu, S. L. 1978. Fat bodies and spermatogenesis in South Indian green frog *Rana hexadactyla*. Bolletino Zoologico 45: 15-22.

Larsen, L. 1963. Fat bodies in *Bufo bufo* with autotransplanted pars distalis. General and Comparative Endocrinology 3: 713-714.

Masood Parveez, U. 1987. Some aspects of reproduction in the female Apodan Amphibian *Ichthyophis*. PhD. Dissertation, Karnatak University, Dharwad, India.

Mauget, C., Cambar, R., Maurice, A. and Baraud, J. 1971. Observations sur les variations cycliques annuelles des corps adipeux chez l'adulte de grenouille verte (*Rana esculenta*). Comptes rendus des Séances de l'Académie des Sciences de Paris 273: 2603-2605.

Paillot, R. 1995. Variations de la calcification osseuse chez *Typhlonectes compressicaudus* (Duméril et Bibron, 1841) au cours de la gestation. Place du tissu osseux des Gymnophiones parmi les Vertébrés inférieurs. Mémoire de fin d'études, année de spécialisation en histologie, Université catholique de Lyon, France.

Pancharatna, M. and Saidapur, S. K. 1985. Ovarian cycle in the frog *Rana cyanophlyctis*: a quantitative study of follicular kinetics in relation to body mass, oviduct and fat body cycles. Journal of Morphology 186: 136-147.

Penhos, J. C. 1953. Rôle des corps adipeux du *Bufo arenarum*. Comptes rendus de la Société de Biologie 47: 1095-1096.

Pierantoni, R., Varriale, B., Simeoli, C., di Matteo, L., Milone, M., Rastogi, R. K. and Chieffi, G. 1983. Fat body autumn recrudescence of the ovary in *Rana esculenta*. Comparative Biochemistry Physiology A 76: 31-35.

Roca, A., Maurice, B., Baraud, J. Mauget, C. and Cambar, R. 1970. Etude des lipides du corps adipeux de *Rana esculenta*. Comptes Rendus des Séances de l'Académie des Sciences de Paris 270: 1278-1281.

Sarasin, P. and Sarasin, F. 1887-1890. Ergebnisse Naturwissenschaftlicher Forschungen auf Ceylon. Zur Entwicklungsgeschichte und Anatomie der Ceylonischen Blindwuhle *Ichthyophis glutinosus*. C. W. Kreidel's Verlag, Wiesbaden.

Sammouri, R., Renous, S., Exbrayat, J.-M. and Lescure, J. 1990. Développement embryonnaire de *Typhlonectes compressicaudus* (Amphibia Gymnophiona). Annales des Sciences Naturelles, Zoologie, 13ème série 11: 135-163.

Schlaghecke, R. and Blüm, R. 1978a. Seasonal variations in fat body metabolism of the green frog *Rana esculenta* (L.). Experientia 34: 1019-1020.

Schlaghecke, R. and Blüm, R. 1978b. Seasonal variations in liver metabolism of the green frog *Rana esculenta* (L.). Experientia 34: 456-459.

Sumitrappa Pramoda and Saidapur, S. K. 1984. Effect of fat body excision on vitellogenic growth of oocytes in *Rana tigrina*. Bolletino Zoologico 51: 329-333.

Toews, D. and Macintyre, D. 1977. Blood respiratory properties of a viviparous Amphibian. Nature 266: 464-465.

Tonutti, E. 1931. Beitrag zur Kenntnis der Gymnophionen. XV. Das Genital-system. Morphologisches Jahrbuch 68: 151-292.

Vilter, V. 1986. La reproduction de la salamandre noire. Pp. 487-495. In P.P. Grassé et M. Delsol (eds), *Traité de Zoologie*, tome XIV, fasc.I B, Masson, Paris.

Vilter, V. and Vilter, A. 1960. Sur la gestation de la salamandre noire des Alpes, *Salamandra atra* Laur. Comptes rendus de la Société de Biologie 154: 290-294.

Wake, M. H. 1968a. The comparative morphology and evolutionary relationships of the urogenital system of Caecilians. Ph.D. Dissertation, University of California, U.S.A.

Wake, M. H. 1968b. Evolutionary morphology of the Caecilian urogenital system. Part I: the gonads and fat bodies. Journal of Morphology 126: 291-332.

Wake, M. H., Exbrayat, J.-M. and Delsol, M. 1985. The development of the chondrocranium of *Typhlonectes compressicaudus* (Gymnophiona), with comparison to other species. Journal of Herpetology 19: 568-577.

Welsch, U. 1981. Fine structural and enzyme histochemical observations on the respiratory epithelium on the Caecilian lungs and gills. A contribution to the understanding of the evolution of the Vertebrate respiratory epithelium. Archivium Histologicum Japonicum 14: 117-133.

Welsch, U. and Storch, V. 1973. Die Feinstruktur verhornter und nichtverhornter ektodermaler Epithelien und der Hautdrüsen, embryonaler und adulter Gymnophionen. Zoologische Jahrbuch, Anatomie 90: 323-342.

CHAPTER 12

Fertilization and Embryonic Development

Jean-Marie Exbrayat

Embryonic development of Gymnophiona has been little studied. The earliest papers, dealing development of *Ichthyophis glutinosus*, came from Sri Lanka (Sarasin and Sarasin 1887-1890). In several papers, development was described in *Hypogeophis rostratus* (Brauer 1897a, b, 1899, 1900, 1902). Further studies were devoted to *Ichthyophis glutinosus* (Breckenridge and de Silva 1973, Breckenridge and Jayasinghe 1979, Breckenridge *et al.* 1987). Viviparity of *Typhlonectes compressicauda* was first recognized in 1841 by Leprieur (Dumeril and Bibron 1841) and new-born of this species described (Peters1874 a, b, 1875). *Typhlonectes compressicauda* embryos were first classified into four categories (Exbrayat *et al.* 1981). A previous table of development was published by Delsol *et al.* (1981). Later, a more detailed normal table was established (Sammouri *et al.* 1990). Recently, embryonic and larval development of *Ichthyophis kohtaoensis*, an oviparous species, has also been described (Dünker *et al.* 2000).

In addition, some observations have been given about eggs and some aspects of development (Seshachar 1942, Ramaswami 1954, Parker 1956, Parker and Dunn 1964, Wake 1967, 1969, 1976, 1978, 1980a, b, 1986, Barrio 1969, Moodie 1978, Hetherington and Wake 1979, Seshachar *et al.* 1982, Balakrishna *et al.* 1983, Billo *et al.* 1985, Wake *et al.* 1985, Delsol *et al.* 1986, Bhatta and Exbrayat 1998, Exbrayat *et al.* 1999).

12.1 EGGS AND FERTILIZATION

At ovulation, the number of oocytes laid is variable according to the species (Table 10.1, chapter **10** of this volume). In oviparous species, eggs are surrounded by several envelopes during the transit through the oviduct. In viviparous species, they are surrounded by a thin envelope and

---

Laboratoire de Biologie générale, Université Catholique de Lyon and Laboratoire de Reproduction et Développement des Vertébrés, Ecole Pratique des Hautes Etudes, 25, rue du Plat, F-69288 Lyon Cedex 02.

development occurs in the anterior region of the oviduct (chapter **10** of this volume). Fertilization is internal in all species. Fertilization has been described only in *Typhlonectes compressicauda* (Hraoui-Bloquet 1995, Hraoui-Bloquet and Exbrayat 1993). In this species, ovulation and fertilization occur in February, the beginning of the breeding period. Mature oocytes are filled with yolk platelets. The nucleus is displaced to the animal pole (Fig. 12.1A). It is surrounded by a thin layer of cytoplasm without yolk (Fig. 12.1B). The size of yolk platelets increases from this part of oocyte to the vegetative pole.

At ovulation, the oocyte surrounded by its vitelline membrane leaves the ovary and reaches the anterior part of the oviduct in which mucous envelopes are elaborated to protect the future egg. Fertilization occurs in this region of the oviduct (Fig. 12.1C).

Several nuclei belonging to spermatozoa are observed at the animal pole of the activated oocyte (Fig. 12.1D, E). A perivitelline gap is now observed between the plasmalemma and the pellucid membrane (Fig. 12.2A). Numerous spermatozoa are found between these two structures (Fig. 12.2B). Polyspermy is observed. Several spermatozoa enter the oocyte (Fig. 12.2C, D) but only one of them unites with the female pronucleus. The other sperm nuclei degenerate quickly (Fig. 12.2E).

12.2 FIRST STAGES OF DEVELOPMENT

The first stages of development have been very little studied, owing to the rarity of material. These stages have been mainly described in *Ichthyophis glutinosus* (Sarasin and Sarasin 1887-1890) and *Hypogeophis rostratus* (Brauer 1899). In *I. glutinosus*, at the beginning of development, eggs are divided into blastomeres situated at the animal pole. The vegetal pole is composed of a residual multinucleate mass of cytoplasm with yolk.

12.3 TABLES OF DEVELOPMENT

12.3.1 Old Tables and Development of *Ichthyophis glutinosus*

Very few works have been devoted to embryonic development in caecilians. Yet, some species have been the object of such studies. *Ichthyophis glutinosus* was the first species for which development was studied (Sarasin and Sarasin 1887-1890) (Fig. 12.3A, B). Their table has been reviewed (Breckenridge and de Silva 1973, Breckenridge and Jayasinghe 1979, Breckenridge *et al.* 1987). Other works concern *Siphonops annulatus* (Goeldi 1899) (Fig. 12.3C-E) and *Hypogeophis rostratus* (Brauer 1897a, b, 1899, 1900, 1902).

At the end of segmentation, the embryo of *Hypogeophis* has the appearance of a germinal disc disposed on the yolk mass. This disc is indented posteriorly. During gastrulation, a cephalic lengthening is

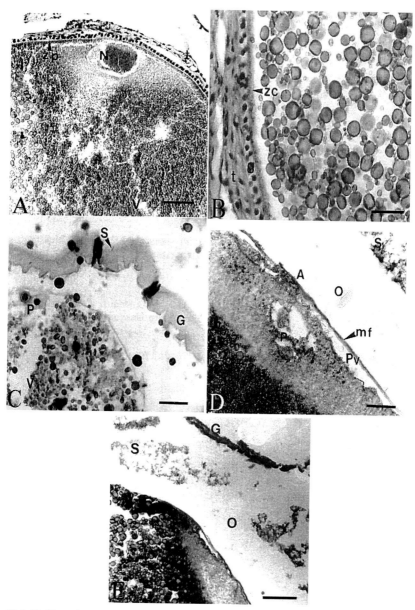

Fig. 12.1. *Typhlonectes compressicauda*. **A.** Animal pole of a mature ovarian follicle. Scale bar = 130 µm. **B.** Details of peripheral zone of a mature follicle. Scale bar = 100 µm. **C.** Fertilization of an oocyte; male and female pronuclei can be observed. Scale bar = 80 µm. **D.** Perivitelline space and fertilization membrane in the egg. Scale bar = 80 µm. **E.** Substances rejected at the periphery of the egg. Scale bar = 100 µm. A, animal pole; g, granulosa; G, mucous envelope; mf, fertilisation membrane; N, nucleus; O, extra-ovular space; P, pronuclei; PV, perivitelline space; S, spermatozoon; t, connective theca; V, yolk; x, spermatic nuclei; Zc, cortical zone; Zp, pellucid zone; Sub, substances rejected by the egg. From Hraoui-Bloquet 1995. Unpublished PhD. Thesis, Ecole Pratique des Hautes Etudes, Paris, France, Pl. I, Figs. 1-3. Pl. II, Figs. 1, 2.

362 Reproductive Biology and Phylogeny of Gymnophiona

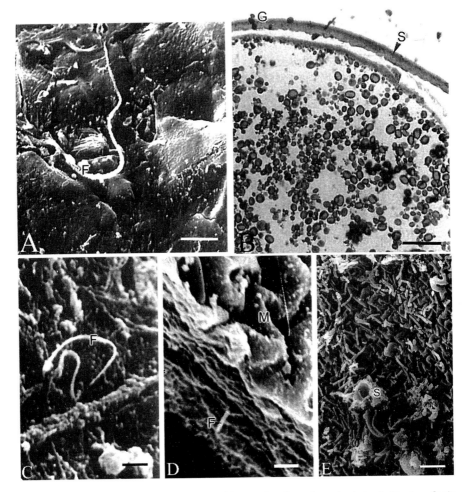

Fig. 12.2. *Typhlonectes compressicauda*. **A.** SEM view of surface of oocyte during fertilization. Scale bar = 10 µm. **B.** Fertilization. Scale bar = 100 µm. **C.** SEM view of the external part of a mucous envelope of the egg during fertilization. Scale bar = 2 µm. **D.** SEM of a spermatozoon traversing the mucous gangue during fertilization. Scale bar = 1 µm. **E.** SEM view of internal surface of mucous gangue of a fertilized egg. Scale bar = 10 µm. Abbreviations: F, flagellum; G, mucous gangue; M, microfolds of the internal face of mucous gangue; S, spermatozoon; s, secretions; t, head of a spermatozoon; From Hraoui-Bloquet, S. 1995. Unpublished PhD. Thesis, Ecole Pratique des Hautes Etudes, Paris, France. Pl. I, Fig. 6. Pl. II, Fig. 6. Pl. III, Fig. 2-4.

observed, after invagination of mesodermal tissues. At this stage, a yolk plug is also observed. At neurulation, somites are observed, then development of the embryo proceeds with the appearance of optic vesicles and multiplication of somites. At this period, each pronephros and Wolffian duct develops. Then, cephalic structure becomes complex, otic and olfactory vesicles develop.

Gill buds develop on each side of the head and later will give rise to a pair of triradiate gills. At the end of development, gills are well reduced in size. The yolk mass around which embryo seems to be rolled throughout development, is progressively resorbed, being used as food by animal. By the end, a residual yolk mass is observed in the posterior part of the body. Finally, the animal lengthens and gills disappear.

In *Siphonops annulatus*, eggs are linked to each other during external development. Embryos have been observed to possess a pair of triradiate gills (Goeldi 1899). These animals develop around the yolk mass. Yolk is progressively resorbed and it is very reduced, at the posterior part of body (Fig. 12.3C, D, E), as in *Hypogeophis*.

Ichthyophis glutinosus has also been studied from individuals caught at different times. The youngest stage looks like a small white disc disposed on the surface of its yolk mass (that is 3 to 5 mm in diameter) (Fig. 12.4A, B). Four days later, the embryo has a well defined shape with a cephalic bulb. The beating heart can be observed. Blood vessels are found on the yolk mass. Gill buds develop. The embryo can move within its mucous envelope (Fig. 12.4, C).

Nine days after the beginning of development, the embryo is 8 to 9 mm in length. The brain develops with distinct vesicles. The neural fold opens posteriorly. Optic vesicles with lenses, and otic vesicles can be observed. Three pairs of gill buds are found. One of the animals possessed 21 somites. Others had 58 and 73 somites. The heart has 40 beats per min. Two weeks later, the yolk mass is conspicuous and two thick blood vessels are observed within it. Gill buds are still poorly developed.

At 16 days, the embryo is 66 mm in length. Three pairs of red gills begin to develop. Two pairs of them are equipped with branches. Eyes are observed. The yolk mass is less abundant than previously and the embryo floats in a fluid contained in the mucous envelope.

At 18 days, the embryo is 20 mm in length. Two lateral buds composed of small black spots extend from the insertion of the gills to the posterior end of animal (Fig. 12.4D). Head pigmentation is lighter than body pigmentation. Gills have increased in size; filaments of the third pair of branches are shorter than the three others. The embryo seems to be coiled throughout its yolk mass.

30 days after commencement of development, the disposition of pigment has changed. The head is always lighter than the body. Ventral and median dorsal parts are unpigmented. Otic vesicles have the appearance of two white spots behind eyes. The embryo is now about 33 mm long.

The animal then becomes darker in color. After 36 or 37 days, the yolk has been strongly resorbed. Numerous vitellin blood vessels are observed on the surface of remaining yolk. The embryo is moving within its envelope.

After 50 days, yolk is almost all resorbed and the remaining yolk is observed in the ventral and posterior parts of animal. On each side of the body, a narrow dark strip is observed from gills to tail. Light circular spots

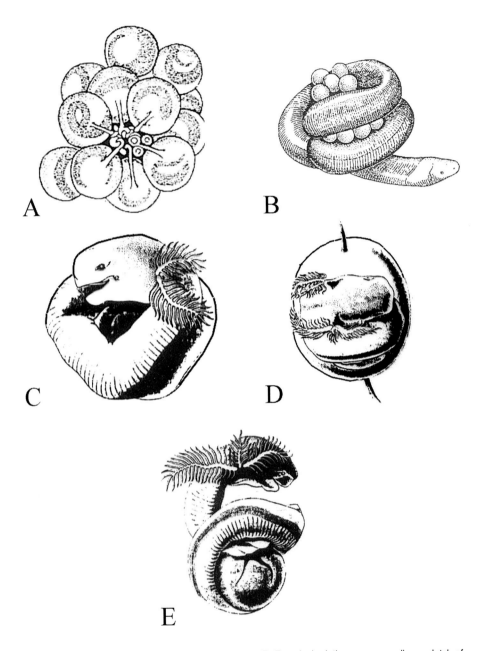

Fig. 12.3. A. Clutch of eggs of *Ichthyophis glutinosus*. **B.** Female *I. glutinosus* surrounding a clutch of eggs. **C, D, E.** Embryos of *Siphonops annulatus*. **A, B**. Modified after Sarasin, F. and Sarasin, P. 1887-1890 Ergebnisse Naturwissenschaftlicher Forschungen auf Ceylon. Zur Entwicklungsgeschichte und Anatomie der Ceylonischen Blindwuhle *Ichthyophis glutinosus*. C.W. Kreidel's Verlag, Wiesbaden. **C, D, E.** Modified after Goeldi, E. 1899. Zoologische Jahrbuch 12: 170-173, Figs. 1, 2, 3.

appear around the eyes, near each gill and on the sides of animal. Organs can be observed through the transparent ventral surface.

Before hatching, the animal is formed and can be observed in its transparent mucous envelope. Eyes are prominent. Triradiate gills are characteristic of Gymnophiona. The reduced yolk mass is observed into the posterior or ventral part of the animal and blood vessels are observed in it (Fig. 12.4E). The animal moves slowly in its envelope. At hatching, movements are more and more rapid and abrupt. By that time, the animal deforms and distends the egg wall to breaking point and the larva is born.

At hatching, the animal is 80 mm in length. A very small yolk mass is still observable (Fig. 12.4F). The animal possesses a tail with a membranous fin. The mouth is at the anterior tip; tentacles are not visible. The three pairs

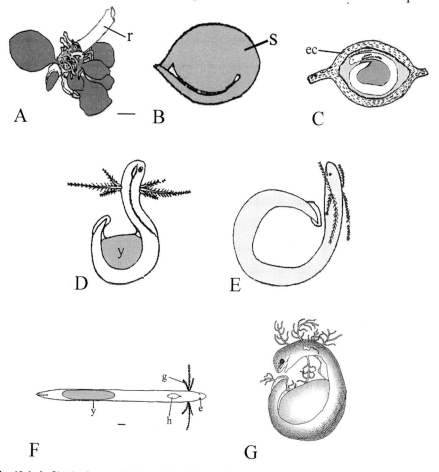

Fig. 12.4. A. Clutch of eggs of *Ichthyophis glutinosus*. **B.** Egg of *I. glutinosus*. **C.** *I. glutinosus* embryo surrounded with a mucous gangue. **D.** Developed *I. glutinosus* embryo. **E.** 10 months old *Ichthyophis glutinosus* embryo. **F.** Free larval *I. glutinosus*. **G.** *I. beddomei* embryo. e, eye; ec, egg case; g, gill; h, heart; r, root; s, stalk; y, yolk; **A** to **F**, from Breckenridge, W.R.and Jayasinghe, S. 1979. Ceylon Journal of Sciences (Biological Sciences) 13: 187-202, **A, B, C**. Fig. 1a, b, d. **D, E, G**. Fig. 2c, d, e. **F**. Original.

of gills are always present, the anterior one being the longest, the posterior one the shortest. These gills will degenerate two days after hatching, probably by reabsorption of cell material. The remaining yolk is progressively resorbed. The tail and fin disappear by 10 months. Tentacles and yellow lateral stripes that are characteristic of the adult also appear at 10 months. This period corresponds to metamorphosis (Breckenridge et al. 1987). These changes are observed in 120 to 140 mm length animals.

Observations of some *Ichthyophis beddomei* embryos reveal close similarties with *Ichthyophis glutinosus* (Bhatta and Exbrayat 1998, Exbrayat et al. 1999) (Fig. 12.4G). A more detailed table of another ichthyophiid, *Ichthyophis kohtaoenesis* has been published recently (Dünker et al. 2000).

12.3.2 Development of *Ichthyophis kohtaoensis*

In *Ichthyophis kohtaoensis* 40 stages are recognised (Dünker et al. 2000). The description begins at stage 21 only because this material is very rare and, like the other species of caecilians, it is difficult to obtain the first stages of development. It is stated that certain stages were represented by a single individual. Length of embryos is variable from one stage to another (see, for instance, stages 23 and 24 at which individuals are shorter at stage 24 than at stage 23).

The authors used the stages previously described by Sammouri et al. (1990) for *Typhlonectes compressicauda*. Below, we give the main facts concerning each stage. A more detailed table is given in the original article (Dünker et al. 2000).

Stage 21 (Fig. 12.5A). The embryo is 16.6 mm in length, rolled around the yolk mass. 125 pairs of somites have been counted. The head and tail are near one another. The neural folds are closed at the posterior end of body and in the forebrain region but not in the hindbrain. Paired mandibular elements touch. Optic vesicles are well observed, with central lens. Otic vesicles and olfactory placodes are present on each side of the head. Three external gills without filaments are observed on each side of the head. At this stage, the tail bud can be observed. The cloacal opening is triangular.

Stage 22 (Fig. 12.5B). The embryo is 18.5 to 23.6 mm in length. 125 somites have been counted. The mandibular elements are now continuous ventrally. The nasal pits are large, lateral, round or oval; and the first and second gills begin to be covered by filaments.

Stage 23 (Fig. 12.5C). The embryo is 25.9 to 29.6 mm in length. 127 or 128 pairs of somites have been counted. The anterior two-third of body is covered by pigment. The optic vesicles and lens project. The eyes are separated from the skin by grooves. The two first gills have lengthened and are covered by more and more filaments. The cloaca is slit shaped, bordered by two lateral depressions.

Stage 24 (Fig. 12.5D). The embryo is 18.3 to 22.2 mm in length. 75% of body is covered with pigment. Resorption of yolk begins, giving the yolk a constricted appearance. Neuromasts appear for the first time. The third gills begin to be covered by filaments. A small fin begins to develop in the tail region of the body.

Stage 25 (Fig. 12.5E). The embryo is 23.3 mm in length and the body is covered by pigment. Neuromasts continue to develop in a nasal, supraorbital and infraorbital situation and are particularly abundant anteriorly. Otic vesicles are no longer observed. A gill chamber begins to develop by outpocketing of the base of second and third gills. Blood vessels are seen for the first time on the yolk surface.

Stage 26 (Fig. 12.5F). The embryo is 24.9 to 30.2 mm in length and the anterior part of the body is elevated from the yolk mass. Neuromasts

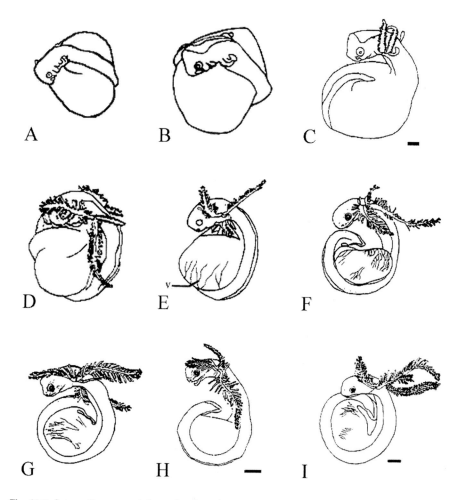

Fig. 12.5. Schematic representations of embryonic stages in *Ichthyophis kohtaoensis*. **A.** Stage 21. **B.** Stage 22. **C.** Stage 23. **D.** Stage 24. **E.** Stage 25. **F.** Stage 26. **G.** Stage 27. **H.** Stage 28. **I.** Stage 29. Scale bar in C, also applied to E, F = 1 mm. Scale bar in I also applied in A, B, D, G, H = 2 mm. From Dünker, N., Wake, M.H. and Olson, W. M. 2000. Journal of Morphology 243: 3-34, Fig. 4A-H; Fig. 5I.

continue to develop. Two rows are observed above the eyes. Neuromasts of the head develop and on the body; elevated neuromasts are observed along 75% of the trunk. Eyes are pigmented. The nasal opening is triangular and enlarges. The tail fin becomes high and continues to develop. Vascularization is more and more developed on the yolk mass.

Stage 27 (Fig. 12.5G). The embryo is 32.5 mm in length. The lower jaw extends posteriorly and the opening of mouth is narrow. Nasal openings are in a more frontal position than previously. Neuromasts continue to develop and appear as white dots on the head, lower jaw, and chin. Three rows of lateral line organs are observed above the eyes. On the trunk, neuromasts are more and more dorsal then lateral. Ampullary organs are observed. The tail bud is rounded and thick and the tail fin broadens.

Stage 28 (Fig. 12.5H). The embryo is 27.4 to 33.2 mm in length. The yolk mass decreases. Rudiments of tentacles appear near the eyes. The lower jaw forms a lip-like structure. Neuromasts begin to appear near the gills and are disposed along an antero-posterior axis on the trunk. In the anterior region, they are more dorsal than lateral. Posteriorly part, the reverse is the true.

Stage 29 (Fig. 12.5I). The embryo is 36.3 mm in length. The yolk mass is elongated with a heart-shaped curvature. The head is elongated. Neuromasts are elevated in the head region and the ampullary organs are observed below the eyes.

Stage 30 (Fig. 12.6A). The embryo is 31.7 to 41.5 mm in length. Bending of the yolk mass gives it a U to S shape. Head outline is straight. Neural folds fuse are closed. A pair of tentacle rudiments appears as small indentations or grooves near the eyes at the frontal border. Two rows of neuromasts and ampullary organs are observed below the eyes and three rows (1 row of neuromasts and 2 rows of ampullary organs) are found behind the eyes. The cloacal opening is elongated and is bordered by oval lengthened buds.

Stage 31 (Fig. 12.6B). The embryo is 40.9 mm in length. Resorption of the yolk mass continues; it is S-shaped. The head is flattened and lengthened. Glands begin to appear on the chin. A second row of ampullary organs begins to develop below the eyes. The mouth takes on the subterminal appearance of the larva.

Stage 32 (Fig. 12.6C). The embryo is 43.8 to 45.9 mm in length and yolk mass is increasingly reduced, it is spirelike, parallel to the longitudinal axis of the body; two symmetrical yolk lobes are separated by the gut. Skin glands are distinct on the chin. Neuromasts are observed as large dots. Ampullary organs are small white dots. Below the eyes one neuromast and two ampullary rows can be observed. The tip of the tail is arrow-shaped, slightly curved ventrally. An incision is observed between the cloaca and the end of the tail fin which broadens. The cloacal opening is more differentiated than previously.

Stage 33 (Fig. 12.6D). The embryo is 49.8 to 53.8 mm in length. The yolk mass is reduced to a spindle or corkscrew shaped mass. The eyes are covered with a thickened skin; eye and tentacle rudiments are overlain by a tear-shaped region of the skin. The cloaca is lengthened and bordered by a crescent or a bean-shaped swelling.

Stage 34 (Fig. 12.6E). Embryos are 54.9 to 78.3 mm in length. The yolk mass is reduced and entirely enclosed in an abdominal fold, the yolk is a zigzag-shaped tube. The anterior neural folds are connected without sutures. Neuromasts of the trunk begin to sink into the skin and the eyes are rarely discernible.

Stage 35 (Fig. 12.6F). Embryos are 74.0 to 80.0 mm in length. The yolk mass is very reduced, enclosed in abdominal folds and the yolk tube begins to segment. Neuromasts are centrally indented. Ampullary organs begin to sink into the skin. Nasal openings enlarge and become deeper.

Stage 36 (Fig. 12.6G). Embryos are 70.1 to 83.0 mm in length. Neuromasts and ampullary organs are barely observable and penetrate into the skin. Only two external gills are now observed. The third gill has degenerated and is contained in the gill chamber that is covered by epithelial fold. The cloaca is elongated and bordered by a wall of thickened tissue lacking pigmentation.

Stage 37 (Fig. 12.6H). Embryos are 65.1 to 76.2 mm in length. This period is the beginning of hatching. The two gills are stripped off immediately after hatching. The opening of the gill chamber lengthens. The tail fin is less broad and tail tip becomes rounded. After hatching, 123 vertebrae can be counted.

Stage 38. Embryos are 78.2 to 120.0 mm in length. A pair of lateral yellow stripes appears. Neuromasts are larger and each one is surrounded by a dark circle. No ampullary organ is found on the skin. The frontal nasal opening increases. The tail fin becomes narrow and dorsally curved.

Stage 39. Embryos are 133.1 to 163.0 mm in length. Yellow lateral stripes are observed on the head. The ventral surface becomes segmented. Skin glands are barely observed. The tail fin is no longer observable and the tail tip resembles the rounded tail of the adults.

Stage 40. The animal is 156.2 mm in length. It is the metamorphic stage. The tentacle sheath is funnel-shaped and a tentacle fold appears in the tentacle opening.

In their description, Dünker *et al.* (2000) emphasized several points of development: neural folds, more particularly lateral line organs, stomodeum, eyes, tentacles, otic vesicles, nasal openings, gills and gill chamber, tail, cloacal region. It seems that total number of somites is already observed by stage 21. We will see that in *Typhlonectes compressicauda*, the total number of somites seems also to be definitive at stage 22 (Delsol *et al.* 1981). At this stage, we could count about 75 somites, a number that varies very little in the older stages. In *Typhlonectes compressicauda*, the number of vertebrae varies around 90. In *Ichthyophis kohtaoensis* they are more numerous.

In *Ichthyophis kohtaoensis*, some characters are linked to the oviparity and to the aquatic life of the larvae: presence of an abundant yolk mass that is used over a long period, presence of neuromasts and ampullary organs,

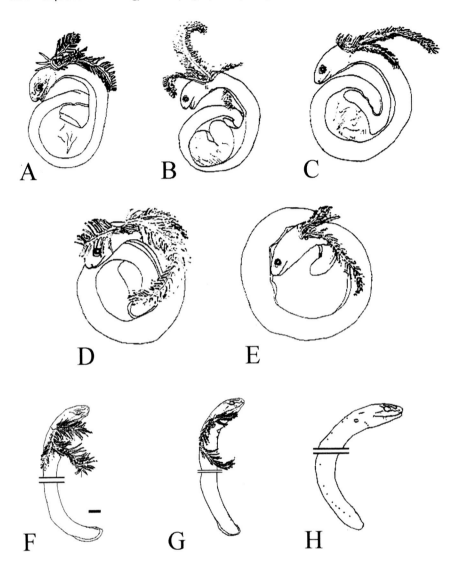

Fig. 12.6. Schematic representations of embryonic stages in *Ichthyophis kohtaoensis*. **A.** Stage 30. **B.** Stage 31. **C.** Stage 32. **D.** Stage 33. **E.** Stage 34. **F.** Stage 35. **G.** Stage 36. **H.** Stage 37. Scale bar = 2 mm. From Dünker, N., Wake, M.H. and Olson, W. M. 2000. Journal of Morphology 243: 3-34, Fig. 5J-N; Fig. 6O-Q.

presence of a tail fin, development of tentacles immediately after hatching. Hatching is also delayed in comparison with *Typhlonectes compressicauda*.

12.3.3 Development of *Typhlonectes compressicauda*

In an initial work, five large stages have been distinguished in this viviparous species (Exbrayat *et al.* 1981). Stage 0 is that of eggs surrounded by a mucous envelope, stages I groups the first phases of development, stage II are embryos with a relatively voluminous yolk mass and always surrounded by a mucous envelope. Stage III is that of fetuses without yolk mass, that can move free in the uterus, stage IV is that of intra-uterine larvae that look like small adults, even if they possess a voluminous pair of vesicular gills. This last stage is the growth phase (chapter **11**, Fig. 11.6A, B, C).

A preliminary short table of development inspired from that of the Anura *Alytes obstetricans* (Cambar and Martin 1959) has been published by Delsol *et al.* (1981). The normal table of *Typhlonectes compressicauda* begins at stage II_{11}, then II_8. It is divided into stages II_8 to II_{16}, III_1 to III_5, IV_1 to IV_3. This last being that of new-born. Subsequently a more detailed table was published (Sammouri *et al.* 1990). Development of *Typhlonectes compressicauda* is now divided into 21 stages numbered 14 to 34. Corresponding stages in the first table are given here. This description is based on the observations of several embryos but, from one stage to another, as in *Ichthyophis kohtaoensis*, the same events may not be synchronised (see for instance, stages 18 and 19: one can observe an increase of somite number from stage 18 to 19, but neural development is more advanced in stage 18 than 19). We give here the main features to aid in an understanding of development. The original paper (Sammouri *et al.* 1990) is more detailed.

Stage 14 or II_8 (Fig. 12.7A). The embryo is 2.5 mm in length. It is narrowly linked to the voluminous yolk mass. 15 pairs of somites have been counted. The neural plate ends anteriorly at an open cephalic vesicle.

Stage 15 or II_9 (Fig. 12.7B). The embryo is 3.2 mm in length. 27 pairs of somites are counted. The cephalic vesicle begins to close but the neural canal is still open at its posterior end. Thickenings of endoderm presage the future optic and otic vesicles. A small stomodeal invagination delimits the upper maxillaries. Mandibular, hyoideal arches and two branchial arches are observed. Three visceral slits make their appearance.

Stage 16 or II_9 (Fig. 12.7C, D). The embryo is 3.6 mm in length. 36 pairs of somites are counted. The yolk mass begins to be resorbed. The stomodeal invagination is particularly marked. Visceral slits are clearly visible.

Stage 17 or II_{10} (Fig. 12.7E). The embryo is 3.8 mm in length. 38 pairs of somites are counted. The neural lobe is closed. Otic vesicles are perfectly individualised. The stomodeal invagination increases. Opening of the third branchial arch is observed. Three branchial slits are well developed; the fourth is only superficially indicated.

Stage 18 or II_{11} (Fig. 12.7F, G). The embryo is 3.5 to 7 mm in length. An average of 38 pairs of somites can be counted (23 to 43 pairs according to the individuals). The yolk mass begins to be greatly resorbed. Absorptive buttons are situated throughout the body. The brain is divided into three vesicles. The neural gutter is wide. Otic placodes are clearly visible. The three expansions of gills are developed and the first visceral slit begins to close.

Stage 19 or II_{11} (Fig. 12.7H). The embryo is 5 mm in length, with 45 pairs of somites. The neural folds bring approach one another. Olfactory placodes begin to appear. The lower jaw projects. Four branchial slits are present.

Stage 20 or II$_{12}$ (Figs. 12.7I, 12.8A). The embryo is 6 to 8 mm in length and it is rolled around the yolk mass. 35 to 60 pairs of somites can be counted. Absorptive buttons persist. Neural fold is still opened anteriorly. Olfactory buds are well indicated but optic placodes are very little visible. Upper jaw is outlined, lower jaw clearly visible.

Stage 21 or II$_{13}$ (Fig. 12.8B). The embryo is 7 to 8.5 mm in length. The head and tail are near one another. The head is elevated above the yolk mass and this has largely been resorbed. 65 pairs of somites have been counted. Absorptive buttons are numerous and very well developed. The brain begins

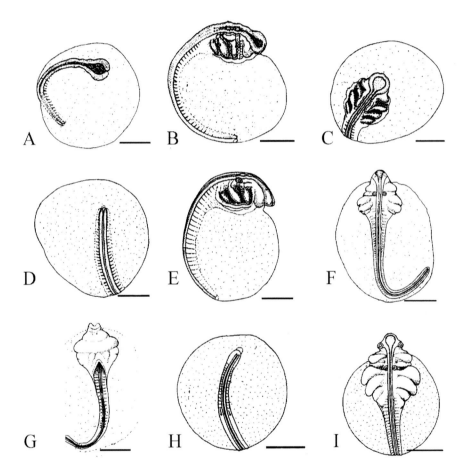

Fig. 12.7. Schematic representations of embryonic stages in *Typhlonectes compressicauda*. **A.** Stage 14. **B.** Stage 15. **C.** Stage 16, view of the head. **D.** Stage 16, view of the posterior part of body. **E.** Stage 17. **F.** Stage 18, view of the anterior and median parts of the body. **G.** Stage 18, view of the posterior part of the body. **H.** Stage 19, view of the posterior part of the body. **I.** Stage 19, view of the anterior party of the body. Scale bars = 0.5 mm. From Sammouri, R., Renous, S., Exbrayat, J.-M. and Lescure, J. 1990. Annales des Sciences Naturelles, Zoologie, 13ème série 11: 135–163, Pl. I, II.

to protrude. Medullary folds of the neural tube are almost fused. The optic placodes can be observed. A small depression appears in the olfactory placode. The upper maxillaries separate and the mandibles join ventrally. A single external gill is observed on each side of the head. Each gill arises from fusion of three lateral expansions. The heart is observed. A small tubular structure links the vertical part of the abdominal cavity to the yolk mass. The cloacal opening is clearly visible. The caudal bud is observed from this stage.

Stage 22 or II_{13} (Fig. 12.8C). The embryo is 8 mm in length with 76 pairs of somites. The yolk mass continues to be resorbed. Absorptive buttons are more and more voluminous anteriorly. At this stage, the ectotrophoblast is observed on the ventral ectoderm (Delsol et al. 1981). The brain is divided into five vesicles. The centre of the olfactory placodes is depressed. The optic vesicles are more and more visible. The upper maxillaries are progressively approximated. The mandibular components have almost merged. Gills are lengthening and are 2 mm in length. The heart is clearly visible. The caudal bud and cloacal opening are increasingly obvious.

Stage 23 or II_{14} (Fig. 12.8D, E). The embryo is 7 to 9 mm. 65 to 80 pairs of somites can be counted. Absorptive buttons are increasingly numerous. Dorsal pigmentation appears. The ectotrophoblast is particularly well developed. The neural tube is closed. The brain is divided into five vesicles. Otic placodes are very well marked. Optic placodes protrude. Maxillaries move to the ventral face. Mandibles are merged. Gills look like blades, they lengthen and they are irrigated by blood vessels. The caudal bud is observed. The cloacal opening becomes triangular.

Stage 24 or $II_{15.}$ The embryo is 7 to 10 mm in length with 65 to 75 pairs of somites. Absorptive buttons are more and more numerous. Pigmentation increases. The yolk mass is increasingly resorbed. Otic placodes are labelled by pigmented cells. External nasal apertures are obvious. Lenses are observed. The lower jaw is now formed. Branchial slits are less visible than previously. Gills develop. The lung outline is more and more conspicuous. The heart is divided into two parts. Gut and liver outlines are now observed.

Stage 25 or II_{16} (Fig. 12.8F). The embryo is 9 to 11 mm. Somites are now hidden by pigmentation. The yolk mass has almost disappeared. Absorptive buttons persist. Ectotrophoblast regresses. Brain hemispheres are fused and the pituitary can be observed. Nostrils are clearly visible. The maxillaries overhang the lower jaw on which dentary buds are observed for the first time. Gills are developed (5.5 mm in length) and highly vascularized. The heart is composed of two auricles and one ventricle. Lungs are lengthening and spread ventrally.

Stage 26 or III_1 (Fig. 12.8G). The embryo is 10 to 17 mm long. Absorptive buttons are always observed by scanning electron microscopy. Future segments begin to appear as small folds on the anterior ventral part of animal. Otic placodes can no longer be seen. Olfactory vesicles and lenses are developed. Maxillaries and olfactory folds are merged. On the lower jaw, three rows of parallel fetal teeth are found. Gills are developed (7 mm in length), branchial slits are less and less visible. Lungs reach the level of posterior end of the liver. Organs are lengthening, kidneys can be seen through the body wall. The cloacal slit is longitudinal.

Stage 27 or III$_1$ (Fig. 12.8H, I). The embryo is 14 mm in length and circular. Ventral segments begin to form. The yolk mass is residual. The ectotrophoblast has completely regressed. Gills continue to develop. Kidneys are increasingly visible. Tentacle outlines begin to be observed.

Stage 28 or III$_2$ (Fig. 13A). The embryo is 11 to 25 mm in length. Pigmentation reaches the caudal region. Annuli are observed throughout the body. Nostrils are small. Each eye is covered by a membrane and it is surrounded by a pigmented ring interrupted ventrally. The mouth begins to open and the lower jaw is more or less protruded. On each side of the snout, the tentacle apparatus extends from each eye to the nostril.

Stage 29 or III$_3$. The embryo is 17 to 30 mm in length. Absorptive buttons decrease. The dorsal face of the body is more pigmented than the ventral face.

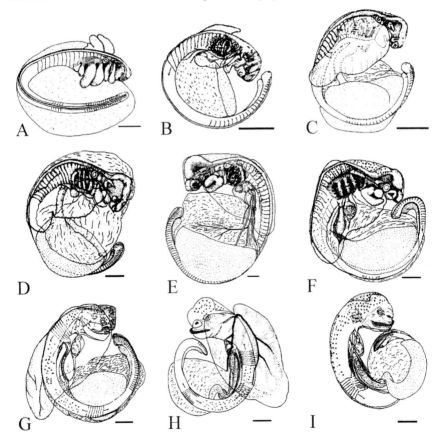

Fig. 12.8. Schematic representations of embryonic stages in *Typhlonectes compressicauda*. **A.** Stage 20. **B.** Stage 21. **C.** Stage 22. **D.** Stage 23. **E.** Stage 24. **F.** Stage 25. **G.** Stage 26. **H.** Stage 27. **I.** Stage 27. Scale bar = 0.5 mm. From Sammouri, R., Renous, S., Exbrayat, J.-M. and Lescure, J. 1990. Annales des Sciences Naturelles, Zoologie, 13ème série 11 : 135–163, Pl. II, III, IV.

Annuli are completely formed. External nostrils seem to be reduced. Tentacles and nostrils come together. Jaws are aligned. Fetal teeth are observed on lower jaw. Gills are well developed. Organs can be observed through the ventral surface of the animal. A fold surrounds the cloacal slit.

Stage 30 or III_4 (Fig. 12.9C). The embryo is 20 to 33 mm in length. The head is swollen in comparison to body. Absorptive buttons continue to regress. Pigmentation still develops. Annuli are clearly visible. Ventral epidermis is covered by pigment except in a narrow median strip. The cloacal papilla forms a projecting button.

Stage 31 or III_5 (Fig. 12.9B, D). The embryo is 32 to 55 mm long and the pigmentation is complete. Epidermal glands begin to appear. Nostrils are developed. The eyes regress. The mouth migrates ventrally. The cloacal papilla becomes narrow. Tentacle ducts have a funnel shape. On the ventral surface, a very narrow white longitudinal stripe is observed.

Stage 32 or IV_1 (Fig. 12.9E). The embryo is 43 to 65 mm in length. It is often folded over in the uterus. Epidermal glands are increasingly numerous, certain of them are disposed in lines delimiting the annuli. 78 to 84 annuli can be counted. Nostrils are small. The mouth has reached its definitive ventral position. Organs are no longer observed. The lengthened cloacal slit is surrounded by a fold.

Stage 33 or IV_2 (Fig. 12.9F). The animal is 65 to 152 mm in length, its maximal size. 78 to 80 annuli are counted. The pigment darkens. Epidermal glands have elongated with a regular alignment in quincunx. The gills are thick and vesicular; they are very well developed and entirely surround the animal, isolating it from the environment. Tentacles and their ducts are clearly visible, with apertures near the nostrils.

Stage 34 or IV_3 (Fig. 12.9G). It is the stage of birth. Animal has the same general aspect and the same size as previously. It looks like a small adult. Gills fall down just before or behind birth.

12.4 COMPARISON OF DEVELOPMENT IN CAECILIANS

Normal tables of development of *Ichthyophis kohtaoensis* and *Typhlonectes compressicauda* have been compared in a profound study (Dünker *et al.* 2000). Development of *I. kohtaoensis* has also been compared with data given in other works for *I. glutinosus* (Sarasin and Sarasin 1887-1890, Breckenridge and Jayasinghe 1979, Breckenridge *et al.* 1987) and *Hypogeophis rostratus* (Brauer 1899).

It appears that development of *Ichthyophis kohtaoensis* closely resembles that of *Ichthyophis glutinosus* after comparison with works of Breckenridge and colleagues and Sarasin and Sarasin. Our own examination of some *I. beddomei* embryos show that stages studied can also be compared to stages of *I. kohtaoensis* or *I. glutinosus*.

Development of *Ichthyophis kohtaoensis* (and therefore Ichthyophiidae) and *Hypogeophis rostratus* present several differences (Dünker *et al.* 2000). Development of otic vesicles occurs at a faster rate relative to other features in *H. rostratus* than in *Ichthyophis kohtaoensis*. In *H. rostratus*, development of

Fig. 12.9. Schematic representations of embryonic stages in *Typhlonectes compressicauda*. **A.** Stage 28. **B.** Stage 31. **C.** Stage 30, upper: general view of the embryo, lower view of the posterior part with organs seen by transparency. **D.** Stage 31 (details). **E.** Stage 32. **F.** Stage 33. **G.** Stage 34. Scale bars = 0.5 mm. From Sammouri, R., Renous, S., Exbrayat, J.-M. and Lescure, J. 1990. Annales des Sciences Naturelles, Zoologie, 13ème série 11: 135 – 163, Pl. IV, V, VI.

gills is slightly different from that in other species. Although embryos of both *H. rostratus* and *I. kohtaoensis* have the same length from stages 21 to 27, *I. kohtaoensis* grows larger and perhaps faster than the other species from this last stage to metamorphosis. Nevertheless, similarities can be observed between the two species, particularly, premetamorphic embryos of *H. rostratus* resemble stage 36 of *I. kohtaoensis*.

Comparison between *Ichthyophis kohtaoensis* and *Typhlonectes compressicauda* reveals important differences. Among differences pointed out by Dünker *et al.* (2000), are a faster somitogenesis in *I. kohtaoensis*, a pigmentation that appears earlier in *I. kohtaoensis* than *T. compressicauda*, an earlier formation of the lens in *I. kohtaoensis*, an earlier and faster tentacle development in *T. compressicauda*, absence of lateral line organs and tail fin, presence of fetal teeth in *T. compressicauda*, mandibular elements fuse earlier in this species.

Otic vesicles appear at the same stage in both species, but they are not pigmented in *I. kohtaoensis*; nasal openings start earlier and develop faster in *I. kohtaoensis*, the cloacal region is similar in early stages but differences in development occur in later stages; the yolk mass disappears earlier in *T. compressicauda* than in *I. kohtaoensis*; gills of *T. compressicauda* have a very different development to reach the characteristic structure found in Typhlonectidae.

In conclusion, the study of comparative development of caecilians show an homogeneity in Ichthyophiidae, some similar points in *Hypogeophis rostratus* (a Caeciliidae); and a very distinct situation in Typhlonectidae, which is certainly linked to viviparity. But it is still difficult to generalise embryonic development of caecilians for only three species have been deeply studied, with establishment for two of them of a normal table based on the same criteria. Studies of development in other species are required to reveal general patterns of development in Gymnophiona, even though capture of these animal and particularly of eggs, embryos and larvae is always very difficult.

12.5 METAMORPHOSIS IN GYMNOPHIONA

Metamorphosis is an important step in the life of Amphibia for it is the transition between aquatic and terrestrial life. During this period, the animal is subjected to important morphological and physiological changes. This phenomenon has been the subject of numerous works in Anura and Urodela. In Gymnophiona, very few studies have been devoted to metamorphosis (Exbrayat and Hraoui-Bloquet 1995). Most of them have a larval life at the end of which they become young animals, as in most other amphibia. Metamorphosis in caecilians is often complicated by an intra-uterine life.

12.5.1 Metamorphosis Features

Most Gymnophiona present only one generation of triradiate or vesicular gills (Sarasin and Sarasin 1887-1890, Brauer 1899, Goeldi 1899, Ramaswami 1954, Wake 1967, 1969, 1977, Taylor 1968, Exbrayat 1986, Breckenridge *et al.* 1987, Exbrayat and Hraoui-Bloquet 1991, 1992, Hraoui-Bloquet and Exbrayat 1994, Dünker *et al.* 2000). In most species, gills disappear by autolysis at metamorphosis (Marcus 1909, Wake 1967, 1969). In *Ichthyophis glutinosus* and *I. kohtaoensis*, gills degenerate two days after hatching. In *Typhlonectes compressicauda*, gills persist to the birth of animal then they degenerate about one day before or after the birth. Gill arteries disappear at metamorphosis.

Metamorphosis also implies locomotion. In *Ichthyophis glutinosus* and *I. kohtaoensis*, an unpaired fin is observed at the posterior end of larval body. It will disappear after about 40 weeks (Breckenridge and Jayasinghe 1979, Breckenridge et al. 1987). At metamorphosis, the outlines of anterior limbs regress in *T. compressicauda* (Renous et al. 1997).

The structure of the larval skin changes at metamorphosis. In larvae, Leydig cells are observed in the skin as in Urodela (Fox 1983a, b, 1985 1986, 1987). In *T. compressicauda*, skin glands are observed at metamorphosis (Delsol et al. 1981, Sammouri et al. 1990). At this period, scales appear in the skin of certain species. Ossification of the skeleton increases (Wake and Hanken 1982, Wake et al. 1985, Wake 1989).

In *Typhlonectes compressicauda*, the brain is subject to two important modifications during development. The first is an increase in size of the cerebral hemispheres that occurs at hatching (Sammouri et al. 1990, Estabel and Exbrayat 1998). The second is the acquisition of final structure at stages 30-31 with, particularly, a rhombencephalic fusion that is characteristic of Gymnophiona (Estabel and Exbrayat 1998). At this time, the pineal organ acquires its final structure (Leclercq 1995, Leclercq et al. 1995). In *T. compressicauda*, the gut is modified at metamorphosis. Gastric glands begin to develop, finally the intestinal epithelium is formed and there is a modification of the intra-uterine larval diet.

12.5.2 Place of Metamorphosis in the Life of Gymnophiona

In primitive species such as *Ichthyophis glutinosus*, branchial buds develop four days after fertilization. At hatching, the embryo possesses a pair of gills, a reduced yolk mass and a tail with a membranous fin. Tentacles are not yet present. Two days after hatching, the gills regress and yolk is resorbed. After 10 months, the membranous fin disappears and tentacles are visible. This slow transformation constitutes metamorphosis (Breckenridge et al. 1987).

For "intermediary" species, very few data are available. In the oviparous *Siphonops annulatus*, embryos that are surrounded by a mucous envelope and still possess gills (Goeldi 1899). Then the yolk mass is progressively resorbed. In *Hypogeophis rostratus*, development is direct and metamorphosis occurs before hatching to give a young animal resembling an adult (Marcus 1909). In some viviparous species (*Dermophis mexicanus*, *Gymnopis multiplicata*), hatching and larval life occur in the uterus. Several authors have reported some signs of metamorphosis during larval life and more particularly resorption of gills (Wake 1977, 1980a, b, Dünker et al. 2000).

In the evolved species that all belong to the family Typhlonectidae, some transformations have been observed at metamorphosis that occurs three to four months after fertilization in *Typhlonectes compressicauda*. In *Chthonerpeton indistinctum*, development resembles that of *Typhlonectes compressicauda*, more particularly concerning the skin, even if data are not so numerous (Fox 1983a, b, 1985, 1986, 1987).

12.5.3 Conclusions on Metamorphosis

Like other amphibia, Gymnophiona undergo metamorphosis. It is more or less spread out in the time, according to the level of evolution of the considered species. In the most primitive forms, the aquatic larval stage is long and metamorphosis is progressive, reminiscent of Urodela. In *Hypogeophis rostratus*, a species with direct development, metamorphosis is rapid and occurs within the mucous membrane of egg. In viviparous species, metamorphosis also seems to be quick and occurs in the uterus after hatching.

Duration of metamorphosis can be compared to the mode of life. The slowness of metamorphosis in *Ichthyophis* can be linked to the long aquatic phase and could be the "basal pattern". In those Urodela that live near rivers, there is also a long aquatic larval life with an unspectacular metamorphosis reminiscent of metamorphosis in *Ichthyophis*. *Siphonops* and *Hypogeophis* are terrestrial, burrowing animals, an existence that is not compatible with a long aquatic larval phase. Metamorphosis is presumed to be quick and just before or after hatching and the new-born may be deposited directly on the ground. In intra-uterine species, eggs and larvae are protected throughout development. In the most evolved species, that are secondarily aquatic (i.e. Typhlonectidae), trends to reduction of larval life and acceleration of metamorphosis are retained.

In conclusion, in Gymnophiona the trend is towards reduction of aquatic larval life and an acceleration of metamorphosis. Metamorphosis is indispensable for terrestrial life, but it is also a handicap.

12.6 EMBRYONIC AND LARVAL GROWTH

12.6.1 In Viviparous Species

In viviparous species, duration of pregnancy and growth are variable. Duration of gestation has been given for several species: *Chthonerpeton indistinctum* (Barrio 1969; Prigioni and Langone 1983a, b), *Gymnopis multiplicata* (Wake 1968), *Dermophis mexicanus* (Wake 1980b) and *Typhlonectes compressicauda* (Moodie 1978, Exbrayat et al. 1981, Exbrayat 1983, 1984, 1986, Billo et al. 1985).

In *Chthonerpeton indistinctum*, a species studied in Uruguay, breeding occurs in August-September and births are observed in January-February. Pregnancy is four months long (Barrio 1969). Adults are 40 cm in length, oocytes are 2 mm in diameter before ovulation; at birth, animals are 10 to 12 cm in length and 7 mm in diameter.

In *Typhlonectes compressicauda,* duration of pregnancy has been established (Exbrayat and Delsol 1985, Exbrayat 1986). Total duration of pregnancy has been calculated by examination of number of pregnant females and appearance of new-born in the field. In February, only some females are pregnant. From April until July, one female in two bears embryos. Then number of pregnant females strongly decreases from July to October. From July, some new-born are observed in the field. They become increasingly numerous. It seems that births occur over three or four months (July till October). From these observations, one can conclude that the first fertilised females (in February) are the first to give birth to young animals, in July; the last females to be fertilised (April or May) would be the last to give birth to young, in October. The gestation is 6 to 7 months long.

A study of the distribution of embryos has allowed estimation of the approximate duration of each general phase (Exbrayat 1986) (Fig. 12.10A). At the beginning of breeding, in February, females bear oocytes, fertilised eggs and young embryos at early stages of development. In April, embryos are very little developed with or without a yolk mass, and several cohorts can be found together in the total population of embryos. The first is composed of embryos at stages 24 to 26, a second of embryos at stage 18. In May, a cohort is composed of young stages (no more than stage 22) and the other by fetuses (stages no more than 30). In June, three cohorts have been found: a group with stages 24 to 26, another with stages 29 to 33 (maximum: stage 31). In July, one cohort is composed of stages 23 and 24, another of stage 29 and a third of stage 33. Between August and October, only stage 33, just before birth, is found. These observations allow one to follow each cohort of embryos and deduce the approximate duration of each stage of development. This method permits us to deduce that the embryonic phase (with yolk mass) is approximately two months long, the fetal phase (fetuses free in the uterus, without any mucous envelope) is also two months long and the larval phase is two to three months long. It is possible that the most developed embryos are found in females that were the first to be fertilised. This could explain the fact that early stages that were found in February reach stages 24 to 26 in April, stage 30 in May, 31 in June, 33 in July. The early stages found in April (stage 18) reach stage 22 in May, 24 to 26 in June and 33 in August or October, etc.

Length and growth have been studied in *Typhlonectes compressicauda* (Exbrayat 1986). Growth curves are exponential for both length and weight (Fig. 12.10B, C). Growth is weak from stage 18 to stage 25, stronger from 25 to 30 with acceleration. During the last two months, growth is the highest. Animals grow from 50 to 150 mm in length. Weight curve resemble length curves, even in Bouin's preserved material. Growth is weak from stages 18 to 26, stronger from 26 to 30. Weight increase is particularly significant after stage 30. During the last two months an animal that was about 200 mg reaches 2000 mg.

In *Dermophis mexicanus* studied by Wake (1980b), pregnancy is a full year long, from June until June. Embryonic growth is about 26 mm a month at the beginning of pregnancy. It is 13.2 mm a month from August until December and 4 mm a month from February until March. Embryos reach 150 mm at birth with a wide variability (108 to 147 mm). In this species, growth is rapid at the beginning of pregnancy, then it slows down after the eighth month (Fig. 12.10D). This growth is different from that of *Typhlonectes compressicauda*.

12.6.2 In Oviparous Species

In *Ichthyophis glutinosus*, weight and length are progressive at the beginning of development. After metamorphosis, growth strongly increases. In 50 weeks, the animal grows from 7 to 120 mm in length and reaches 3000 mg, then in 10 weeks, it grows from 120 mm to 300 mm and from 3 g to 35 g (Breckenridge *et al.* 1987). Although there is no indication of time, it seems that, in *Ichthyophis kohtaoensis*, growth in length and weight varies according to the same curve (Dünker *et al.* 2000). In these species, two phases of growth are observed, resembling growth curves in *Typhlonectes compressicauda*. In *Boulengerula taitanus*, Malonza and Measey (2005) suggest that the ontogenetic development from juvenile to subadult to adult takes about two years.

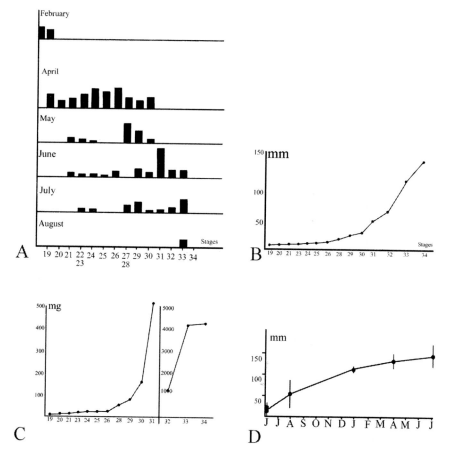

Fig. 12.10. A. Distribution of embryos in *Typhlonectes compressicauda*. **B.** Length growth of *Typhlonectes compressicauda* during development. **C.** Weight growth of *Typhlonectes compressicauda* during development. **D.** Length growth of *Dermophis mexicanus* during development. **A, B, C,** from Exbrayat, J.-M. 1986, Unpublished D. Sci. Thesis, Université Paris VI, France. Figs. 101, 102, 103. **D,** after data in Wake, M.H. 1980b. Herpetologica, 36: 244-256.

12.7 CONCLUSIONS

Embryonic development of caecilians is still little known. Only two normal tables of development have been recently published (*Ichthyophis kohtaoensis* and *Typhlonectes compressicauda*) with some precise observations concerning *Ichthyophis glutinosus* (Sarasin and Sarasin 1887-1890, Breckenridge and Jayasinghe 1979, Breckenridge *et al*. 1987). Growth of embryos and larvae is also very little known and only in a few species (*Ichthyophis glutinosus*, *I. kohtaoensis*, *Typhlonectes compressicauda*, *Dermophis mexicanus*) with some additional measures (*Gymnopis multiplicata*, *Chthonerpeton indistinctum*, *Hypogeophis rostratus*). It appears that oviparous species have several similar points of development. The viviparous *Typhlonectes compressicauda* has a peculiar development relative to the other species, that is certainly linked to the intra-uterine life of embryos. We have seen in another chapter that this animal presents special characters in direct relation to viviparity (chapter 11, this volume). It is also interesting to observe the similarity of embryonic growth curves in *Typhlonectes compressicauda*, *Ichthyophis glutinosus* and *I. kohtaoensis*, and that the embryonic growth curve of *Dermophis mexicanus* is different. In their important work, Dünker *et al*. (2000) made comparisons of *I. kohtaoensis* development with that of the other amphibians, and they found similarities with certain Urodela and Anura especially direct-developing species.

In spite of these interesting data, we need further data in order to understand embryonic development in caecilians, in both oviparous and viviparous forms, to obtain an evolutionary perspective.

12.8 ACKNOWLEDGMENTS

The author thanks Catholic University of Lyon (U.C.L.) and Ecole Pratique des Hautes Etudes (E.P.H.E.) that allowed him freedom in organizing his research fields and supported the works on Gymnophiona. He also thanks Singer-Polignac Foundation that supported the missions necessary to collect the material of study on *Typhlonectes compressicauda*. The author also thanks Michel Delsol, Honorary Professor at he U.C.L. and at E.P.H.E. who is at the origin of these works in Gymnophiona, Marie-Thérèse Laurent, who made thousand and thousand of sections. He also thanks Barrie Jamieson, who invited him to contribute to this collection devoted to reproduction and phylogeny of Amphibia and also for his patience in reviewing the manuscript. He thanks his son, Jean-François Exbrayat for his help in preparing the illustrations.

12.9 LITERATURE CITED

Balakrishna, T. A., Gundappa, K. R. and Katre Shakuntala. 1983. Observations on the eggs and embryo of *Ichthyophis malabarensis* (Taylor) (Apoda: Amphibia). Current Science 52: 990-991.

Barrio, A. 1969. Observaciones sobre *Chthonerpeton indistinctum* (Gymnophiona, Caecilidae) y su reproduccion. Physis 28: 499-503.

Bhatta, G. K. and Exbrayat, J.-M. 1998. Premières observations sur le développement embryonnaire d'*Ichthyophis beddomei*, Amphibien Gymnophione ovipare. Bulletin de la Société Zoologique de France. 124: 117-118.

Billo, R. R., Straub, J. O. and Senn, D. G. 1985. Vivipare Apoda (Amphibia : Gymnophiona) *Typhlonectes compressicaudus* (Duméril et Bibron, 1841): Kopulation, Tragzeit und Geburt. Amphibia-Reptilia 6: 1-9.

Brauer, A. 1897a. Beitrage zur kenntniss der Entwicklungsgeschichte und der Anatomie der Gymnophionen. Zoologisches Jahrbuch für Anatomie 10: 277-279.

Brauer, A. 1897b. Beitrage zur kenntniss der Entwicklungsgeschichte und der Anatomie der Gymnophionen. Zoologisches Jahrbuch für Anatomie 10: 389-472.

Brauer, A. 1899. Beitrage zur kenntniss der Entwicklung und Anatomie der Gymnophionen.II. Die Entwicklung der aüssern Form. Zoologisches Jahrbuch für Anatomie 12: 477-508.

Brauer, A. 1900. Zur kenntniss der Entwicklung der Excretionsorgane der Gymnophionen. Zoologischer Anzeiger 23: 353-358.

Brauer, A. 1902. Beitrage zur kenntniss der Entwicklung und Anatomie der Gymnophionen. III. Die Entwicklung der Excretionsorgane. Zoologisches Jahrbuch für Anatomie 16: 1-176.

Breckenridge, W. R. and De Silva, G. I. S. 1973. The egg case of *Ichthyophis glutinosus* (Amphibia, Gymnophiona). Proceedings of Ceylon Association for Advancement of Sciences 29: 107.

Breckenridge, W. R. and Jayasinghe, S. 1979. Observations on the eggs and larvae of *Ichthyophis glutinosus*. Ceylon Journal of Sciences (Biological Sciences) 13: 187-202.

Breckenridge, W. R., Shirani, N. and Pereira, L. 1987. Some aspects of the biology and development of *Ichthyophis glutinosus* (Amphibia, Gymnophiona). Journal of Zoology 211: 437-449.

Cambar, R. and Martin, S. 1959. Table chronologique du développement embryonnaire et larvaire du Crapaud accoucheur (*Alytes obstetricans* Laur.). Actes de la Société Linnéenne de Bordeaux 98: 1-20.

Delsol, M., Flatin, J., Exbrayat, J.-M. and Bons, J. 1981. Développement de *Typhlonectes compressicaudus* Amphibien Apode vivipare. Hypothèse sur sa nutrition embryonnaire et larvaire par un ectotrophoblaste. Comptes Rendus des Séances de l'Académie des Sciences de Paris, série III 293: 281-285.

Delsol, M., Exbrayat, J.-M., Flatin, J. and Gueydan-Baconnier, M. 1986. Nutrition embryonnaire chez *Typhlonectes compressicaudus* (Duméril et Bibron, 1841) Amphibien Apode vivipare. Mémoires de la Société Zoologique de France 43: 39-54.

Dumeril, A. M. C. and Bibron, G. 1841. *Erpétologie générale ou histoire naturelle des Reptiles*. Tome 8, Librairie encyclopédique de Roret, Paris.

Dünker, N., Wake, M. H. and Olson, W. M. 2000. Embryonic and larval development in the Caecilian *Ichthyophis kohtaoensis* (Amphibia, Gymnophiona). A staging table. Journal of Morphology 243: 3-34.

Estabel, J. and Exbrayat, J.-M. 1998. Brain development of *Typhlonectes compressicaudus* (Dumeril and Bibron, 1841). Journal of Herpetology 32: 1-10.

Exbrayat, J.-M. 1986. Quelques aspects de la biologie de la reproduction chez *Typhlonectes compressicaudus* (Duméril et Bibron, 1841), Amphibien Apode. Doctorat ès Sciences Naturelles, Université Paris VI, France.

Exbrayat, J.-M. and Delsol, M. 1985. Reproduction and growth of *Typhlonectes compressicaudus*, a viviparous Gymnophione. Copeia 1985: 950-955.

Exbrayat, J.-M. and Hraoui-Bloquet, S. 1991. Morphologie de l'épithélium branchial des embryons de *Typhlonectes compressicaudus* (Amphibien, Gymnophione) étudié en microscopie électronique à balayage. Bulletin de la Société Herpétologique de France 57: 45-52.

Exbrayat, J.-M. and Hraoui-Bloquet, S. 1992. Evolution de la surface branchiale des embryons de *Typhlonectes compressicaudus*, Amphibien Gymnophione vivipare, au cours du développement. Bulletin de la Société Zoologique de France 117: 340.

Exbrayat, J.-M. and Hraoui-Bloquet, S. 1995 . Evolution of reproductive patterns in Gymnophiona Amphibia. Pp. 48-52. In G. A. Llorente, A. Montori, X. Santos and M. A. Carretero (eds), *Scientia Herpetologica*, Proceedings of the 7th General Meeting of the Societas Europaea Herpetologica, Barcelona, Spain.

Exbrayat, J.-M., Delsol, M. and Flatin, J. 1981. Premières remarques sur la gestation chez *Typhlonectes compressicaudus* (Duméril et Bibron, 1841) Amphibien Apode vivipare. Comptes Rendus des Séances de l'Académie des Sciences de Paris 292: 417-420.

Exbrayat, J.-M., Bhatta, G. K., Estabel, J. and Paillot, R. 1999. First observations on embryonic development of *Ichthyophis beddomei*, an oviparous Gymnophionan Amphibia. Pp. 113-120. In C. Miaud and R. Guyetant (eds), *Current Studies in Herpetology*, Proceedings of the 9th General Meeting of the Societas Europaea Herpetologica, Le Bourget du Lac, France.

Fox, H. 1983a. *Amphibian Morphogenesis*. Humana Press, Clifto, New Jersey, 301 pp.

Fox, H. 1983b. The skin of *Ichthyophis* (Amphibia: Caecilia): an ultrastructural study. Journal of Zoology 199: 223-248.

Fox, H. 1985. The tentacles of *Ichthyophis* (Amphibia: Caecilia) with special reference to the skin. Journal of Zoology 205: 223-234.

Fox, H. 1986. Early development of caecilian skin with special reference to the epidermis. Journal of Herpetology 20: 154-157.

Fox, H. 1987. On the fine structure of the skin of larval juvenile and adult *Ichthyophis* (Amphibia: Caecilia). Zoomorphology 107: 67-76.

Goeldi, E. 1899. Ueber die Entwicklung von *Siphonops annulatus*. Zoologische Jahrbuch 12: 170-173.

Hetherington, T. E. and Wake, M. H. 1979. The lateral line system in larval *Ichthyophis* (Amphibian: Gymnophiona). Zoomorphology 93: 209-225.

Hraoui-Bloquet, S. 1995. Nutrition embryonnaire et relations materno-fetales chez *Typhlonectes compressicaudus* (Duméril et Bibron, 1841), Amphibien Gymnophione vivipare. Thèse de Doctorat E.P.H.E., Lyon, France.

Hraoui-Bloquet, S. and Exbrayat, J.-M. 1993. La fécondation chez *Typhlonectes compressicaudus* (Duméril et Bibron, 1841), Amphibien Gymnophione. Bulletin de la Société Zoologique de France 118: 356-357.

Hraoui-Bloquet, S. and Exbrayat, J.-M. 1994. Développement des branchies chez les embryons de *Typhlonectes compressicaudus*, Amphibien Gymnophione vivipare. Annales des Sciences naturelles, Zoologie, 13éme série 15: 33-46.

Leclercq, B. 1995. Contribution à l'étude du complexe pinéal des Amphibiens actuels. Diplôme de l'E.P.H.E., Lille, France.

Leclercq, B., Martin-Bouyer, L. and Exbrayat, J.-M. 1995. Embryonic development of pineal organ in *Typhlonectes compressicaudus* (Dumeril and Bibron, 1841), a viviparous Gymnophionan Amphibia. Pp. 107-111. In G. A. Llorente, A. Montori, X. Santos and M. A. Carretero (eds), *Scientia Herpetologica*, Proceedings of the 7th General Meeting of the Societas Europaea Herpetologica, Barcelona, Spain.

Malonza, P.K. and Measey, G.J. 2005. Life history of an African caecilian: *Boulengerula taitanus* Loveridge 1935 (Caeciilidae: Amphibia: Gymnophiona). Tropical Zoology 18: 49-66.

Marcus, H. 1909. Beiträge zur Kenntniss der Gymnophionen. III. Zur Entwicklungsgeschichte des Kopfes. Morphologisches Jahrbuch 40: 105-183.

Moodie, G. E. E. 1978. Observations on the life history of the caecilian *Typhlonectes compressicaudus* (Dumeril and Bibron) in the Amazon Basin. Canadian Journal of Zoology 56: 1005-1008.

Parker, H. W. 1956. Viviparous caecilians and amphibian phylogeny. Nature 178: 250-252.

Parker, H. W. and Dunn, E. R. 1964. Dentitional metamorphosis in the Amphibia. Copeia 1964: 75-86.

Peters, W. 1874a. Observations sur le développement du *Caecilia compressicauda*. Annales des Sciences Naturelles, Zoologie, série 5: article 13.

Peters, W. 1874b. Derselbe las ferner über die Entwicklung der Caecilien und besonders der *Caecilia compressicauda*. Monatsberichte der Deutschen Akademie der wissenschaften zu Berlin 1874: 45-49.

Peters, W. 1875. Uber die Entwicklung der Caecilien. Monatsberichte der Deutschen Akademie der wissenschaften zu Berlin 1875: 483-486.

Prigioni, M. Y. and Langone, J. A. 1983a. Notas sobre *Chthonerpeton indistinctum* (Amphibia, Typhlonectidae). IV Notas complementarias. Jornadas de Ciencias naturales, Montevideo 19-24 set. 1983, 3: 81-83.

Prigioni, M. Y. and Langone, J. A. 1983b. Notas sobre *Chthonerpeton indistinctum* (Amphibia, Typhlonectidae). V Notas complementarias. Jornadas de Ciencias naturales, Montevideo 19-24 set. 1983, 3: 97-99.

Ramaswami, L. S. 1954. The external gills of *Gegenophis* embryos. Anatomischer Anzeiger 101 : 120-122.

Renous, S., Exbrayat, J.-M. and Estabel, J. 1997. Recherche d'indices de membres chez les Amphibiens Gymnophiones. Annales des Sciences Naturelles, Zoologie, 13ème série 18: 11-26.

Sammouri, R., Renous, S., Exbrayat, J.-M. and Lescure, J. 1990. Développement embryonnaire de *Typhlonectes compressicaudus* (Amphibia Gymnophiona). Annales des Sciences Naturelles, Zoologie, 13ème série 11: 135-163.

Sarasin, P. and Sarasin, F. 1887-1890. Ergebnisse Naturwissenschaftlicher Forschungen auf Ceylon. Zur Entwicklungsgeschichte und Anatomie der Ceylonischen Blindwuhle *Ichthyophis glutinosus*. C. W. Kreidel's Verlag, Wiesbaden.

Seshachar, B. R. 1942. Origin of intralocular oocytes in male Apode. Proceedings of Indian Academy of Sciences 15: 278-279.

Seshachar, B. R., Balakrishna, T. A., Katre Shakuntala and Gundappa, K. R. 1982. Some unique feature of egg laying and reproduction in *Ichthyophis malabarensis* (Taylor) (Apoda: Amphibia). Current Science 51: 32-34.

Taylor, E. H. 1968. *The Caecilians of the World. A Taxonomic Review*. University of Kansas Press. Lawrence, Kansas, U.S.A., 848 pp.

Wake, M. H. 1967. Gill structure in the Caecilian genus *Gymnopis*. Bulletin of South California Academy of Sciences 66: 109-116.

Wake, M. H. 1968. Evolutionary morphology of the Caecilian urogenital system. Part I: the gonads and fat bodies. Journal of Morphology 126: 291-332.

Wake, M. H. 1969. Gill ontogeny in embryos of *Gymnopis* (Amphibia : Gymnophiona). Copeia 1969: 183-184.

Wake, M. H. 1976. The development and replacement of teeth in viviparous Caecilians. Journal of Morphology 148: 33-63.

Wake, M. H. 1977. Fetal maintenance and its evolutionary significance in the Amphibia: Gymnophiona. Journal of Herpetology 11: 379-386.

Wake, M. H. 1978. Comments on the ontogeny of *Typhlonectes obesus* particulary its dentition and feeding. Papeis avulsos de Zoologia 32: 1-13.

Wake, M. H. 1980a. Fetal tooth development and adult replacement in *Dermophis mexicanus* (Amphibia : Gymnophiona) : Fields versus clones. Journal of Morphology 166: 203-216.

Wake, M. H. 1980b. Reproduction, growth and population structure of the Central American Caecilian *Dermophis mexicanus*. Herpetologica 36: 244-256.

Wake, M. H. 1986. A perspective on the systematics and morphology of the Gymnophiona (Amphibia). Mémoires de la Société Zoologique de France 43: 21-35.

Wake, M. H. 1989. Metamorphosis of the hyobranchial apparatus in *Epicrionops*, Amphibia, Gymnophiona, Rhinatrematidae: replacement of bone by cartilage. Annales des Sciences Naturelles, Zoologie, 13ème série 10: 171-182.

Wake, M. H., Exbrayat, J.-M. and Delsol, M. 1985. The development of the chondrocranium of *Typhlonectes compressicaudus* (Gymnophiona), with comparison to other species. Journal of Herpetology 19: 568-577.

Wake, M. H. and Hanken, J. 1982. Development of the skull of *Dermophis mexicanus* (Amphibia: Gymnophiona) with comments on skull kinesis and amphibian relationships. Journal of Morphology 173: 203-223.

Index

17β Estradiol 202

A

A1 Cell 205
A2 Cell 205
Absorptive Button 316, 317, 371, 372, 373, 374, 375
Acth 207, 208, 213
Adelphophagy 307, 308, 317, 331
Adenohypophysis 183, 205, 210
Adrenal Cell 217
Adrenaline 217
Adrenocortical Cell 213, 215
Adult 7, 11, 14, 15, 16, 18, 45, 53, 55, 59, 61, 63, 66, 70, 71, 79, 82, 83, 85, 88, 91, 92, 93, 94, 96, 97, 99, 102, 103, 105, 108, 109, 113, 114, 119, 124, 131, 134, 135, 157, 186, 188, 191, 192, 193, 196, 198, 199, 203, 204, 208, 209, 212, 213, 214, 278, 284, 292, 296, 299, 308, 309, 310, 314, 335, 366, 369, 371, 375, 378, 379, 381
Afrocaecilia 43, 59, 124, 248, 250, 304
Alveola 96, 318
Ampullary Organ 18, 92, 368, 369
Anatomy 2, 4, 5, 7, 8, 79, 82, 94, 96, 97, 105, 120, 124, 125, 135, 158, 159, 187, 194, 204, 215, 232, 233, 234
Animal Pole 360, 361
Annuli 45, 47, 53, 55, 56, 57, 59, 64, 69, 79, 99, 374, 375
Anticortisol Serum 215
Anti-GH 207
Anura 41, 82, 85, 104, 107, 122, 127, 132, 133, 135, 157, 189, 217, 219, 220, 221, 235, 239, 240, 261, 263, 267, 269, 270, 271, 298, 299, 318, 351, 352, 371, 377, 382
Aorta 97, 213, 294, 295, 296, 297, 298
Arterial Cone 97
Ascaphus 133, 157
Atresia 285, 287
Atretic Bodies 202, 220
Atretic Follicle 194, 202, 203, 204, 277, 279, 280, 281, 282, 283, 284, 285, 287, 288

Atretochoana 40, 48, 69, 96, 97
Auricle 83, 97, 98, 373
Avt 87

B

B1 Cells 205
B3 Cell 205
Basal Cells 94, 100, 163, 166, 167, 168, 169, 170, 171, 187, 189, 328, 340, 344, 345, 346
Behavior 2, 5, 6, 217, 221
Biennial 14, 15, 131, 193, 197, 210, 221, 284, 288
Biennial Cycle 193, 197, 210
Bile 94
Biogeography 3
Bipotential 292, 293, 295, 298
Birth 8, 10, 12, 13, 14, 97, 113, 117, 119, 120, 122, 124, 134, 190, 203, 219, 275, 276, 278, 288, 292, 296, 298, 307, 308, 309, 310, 314, 317, 318, 325, 326, 351, 352, 375, 377, 379, 380
Bladder 83, 94, 99, 101, 120, 121, 123, 134, 187, 197, 336
Blastomere 360
Blind Sac 91, 121, 123
Blood Cell 98, 99, 197, 201, 316, 347
Bone Marrow 99
Bony Tissue 352
Boulengerula 12, 14, 43, 49, 59, 61, 63, 99, 105, 111, 114, 121, 124, 185, 248, 249, 250, 252, 253, 257, 258, 259, 260, 268, 270, 271, 303, 304, 305, 319, 381
Brain 11, 17, 84, 85, 87, 88, 89, 91, 208, 363, 371, 372, 373, 378
Branchial Slit 371, 372, 373
Breeding 108, 111, 115, 116, 120, 122, 127, 128, 131, 132, 134, 135, 185, 186, 187, 188, 189, 190, 191, 192, 194, 195, 196, 197, 200, 205, 206, 207, 210, 215, 217, 218, 219, 220, 221, 232, 233, 235, 281, 284, 285, 287, 288, 307, 348, 349, 350, 351, 352, 353, 360, 379, 380
Breeding Cycle 108, 111, 120, 132, 135, 185, 192, 235, 281, 348, 350, 351, 352, 353

Breeding Season 185, 186, 187, 189, 190, 192, 195, 200, 205, 206, 207, 210, 215, 217, 218, 219, 221
Brünner Gland 94
Buccopharyngeal 43, 97
Bufo 122, 217, 263, 351, 352

C

Caecilia 2, 9, 12
Calcification 316, 352, 353
Calvaria 352, 353
Capsule 12, 85, 91, 100, 121, 122, 132, 215
Carotid 44, 98
Caudal Bud 292, 373
Cell Nests 234, 235, 238, 239, 240
Centrum 82, 84
Cephalic Vesicle 371
Chemosensory Organ 17
Chin 368
Choane 135
Chondrocranium 16, 352
Chromaffin Cell 215, 216, 217
Chthonerpeton 8, 9, 10, 13, 48, 61, 70, 96, 103, 104, 105, 114, 120, 124, 127, 129, 185, 192, 202, 205, 232, 233, 238, 247, 248, 249, 251, 252, 253, 257, 259, 263, 270, 271, 272, 275, 278, 279, 280, 288, 307, 308, 310, 314, 330, 347, 378, 379, 382
Ciliated Cells 92, 96, 114, 115, 116, 129, 131, 163, 167, 189, 195, 196, 197, 199, 326, 333, 338, 347
Circulatory System 97
Classification 2, 39, 40, 41, 51, 52, 53, 239, 240
Cloaca 8, 9, 10, 11, 52, 53, 69, 70, 71, 79, 82, 83, 93, 94, 100, 101, 105, 113, 114, 117, 120, 121, 122, 123, 124, 125, 127, 129, 131, 134, 157, 158, 159, 160, 175, 186, 187, 188, 189, 192, 194, 195, 197, 204, 217, 218, 219, 243, 335, 336, 337, 349, 366, 368, 369, 373, 374, 375, 377
Cloacal Ampulla 121
Cloacal Opening 53, 83, 127, 337, 366, 368, 373
Cloacal Papilla 375
Cloacal Slit 374, 375
Clutch 12, 14, 15, 63, 364, 365
Clutch Size 12, 14, 15
Coecilia 7, 56, 65
Coelomic Cavity 293, 326, 328, 330, 335, 337
Coelomic Wall 127
Collar 99, 100, 309
Columella 91

Copulation 10, 11, 122, 132, 158, 192
Cord
— nerve 331
— umbilical 347
— spinal 88
Corpora Lutea 11, 125, 128, 183, 193, 201, 202, 211, 219, 220, 221, 275, 280, 282, 283, 284, 287, 288, 306
Cortex 99, 163, 264, 296, 297, 298
Cortical Granule 281, 287
Corticotropic Cell 205, 207, 209, 210, 213, 214, 215, 217, 218, 221
Courtship 10, 19
Crown 52, 93, 195, 309, 312
Cutaneous Gland 85, 330

D

Density 116, 135, 159, 163, 207, 348, 349, 350
Dentary Bud 373
Dermis 85, 86, 89, 99, 122, 328, 330, 339, 341, 375
Dermophis 9, 11, 14, 15, 16, 18, 43, 49, 63, 66, 68, 81, 82, 85, 89, 91, 105, 111, 113, 114, 116, 124, 125, 127, 129, 174, 185, 187, 190, 192, 193, 204, 232, 234, 238, 248, 250, 275, 303, 305, 307, 308, 309, 314, 315, 317, 378, 379, 380, 381, 382
Development 1, 2, 3, 5, 7, 8, 11, 12, 13, 14, 15, 16, 17, 18, 19, 63, 82, 87, 89, 92, 93, 95, 96, 99, 101, 103, 108, 113, 125, 132, 133, 135, 157, 186, 189, 190, 191, 200, 201, 202, 204, 213, 217, 218, 219, 220, 221, 232, 235, 237, 275, 278, 280, 284, 287, 288, 289, 291, 292, 293, 296, 299, 305, 306, 307, 308, 309, 310, 312, 313, 314, 315, 316, 317, 318, 319, 325, 326, 327, 331, 332, 334, 335, 336, 337, 338, 339, 341, 342, 345, 347, 349, 351, 352, 353, 359, 360, 362, 363, 366, 369, 370, 371, 373, 375, 377, 378, 379, 380, 381, 382
Developmental Biology 5, 15, 18
Diencephalon 85, 88, 101
Diffusive Exchange 316, 318, 348
Digestive Tract 92, 93, 95, 96, 125, 134, 314, 318, 330, 331, 332, 338
Direct Developer 12, 14
Distal ductule 100
Distal Lobe (Pars Distalis) 205
Distal Tubule 100
Dorsal Crest 292
Duct Epithelium of Tubular Glands 170
Ductus Epididymidis 178, 179
Duodenum 83, 93, 94

Index **389**

E

Ear 15, 17, 91
Ecology 2, 3, 5, 6, 47
Ectoderm 292, 326, 327, 373
Ectotrophoblast 316, 317, 326, 327, 328, 329, 330, 335, 339, 347, 373, 374
Efferent Ductule 108, 111
Egg 12, 15, 33, 128, 129, 132, 133, 185, 192, 194, 201, 232, 257, 264, 275, 282, 285, 287, 303, 304, 305, 306, 307, 308, 317, 318, 319, 325, 326, 330, 359, 360, 361, 362, 363, 364, 365, 370, 377, 379, 380
Egg Case 12, 365
Egg Envelope 132, 133, 194
Egg-layer 12, 303
Egg-laying 192
Ellipsoid 103, 125
Embryo 13, 14, 15, 83, 84, 88, 102, 129, 133, 134, 197, 201, 219, 299, 305, 306, 307, 308, 310, 312, 314, 315, 316, 317, 318, 326, 330, 331, 335, 338, 347, 348, 349, 351, 352, 353, 360, 362, 363, 365, 366, 367, 368, 371, 372, 373, 374, 375, 376, 378
Embryogenesis 132, 308
Embryonic Development 82, 87, 101, 103, 113, 132, 133, 135, 190, 327, 334, 336, 337, 341, 342, 349, 353, 359, 360, 377, 382
Enamel 309, 310, 311, 312
Endocrine Gonad 101
Endocrine Organ 8, 101, 183, 218
Endocrine Ovary 197, 203
Endocrine Regulation 125, 183, 217, 218
Endocrinology 11, 179, 183, 200, 219, 221
Endoderm 292, 293, 294, 296, 298, 332, 371
Endodermal Cell 292, 293, 294
Endodermal Roof 293
Endolymphatic Duct 91
Enterocyte 94, 336
Epicrionops 18, 43, 44, 53, 248, 251
Epidermal Gland 375
Epidermis 41, 83, 84, 85, 86, 89, 92, 122, 309, 310, 314, 326, 328, 330, 339, 341, 375
Epididymis 158, 167, 174
Epiphysis 101, 102, 103
Erythrocyte 98, 99
Esophagus 43, 83, 92, 93, 94, 95, 97, 332, 333, 334
Estriol 202
Estrogenic Hormone 202, 219
Euviviparity 13
Excretion 101
Eye 15, 16, 17, 41, 47, 49, 53, 55, 59, 61, 63, 64, 65, 66, 67, 68, 69, 70, 82, 87, 88, 89, 363, 365, 366, 368, 369, 374, 375

F

Fat Bodies 8, 41, 105, 122, 124, 134, 135, 232, 275, 291, 293, 296, 297, 348, 349, 350, 351, 352
Fat Body 83, 124, 125, 127, 275, 295, 349, 351
Female 5, 6, 7, 8, 9, 10, 12, 14, 61, 63, 79, 83, 99, 100, 124, 125, 127, 129, 131, 134, 135, 157, 158, 178, 183, 185, 192, 193, 194, 196, 197, 198, 199, 201, 203, 204, 207, 208, 210, 211, 212, 213, 214, 215, 217, 219, 221, 232, 275, 278, 280, 281, 282, 283, 284, 288, 296, 299, 303, 305, 306, 307, 315, 316, 317, 319, 331, 348, 349, 350, 351, 352, 353, 360, 361, 364, 379, 380
Female Genital Duct 204, 353
Female Genital Tract 124, 125, 193, 194, 275
Female Pronucleus 360
Female Reproductive System 5, 8
Fenestra Ovalis 91
Fertilization 9, 13, 19, 105, 120, 124, 132, 135, 157, 158, 167, 177, 187, 192, 194, 231, 287, 326, 359, 360, 361, 362, 378
Fetal Dentition 13, 93, 308, 309, 310, 314
Fetal Teeth 129, 132, 305, 307, 308, 309, 310, 312, 313, 314, 315, 317, 331, 332, 373, 375, 377
Fetuses 12, 14, 15, 48, 97, 292, 306, 308, 314, 315, 317, 325, 330, 331, 347, 371, 380
Fin 70, 365, 366, 367, 368, 369, 370, 377, 378
First Reproduction 14, 15
First Stage 132, 287, 291, 325, 331, 332, 345, 360, 366
FMR Amide 87
Follicle 103, 104, 109, 125, 128, 183, 190, 192, 193, 194, 200, 201, 202, 203, 204, 210, 219, 220, 275, 276, 277, 278, 279, 280, 281, 282, 283, 284, 285, 287, 288, 297, 298, 361
Follicle Cell 190, 192, 200, 201, 204, 276, 277, 278, 280, 281, 282, 284, 285, 287, 297, 298
Follicular Cell 103, 107, 108, 111, 113, 200, 201, 204, 235, 279, 296
Folliculogenesis 125, 275
Forebrain 366
Frogs 12, 13, 18, 41, 243, 247, 263, 268
Frontal Organ 101
Funnel 71, 79, 113, 127, 128, 194, 195, 196, 369, 375

G

Gaba 89

Gall-bladder 336
Gastric Gland 94, 334, 335, 378
Gastrulation 291, 325, 360
Gegeneophis 11, 12, 16, 49, 63, 93, 124, 205, 232, 233, 247, 248, 249, 250, 252, 253, 254, 257, 259, 265, 266, 270, 271, 303, 304
Genital Gland 293, 294, 295, 297, 298
Geotrypetes 12, 14, 16, 63, 64, 68, 69, 86, 105, 114, 121, 124, 127, 187, 204, 233, 248, 250, 303, 304, 308, 314, 319
Germ Cell 107, 108, 110, 111, 112, 113, 114, 185, 186, 190, 192, 206, 232, 234, 235, 237, 240, 275, 276, 277, 280, 281, 284, 287, 292, 293, 295, 296, 297
Germinal Area 281, 284
Germinal Cyst 194
Germinal Disc 360
Germinal Nest 204, 276, 277, 280, 281, 284, 287, 288, 296, 298
Gestation 8, 11, 13, 14, 15, 129, 131, 135, 193, 202, 211, 219, 220, 221, 306, 317, 352, 379, 380
GFAP 87
Gill 70, 71, 306, 314, 315, 316, 338, 339, 340, 341, 342, 343, 344, 345, 346, 347, 348, 363, 365, 367, 369, 373, 377
Gill Chamber 367, 369
Girdle 41, 52, 82, 83, 135
Gland 5, 11, 14, 85, 86, 89, 90, 92, 101, 102, 103, 104, 114, 115, 116, 117, 118, 119, 128, 132, 157, 158, 159, 160, 161, 162, 163, 164, 165, 166, 167, 168, 169, 170, 171, 172, 173, 174, 175, 176, 177, 178, 179, 183, 186, 187, 188, 189, 195, 197, 199, 206, 207, 215, 216, 220, 293, 294, 295, 297, 334, 337
Glomerulus 100, 105, 216
Golgi Apparatus 103, 104, 116, 131, 163, 165, 166, 167, 168, 169, 240, 328, 329, 340, 342, 343, 344, 346
Gonad 16, 125, 221, 296, 298
Gonadotropic cell 205, 207, 208, 209, 210, 214, 218, 219, 220, 221, 352
Gonadotropic Hormones 221
Gonadotropin-releasing Hormone 87
Gonia 293, 296, 297
Grandisonia 7, 12, 15, 49, 64, 92, 248, 250
Granulosa 200, 201, 202, 203, 204, 219, 276, 281, 282, 285, 287, 361
Granulosa Cell 200, 201, 202, 219
Growth 15, 85, 113, 117, 119, 120, 122, 125, 132, 133, 134, 186, 200, 208, 219, 275, 277, 285, 292, 317, 318, 338, 351, 353, 371, 379, 380, 381, 382
Gut 82, 93, 94, 105, 187, 275, 292, 293, 308, 317, 318, 331, 332, 335, 336, 338, 353, 368, 373, 378

Gymnopis 7, 43, 49, 59, 64, 65, 68, 97, 101, 105, 111, 114, 121, 127, 129, 159, 185, 190, 193, 233, 241, 248, 250, 303, 304, 305, 309, 310, 315, 378, 379, 382

H

Harderian Gland 89
Hatching 12, 87, 88, 102, 103, 192, 292, 305, 310, 325, 326, 338, 365, 366, 369, 370, 377, 378, 379
Head 16, 71, 81, 88, 90, 91, 92, 97, 99, 127, 176, 249, 250, 251, 252, 260, 270, 309, 315, 316, 338, 339, 347, 362, 363, 366, 368
Heart 18, 55, 56, 57, 69, 88, 97, 98, 100, 330, 363, 365, 368, 373
Hematopoietic Layer 94, 99
Hemoglobin 98
Hepatocyte 94, 336
Hermaphroditism 291
Herpele 49, 57, 65, 105, 114, 121, 187, 233
Hindbrain 366
Hydromineral Balance 215
Hynobius 298
Hyobranchial Apparatus 16, 17, 43
Hypogeophis 7, 9, 12, 15, 16, 49, 64, 65, 85, 91, 92, 93, 95, 96, 97, 99, 101, 114, 120, 121, 122, 124, 127, 128, 194, 201, 213, 232, 248, 250, 304, 316, 331, 332, 359, 360, 363, 375, 377, 378, 379, 382
Hypophyseal Hormone 217
Hypophysis 88, 101, 183, 197, 204, 205, 207, 208, 209, 211, 212, 213, 214, 215, 217, 218, 220

I

Ichthyophis 3, 5, 6, 7, 8, 9, 12, 14, 15, 16, 18, 40, 45, 51, 55, 56, 63, 79, 81, 83, 86, 87, 88, 91, 92, 93, 96, 97, 99, 101, 102, 103, 104, 105, 108, 111, 113, 114, 121, 124, 125, 127, 128, 129, 131, 132, 133, 185, 187, 192, 193, 194, 200, 201, 202, 204, 205, 207, 210, 213, 215, 216, 217, 218, 220, 232, 233, 234, 235, 237, 238, 239, 240, 241, 242, 243, 247, 248, 249, 250, 251, 252, 253, 255, 257, 259, 261, 265, 266, 267, 268, 270, 271, 275, 278, 280, 284, 285, 287, 288, 291, 303, 304, 307, 319, 330, 338, 347, 359, 360, 363, 364, 365, 366, 367, 369, 370, 371, 375, 376, 377, 378, 379, 381, 382
Ichthyophis glutinosus 3, 5, 7, 9, 12, 14, 15, 16, 18, 81, 185, 194, 205, 213, 217, 232, 234, 235, 237, 247, 248, 250, 270, 271, 303, 304, 307, 319, 330, 338, 359, 360,

363, 364, 365, 366, 375, 377, 378, 381, 382
Idiocranium 10, 12, 14, 65, 105, 124, 127, 128, 194, 233, 248, 250, 303
Immune System 98
Immunological Relationship 316, 352
Immunology 11
Integument 16, 86, 88, 327, 352
Intermediate Lobe (Pars Intermedia) 205
Internal Fertilisation 120, 124, 197, 192, 287
Interrenal 82, 101, 183, 207, 213, 215, 216, 217, 218, 221
Interrenal Gland 215, 216
Interstitial Cells 217, 234
Interstitial Tissue 189, 190, 217, 218, 234
Intestine 18, 83, 92, 93, 94, 95, 120, 121, 189, 197, 292, 293, 294, 295, 296, 297, 326, 330, 331, 332, 333, 335, 336, 337, 339
Intrauterine Feeding 317
Intra-uterine Medium 318, 326
Intromittent Organ 6, 9, 10, 120, 135, 187, 241
Iridophore 85
Islet of Langerhans 95

K

Kidney 18, 83, 98, 100, 105, 125, 158, 160, 215, 216, 241, 243, 275, 295, 297, 298
Küppfer Cell 94

L

Labyrinth 91
Lactotropic cell 205, 206, 208, 209, 210, 211, 214, 218, 221, 352
Lagena 91
Lamellar Bodies 96, 347
Laminophore 85
Larvae 14, 16, 41, 44, 53, 55, 63, 90, 91, 92, 93, 94, 96, 97, 99, 103, 292, 306, 307, 308, 313, 314, 316, 317, 318, 325, 330, 331, 338, 348, 351, 365, 368, 369, 371, 377, 378, 379, 382
Lateral Line 17, 18, 92, 368, 369, 377
Lecithotrophy 13
Lens 41, 87, 88, 366, 377
Leukocytes 169
Leydig 5, 7, 9, 17, 87, 89, 107, 136, 183, 189, 190, 191, 192, 206, 217, 218, 234, 296, 299, 378
Leydig Cell 107, 183, 189, 190, 192, 206, 217, 234, 378
Leydig-like Cell 9, 191, 299
LH 205, 208, 210
Life Cycles 231

Life History 6, 12, 14, 47
Limb 19, 82, 84
Live-bearing 8, 12, 13, 305
Liver 82, 92, 94, 96, 99, 113, 114, 124, 135, 158, 211, 336, 337, 348, 349, 350
Lobules 95, 99, 107, 108, 111, 113, 159, 185, 189, 190, 192, 234, 235, 237, 239, 243
Locules 232, 234, 235, 239, 241
Lower Jaw 93, 308, 309, 310, 312, 314, 331, 352, 368, 372, 373, 374, 375
Lung 18, 43, 70, 71, 83, 96, 97, 297, 334, 373
Lungenbronchus 96
Lymphatic System 99
Lymphocyte 99, 100
Lymphoid Cell 336
Lysosomes 104, 116, 168, 169, 171, 172, 240, 345

M

Macrophage 99, 170
Male 5, 6, 7, 8, 9, 10, 11, 12, 14, 52, 61, 63, 79, 83, 99, 100, 105, 111, 113, 114, 119, 120, 121, 122, 124, 127, 129, 131, 134, 135, 157, 158, 159, 160, 165, 175, 177, 178, 183, 185, 186, 187, 188, 189, 190, 191, 192, 193, 194, 196, 197, 198, 199, 201, 203, 204, 205, 207, 208, 209, 210, 211, 212, 213, 214, 215, 217, 218, 219, 221, 231, 232, 234, 237, 275, 278, 280, 281, 282, 283, 284, 288, 296, 299, 303, 305, 306, 307, 315, 316, 317, 319, 331, 348, 349, 350, 351, 352, 353, 360, 361, 364, 379, 380, 390
Male Genital Tract 105, 124, 189, 190, 193, 194, 275, 390
Maternal Serum 11
Materno-fetal Exchange 307, 316, 317, 318, 338, 348
Materno-fetal relationships 352
Matrix 109, 110, 112, 234, 235
Maturation 87, 131, 175, 178, 204, 235, 237, 275, 298
Maturity 183, 190, 192, 204, 220
Median Eminence 205, 208
Medulla 88, 99, 165, 264, 278, 296, 297, 298, 372
Medullary Fold 372
Melanophore 85
Merckel Cells 85
Mertensiella 133
Mesencephalon 85, 88
Mesenterium 293
Mesentery 292
Mesoblast 292
Mesodermal Tissue 293, 362
Mesotocin 87

Mesovarium 125, 275
Metamorphosis 13, 18, 88, 292, 296, 298, 299, 325, 326, 345, 366, 376, 377, 378, 379, 381
Microcaecilia 61, 66, 67, 79, 95, 121, 127, 128, 132, 134, 187, 194, 248, 250
Microvilli 163, 165, 167, 172, 173, 197, 200, 327, 341
Mid Intestine 335, 336
Mitochondria 40, 44, 100, 103, 104, 110, 116, 129, 163, 165, 166, 167, 168, 169, 172, 175, 176, 177, 240, 241, 249, 257, 258, 259, 260, 261, 263, 265, 266, 267, 268, 270, 271, 328, 329, 338, 340, 342, 344, 345, 346
Monocyte 99, 170
Morphogenesis 208, 209, 213, 214, 325
Morphology 2, 3, 4, 5, 6, 7, 8, 9, 10, 11, 12, 19, 106, 119, 158, 160, 162, 165, 167, 169, 170, 171, 173, 177, 179, 194, 247, 254, 255, 270, 310, 367, 370
Mother 133, 305, 308, 316, 317, 318, 319, 330, 345, 347, 348, 353
Motility of Sperm 174, 177, 178
Mouth 46, 53, 55, 56, 67, 68, 82, 91, 92, 93, 96, 133, 308, 310, 311, 314, 331, 338, 339, 365, 368, 374, 375
Mucous Envelope 291, 292, 305, 306, 307, 308, 315, 325, 326, 360, 361, 362, 363, 365, 370, 371, 378, 380
Mullerian Duct 11, 83, 100, 105, 114, 115, 116, 117, 119, 120, 121, 123, 135, 157, 158, 159, 160, 177, 178, 186, 187, 188, 189, 206, 212, 213, 218, 299
Mullerian Gland 5, 9, 10, 11, 114, 116, 117, 118, 119, 123, 157, 158, 159, 160, 162, 163, 165, 167, 170, 171, 172, 173, 174, 175, 176, 177, 178, 186, 187, 188, 192, 207, 217, 218, 299
Mullerian Gland Secretory Proteins 174
Muscle 44, 48, 64, 69, 82, 83, 87, 89, 96, 97, 120, 121, 131, 159, 160, 162, 167, 169, 171, 333, 335, 339
Musculus Retractor Cloacae 120, 123, 134

N

Nasal Cavity 89
Nasal Opening 368, 369, 377
Nasal Pit 366
Nectophrynoides 133, 219, 220, 221, 318
Nephron 100
Nephrostome 100
Nephrostomial Tubule 16
Nervous Lobe 205
Nervous System 18, 85
Nest 63, 91, 97, 163, 194, 200, 204, 234, 235, 238, 239, 240, 276, 277, 280, 281, 284, 287, 288, 296, 298, 387, 390, 391, 395
Neural Fold 363, 366, 368, 369, 371, 372
Neural Gutter 371
Neural Plate 371
Neural Tube 84, 294
Neuroanatomical Character 85
Neuroendocrinology 5
Neurohypophysial Peptide 11
Neurohypophysis 205
Neuromast 44, 92
Neurulation 325, 291
New-born 307, 309, 310, 313, 314, 315, 317, 325, 331, 61, 90, 113, 118, 204, 277
Noradrenaline 217
Noradrenaline Cell 217
Normal Table 15, 16
Nostril 55, 56, 82, 84, 88, 89
Notochord 16, 17, 294, 295, 297, 331

O

Ocular Muscle 87
Odontoblast 309, 311, 312, 313
Olfactory 17, 85, 89
Olfactory Bud 372
Olfactory Bulb 89
Olfactory Lobe 85
Olfactory Nerve 85
Olfactory Placode 366, 371, 373
Olfactory Vesicle 362, 373
Oocytes 125, 127, 128, 129, 131, 133, 136, 192, 193, 194, 195, 196, 200, 201, 202, 204, 219, 220, 221, 275, 276, 277, 278, 280, 281, 284, 285, 287, 288, 303, 304, 305, 306, 307, 308, 317, 330, 351, 352
Oogenesis 125, 275, 278, 284, 288
Oogonia 192, 194, 200, 204, 276, 277, 278, 280, 281, 283, 284, 285, 287, 288
Oophagy 307, 308, 330, 331
Operculum 135
Optic Chiasma 88
Optic Nerve 87, 88
Optic Vesicle 362, 363, 366, 373
Oscaecilia 43, 49, 67, 114, 124
Osmiophilic Bodies 96
Osmoregulation 101
Osteology 67
Ostium 8, 127
Otic Capsule 91
Otic Placode 371, 373
Otic Vesicle 328, 340, 346, 99
Otolith 91
Ova 39, 43, 44, 46, 47, 57, 70, 82, 91, 104, 124, 125, 127, 128, 133, 135
Ovarian Epithelium 278, 280
Ovarian Follicle 183, 193, 278, 284, 361

Index **393**

Ovaries 6, 7, 8, 82, 124, 125, 127, 128, 135, 192, 193, 194, 200, 201, 204, 211, 275, 276, 279, 280, 281, 282, 284, 287, 288, 296, 298, 351
Ovary 360
Oviduct 359, 360
Oviparity 369
Oviparous Species 359, 378, 381, 382
Ovoviviparity 13
Ovulation 359, 360, 379

P

Palate 18, 91
Pancreas 82, 83, 92, 94, 95, 99
Papilla Amphibia 91
Papilla Basilia 91
Papilla Neglecta 91
Parapineal Vesicle 101
Parathyroid Gland 103, 104, 183
Parental Care 12, 319
Parity 359, 369, 377, 382
Pars Convoluta 131, 194, 195
Pars Nervosa 205, 208
Pars Recta 128, 131, 194, 195
Pars Tuberalis 205, 208
Pars Utera 128, 195
Parturition 129, 132, 193, 194, 197, 199, 202, 210, 211, 213, 219, 282, 283, 284
Pedicle 101, 103
Phallodeum 10, 105, 120, 121, 122, 123, 134, 158, 187, 189
Pharynx 103, 331, 333
Photoreceptor 102
Pigment 363, 366, 367, 368, 369, 373, 374, 375, 377
Pigmentation 363, 369, 373, 374, 375, 377
Pineal 378
Pineal Gland 183
Pineal Organ 378
Pineal Pedicle 103
Pipa 318
Pituitary 373
Placenta 13, 316, 317, 318, 347, 348, 353, 97, 133, 197
Pneumocyte 96
Podocyte 100
Poisonous Gland 85
Polyribosomes 172
Polyspermy 360
Population 380
Postovulatory Follicle 201, 285, 287
Predentine 311
Pregnancy 379, 380
Pregnant 379
Pregnant Female 11, 83, 99, 129, 193, 194, 204, 210, 214, 280, 281, 282, 283, 284, 288, 306, 348, 349, 350, 351, 352, 353, 379
Pregnenelone 201
Preovulatory Follicle 285
Previtellogenic Follicle 282, 284
Previtellogenic Oocyte 287
Previtellogenic Oocyte 204, 219, 277, 278, 281, 285, 287, 288
Primary Gonad 291
Primordial Genital Gland 298
Primordial Median Ridge 293
PRL 206, 208
Progesterone 201, 219
Prolactin Receptor 11, 183, 206, 211, 212, 213, 218, 221
Proliferative Area 277
Pronephritic Mesoderm 291
Pronephros 362
Prostate 116, 158, 165, 167, 174, 177, 178, 187
Prostate Gland 11
Pulmonary Epithelium 347
Pyloric Sphincter 94

R

Radial Cell 87
Rana 351, 122, 217, 239
Reproduction 359, 382
Reproductive Biology 1, 2, 3, 5, 6, 15, 19
Reproductive Cycle 14, 131, 200, 210, 213, 218, 221, 233, 283, 291, 348
Reproductive Mode 5, 6, 11, 12, 13, 19, 128, 304
Reptile 4, 6, 39, 85, 158, 165, 167, 174, 178
Reserve Organ 211, 220, 221, 350
Respiration 48, 347, 348
Respiratory Epithelium 89
Respiratory System 82, 96
Rete Testis 100, 105, 106
Reticulocyte 99
Retina 61, 87, 88, 89, 135
Retroarticular Process 52, 53, 82
Rhacophora 298
Rhinatrema 7, 41, 42, 43, 44, 47, 53, 55, 105, 124, 125, 128, 194, 233, 241, 248, 251

S

Saccula 91
Salamanders 12, 13, 18, 41, 243, 247, 265, 267, 268
Salamandra 133, 219, 220, 317, 318, 331, 347
Scale 361, 362, 367, 370, 372, 374, 376, 378
Schistometopum 49, 59, 61, 64, 68, 105, 114, 124, 127, 129, 205, 233, 304, 307, 308, 309, 315

Scolecomorphus 7, 10, 47, 57, 61, 82, 89, 105, 114, 120, 121, 124, 129, 187, 233, 248, 251, 309, 315
Secondary Gonad 291
Secretory Cells 163, 164, 165, 166, 167, 169, 170, 171, 172, 176, 178
Secretory Material of Mullerian Gland 173
Segmentation 360
Semicircular Duct 91
Semicircular Vesicle 91
Seminiferous Ductule 107
Seminiferous Epithelium 239
Seminiferous Tubule 239
Sensory Organ 17, 87, 387
Sertoli Cell 9, 107, 108, 109, 110, 111, 112, 113, 135, 157, 190, 206, 234, 235, 237, 238, 239, 240, 296, 298, 299
Sex 79, 82, 107, 111, 113, 114, 115, 116, 120, 121, 122, 123, 124, 127, 128, 131, 132, 134, 158, 178, 183, 185, 186, 187, 188, 189, 190, 191, 192, 193, 194, 195, 197, 198, 204, 206, 210, 211, 212, 213, 217, 218, 219, 220, 221, 275, 281, 283, 284, 285, 288, 292, 296, 298, 299, 349, 351
Sexual Cycle 107, 111, 114, 115, 120, 121, 122, 124, 128, 131, 183, 185, 186, 187, 189, 192, 193, 194, 195, 197, 204, 210, 211, 217, 218, 219, 220, 275, 281, 284, 285, 288
Sexual Differentiation 113, 292, 296, 298, 299
Sexual Quiescence 123, 127, 131, 185, 186, 188, 189, 191, 193, 195, 197, 198, 206, 211, 212, 213, 218, 219, 283, 349
Shield 59, 61, 63, 64, 65, 66, 67, 68, 69, 79
Siphonops 360, 363, 364, 378, 379
Skeletogenesis 16
Skeleton 378
Skin 366, 368, 369, 378
Skin Gland 368, 369, 378
Skull 16, 17, 42, 61, 82, 87, 91
Snake 2, 41
Somatopleure 292, 326, 327, 328
Somatotropic Cell 205, 207, 208, 209, 211, 214
Somite 362, 363, 366, 369, 371, 372, 373
Sperm 360, 361, 362
Spermatid 107, 108, 109, 111, 112, 113, 119, 185, 186, 192, 235, 237, 238, 239, 240, 241, 248, 259, 261, 267
Spermatocytes 107, 109, 111, 112, 113, 114, 185, 186, 192, 235, 237, 238, 239, 240
Spermatogenesis 5, 9, 11, 111, 113, 160, 172, 185, 186, 190, 207, 231, 232, 233, 234, 235, 237, 239, 240, 243, 278
Spermatogenetic Cycle 185
Spermatogenic Cycle 233

Spermatogonia 109, 185, 186, 190, 192, 207, 237
Spermatozoa 360
Spermiogenesis 111, 240
Spicule 10, 57, 120, 122
Splanchnopleure 292, 326, 335
Stapes 47, 53, 55, 56, 57, 69
Stomach 83, 92, 93, 94, 95, 334, 337
Stomodeum 369
Superfactant 96
Systematics 1, 2, 5, 6, 41
Systemic Arch 97
Systemico-pulmonary Arch 97

T

Table of Development 359, 371
Tail 359, 361, 363, 365, 366, 367, 368, 369, 370, 371, 372, 376, 377, 378
Taste Bud 17, 91
Teeth 373, 375, 377
Telencephalon 85, 88
Telolecithal Egg 307
Tentacle 365, 366, 368, 369, 370, 374, 375, 377, 378
Tentacle Rudiment 368
Testes 5, 9, 11, 82, 105, 106, 107, 108, 111, 113, 114, 117, 135, 183, 185, 186, 187, 189, 190, 205, 206, 207, 217, 218, 233, 271, 296, 298
Testis 6, 9, 11, 83, 100, 105, 106, 107, 109, 110, 111, 112, 113, 157, 158, 159, 160, 169, 172, 173, 175, 178, 185, 190, 191, 237, 242, 243, 287, 291, 296, 399
Testis Lobes 106, 158, 159, 160, 232, 233, 234, 241, 243
Testosterone 190, 191, 192, 217, 297, 298
Thalamus 85, 208
Theca 361
Theca Cell 200
Thymus 17, 99
Thyroid 17, 102, 183, 397
Thyroid Gland 103, 104, 183, 397
Thyrotropic Cell 205
Tongue 18, 63, 64, 66, 90, 312
Tooth 17, 90, 93, 309, 310, 311, 312, 313, 327
Trachea 83
Tubular Glands 11, 116, 160, 162, 165, 171, 176, 389
Typhlonectes 359, 360, 361, 362, 366, 369, 370, 371, 372, 374, 375, 376, 377, 378, 379, 380, 381, 382

U

Ultimobranchial Bodies 103, 104, 183

Upper Jaw 93, 308, 309, 310, 312, 313, 314, 372
Uraeotyphlus 7, 9, 14, 45, 46, 56, 59, 105, 114, 116, 121, 124, 127, 128, 129, 157, 158, 159, 160, 170, 172, 174, 175, 186, 187, 194, 232, 233, 234, 235, 237, 238, 239, 240, 241, 243, 247, 248, 249, 251, 252, 253, 256, 257, 259, 265, 267, 268, 270, 271, 304
Urinogenital System 158, 159, 160
Urodela 82, 85, 104, 127, 132, 133, 135, 217, 219, 220, 221, 235, 239, 261, 263, 265, 267, 268, 269, 270, 271, 298, 377, 378, 379, 382
Urogenital Duct 8, 124, 275
Urogenital System 5, 7, 8, 9, 232, 243,
Urogenital Tract 127, 291
Uterine Fluid 133, 308, 316, 326, 330
Uterine Milk 317, 330
Uterine Wall 97, 132, 133, 134, 197, 306, 314, 316, 317, 318, 331, 338, 339, 340, 346, 347, 353
Uterus 83, 97, 124, 125, 131, 132, 133, 194, 195, 197, 198, 199, 213, 219, 220, 292, 303, 306, 307, 308, 314, 315, 317, 318, 325, 330, 331, 338, 339, 347, 348, 353, 371, 375, 378, 379, 380

V

Vasotocin 87
Vegetal Pole 360
Venous Sinus 97
Ventricle 44, 83, 88, 97, 98, 373
Vertebrae 47, 52, 53, 79, 82, 84, 352, 369
Vertebrogenesis 16, 17
Vesicle 91, 99, 101, 102, 104, 158, 163, 165, 166, 167, 168, 169, 171, 172, 174, 175, 176, 177, 178, 240, 243, 249, 250, 252, 253, 257, 258, 260, 263, 264, 266, 267, 268, 270, 328, 329, 340, 345, 346, 362, 363, 366, 367, 369, 371, 373, 375, 377
Visceral Slit 371
Vision 13, 19, 39, 43, 46, 51, 87, 186, 192, 238, 239, 268, 288, 292, 325
Vitellin Blood Vessel 363
Vitelline Envelope 200

Vitelline Membrane 128, 133, 200, 201, 204, 360
Vitellogenesis 185, 194, 200, 204, 211, 212, 220, 221, 283, 284, 351
Vitellogenic Follicle 203, 282, 283, 284, 285, 287
Vitellogenic Oocyte 192, 193, 200, 202, 203, 204, 219, 277, 278, 281, 284, 285, 287, 288, 351
Viviparity 6, 8, 12, 13, 15, 16, 18, 61, 69, 93, 97, 124, 125, 127, 128, 129, 132, 133, 134, 135, 183, 192, 194, 195, 200, 201, 202, 207, 219, 220, 221, 231, 275, 280, 287, 291, 303, 304, 305, 307, 308, 314, 316, 317, 318, 319, 325, 326, 327, 330, 338, 347, 348, 352, 353, 359, 370, 377, 378, 379, 382
Viviparous Species 13, 93, 97, 124, 125, 128, 129, 133, 183, 192, 194, 195, 201, 202, 219, 220, 275, 280, 287, 291, 303, 305, 307, 308, 314, 317, 326, 330, 352, 359, 370, 378, 379
Vomeronasal Nerve 89
Vomeronasal Organ 85, 89
Vomeronasal System 17

W

Wolffian Duct 100, 101, 105, 123, 134, 197, 293, 296, 297, 362

Y

Yolk 12, 13, 128, 133, 192, 197, 198, 200, 202, 203, 221, 278, 279, 280, 281, 282, 285, 287, 291, 292, 294, 306, 307, 308, 309, 310, 314, 316, 317, 318, 325, 326, 327, 330, 332, 335, 336, 337, 338, 340, 345, 351, 353, 360, 361, 362, 363, 365, 366, 367, 368, 369, 371, 372, 373, 374, 377, 378, 380
Yolk Platelet 128, 192, 197, 198, 200, 202, 203, 281, 282, 285, 287, 292, 326, 330, 335, 337, 340, 345, 360

Z

Zona Pellucida 276, 277, 278, 279, 280, 281, 287